T0211355

GEOMETRIC FUNCTION THEORY IN ONE AND HIGHER DIMENSIONS

PURE AND APPLIED MATHEMATICS

A Program of Monographs, Textbooks, and Lecture Notes

EXECUTIVE EDITORS

Earl J. Taft
Rutgers University
New Brunswick, New Jersey

Zuhair Nashed
University of Central Florida
Orlando, Florida

EDITORIAL BOARD

MONOGRAPHS AND TEXTBOOKS IN
PURE AND APPLIED MATHEMATICS

1. *K. Yano*, Integral Formulas in Riemannian Geometry (1970)
2. *S. Kobayashi*, Hyperbolic Manifolds and Holomorphic Mappings (1970)
3. *V. S. Vladimirov*, Equations of Mathematical Physics (A. Jeffrey, ed.; A. Littlewood, trans.) (1970)
4. *B. N. Pshenichnyi*, Necessary Conditions for an Extremum (L. Neustadt, translation ed.; K. Makowski, trans.) (1971)
5. *L. Narici et al.*, Functional Analysis and Valuation Theory (1971)
6. *S. S. Passman*, Infinite Group Rings (1971)
7. *L. Dornhoff*, Group Representation Theory. Part A: Ordinary Representation Theory. Part B: Modular Representation Theory (1971, 1972)
8. *W. Boothby and G. L. Weiss, eds.*, Symmetric Spaces (1972)
9. *Y. Matsushima*, Differentiable Manifolds (E. T. Kobayashi, trans.) (1972)
10. *L. E. Ward, Jr.*, Topology (1972)
11. *A. Babakhanian*, Cohomological Methods in Group Theory (1972)
12. *R. Gilmer*, Multiplicative Ideal Theory (1972)
13. *J. Yeh*, Stochastic Processes and the Wiener Integral (1973)
14. *J. Barros-Neto*, Introduction to the Theory of Distributions (1973)
15. *R. Larsen*, Functional Analysis (1973)
16. *K. Yano and S. Ishihara*, Tangent and Cotangent Bundles (1973)
17. *C. Procesi*, Rings with Polynomial Identities (1973)
18. *R. Hermann*, Geometry, Physics, and Systems (1973)
19. *N. R. Wallach*, Harmonic Analysis on Homogeneous Spaces (1973)
20. *J. Dieudonné*, Introduction to the Theory of Formal Groups (1973)
21. *I. Vaisman*, Cohomology and Differential Forms (1973)
22. *B.-Y. Chen*, Geometry of Submanifolds (1973)
23. *M. Marcus*, Finite Dimensional Multilinear Algebra (in two parts) (1973, 1975)
24. *R. Larsen*, Banach Algebras (1973)
25. *R. O. Kujala and A. L. Vitter, eds.*, Value Distribution Theory: Part A; Part B: Deficit and Bezout Estimates by Wilhelm Stoll (1973)
26. *K. B. Stolarsky*, Algebraic Numbers and Diophantine Approximation (1974)
27. *A. R. Magid*, The Separable Galois Theory of Commutative Rings (1974)
28. *B. R. McDonald*, Finite Rings with Identity (1974)
29. *J. Satake*, Linear Algebra (S. Koh et al., trans.) (1975)
30. *J. S. Golan*, Localization of Noncommutative Rings (1975)
31. *G. Klambauer*, Mathematical Analysis (1975)
32. *M. K. Agoston*, Algebraic Topology (1976)
33. *K. R. Goodearl*, Ring Theory (1976)
34. *L. E. Mansfield*, Linear Algebra with Geometric Applications (1976)
35. *N. J. Pullman*, Matrix Theory and Its Applications (1976)
36. *B. R. McDonald*, Geometric Algebra Over Local Rings (1976)
37. *C. W. Groetsch*, Generalized Inverses of Linear Operators (1977)
38. *J. E. Kuczkowski and J. L. Gersting*, Abstract Algebra (1977)
39. *C. O. Christenson and W. L. Voxman*, Aspects of Topology (1977)
40. *M. Nagata*, Field Theory (1977)
41. *R. L. Long*, Algebraic Number Theory (1977)
42. *W. F. Pfeffer*, Integrals and Measures (1977)
43. *R. L. Wheeden and A. Zygmund*, Measure and Integral (1977)
44. *J. H. Curtiss*, Introduction to Functions of a Complex Variable (1978)
45. *K. Hrbacek and T. Jech*, Introduction to Set Theory (1978)
46. *W. S. Massey*, Homology and Cohomology Theory (1978)
47. *M. Marcus*, Introduction to Modern Algebra (1978)
48. *E. C. Young*, Vector and Tensor Analysis (1978)
49. *S. B. Nadler, Jr.*, Hyperspaces of Sets (1978)
50. *S. K. Segal*, Topics in Group Kings (1978)
51. *A. C. M. van Rooij*, Non-Archimedean Functional Analysis (1978)
52. *L. Corwin and R. Szczarba*, Calculus in Vector Spaces (1979)
53. *C. Sadosky*, Interpolation of Operators and Singular Integrals (1979)
54. *J. Cronin*, Differential Equations (1980)
55. *C. W. Groetsch*, Elements of Applicable Functional Analysis (1980)

56. *I. Vaisman*, Foundations of Three-Dimensional Euclidean Geometry (1980)
57. *H. I. Freedan*, Deterministic Mathematical Models in Population Ecology (1980)
58. *S. B. Chae*, Lebesgue Integration (1980)
59. *C. S. Rees et al.*, Theory and Applications of Fourier Analysis (1981)
60. *L. Nachbin*, Introduction to Functional Analysis (R. M. Aron, trans.) (1981)
61. *G. Orzech and M. Orzech*, Plane Algebraic Curves (1981)
62. *R. Johnsonbaugh and W. E. Pfaffenberger*, Foundations of Mathematical Analysis (1981)
63. *W. L. Voxman and R. H. Goetschel*, Advanced Calculus (1981)
64. *L. J. Corwin and R. H. Szczarba*, Multivariable Calculus (1982)
65. *V. I. Istrățescu*, Introduction to Linear Operator Theory (1981)
66. *R. D. Järvinen*, Finite and Infinite Dimensional Linear Spaces (1981)
67. *J. K. Beem and P. E. Ehrlich*, Global Lorentzian Geometry (1981)
68. *D. L. Armacost*, The Structure of Locally Compact Abelian Groups (1981)
69. *J. W. Brewer and M. K. Smith, eds.*, Emmy Noether: A Tribute (1981)
70. *K. H. Kim*, Boolean Matrix Theory and Applications (1982)
71. *T. W. Wieting*, The Mathematical Theory of Chromatic Plane Ornaments (1982)
72. *D. B.Gauld*, Differential Topology (1982)
73. *R. L. Faber*, Foundations of Euclidean and Non-Euclidean Geometry (1983)
74. *M. Carmeli*, Statistical Theory and Random Matrices (1983)
75. *J. H. Carruth et al.*, The Theory of Topological Semigroups (1983)
76. *R. L. Faber*, Differential Geometry and Relativity Theory (1983)
77. *S. Barnett*, Polynomials and Linear Control Systems (1983)
78. *G. Karpilovsky*, Commutative Group Algebras (1983)
79. *F. Van Oystaeyen and A. Verschoren*, Relative Invariants of Rings (1983)
80. *I. Vaisman*, A First Course in Differential Geometry (1984)
81. *G. W. Swan*, Applications of Optimal Control Theory in Biomedicine (1984)
82. *T. Petrie and J. D. Randall*, Transformation Groups on Manifolds (1984)
83. *K. Goebel and S. Reich*, Uniform Convexity, Hyperbolic Geometry, and Nonexpansive Mappings (1984)
84. *T. Albu and C. Năstăsescu*, Relative Finiteness in Module Theory (1984)
85. *K. Hrbacek and T. Jech*, Introduction to Set Theory: Second Edition (1984)
86. *F. Van Oystaeyen and A. Verschoren*, Relative Invariants of Rings (1984)
87. *B. R. McDonald*, Linear Algebra Over Commutative Rings (1984)
88. *M. Namba*, Geometry of Projective Algebraic Curves (1984)
89. *G. F. Webb*, Theory of Nonlinear Age-Dependent Population Dynamics (1985)
90. *M. R. Bremner et al.*, Tables of Dominant Weight Multiplicities for Representations of Simple Lie Algebras (1985)
91. *A. E. Fekete*, Real Linear Algebra (1985)
92. *S. B. Chae*, Holomorphy and Calculus in Normed Spaces (1985)
93. *A. J. Jerri*, Introduction to Integral Equations with Applications (1985)
94. *G. Karpilovsky*, Projective Representations of Finite Groups (1985)
95. *L. Narici and E. Beckenstein*, Topological Vector Spaces (1985)
96. *J. Weeks*, The Shape of Space (1985)
97. *P. R. Gribik and K. O. Kortanek*, Extremal Methods of Operations Research (1985)
98. *J.-A. Chao and W. A. Woyczynski, eds.*, Probability Theory and Harmonic Analysis (1986)
99. *G. D. Crown et al.*, Abstract Algebra (1986)
100. *J. H. Carruth et al.*, The Theory of Topological Semigroups, Volume 2 (1986)
101. *R. S. Doran and V. A. Belfi*, Characterizations of C*-Algebras (1986)
102. *M. W. Jeter*, Mathematical Programming (1986)
103. *M. Altman*, A Unified Theory of Nonlinear Operator and Evolution Equations with Applications (1986)
104. *A. Verschoren*, Relative Invariants of Sheaves (1987)
105. *R. A. Usmani*, Applied Linear Algebra (1987)
106. *P. Blass and J. Lang*, Zariski Surfaces and Differential Equations in Characteristic *p* > 0 (1987)
107. *J. A. Reneke et al.*, Structured Hereditary Systems (1987)
108. *H. Busemann and B. B. Phadke*, Spaces with Distinguished Geodesics (1987)
109. *R. Harte*, Invertibility and Singularity for Bounded Linear Operators (1988)
110. *G. S. Ladde et al.*, Oscillation Theory of Differential Equations with Deviating Arguments (1987)
111. *L. Dudkin et al.*, Iterative Aggregation Theory (1987)
112. *T. Okubo*, Differential Geometry (1987)

113. *D. L. Stancl and M. L. Stancl*, Real Analysis with Point-Set Topology (1987)
114. *T. C. Gard*, Introduction to Stochastic Differential Equations (1988)
115. *S. S. Abhyankar*, Enumerative Combinatorics of Young Tableaux (1988)
116. *H. Strade and R. Farnsteiner*, Modular Lie Algebras and Their Representations (1988)
117. *J. A. Huckaba*, Commutative Rings with Zero Divisors (1988)
118. *W. D. Wallis*, Combinatorial Designs (1988)
119. *W. Wiesław*, Topological Fields (1988)
120. *G. Karpilovsky*, Field Theory (1988)
121. *S. Caenepeel and F. Van Oystaeyen*, Brauer Groups and the Cohomology of Graded Rings (1989)
122. *W. Kozlowski*, Modular Function Spaces (1988)
123. *E. Lowen-Colebunders*, Function Classes of Cauchy Continuous Maps (1989)
124. *M. Pavel*, Fundamentals of Pattern Recognition (1989)
125. *V. Lakshmikantham et al.*, Stability Analysis of Nonlinear Systems (1989)
126. *R. Sivaramakrishnan*, The Classical Theory of Arithmetic Functions (1989)
127. *N. A. Watson*, Parabolic Equations on an Infinite Strip (1989)
128. *K. J. Hastings*, Introduction to the Mathematics of Operations Research (1989)
129. *B. Fine*, Algebraic Theory of the Bianchi Groups (1989)
130. *D. N. Dikranjan et al.*, Topological Groups (1989)
131. *J. C. Morgan II*, Point Set Theory (1990)
132. *P. Biler and A. Witkowski*, Problems in Mathematical Analysis (1990)
133. *H. J. Sussmann*, Nonlinear Controllability and Optimal Control (1990)
134. *J.-P. Florens et al.*, Elements of Bayesian Statistics (1990)
135. *N. Shell*, Topological Fields and Near Valuations (1990)
136. *B. F. Doolin and C. F. Martin*, Introduction to Differential Geometry for Engineers (1990)
137. *S. S. Holland, Jr.*, Applied Analysis by the Hilbert Space Method (1990)
138. *J. Okniński*, Semigroup Algebras (1990)
139. *K. Zhu*, Operator Theory in Function Spaces (1990)
140. *G. B. Price*, An Introduction to Multicomplex Spaces and Functions (1991)
141. *R. B. Darst*, Introduction to Linear Programming (1991)
142. *P. L. Sachdev*, Nonlinear Ordinary Differential Equations and Their Applications (1991)
143. *T. Husain*, Orthogonal Schauder Bases (1991)
144. *J. Foran*, Fundamentals of Real Analysis (1991)
145. *W. C. Brown*, Matrices and Vector Spaces (1991)
146. *M. M. Rao and Z. D. Ren*, Theory of Orlicz Spaces (1991)
147. *J. S. Golan and T. Head*, Modules and the Structures of Rings (1991)
148. *C. Small*, Arithmetic of Finite Fields (1991)
149. *K. Yang*, Complex Algebraic Geometry (1991)
150. *D. G. Hoffman et al.*, Coding Theory (1991)
151. *M. O. González*, Classical Complex Analysis (1992)
152. *M. O. González*, Complex Analysis (1992)
153. *L. W. Baggett*, Functional Analysis (1992)
154. *M. Sniedovich*, Dynamic Programming (1992)
155. *R. P. Agarwal*, Difference Equations and Inequalities (1992)
156. *C. Brezinski*, Biorthogonality and Its Applications to Numerical Analysis (1992)
157. *C. Swartz*, An Introduction to Functional Analysis (1992)
158. *S. B. Nadler, Jr.*, Continuum Theory (1992)
159. *M. A. Al-Gwaiz*, Theory of Distributions (1992)
160. *E. Perry*, Geometry: Axiomatic Developments with Problem Solving (1992)
161. *E. Castillo and M. R. Ruiz-Cobo*, Functional Equations and Modelling in Science and Engineering (1992)
162. *A. J. Jerri*, Integral and Discrete Transforms with Applications and Error Analysis (1992)
163. *A. Charlier et al.*, Tensors and the Clifford Algebra (1992)
164. *P. Biler and T. Nadzieja*, Problems and Examples in Differential Equations (1992)
165. *E. Hansen*, Global Optimization Using Interval Analysis (1992)
166. *S. Guerre-Delabrière*, Classical Sequences in Banach Spaces (1992)
167. *Y. C. Wong*, Introductory Theory of Topological Vector Spaces (1992)
168. *S. H. Kulkarni and B. V. Limaye*, Real Function Algebras (1992)
169. *W. C. Brown*, Matrices Over Commutative Rings (1993)
170. *J. Loustau and M. Dillon*, Linear Geometry with Computer Graphics (1993)
171. *W. V. Petryshyn*, Approximation-Solvability of Nonlinear Functional and Differential Equations (1993)

226. *R. Li et al.*, Generalized Difference Methods for Differential Equations: Numerical Analysis of Finite Volume Methods (2000)
227. *H. Li and F. Van Oystaeyen*, A Primer of Algebraic Geometry (2000)
228. *R. P. Agarwal*, Difference Equations and Inequalities: Theory, Methods, and Applications, Second Edition (2000)
229. *A. B. Kharazishvili*, Strange Functions in Real Analysis (2000)
230. *J. M. Appell et al.*, Partial Integral Operators and Integro-Differential Equations (2000)
231. *A. I. Prilepko et al.*, Methods for Solving Inverse Problems in Mathematical Physics (2000)
232. *F. Van Oystaeyen*, Algebraic Geometry for Associative Algebras (2000)
233. *D. L. Jagerman*, Difference Equations with Applications to Queues (2000)
234. *D. R. Hankerson et al.*, Coding Theory and Cryptography: The Essentials, Second Edition, Revised and Expanded (2000)
235. *S. Dăscălescu et al.*, Hopf Algebras: An Introduction (2001)
236. *R. Hagen et al.*, C*-Algebras and Numerical Analysis (2001)
237. *Y. Talpaert*, Differential Geometry: With Applications to Mechanics and Physics (2001)
238. *R. H. Villarreal*, Monomial Algebras (2001)
239. *A. N. Michel et al.*, Qualitative Theory of Dynamical Systems: Second Edition (2001)
240. *A. A. Samarskii*, The Theory of Difference Schemes (2001)
241. *J. Knopfmacher and W.-B. Zhang*, Number Theory Arising from Finite Fields (2001)
242. *S. Leader*, The Kurzweil-Henstock Integral and Its Differentials (2001)
243. *M. Biliotti et al.*, Foundations of Translation Planes (2001)
244. *A. N. Kochubei*, Pseudo-Differential Equations and Stochastics over Non-Archimedean Fields (2001)
245. *G. Sierksma*, Linear and Integer Programming: Second Edition (2002)
246. *A. A. Martynyuk*, Qualitative Methods in Nonlinear Dynamics: Novel Approaches to Liapunov's Matrix Functions (2002)
247. *B. G. Pachpatte*, Inequalities for Finite Difference Equations (2002)
248. *A. N. Michel and D. Liu*, Qualitative Analysis and Synthesis of Recurrent Neural Networks (2002)
249. *J. R. Weeks*, The Shape of Space: Second Edition (2002)
250. *M. M. Rao and Z. D. Ren*, Applications of Orlicz Spaces (2002)
251. *V. Lakshmikantham and D. Trigiante*, Theory of Difference Equations: Numerical Methods and Applications, Second Edition (2002)
252. *T. Albu*, Cogalois Theory (2003)
253. *A. Bezdek*, Discrete Geometry (2003)
254. *M. J. Corless and A. E. Frazho*, Linear Systems and Control: An Operator Perspective (2003)
255. *I. Graham and G. Kohr*, Geometric Function Theory in One and Higher Dimensions (2003)

Additional Volumes in Preparation

GEOMETRIC FUNCTION THEORY IN ONE AND HIGHER DIMENSIONS

IAN GRAHAM
University of Toronto
Toronto, Ontario, Canada

GABRIELA KOHR
Babeş-Bolyai University
Cluj-Napoca, Romania

CRC Press
Taylor & Francis Group
Boca Raton London New York

CRC Press is an imprint of the
Taylor & Francis Group, an **informa** business

CRC Press
Taylor & Francis Group
6000 Broken Sound Parkway NW, Suite 300
Boca Raton, FL 33487-2742

First issued in paperback 2019

© 2003 by Taylor Francis Group, LLC
CRC Press is an imprint of Taylor & Francis Group, an Informa business

No claim to original U.S. Government works

ISBN-13: 978-0-8247-0976-1 (hbk)
ISBN-13: 978-0-367-39533-9 (pbk)

Library of Congress Cataloging-in-Publication Data
A catalog record for this book is available from the Library of Congress.

Visit the Taylor & Francis Web site at
http://www.taylorandfrancis.com

and the CRC Press Web site at
http://www.crcpress.com

To my twin sister, Mirela
Gabriela Kohr

To Norberto Kerzman
Ian Graham

Preface

In this book we give a combined treatment of classical results in univalent function theory (as well as newer results in geometric function theory in one variable) and generalizations of these results to higher dimensions, in which there has been much recent progress.

The one-variable topics treated include the class S of normalized univalent functions on the unit disc and various subclasses, the theory of Loewner chains and applications, Bloch functions and the Bloch constant, and linear-invariant families. Our treatment of these topics is designed to prepare the ground for the several-variables material.

The second part of the book begins with a concise introduction to those aspects of the theory of several complex variables and complex analysis in infinite dimensions which are needed. We then study the class $S(B)$ of normalized biholomorphic mappings from the unit ball B of $\mathbb{C}^n$ into $\mathbb{C}^n$. We consider growth, covering, and distortion theorems and coefficient estimates for various subclasses of $S(B)$, some of which are direct generalizations of familiar subclasses of S, and some of which are not. We give a detailed exposition of the theory of Loewner chains in several variables with applications. We also consider Bloch mappings and analogs of the Bloch constant problem, and the theory of linear-invariant families in several variables. Finally we study extension operators such as the Roper-Suffridge operator which can be used to construct biholomorphic mappings of the unit ball with certain geometric properties using univalent functions of the unit disc with related properties.

The book is intended for both graduate students and research mathematicians. The prerequisites are a good first course in complex analysis, including the Riemann mapping theorem, a course in measure theory, and some basic notions of functional analysis. A course in several complex variables is not a prerequisite (though we hope that one-variable readers will be led to explore other aspects of this subject); the necessary background is given in the first section of Chapter 6. In fact, the book can be used as an introduction to several complex variables. Numerous exercises are given throughout.

A more detailed description of the contents appears in the Introduction.

We would like to acknowledge a number of people.

Gabriela Kohr wishes to express her gratitude to Professor Petru T. Mocanu for his help and encouragement and for all that she learned from him over many years. She particularly wishes to thank Hidetaka Hamada for his great help throughout a long and valuable collaboration. Professor Ted Suffridge has provided much useful advice over the years. Professor John Pfaltzgraff has given some much appreciated encouragement and ideas.

Ian Graham wishes to thank David Minda for discussions about geometric function theory in one variable, and Dror Varolin for discussions about covering theorems. Among earlier mathematical influences, he would like to mention the advice and enthusiasm of Norberto Kerzman and his collaboration with H. Wu.

We also thank Professor Sheng Gong for discussions about geometric function theory of several complex variables.

We would like to thank all those who assisted with the preparation of the manuscript, especially Georgeta Bonda of Babeş-Bolyai University and Ida Bulat of the University of Toronto. The figures were made with the help of Nadia Villani and Miranda Tang of the University of Toronto and Radu Trîmbiţaş of Babeş-Bolyai University. We would also like to acknowledge the hospitality of each other's university and the support of the Natural Sciences and Engineering Research Council of Canada.

Finally we would like to thank the staff at Marcel Dekker Inc., including Maria Allegra, for their help with the publication of this book.

Ian Graham and Gabriela Kohr

Contents

Introduction

The theory of univalent functions is one of the most beautiful topics in one complex variable. There are many remarkable theorems dealing with extremal problems for the class S of normalized univalent functions on the unit disc, from the Bieberbach conjecture which was solved by de Branges in 1985, to others of a purely geometrical nature. A great variety of methods was developed to study these problems.

The study of the direct analog of the class S in several variables, i.e. the class $S(B)$ of normalized biholomorphic mappings of the unit ball B in $\mathbb{C}^n$, was comparatively slow to develop, although it was suggested by H. Cartan in 1933 [Cart2]. Perhaps this was because of the failure of the Riemann mapping theorem in higher dimensions, and its replacement by many new types of mapping questions. Moreover, some of the most obvious questions that one can formulate about the class $S(B)$ lead to counterexamples rather than generalizations of one-variable theorems. However, in recent years there have been many developments in univalent mappings in higher dimensions, and this subject now includes a significant body of results.

It is our belief that a book which combines both classical results in univalent function theory and analogous recent results in higher dimensions will be useful at this time. Indeed, it is our hope that the book will lead to increased interaction between specialists in one and in several complex variables.

The book begins with the classical growth, covering and distortion theorems for the class S.

In Chapter 2 we consider various subclasses of S, including not only starlike and convex functions but also functions which are spirallike, close-to-convex, starlike of order α, or α-convex. (Part of the reason for doing so is that in several variables it is necessary to consider proper subclasses of the normalized univalent mappings on the unit ball in $\mathbb{C}^n$ in order to obtain nontrivial theorems.) These subclasses are defined by geometric conditions which can be reformulated as analytic conditions, which in turn lead to interesting theorems. Our intention is not to give an exhaustive treatment of subclasses of S, but to give a number of applications which are typical of the results which can be found in the one-variable literature.

The study of Loewner chains (Chapter 3) will be of special interest, partly by way of comparison with recent results in this area in several variables. We shall give some well-known and beautiful applications of this method in one variable, including the radius of starlikeness, the rotation theorem, the bound for the third coefficient of functions in S, alternative characterizations of starlikeness, convexity, spirallikeness, and close-to-convexity, and univalence criteria. We have omitted the proof of de Branges' theorem, since some of the methods do not generalize to several variables. (Proofs can be found in the books of Conway, Hayman, Henrici, or Rosenblum and Rovnyak.)

Some of the ideas from univalent function theory extend naturally to the study of certain classes of non-univalent functions. In Chapter 4 we study Bloch functions, including Bonk's distortion theorem and estimates for the Bloch constant. In Chapter 5 we consider linear-invariant families, introduced by Pommerenke, in which the study of estimates for the second coefficient is extended to families of locally univalent functions.

The second part of the book begins with a summary of results from the general theory of several complex variables which will be needed, and some examples which show that not all of the results of classical univalent function theory can be expected to carry over to higher dimensions. We then treat particular subclasses of normalized univalent mappings on the unit ball (and in some cases on more general domains and in infinite dimensions). The convex and starlike mappings are of course analogs of well-known subclasses of S, but other new classes in several variables are introduced. As in one variable, the focus is on growth, distortion, and covering theorems and on coefficient estimates. Among the recent results treated here, we mention the compactness of the class $\mathcal{M}$ which plays the role of the Carathéodory class in several variables. This is the subject of Chapters 6 and 7.

The theory of Loewner chains in several variables (Chapter 8) is one of the main themes of the second part of the book. There are many recent results in this area, including improvements in the existence theorems resulting from the compactness of the class $\mathcal{M}$, and new applications. Of particular importance is the subclass $S^0(B)$ of $S(B)$ consisting of mappings which have parametric representation, because many of the results for the class S in one variable can be generalized to this class, and many useful subclasses of $S(B)$ are also subclasses of $S^0(B)$. Surprisingly, in higher dimensions $S^0(B)$ turns out to be a proper subclass of the class of normalized holomorphic mappings of B which can be embedded as the first element of a Loewner chain.

We also consider Bloch mappings in higher dimensions (Chapter 9), and we give a detailed exposition of the theory of linear-invariant families on the Euclidean unit ball and the polydisc in Chapter 10. The book concludes with a study of the Roper-Suffridge extension operator (Chapter 11), a particularly interesting way of constructing mappings of the unit ball in $\mathbb{C}^n$ which extend univalent functions on the disc, preserving certain properties. Many of the results and methods of previous chapters are tied together in this chapter.

GEOMETRIC FUNCTION THEORY IN ONE AND HIGHER DIMENSIONS

Part I

Univalent functions

Chapter 1

Elementary properties of univalent functions

The theory of univalent functions is one of the most beautiful subjects in geometric function theory. Its origins (apart from the Riemann mapping theorem) can be traced to the 1907 paper of Koebe [Koe], to Gronwall's proof of the area theorem in 1914 [Gro], and to Bieberbach's estimate for the second coefficient of a normalized univalent function in 1916 and its consequences [Bie1]. By then, univalent function theory was a subject in its own right.

We begin the one-variable part of the book with the study of basic notions about the class S of normalized univalent functions on the unit disc, including growth, covering, and distortion theorems. Most of the results in the theory of univalent functions that we present here are classical, but there are some which are relatively new and provide a slightly different viewpoint of older results.

1.1 Univalence in the complex plane

1.1.1 Elementary results in the theory of univalent functions. Examples of univalent functions

Let $\mathbb{C}$ be the complex plane. If $z_0 \in \mathbb{C}$ and $r > 0$, we let $U(z_0, r) = \{z \in \mathbb{C} : |z - z_0| < r\}$ be the open disc of radius r centered at z_0. The closure of

3

$U(z_0, r)$ will be denoted by $\overline{U}(z_0, r)$ and its boundary by $\partial U(z_0, r)$. The open disc $U(0, r)$ will be denoted by U_r, and the unit disc U_1 will be denoted by U.

If G is an open subset of $\mathbb{C}$, let $H(G)$ denote the set of holomorphic functions on G with values in $\mathbb{C}$. With the topology of local uniform convergence (or uniform convergence on compact subsets), $H(G)$ becomes a topological space.

Let D be a domain in $\mathbb{C}$. A function $f : D \to \mathbb{C}$ is called *univalent* if $f \in H(D)$ and f is one-to-one on D. We shall be interested in the study of the class $H_u(D)$ of univalent functions on D. It is well known that the class $H_u(\mathbb{C})$ contains only functions of the form $f(z) = az + b$, $z \in \mathbb{C}$, where $a, b \in \mathbb{C}$, $a \neq 0$. However for a general domain D, $H_u(D)$ contains many other functions.

A function $f \in H(D)$ is called *locally univalent* if each point $z \in D$ has a neighbourhood V such that $f|_V$ is univalent. Since f is holomorphic, local univalence is equivalent to the condition that $f'(z) \neq 0$, $z \in D$.

If $f \in H(D)$ is locally univalent and $z \in D$, the derivative $f'(z)$ determines the local geometric behaviour of f at z. The quantity $|f'(z)|$ gives the local magnification factor for lengths, and $\arg f'(z)$ is the local rotation factor. Moreover, if f is viewed as a transformation from a domain $D \subset \mathbb{R}^2$ to $\mathbb{R}^2$, the Jacobian determinant of this transformation is given by $|f'(z)|^2$.

A locally univalent function therefore preserves angles and orientation. For this reason it is customary to refer to a univalent function as a *conformal mapping* or a *conformal equivalence*.

The condition that $f'(z) \neq 0$ on a domain $D \subset \mathbb{C}$ is necessary but of course not sufficient for the univalence of f on D. For example, $f(z) = e^{kz}$ is locally univalent on U for all $k \in \mathbb{C}$, but is not globally univalent on U if $|k| > \pi$. It is more difficult to give conditions for global univalence than for local univalence, but as we shall see there are many such conditions, some of them quite remarkable. One of the most easily stated and proved is the following criterion of Noshiro [Nos], Warschawski [War], and Wolff [Wol] (see Lemma 2.4.1): If f is holomorphic on a convex domain $D \subset \mathbb{C}$ and $\operatorname{Re} f'(z) \neq 0$, $z \in D$, then f is univalent on D.

One of the most basic results in the theory of univalent functions in one variable is the Riemann mapping theorem. Its failure in several variables is one of the key differences between complex analysis in one variable and in higher dimensions.

Riemann mapping theorem. *Every simply connected domain D, which is a proper subset of $\mathbb{C}$, can be mapped conformally onto the unit disc. Moreover, if $z_0 \in D$, there is a unique conformal map of D onto U such that $f(z_0) = 0$ and $f'(z_0) > 0$.*

The entire complex plane cannot be conformally equivalent to the unit disc U by Liouville's theorem, although these domains are homeomorphic.

We note that there is a stronger version of the Riemann mapping theorem in the case when the boundary of D is a closed Jordan curve, due to Carathéodory:

Carathéodory's theorem. *Let $D \subset \mathbb{C}$ be a simply connected domain bounded by a closed Jordan curve. Then any conformal map of D onto U extends to a homeomorphism of $\overline{D}$ onto $\overline{U}$.*

In view of the Riemann mapping theorem, it suffices to study many questions involving univalence on the unit disc U rather than on a general simply connected domain. For this purpose, we introduce the class S of functions $f \in H_u(U)$ which are *normalized* by the condition $f(0) = f'(0) - 1 = 0$. (Any holomorphic function f on U which satisfies $f(0) = f'(0) - 1 = 0$ will be said to be normalized.) If g is any univalent function on U and $h = (g - g(0))/g'(0)$, then $h \in S$, so the study of the class S provides information about any univalent function on U. Similarly, for $0 < r < 1$ we shall sometimes consider the class $S(U_r)$ of functions $f \in H_u(U_r)$ which are normalized.

A function f in the class S has a Taylor series expansion of the form

$$(1.1.1) \qquad f(z) = z + a_2 z^2 + \ldots + a_n z^n + \ldots, \quad z \in U.$$

We also introduce the class Σ of functions φ which are univalent on $\Delta = \{\zeta \in \mathbb{C} : |\zeta| > 1\}$, with a simple pole at ∞, and which are normalized so that the Laurent series expansion of φ at ∞ has the form

$$(1.1.2) \qquad \varphi(\zeta) = \zeta + \alpha_0 + \frac{\alpha_1}{\zeta} + \ldots + \frac{\alpha_n}{\zeta^n} + \ldots, \quad |\zeta| > 1.$$

Any such function $\varphi \in \Sigma$ maps Δ onto the complement of a connected compact set.

The classes S and Σ are closely related, in fact we have the following properties:

- $f \in S$ and $\beta \in \mathbb{C} \Rightarrow \varphi \in \Sigma$, where $\varphi(\zeta) = \dfrac{1}{f(1/\zeta)} + \beta$.

Indeed, since f vanishes only at $z = 0$, it follows that φ is holomorphic on Δ and obviously φ has a Laurent series expansion at ∞ of the form (1.1.2). Moreover, the univalence of f implies the univalence of φ as well.

Conversely,

- $\varphi \in \Sigma$ and $\beta \in \mathbb{C} \setminus \varphi(\Delta) \Rightarrow f \in S$, where $f(z) = \dfrac{1}{\varphi(1/z) - \beta}$.

Indeed, since $\beta \notin \varphi(\Delta)$, f is holomorphic on U. It is elementary to see that f is univalent, and the normalization of φ implies that f is normalized.

Using the above relations, we may derive properties of the class S from corresponding results for the class Σ.

Next we consider some elementary transformations which leave invariant the class S. We have

Theorem 1.1.1. *Let* $f \in S$, $f(z) = z + \displaystyle\sum_{k=2}^{\infty} a_k z^k$, $z \in U$. *The following assertions hold:*

(i) If $\theta \in \mathbb{R}$, *the function given by*

$$e^{-i\theta} f(e^{i\theta} z) = z + \sum_{k=2}^{\infty} a_k e^{i(k-1)\theta} z^k, \ z \in U,$$

belongs to S *and is called a rotation of* f.

(ii) If $r \in (0,1)$, *the function given by*

$$\frac{1}{r} f(rz) = z + \sum_{k=2}^{\infty} a_k r^{k-1} z^k, \ z \in U,$$

belongs to S *and is called a dilation of* f.

(iii) Suppose $w \notin f(U)$ *and* g *is defined by*

$$g(z) = \frac{f(z)}{1 - \dfrac{f(z)}{w}}, \quad z \in U.$$

Then $g \in S$ and is called an omitted value transform of f.

(iv) If $z_0 \in U$ and g is defined by

$$g(z) = \frac{f\left(\dfrac{z_0 + z}{1 + \overline{z}_0 z}\right) - f(z_0)}{(1 - |z_0|^2)f'(z_0)}, \quad z \in U,$$

then $g \in S$ and is called a Koebe transform of f.

(v) If $n = 2, 3, \ldots,$ and h is defined by

$$h(z) = \sqrt[n]{f(z^n)} = z\left(\frac{f(z^n)}{z^n}\right)^{1/n}, \quad z \in U,$$

then $h \in S$ and is called the nth-root transform of f. We choose the branch of the power function such that

$$\left.\left(\frac{f(z^n)}{z^n}\right)^{1/n}\right|_{z=0} = 1.$$

Proof. It suffices to give the proofs of properties (iii), (iv) and (v), since the others are obvious.

(iii) Since $w \notin f(U)$, the function g is holomorphic on U, and the normalization of f implies that g is normalized as well. The univalence of g is a simple consequence of the univalence of f.

(iv) The Möbius transformation $q(z) = \dfrac{z + z_0}{1 + \overline{z}_0 z}$ gives a conformal mapping of U onto U, and hence $f \circ q$ is univalent on U. Since g is normalized by the construction, it follows that $g \in S$.

(v) Clearly the function h is well defined and holomorphic on the unit disc, since $f \in S$. Moreover, if β is an nth root of 1, then $h(\beta z) = \beta h(z)$, $z \in U$. We show that h is univalent. To this end, let $z_1, z_2 \in U$ be such that $h(z_1) = h(z_2)$. Then $h^n(z_1) = h^n(z_2)$, and hence $f(z_1^n) = f(z_2^n)$. Since f is univalent on U, we deduce that there exists $\beta \in \mathbb{C}$, $\beta^n = 1$, such that $z_2 = \beta z_1$. Because $h(\beta z_1) = \beta h(z_1)$, we must have

$$h(z_2) = h(\beta z_1) = \beta h(z_1) = \beta h(z_2),$$

and thus $\beta = 1$ or $h(z_2) = 0$. If $\beta = 1$ then $z_2 = z_1$, and if $h(z_2) = 0$, we deduce that $z_2 = z_1 = 0$. This completes the proof.

Examples of univalent functions. One of the most important examples of a function in S is the *Koebe function*

$$k(z) = \frac{z}{(1-z)^2}, \quad z \in U.$$

This function maps the unit disc U conformally onto the complex plane except for a slit along the negative real axis from $-\infty$ to $-1/4$, and plays an extremal role in many problems in the theory of univalent functions.

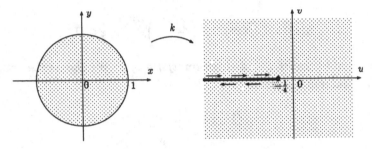

Figure 1.1: The Koebe function

In the case of the Koebe function, the corresponding function $\varphi(\zeta) = 1/k(1/\zeta) \in \Sigma$ is given by $\varphi(\zeta) = \zeta - 2 + \dfrac{1}{\zeta}$. This function maps Δ onto the doubly-connected domain consisting of the entire complex plane minus a slit along the line segment $[-4, 0]$.

The rotation of the Koebe function, i.e. the function given by

$$k_\theta(z) = \frac{z}{(1 - e^{i\theta} z)^2}, \quad z \in U,$$

belongs to the class S, for each $\theta \in \mathbb{R}$. The image of the unit disc is the complex plane except for a radial slit from ∞ to $-e^{-i\theta}/4$.

The function $f(z) = \dfrac{1}{2\alpha}\left[\left(\dfrac{1+z}{1-z}\right)^\alpha - 1\right]$, $z \in U$, where $\alpha \in (0, 2]$, is called the *generalized Koebe function*, and belongs to S (see Problem 1.1.7).

The function $f(z) = \dfrac{z}{1 - z^2}$ is the square root transform of the Koebe function. It maps the unit disc onto the complement of two collinear radial slits, one extending from $i/2$ to ∞, and the other extending from $-i/2$ to ∞.

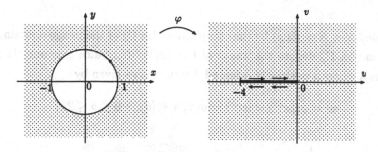

Figure 1.2: The function $\varphi(\zeta) = \zeta - 2 + \dfrac{1}{\zeta}$

The function $f(z) = \dfrac{z}{1-z}$ is a linear fractional transformation which maps the unit disc U onto the half plane Re $w > -1/2$. Since it is normalized it belongs to S. This function plays an extremal role in many problems for the subclass of S consisting of functions with convex image.

The function $f(z) = \dfrac{z}{(1-z)^{2e^{-i\alpha}\cos\alpha}}$, where $\alpha \in (-\pi/2, \pi/2)$, is in the class S. The image of the unit disc is the complement of an arc of a logarithmic α-spiral.

The function $f(z) = \dfrac{z}{(1-z)^3}$, $z \in U$, is holomorphic on U, but is not univalent on the whole disc U, since $f'(-1/2) = 0$. However, this function is univalent on the disc $U_{1/2}$ and this is the largest disc centered at 0 on which f is univalent.

1.1.2 The area theorem

Our subject begins with a theorem about the Laurent series coefficients of functions in Σ, called the *area theorem*. It was proved by Gronwall [Gro] in 1914, and plays a key role in the study of elementary properties of the classes Σ and S.

Theorem 1.1.2. *If $\varphi \in \Sigma$ is given by (1.1.2), then*

(1.1.3)
$$\sum_{n=1}^{\infty} n|\alpha_n|^2 \leq 1.$$

Proof. Let $E = E(\varphi)$ be the complement (in $\mathbb{C}$) of the image domain. Fix $\rho > 1$ and let Γ_ρ denote the image of the circle ∂U_ρ. Since φ is univalent on Δ, Γ_ρ is a smooth positively oriented Jordan curve given by

$$w = \varphi(\rho e^{i\theta}) = w(\theta) = u(\theta) + iv(\theta), \quad 0 \le \theta \le 2\pi.$$

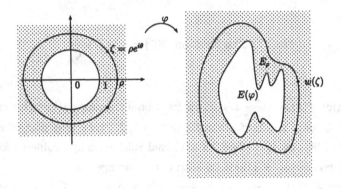

Figure 1.3: $E(\varphi) = $ the complement of the image domain

If E_ρ is the domain enclosed by Γ_ρ then $E_\rho \supset E$, and an application of Green's theorem shows that the area A_ρ of E_ρ is given by

$$A_\rho = \frac{1}{2} \int_{\Gamma_\rho} (u\,dv - v\,du) = \frac{1}{2i} \int_{\Gamma_\rho} \overline{w}\,dw = \frac{1}{2i} \int_{\partial U_\rho} \overline{\varphi(\zeta)}\varphi'(\zeta)d\zeta.$$

Introducing the Laurent series expansion of φ, we obtain

$$
\begin{aligned}
A_\rho &= \frac{1}{2i} \int_{|\zeta|=\rho} \left\{ \overline{\zeta} + \sum_{n=0}^{\infty} \frac{\overline{a}_n}{\overline{\zeta}^n} \right\} \left\{ 1 - \sum_{n=1}^{\infty} \frac{na_n}{\zeta^{n+1}} \right\} d\zeta \\
&= \frac{1}{2i} \int_{|\zeta|=\rho} \left\{ \frac{\rho^2}{\zeta} + \sum_{n=0}^{\infty} \frac{\overline{a}_n \zeta^n}{\rho^{2n}} \right\} \left\{ 1 - \sum_{n=1}^{\infty} \frac{na_n}{\zeta^{n+1}} \right\} d\zeta \\
&= \pi\left(\rho^2 - \sum_{n=1}^{\infty} \frac{n|a_n|^2}{\rho^{2n}} \right).
\end{aligned}
$$

Here we have used the facts that $\int_{|\zeta|=\rho} \zeta^k d\zeta = 0$, $k \in \mathbb{Z}$, except when $k = -1$, in which case the integral is equal to $2\pi i$, and that we can integrate term by term because of the uniform convergence of the above series for fixed ρ. Since $A_\rho \geq 0$, we conclude that

$$\sum_{n=1}^{\infty} \frac{n|\alpha_n|^2}{\rho^{2n}} \leq \rho^2,$$

and letting $\rho \searrow 1$, we deduce that $\sum_{n=1}^{\infty} n|\alpha|^2$ converges and (1.1.3) follows.

In the above proof we have derived the formula

$$A_\rho = \pi\left(\rho^2 - \sum_{n=1}^{\infty} \frac{n|\alpha_n|^2}{\rho^{2n}}\right)$$

for the area of E_ρ. We note that the sets E_ρ are decreasing as $\rho \searrow 1$ and that $E(\varphi) = \bigcap_{\rho>1} E_\rho$. Hence if A is the area (i.e. Lebesgue measure) of $E(\varphi)$, we deduce that

$$A = \pi\left(1 - \sum_{n=1}^{\infty} n|\alpha_n|^2\right).$$

Thus Theorem 1.1.2 is equivalent to the statement that $A \geq 0$. This explains the term "area theorem". Moreover, equality holds in (1.1.3) precisely when $A = 0$.

The following coefficient bounds for functions in the class Σ are immediate consequences of the area theorem.

Corollary 1.1.3. *Let $\varphi \in \Sigma$ be given by (1.1.2). Then $|\alpha_1| \leq 1$. Equality $|\alpha_1| = 1$ occurs if and only if $\varphi(\zeta) = \zeta + \alpha_0 + e^{i\theta}\zeta^{-1}$, where $\theta \in \mathbb{R}$. Moreover, if $\varphi(\zeta) \neq 0$, $|\zeta| > 1$, then $|\alpha_0| \leq 2$. Equality occurs if and only if $\varphi(\zeta) = \zeta + 2e^{i\sigma} + e^{2i\sigma}\zeta^{-1}$, where $\sigma \in \mathbb{R}$.*

Proof. The inequality $|\alpha_1| \leq 1$ is obvious in view of (1.1.3). Furthermore, if $|\alpha_1| = 1$, then from (1.1.3) we deduce that $\alpha_k = 0$ for $k \geq 2$.

Next, assume that $\varphi(\zeta) \neq 0$ for $|\zeta| > 1$. Then $f(z) = 1/\varphi(1/z)$, $|z| < 1$, is a function in S, and letting $g(z) = \sqrt{f(z^2)}$, we also have $g \in S$ by Theorem 1.1.1 (v). Therefore, the function ψ defined by

$$\psi(\zeta) = \frac{1}{g(1/\zeta)} = \zeta\left(\frac{\varphi(\zeta^2)}{\zeta^2}\right)^{1/2}, \quad |\zeta| > 1,$$

belongs to Σ. Moreover, $[\psi(\zeta)]^2 = \varphi(\zeta^2)$, $|\zeta| > 1$. Suppose ψ has the Laurent series

$$\psi(\zeta) = \zeta + \beta_0 + \frac{\beta_1}{\zeta} + \cdots, \quad |\zeta| > 1.$$

Then

$$[\psi(\zeta)]^2 = \zeta^2 + 2\beta_0\zeta + (\beta_0^2 + 2\beta_1) + \cdots$$
$$= \varphi(\zeta^2) = \zeta^2 + \alpha_0 + \frac{\alpha_1}{\zeta^2} + \cdots, \quad |\zeta| > 1.$$

Comparing the coefficients, we see that $\beta_0 = 0$ and $\alpha_0 = 2\beta_1$. Since $\psi \in \Sigma$, we deduce from the first part of the proof that $|\alpha_0/2| \leq 1$, i.e. $|\alpha_0| \leq 2$.

The equality $|\alpha_0| = 2$ or $|\beta_1| = 1$ occurs if and only if $\psi(\zeta) = \zeta + \dfrac{e^{i\sigma}}{\zeta}$, for some $\sigma \in \mathbb{R}$. Then $\varphi(\zeta^2) = [\psi(\zeta)]^2 = \zeta^2 + \dfrac{e^{2i\sigma}}{\zeta^2} + 2e^{i\sigma}$, and thus $\varphi(\zeta) = \zeta + \dfrac{e^{2i\sigma}}{\zeta} + 2e^{i\sigma}$. This completes the proof.

We now consider the implications of these results for the class S. The first statement in Corollary 1.1.3 leads to an estimate for the second order coefficient in the Taylor series expansion of a function in S which was obtained by Bieberbach in 1916 [Bie1]. Another proof of this result, based on an application of Pick's theorem, may be found in [Ros-Rov, p.160].

Theorem 1.1.4. *If $f \in S$ is given by $f(z) = z + \sum_{k=2}^{\infty} a_k z^k$, $z \in U$, then*

(1.1.4) $|a_2| \leq 2.$

Equality occurs in (1.1.4) if and only if f is a rotation of the Koebe function.

Proof. Since $f \in S$, the function φ defined by $\varphi(\zeta) = 1/f(1/\zeta)$, $|\zeta| > 1$, belongs to Σ. Moreover, $\varphi(\zeta) \neq 0$ for $|\zeta| > 1$. A straightforward computation shows that φ has the Laurent series

$$\varphi(\zeta) = \zeta - a_2 + \sum_{k=2}^{\infty} \frac{\alpha_k}{\zeta^k}, \quad |\zeta| > 1.$$

It follows from Corollary 1.1.3 that $|a_2| \leq 2$ and if equality occurs, then $\varphi(\zeta) = \zeta - 2e^{i\sigma} + \dfrac{e^{2i\sigma}}{\zeta}$, for some $\sigma \in \mathbb{R}$. In this case

$$f(z) = \frac{1}{\varphi(1/z)} = \frac{z}{(1 - e^{i\sigma}z)^2}, \quad z \in U.$$

Conversely, if f is a rotation of the Koebe function, then it is clear that the second coefficient a_2 satisfies the condition $|a_2| = 2$. This completes the proof.

On the basis of this estimate and of the fact that $|a_k| = k$ for the rotations of the Koebe function, Bieberbach [Bie1] formulated his famous conjecture:

Bieberbach's Conjecture. If $f \in S$, $f(z) = z + \sum_{k=2}^{\infty} a_k z^k$, $z \in U$, then $|a_k| \leq k$, for $k = 2, 3, \dots$. Equality $|a_k| = k$ for a given $k \geq 2$ holds if and only if f is a rotation of the Koebe function.

This conjecture remained unsolved until 1985, when de Branges [DeB] gave a remarkable proof. Many partial results were obtained in the intervening years, including results for special subclasses of S and for particular coefficients, as well as asymptotic estimates and estimates for general n. For historical comments concerning the Bieberbach conjecture, the reader may consult the books of Duren [Dur], Pommerenke [Pom5], Conway [Con], Goluzin [Gol4], Gong [Gon5], Hayman [Hay], Henrici [Hen], Milin [Mili2], Rosenblum and Rovnyak [Ros-Rov], or the papers in [Bae-Dra-Dur-Mar].

We shall return to this result in Chapter 3.

1.1.3   Growth, covering and distortion results in the class S

The estimate $|a_2| \leq 2$ is the basis for some of the key theorems about functions in the class S. These theorems determine the basic nature of univalent function theory, at least in its elementary aspects. It should be noted that the results in this section cannot be extended to higher dimensions for the full class $S(B)$ of normalized univalent mappings of the unit ball B in $\mathbb{C}^n$ (see Chapter 6). We also remark that there are related results for functions which are not necessarily univalent on U, such as distortion theorems for linear-invariant families and for Bloch functions, and covering theorems of Bloch type. We shall discuss these results in Chapters 4 and 5.

We begin with the Koebe 1/4-theorem.

Theorem 1.1.5. *If $f \in S$ then $f(U) \supseteq U_{1/4}$. This result is sharp for rotations of the Koebe function. Moreover, $\bigcap_{f \in S} f(U) = U_{1/4}$.*

Proof. We shall give a proof based on the omitted value transformation. A second proof may be given using the lower estimate in (1.1.5).

It suffices to show that if $w_0 \in \mathbb{C}$ is such that $w_0 \notin f(U)$, then $|w_0| \geq 1/4$, and that equality $|w_0| = 1/4$ can hold if and only if f is a rotation of the Koebe function. For this purpose, let $g : U \to \mathbb{C}$ be given by

$$g(z) = \frac{w_0 f(z)}{w_0 - f(z)}, \quad z \in U.$$

Then $g \in S$ by Theorem 1.1.1 (iii), and a simple computation shows that the Taylor series of g has the form

$$g(z) = z + \left(a_2 + \frac{1}{w_0} \right) z^2 + \dots, \quad z \in U.$$

From Theorem 1.1.4 we deduce that

$$\left| a_2 + \frac{1}{w_0} \right| \leq 2$$

and since $|a_2| \leq 2$, we obtain

$$\left| \frac{1}{w_0} \right| \leq \left| a_2 + \frac{1}{w_0} \right| + |a_2| \leq 4.$$

Hence $|w_0| \geq 1/4$, as claimed.

On the other hand, using the above reasoning, we see that $|w_0| = 1/4$ if and only if $|a_2| = 2$ and $\left| a_2 + \frac{1}{w_0} \right| = 2$. Hence in this case f must be a rotation of the Koebe function.

To show that $\bigcap_{f \in S} f(U) = U_{1/4}$, consider the rotations of the Koebe function given by

$$k_\theta(z) = \frac{z}{(1 - e^{i\theta} z)^2}, \quad z \in U,$$

where $\theta \in \mathbb{R}$. Then each point of the circle $\partial U_{1/4}$ is a boundary point of one of the domains $k_\theta(U)$, and $\bigcap_{\theta \in \mathbb{R}} k_\theta(U) = U_{1/4}$. It follows that we must also have $\bigcap_{f \in S} f(U) = U_{1/4}$. This completes the proof.

Thus the disc $U_{1/4}$ is the largest disc centered at the origin which is contained in the image of each function in S.

Another very important consequence of Bieberbach's theorem for the second order coefficient a_2 of a function in S is the *Koebe distortion theorem* given in (1.1.6). (The precise form of this estimate is due to Bieberbach [Bie1].) From it one may deduce the *growth theorem*, i.e. the estimate (1.1.5), as well as an estimate for the quantity $zf'(z)/f(z)$.

Theorem 1.1.6. *If $f \in S$ then the following sharp estimates hold for all $z \in U$:*

(1.1.5)
$$\frac{|z|}{(1+|z|)^2} \le |f(z)| \le \frac{|z|}{(1-|z|)^2},$$

(1.1.6)
$$\frac{1-|z|}{(1+|z|)^3} \le |f'(z)| \le \frac{1+|z|}{(1-|z|)^3},$$

and

(1.1.7)
$$\frac{1-|z|}{1+|z|} \le \left|\frac{zf'(z)}{f(z)}\right| \le \frac{1+|z|}{1-|z|}.$$

Equality occurs in one of these estimates at a given point $z \ne 0$ if and only if f is a suitable rotation of the Koebe function.

Proof.

Step 1. It suffices to prove all of these estimates when $z \ne 0$. Fix z, $|z| = r \in (0,1)$, and consider the Koebe transform g of f given by

(1.1.8)
$$g(\zeta) = \frac{f\left(\dfrac{\zeta+z}{1+\bar{z}\zeta}\right) - f(z)}{(1-|z|^2)f'(z)} = \zeta + b_2\zeta^2 + \dots, \quad \zeta \in U.$$

This function belongs to S and a simple computation shows that

$$b_2 = \frac{1}{2}\left[(1-|z|^2)\frac{f''(z)}{f'(z)} - 2\bar{z}\right].$$

We have $|b_2| \le 2$ by Theorem 1.1.4, and hence

(1.1.9)
$$\left|(1-|z|^2)\frac{f''(z)}{f'(z)} - 2\bar{z}\right| \le 4.$$

Now, (1.1.9) leads to

$$\left|\frac{zf''(z)}{f'(z)} - \frac{2r^2}{1-r^2}\right| \le \frac{4r}{1-r^2}$$

and in particular,

$$\frac{2r^2 - 4r}{1 - r^2} \leq \mathrm{Re} \left[\frac{z f''(z)}{f'(z)} \right] \leq \frac{2r^2 + 4r}{1 - r^2}.$$

On the other hand, since $f'(z) \neq 0$ and $f'(0) = 1$, there exists an analytic branch of $\log f'(z)$ such that $\log f'(z)|_{z=0} = 0$. For $z = re^{i\theta}$ we deduce that

$$\frac{\partial}{\partial r} \log |f'(z)| = \frac{\partial}{\partial r} \mathrm{Re} \, \{\log f'(z)\} = \frac{1}{r} \mathrm{Re} \left[\frac{z f''(z)}{f'(z)} \right].$$

Using (1.1.9) and this equality, we obtain

$$(1.1.10) \qquad \frac{2r - 4}{1 - r^2} \leq \frac{\partial}{\partial r} \log |f'(re^{i\theta})| \leq \frac{2r + 4}{1 - r^2}.$$

Integrating with respect to r while keeping θ fixed, and noting that $f'(0) = 1$, we deduce that

$$\log \left[\frac{1 - r}{(1 + r)^3} \right] \leq \log |f'(re^{i\theta})| \leq \log \left[\frac{1 + r}{(1 - r)^3} \right],$$

which yields (1.1.6) upon exponentiating. Furthermore, these bounds are sharp. Indeed, if equality holds in one of inequalities (1.1.6) for some $z = re^{i\theta}$, then we must have equality in the corresponding inequality in (1.1.10) for all ρ, $0 \leq \rho \leq r$. Letting $\rho \to 0$, we obtain

$$4 = \left| \frac{\partial}{\partial \rho} \log |f'(\rho e^{i\theta})| \right|_{\rho=0} = \left| \mathrm{Re} \left[\frac{e^{i\theta} f''(0)}{f'(0)} \right] \right|$$

and since $f'(0) = 1$, we must have $|\mathrm{Re} \, [e^{i\theta} f''(0)]| = 4$. Therefore $|f''(0)| \geq 4$. By Theorem 1.1.4 we deduce that $|f''(0)| = 4$, and hence f must be a rotation of the Koebe function.

Conversely, it is easy to see that if f is a rotation of the Koebe function, then equality holds in each of the upper and lower estimates in (1.1.6) along a ray, and these rays point in opposite directions.

Step 2. Next we prove the upper bound in (1.1.5).

Let $z = re^{i\theta}$, $0 < r < 1$. If $[0, z]$ denotes the closed line segment between 0 and z, we have

$$f(z) = \int_{[0,z]} f'(\zeta) d\zeta = \int_0^r f'(\rho e^{i\theta}) e^{i\theta} d\rho.$$

Using the upper estimate in (1.1.6), we obtain

$$|f(z)| \le \int_0^r |f'(\rho e^{i\theta})| d\rho \le \int_0^r \frac{1+\rho}{(1-\rho)^3} d\rho = \frac{r}{(1-r)^2},$$

as desired.

In order to prove the lower bound in (1.1.5), we let

$$m(r) = \min\left\{|f(z)| : |z| = r\right\}.$$

Then it is clear that the closed disc $\overline{U}_{m(r)}$ is contained in the image of the disc $\overline{U}_r$. Let $z_1 \in U$, $|z_1| = r$, be such that $|f(z_1)| = m(r)$. Then the closed line segment Γ between 0 and $f(z_1)$ lies entirely in the closed disc $\overline{U}_{m(r)}$. Let γ be the inverse image of Γ (see Figure 1.4). Then γ is a simple arc contained in $|z| \le r$, and using the fact that $f'(\zeta)d\zeta$ has constant argument on γ and the lower bound in (1.1.6), we have

$$m(r) = |f(z_1)| = \int_\Gamma |dw| = \int_\gamma |f'(\zeta)||d\zeta| \ge \int_\gamma \frac{1-|\zeta|}{(1+|\zeta|)^3} |d\zeta|$$

$$\ge \int_\gamma \frac{1-|\zeta|}{(1+|\zeta|)^3} d|\zeta| = \int_0^r \frac{1-t}{(1+t)^3} dt = \frac{r}{(1+r)^2}.$$

Here we have used the fact that $|d\zeta| \ge d|\zeta|$ on γ.

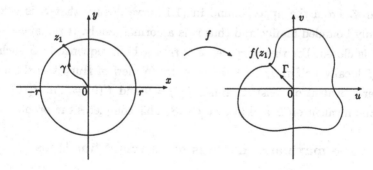

Figure 1.4: $\gamma = f^{-1}(\Gamma)$

Equality for some $z \neq 0$ in (1.1.5) implies equality in (1.1.6) along the ray joining this point z to 0, and hence f must be a rotation of the Koebe function. Conversely, if f is a rotation of the Koebe function, it is obvious that equality holds in the growth theorem (1.1.5) along two rays which point in opposite directions.

Step 3. It remains to prove the estimate (1.1.7). For this purpose, it suffices to consider the growth theorem for the Koebe transform g of f given by (1.1.8), i.e.

$$\frac{|\zeta|}{(1 + |\zeta|)^2} \leq |g(\zeta)| \leq \frac{|\zeta|}{(1 - |\zeta|)^2},$$

and to set $\zeta = -z$. This gives

$$\frac{|z|}{(1 + |z|)^2} \leq \frac{|f(z)|}{(1 - |z|^2)|f'(z)|} \leq \frac{|z|}{(1 - |z|)^2},$$

which leads to (1.1.7).

Again if equality holds in either part of (1.1.7) at some $z \in U \setminus \{0\}$, then the Koebe transform g, given by (1.1.8), satisfies

$$(1.1.11) \qquad |g(-z)| = \frac{|z|}{(1 - |z|)^2} \quad \text{or} \quad |g(-z)| = \frac{|z|}{(1 + |z|)^2}.$$

From either equality in (1.1.11), we conclude that g must be a rotation of the Koebe function and this implies that f is also a rotation of the Koebe function (see Problem 1.1.8). The converse is clear. This completes the proof.

Corollary 1.1.7. *The class S is compact as a subclass of $H(U)$.*

Proof. From the upper bound in (1.1.5) we deduce that S is a locally uniformly bounded family, and thus it is a normal family. It remains to show that S is closed. For this purpose, let $\{f_k\}_{k \in \mathbb{N}}$ be a sequence in S such that $f_k \to f$ locally uniformly on U as $k \to \infty$. In view of Hurwitz's theorem, f is either univalent or constant. Since $f(0) = 0$ and $f'(0) = \lim_{k \to \infty} f_k'(0) = 1$, f must be univalent on U and in fact $f \in S$. This completes the proof.

1.1.4   The maximum modulus of univalent functions

We mention some refinements of Theorem 1.1.6 due to Hayman (see e.g. [Hay]) and Krzyz [Krz1]. Hayman's result describes the growth of the maximum modulus of any function in the class S. The limit $\alpha = \alpha(f)$ which occurs in this theorem is known as the *Hayman index* of f. There is a deeper result, also due to Hayman, called *Hayman's regularity theorem*, which states that if $f \in S$ and $f(z) = z + \sum_{k=2}^{\infty} a_k z^k$, $z \in U$, then $\alpha(f) = \lim_{k \to \infty} |a_k|/k \leq 1$, with strict inequality unless f is a rotation of the Koebe function. This result was an important step in the study of the Bieberbach conjecture, because it showed that $|a_k| \leq k$ for k sufficiently large.

Theorem 1.1.8. *Let $f \in S$ and $M_\infty(r, f) = \max_{|z|=r} |f(z)|$, $0 < r < 1$. If f is not a rotation of the Koebe function, then the function $\psi(r) = \dfrac{1}{r}(1 - r)^2 M_\infty(r, f)$ is strictly decreasing on $(0,1)$ and hence tends to a limit $\alpha = \alpha(f) \in [0, 1)$ as $r \nearrow 1$.*

Proof. From (1.1.7) we have

$$\frac{\partial}{\partial r} \log |f(re^{i\theta})| = \operatorname{Re}\left[\frac{\partial}{\partial r} \log f(re^{i\theta})\right] \leq \left|\frac{f'(re^{i\theta})}{f(re^{i\theta})}\right| \leq \frac{1+r}{r(1-r)},$$

for $|z| = r \in (0,1)$. If f is not a rotation of the Koebe function, then Theorem 1.1.6 implies that strict inequality holds in the above. Integrating this inequality from r_1 to r_2, where $0 < r_1 < r_2 < 1$, we obtain

$$\log \left|\frac{f(r_2 e^{i\theta})}{f(r_1 e^{i\theta})}\right| < \int_{r_1}^{r_2} \frac{1+r}{r(1-r)} dr = \log\left[\frac{(1-r_1)^2 r_2}{(1-r_2)^2 r_1}\right].$$

Therefore

$$\frac{(1-r_2)^2}{r_2} |f(r_2 e^{i\theta})| < \frac{(1-r_1)^2}{r_1} |f(r_1 e^{i\theta})|,$$

for $0 < r_1 < r_2 < 1$ and for all $\theta \in \mathbb{R}$. If we choose $\theta \in \mathbb{R}$ so that $|f(r_2 e^{i\theta})| = M_\infty(r_2, f)$, then the above inequality yields

$$\frac{(1-r_2)^2}{r_2} M_\infty(r_2, f) < \frac{(1-r_1)^2}{r_1} |f(r_1 e^{i\theta})| \leq \frac{(1-r_1)^2}{r_1} M_\infty(r_1, f).$$

Hence we have proved that unless $f = k_\phi$ (a rotation of the Koebe function), the function $\psi(r)$ is strictly decreasing for $r \in (0, 1)$. Using the upper bound in (1.1.5), we deduce that

$$\alpha = \lim_{r \to 1} \frac{(1-r)^2}{r} M_\infty(r, f) < 1.$$

Of course, if f is a rotation of the Koebe function, then it is easily seen that $\alpha = 1$. This completes the proof.

Krzyz [Krz1] extended Hayman's theorem in another direction. He proved a corresponding result for the derivative of a function in S.

Theorem 1.1.9. *If $f \in S$ then $M_\infty(r, f')(1 - r)^3/(1 + r)$ is a strictly decreasing function of $r \in (0, 1)$ unless f is a rotation of the Koebe function. Moreover, the limit*

$$\lim_{r \nearrow 1} M_\infty(r, f')(1 - r)^3 = \beta(f)$$

exists and $\beta(f) \in [0, 2]$. Equality $\beta(f) = 2$ occurs if and only if f is a rotation of the Koebe function.

Proof. Taking into account (1.1.9), we obtain

$$\left| \frac{f''(z)}{f'(z)} \right| \leq \frac{2r + 4}{1 - r^2} = \frac{d}{dr} \log \left[\frac{1 + r}{(1 - r)^3} \right], \quad |z| = r \in (0, 1).$$

Moreover, equality occurs if and only if f is a rotation of the Koebe function by Theorem 1.1.6. From Problem 1.1.16 we deduce that $|f'(re^{i\theta})|(1 - r)^3/(1 + r)$ (θ fixed) and $M_\infty(r, f')(1 - r)^3/(1 + r)$ are both strictly decreasing functions of $r \in (0, 1)$ unless f is a rotation of the Koebe function. Since

$$\lim_{r \searrow 0} M_\infty(r, f')(1 - r)^3/(1 + r) = 1,$$

we deduce that

$$\lim_{r \nearrow 1} M_\infty(r, f') \frac{(1 - r)^3}{1 + r} = \frac{1}{2} \lim_{r \nearrow 1} M_\infty(r, f')(1 - r)^3 = \frac{1}{2}\beta(f) \leq 1.$$

Clearly equality $\beta(f) = 2$ holds if and only if f is a rotation of the Koebe function. This completes the proof.

The proof of Theorem 1.1.8 can also be based on Problem 1.1.16. Indeed, in view of (1.1.7), we have

$$\left| \frac{f'(re^{i\theta})}{f(re^{i\theta})} \right| \leq \frac{1 + r}{r(1 - r)} = \frac{d}{dr} \log \left[\frac{(1 - r)^2}{r} \right].$$

Hence from Problem 1.1.16 we conclude that $|f(re^{i\theta})|(1 - r)^2/r$ (with θ fixed) and $M_\infty(r, f)(1 - r)^2/r$ are strictly decreasing functions of $r \in (0, 1)$ unless f is a rotation of the Koebe function.

We remark that Krzyz [Krz1] proved in addition that if $f \in S$, then

$$2 \lim_{r \nearrow 1}(1-r)^2 M_\infty(r,f) = \lim_{r \nearrow 1}(1-r)^3 M_\infty(r,f').$$

1.1.5   Two-point distortion results for the class S

It is well known that the classical growth theorem stated in (1.1.5) gives a necessary but not sufficient condition for univalence (Problem 1.1.15). Blatter [Bla] considered the question of whether there exists a version of the growth theorem which is sufficient for univalence. He was led to formulate a two-point distortion theorem, stated in terms of the hyperbolic distance and invariant under pre-composition with automorphisms of U and post-composition with automorphisms of $\mathbb{C}$. More recently, Kim and Minda [Kim-Min] obtained a one-parameter family of such theorems which contains Blatter's as a special case, and which also contains an invariant version of the classical distortion theorem as a special case. Their results were further extended by Jenkins [Jen2].

We begin with some basic facts about the hyperbolic metric and the hyperbolic distance on U.

The hyperbolic metric on U, also called the Poincaré metric, is the Riemannian metric whose element of arc length is

$$ds = \frac{|dz|}{1-|z|^2}.$$

The distance function induced by the hyperbolic metric is

$$d_h(a,b) = \inf \int_\gamma ds,$$

where the infimum is taken over all piecewise C^1 curves γ in U joining a and b. It is given by the formula

$$(1.1.12) \qquad d_h(a,b) = \operatorname{arctanh}\left|\frac{b-a}{1-\overline{a}b}\right|, \quad a,b \in U.$$

Both the hyperbolic distance and the hyperbolic metric on U are invariant under conformal automorphisms of U. The geodesics of the hyperbolic metric on U are circular arcs and line segments which are orthogonal to the unit circle.

For further details about the hyperbolic metric and related invariant metrics in complex analysis, the reader may consult the books of Ahlfors [Ahl2], Jarnicki and Pflug [Jar-Pf], Kobayashi [Kob], or Franzoni and Vesentini [Fra-Ve].

First, we present an invariant distortion theorem obtained by Kim and Minda [Kim-Min], the most elementary result of this type. Actually, this theorem also contains the lower estimate in the classical growth theorem. It gives a necessary and sufficient condition for univalence. A related but weaker estimate appears in [Pom10, Corollary 1.5]. (See also Problem 1.1.14.)

We define the invariant differential operator D_1 on holomorphic functions f on U by

$$D_1 f(z) = (1 - |z|^2) f'(z), \quad z \in U.$$

This operator has the property that $D_1 f(z) = (f \circ T)'(0)$, where T is the disc automorphism given by $T(\zeta) = (\zeta + z)/(1 + \overline{z}\zeta), \zeta \in U$.

Theorem 1.1.10. *Suppose f is univalent on U and $a, b \in U$. Then*

$$(1.1.13) \quad |f(a) - f(b)| \geq \frac{\sinh(2d_h(a,b))}{2 \exp(2d_h(a,b))} \max \left\{ |D_1 f(a)|, |D_1 f(b)| \right\}.$$

There exist points $a, b \in U$, $a \neq b$, for which equality holds if and only if $f = \Phi \circ k \circ T$, where Φ is an automorphism of $\mathbb{C}$, k is the Koebe function, and T is an automorphism of the unit disc. Conversely, if f is a nonconstant holomorphic function on U satisfying (1.1.13), then f is univalent on U.

Proof. Let T be the disc automorphism given by $T(z) = \dfrac{z + a}{1 + \overline{a}z}, z \in U$. Also consider the Koebe transform g of f defined by

$$g(z) = \frac{f(T(z)) - f(T(0))}{(f \circ T)'(0)} = \frac{f\left(\dfrac{z + a}{1 + \overline{a}z}\right) - f(a)}{(1 - |a|^2) f'(a)}, \quad z \in U.$$

Then $g \in S$ and from (1.1.5) we have

$$|g(z)| \geq \frac{|z|}{(1 + |z|)^2} = \frac{\sinh(2d_h(0, z))}{2 \exp(2d_h(0, z))}, \quad z \in U.$$

Let $z \in U$ be such that $T(z) = b$, i.e. $z = \dfrac{b-a}{1-\overline{a}b}$. Taking into account the fact that the hyperbolic distance is invariant under the map T, the above relation is equivalent to

$$|f(a) - f(b)| \geq \frac{\sinh(2d_h(a,b))}{2\exp(2d_h(a,b))}|D_1 f(a)|.$$

Interchanging the roles of a and b leads to a second inequality of the same type, and taking the maximum of the lower bounds on $|f(a) - f(b)|$ gives (1.1.13).

The conditions under which equality holds in (1.1.13) can be deduced from the fact that equality in the lower estimate of the classical growth theorem holds only for rotations of the Koebe function.

Now we show that the condition (1.1.13) implies the univalence of f. Let f be a nonconstant holomorphic function in U and suppose that $f(a) = f(b)$ for distinct points $a, b \in U$. Then (1.1.13) implies that $f'(a) = f'(b) = 0$, and hence f is not univalent in any neighbourhood of a (or of b). Hence, we can find two sequences $\{c_n\}_{n\in\mathbb{N}}$, $\{d_n\}_{n\in\mathbb{N}}$ of distinct points in U such that $\lim_{n\to\infty} c_n = a$, $\lim_{n\to\infty} d_n = a$ and $f(c_n) = f(d_n)$, for all $n \in \mathbb{N}$. Again using (1.1.13), we conclude that $f'(c_n) = 0$, for all $n \in \mathbb{N}$, and hence f must be constant. However this is a contradiction with the hypothesis, and we conclude that f is univalent. This completes the proof.

Blatter's original result [Bla] is the following:

Theorem 1.1.11. *Let $f : U \to \mathbb{C}$ be a univalent function and let $a, b \in U$. Then*

$$|f(a) - f(b)|^2 \geq \frac{\sinh^2(2d_h(a,b))}{8\cosh(4d_h(a,b))}\left\{|D_1 f(a)|^2 + |D_1 f(b)|^2\right\}.$$

There exist points $a, b \in U$, $a \neq b$, such that equality holds if and only if f has the form $f = \Phi \circ k \circ T$, where Φ is an automorphism of $\mathbb{C}$, k is the Koebe function, and T is an automorphism of U. Conversely, let f be a nonconstant holomorphic function on U which satisfies the above relation. Then f is univalent on U.

We shall not give the proof, but we remark that it requires the following coefficient estimates that are satisfied for each $f \in S$, $f(z) = z + \sum_{k=2}^{\infty} a_k z^k$,

$z \in U$:

$$|a_2| \leq 2, \quad |a_3| \leq 3, \quad |a_3 - a_2^2| \leq 1.$$

In 1994 Kim and Minda [Kim-Min] showed that both Blatter's result and Theorem 1.1.10 were special cases of a more general theorem in which a parameter appears. The range of permissible parameters was extended by Jenkins [Jen2].

Theorem 1.1.12. *Let f be a univalent function on U. Then for all $a, b \in U$ and for any $p \geq 1$,*

$$|f(a) - f(b)| \geq \frac{\sinh(2d_h(a,b))}{2[2\cosh(2pd_h(a,b))]^{1/p}}\left[|D_1 f(a)|^p + |D_1 f(b)|^p\right]^{1/p}.$$

Equality holds under the same conditions as in Theorems 1.1.10 and 1.1.11. Conversely, if f is a nonconstant holomorphic function on U, satisfying the above inequality, then f is univalent on U.

The case $p = 2$ gives Blatter's result and the case $p = \infty$ gives Theorem 1.1.10. The lower bound decreases with p, and hence the strongest result of this type is the case $p = 1$, while Theorem 1.1.10 is the weakest. Corresponding upper estimates for $|f(a) - f(b)|$ were given by both Jenkins [Jen2] and Ma and Minda [Ma-Min7].

Notes. Most of the material in this chapter is well known. The basic sources used here are [Dur], [Pom5] and [Goo1]. But there are other excellent sources, including [Ahl2], [Con], [Gol4], [Gon5], [Hal-MG], [Hen], [Hil2], [Jen1], [Mili2], [Moc-Bu-Să], [Mon], [Neh2], [Ros-Rov], [Scho].

Problems

1.1.1. Let $f \in S$ be given by (1.1.1) and suppose that $|f(z)| < M$ on U. Show that $|a_2| \leq 2\left(1 - M^{-1}\right)$.
(Pick, 1917 [Pic].)

1.1.2. Suppose that $f \in S$ has the Taylor series expansion (1.1.1). Show that $|a_3 - a_2^2| \leq 1$. Study the case of equality.

1.1.3. Let k be a positive integer and $f \in S$ be a k-fold symmetric function (i.e. $e^{-2\pi i/k} f(e^{2\pi i/k} z) = f(z)$, $z \in U$). Prove the following sharp estimates:

$$\frac{|z|}{(1+|z|^k)^{2/k}} \leq |f(z)| \leq \frac{|z|}{(1-|z|^k)^{2/k}}, \quad z \in U.$$

1.1.4. Let $f \in H(U)$ be given by $f(z) = z + \sum_{k=2}^{\infty} a_k z^k$, $z \in U$. Show that if $\sum_{k=2}^{\infty} k|a_k| \leq 1$, then f is univalent on U.

1.1.5. Find the necessary and sufficient condition on a_2 such that $f(z) = z + a_2 z^2$ is univalent on U.

1.1.6. Let $f \in H_u(U)$ and $\Omega = f(U)$. Prove that

$$\frac{1}{4}(1 - |z|^2)|f'(z)| \leq \delta_\Omega(f(z)) \leq (1 - |z|^2)|f'(z)|, \quad z \in U,$$

where $\delta_\Omega(f(z))$ is the Euclidean distance from $f(z)$ to $\partial\Omega$. These estimates are sharp. In particular, if $f \in S$ then

$$\frac{1}{4} \leq \delta_\Omega(0) \leq 1.$$

1.1.7. Show that if $0 < \alpha \leq 2$ and

$$f(z) = \frac{1}{2\alpha}\left[\left(\frac{1+z}{1-z}\right)^\alpha - 1\right], \quad z \in U,$$

then $f \in S$. Describe the image of f.

1.1.8. Determine which Koebe transforms of the Koebe function are equal to rotations of the Koebe function.

1.1.9. Let $\varphi \in \Sigma$. Derive the inequality

$$|\varphi'(\zeta)| \leq \frac{|\zeta|^2}{|\zeta|^2 - 1}, \quad |\zeta| > 1.$$

Hint. Use the area theorem and the Cauchy-Schwarz inequality.

1.1.10. Let $\varphi \in \Sigma$ be given by (1.1.2) and $E = E(\varphi)$ be the complement of the image domain. Show that $|w - \alpha_0| \leq 2$ for $w \in E$, and that equality holds if and only if E is a line segment of length 4.

1.1.11. Suppose $f \in S$ and $|z| = r < 1$. Show that

$$|\arg f'(z)| \leq 2\log\left[\frac{1+r}{1-r}\right].$$

Hint. Consider the imaginary part of the quantity whose modulus appears in (1.1.9), and integrate along the line segment joining 0 to r.

(This elementary estimate for $|\arg f'(z)|$ is not sharp for any $z \neq 0$. The sharp estimate for $|\arg f'(z)|$ for $f \in S$ will be discussed in Chapter 3.)

1.1.12. Let $f \in S$. Show that

$$\frac{1}{2\pi}\int_0^{2\pi} |f(re^{i\theta})| d\theta \leq \frac{r}{1-r}, \quad r \in [0,1).$$

1.1.13. Let $f(z) = a_1 z + a_2 z^2 + \ldots$ be a univalent function on the unit disc such that $|f(z)| < 1$, $|z| < 1$. Prove the sharp inequality $|a_2| \leq 2|a_1|(1 - |a_1|)$. (Pick, 1917 [Pic].)

1.1.14. Let $f : U \to \mathbb{C}$ be a univalent function and let $a, b \in U$. Show that

$$\exp\{-6d_h(a,b)\} \leq \left|\frac{f'(a)}{f'(b)}\right| \leq \exp\{6d_h(a,b)\}.$$

(Pommerenke, 1992 [Pom10].)

1.1.15. Give an example of a normalized holomorphic function on the unit disc U which satisfies the growth theorem but is not univalent.

1.1.16. Let $g : U \to \mathbb{C}$ be a holomorphic function and h be a positive differentiable function on $(0,1)$. Assume that

$$(1.1.14) \qquad \left|\frac{g'(z)}{g(z)}\right| \leq \frac{h'(|z|)}{h(|z|)}, \quad z \in U \setminus \{0\}.$$

Prove that $|g(re^{i\theta})|/h(r)$ (θ fixed) and $M_\infty(r, g)/h(r)$ are non-increasing functions of $r \in (0,1)$. Moreover, if strict inequality holds in (1.1.14), then the above functions are strictly decreasing.

(Krzyz, 1955 [Krz1].)

1.1.17. Prove the following principle of univalence on the boundary: Let f be a holomorphic function on $\overline{U}$ which is injective on ∂U. Then f is univalent on U.

Chapter 2

Subclasses of univalent functions in the unit disc

In this chapter we shall discuss the basic properties of certain subclasses of S which are of special interest in their own right. These include the starlike, convex, close-to-convex, alpha-convex, spirallike and Φ-like functions. Most of these subclasses have both an analytic and a geometric characterization. In some cases there are more restrictive growth, covering, and distortion theorems than for the full class S. These classes are closely related with functions of positive real part and with subordination.

2.1 Functions with positive real part. Subordination and the Herglotz formula

2.1.1 The Carathéodory class. Subordination

In the following we shall give the basic properties of functions with positive real part in the unit disc U. Also we shall discuss the concept of subordination in the complex plane.

Let $\mathcal{P}$ denote the class of holomorphic functions p in U such that $p(0) = 1$ and Re $p(z) > 0$, $z \in U$.

This class is usually called the *Carathéodory class*.

For example, the function $p(z) = (1 + z)/(1 - z)$, $z \in U$, belongs to $\mathcal{P}$. This function gives a conformal map of U onto the right-half plane, and consequently it plays a fundamental role in the class $\mathcal{P}$, similar to the Koebe function for the class S.

We also note that $\mathcal{P}$ is a convex set and later we will see that $\mathcal{P}$ is also a compact subset of $H(U)$.

Now let $\mathcal{V}$ denote the class of *Schwarz functions*, i.e. $\varphi \in \mathcal{V}$ if and only if $\varphi \in H(U)$, $\varphi(0) = 0$ and $|\varphi(z)| < 1$ on U. In other words, $\mathcal{V}$ consists precisely of those holomorphic functions on U which satisfy the hypotheses of the Schwarz lemma.

Obviously, there is the following correspondence between the classes $\mathcal{P}$ and $\mathcal{V}$:

$$p \in \mathcal{P} \text{ if and only if } p = \frac{1 + \varphi}{1 - \varphi}, \quad \varphi \in \mathcal{V}.$$

Because of this correspondence, certain properties of $\mathcal{P}$ can be inferred from those of the class $\mathcal{V}$ and conversely.

Given two functions $f, g \in H(U)$, we say that f *is subordinate to* g (written $f \prec g$) if there exists a function $\varphi \in \mathcal{V}$ such that $f = g \circ \varphi$.

Thus if $f \prec g$, then $f(0) = g(0)$ and $f(U) \subseteq g(U)$. It follows from the Schwarz lemma that $|f'(0)| \leq |g'(0)|$ and $f(U_r) \subseteq g(U_r)$ for all $r \in (0, 1)$.

In particular, if $f \prec g$ we have

$$\max_{|z| \leq r} |f(z)| \leq \max_{|z| \leq r} |g(z)|, \quad r \in (0, 1).$$

Moreover, using the Schwarz-Pick lemma (see Problem 2.1.5), we deduce that if $f \prec g$ then

$$\max_{|z| \leq r} (1 - |z|^2)|f'(z)| \leq \max_{|z| \leq r} (1 - |z|^2)|g'(z)|, \quad 0 \leq r < 1.$$

If g is univalent on U, then it is obvious that $f \prec g$ if and only if $f(0) = g(0)$ and $f(U) \subseteq g(U)$. The fact that $f(U_r) \subseteq g(U_r)$ for all $r \in (0, 1)$ under these conditions is known as the *subordination principle*.

In particular, a holomorphic function p on U with $p(0) = 1$ belongs to $\mathcal{P}$ if and only if $p(z) \prec \dfrac{1 + z}{1 - z}$ on U. Also, if $\varphi \in H(U)$, $\varphi(0) = 0$, then φ belongs to $\mathcal{V}$ if and only if $\varphi(z) \prec z$ on U.

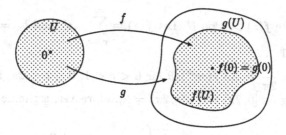

Figure 2.1: $f \prec g$

For a detailed study of the concept of subordination in the complex plane and many applications, see the recent book of Miller and Mocanu [Mill-Moc2].

We now turn to the class of functions with positive real part and we derive the Herglotz representation formula [Her]. This fundamental result leads to integral representation theorems for several subclasses of S. An alternative representation for such functions may be obtained by taking the integral in (2.1.1) to be a Lebesgue integral with respect to a finite positive Borel measure instead of a Riemann-Stieltjes integral with respect to a non-decreasing function (see [Hal-MG], [Rud3]). The integrand in (2.1.1) is the analytic completion of the Poisson kernel.

Theorem 2.1.1. *Let $f \in H(U)$. Then* Re $f(z) \geq 0$ *on U if and only if there exists a non-decreasing function μ on $[0, 2\pi]$ such that $\mu(2\pi) - \mu(0) =$* Re $f(0)$ *and*

$$(2.1.1) \qquad f(z) = \int_0^{2\pi} \frac{1 + ze^{-it}}{1 - ze^{-it}} d\mu(t) + i\text{Im } f(0), \quad z \in U.$$

Proof. First, assume f satisfies (2.1.1). Then it is clear that

$$\text{Re } f(z) = \int_0^{2\pi} \text{Re } \frac{1 + ze^{-it}}{1 - ze^{-it}} d\mu(t) \geq 0,$$

since μ is non-decreasing on $[0, 2\pi]$ and the integrand has positive real part.

Conversely, assume Re $f(z) \geq 0$ on U. Without loss of generality we may

assume that Re $f(z) > 0$ on U. Let $f(z) = \sum_{n=0}^{\infty} a_n z^n$ and $b_n = \text{Re } a_n$, $c_n =$
Im a_n, $n = 0, 1, \ldots$.

If $\mu(r, t) = \dfrac{1}{2\pi} \displaystyle\int_0^t \text{Re } f(re^{i\theta}) d\theta$ for $0 < r < 1$, $0 \le t \le 2\pi$, then $\mu(r, \cdot)$ is
non-decreasing on $[0, 2\pi]$ and $\mu(r, 2\pi) = b_0$. Moreover, a simple computation
shows that

$$\int_0^{2\pi} e^{-int} d\mu(r, t) = \begin{cases} a_n \dfrac{r^n}{2}, & n = 1, 2, \ldots \\ b_0, & n = 0. \end{cases}$$

Hence

$$f(z) = \int_0^{2\pi} d\mu(r, t) + \sum_{n=1}^{\infty} \frac{2}{r^n} \int_0^{2\pi} e^{-int} d\mu(r, t) z^n + ic_0.$$

For $|z| < r$, we may write this as

$$f(z) = \int_0^{2\pi} \left[1 + 2 \sum_{n=1}^{\infty} \left(\frac{e^{-it} z}{r} \right)^n \right] d\mu(r, t) + ic_0,$$

since the series in the last integrand converges uniformly in t. Summing the
series gives

(2.1.2) $$f(z) = \int_0^{2\pi} \frac{re^{it} + z}{re^{it} - z} d\mu(r, t) + ic_0.$$

Now let $\{\rho_n\}_{n \in \mathbb{N}}$ be an increasing sequence in $(0, 1)$ such that $\lim_{n \to \infty} \rho_n = 1$,
and let $\mu_n(t) = \mu(\rho_n, t)$, $t \in [0, 2\pi]$. Then $\{\mu_n\}_{n \in \mathbb{N}}$ is a sequence of non-
decreasing functions on $[0, 2\pi]$, and by the *Helly selection theorem* ([Nat,
p.233]; see also [Dur, p.22-23]) we can find a subsequence $\{\mu_{n_k}\}_{k \in \mathbb{N}}$ and a
non-decreasing function μ on $[0, 2\pi]$ such that $\mu_{n_k} \to \mu$ as $k \to \infty$ and

$$\lim_{k \to \infty} \int_0^{2\pi} h(t) d\mu_{n_k}(t) = \int_0^{2\pi} h(t) d\mu(t),$$

for each continuous function h on $[0, 2\pi]$.

Together with the fact that

$$\frac{\rho_{n_k} e^{it} + z}{\rho_{n_k} e^{it} - z} \to \frac{e^{it} + z}{e^{it} - z}$$

uniformly in t for fixed z as $k \to \infty$, this implies

$$\lim_{k \to \infty} \int_0^{2\pi} \frac{\rho_{n_k} e^{it} + z}{\rho_{n_k} e^{it} - z} d\mu_{n_k}(t) = \int_0^{2\pi} \frac{e^{it} + z}{e^{it} - z} d\mu(t).$$

From (2.1.2) and the above equality we deduce (2.1.1), as desired. This completes the proof.

A direct consequence of the previous theorem is the Herglotz integral formula for functions in the class $\mathcal{P}$ [Her].

Corollary 2.1.2. *Let $p \in H(U)$ satisfy $p(0) = 1$. Then $p \in \mathcal{P}$ if and only if there exists a non-decreasing function μ on $[0, 2\pi]$ with $\mu(2\pi) - \mu(0) = 1$ and such that*

$$(2.1.3) \qquad p(z) = \int_0^{2\pi} \frac{1 + ze^{-it}}{1 - ze^{-it}} d\mu(t) \text{ on } U.$$

The Herglotz formula leads to the following growth and distortion result for functions with positive real part:

Theorem 2.1.3. *If $p \in \mathcal{P}$ and $|z| = r < 1$ then*

$$(2.1.4) \qquad \frac{1 - r}{1 + r} \le |p(z)| \le \frac{1 + r}{1 - r},$$

$$(2.1.5) \qquad \frac{1 - r}{1 + r} \le \operatorname{Re} p(z) \le \frac{1 + r}{1 - r},$$

$$(2.1.6) \qquad |p'(z)| \le \frac{2\operatorname{Re} p(z)}{1 - r^2} \le \frac{2}{(1 - r)^2}.$$

These estimates are sharp.

Proof. The first and second inequalities are simple consequences of the relation (2.1.3). To prove the estimate (2.1.6), it suffices to differentiate both sides of (2.1.3) with respect to z, to obtain

$$|p'(z)| \le \int_0^{2\pi} \frac{2d\mu(t)}{|1 - ze^{-it}|^2} = \frac{2}{1 - r^2} \int_0^{2\pi} \operatorname{Re} \frac{1 + ze^{-it}}{1 - ze^{-it}} d\mu(t) = \frac{2\operatorname{Re} p(z)}{1 - r^2}.$$

Finally, it is obvious to see that equality holds in each of these relations for $p(z) = \frac{1 + \lambda z}{1 - \lambda z}$, $z \in U$, for some $\lambda \in \mathbb{C}$, $|\lambda| = 1$.

Remark 2.1.4. Using formula (2.1.3), it is easy to deduce that for each fixed z, $|z| = r < 1$, and $p \in \mathcal{P}$, $p(z)$ lies in the closed disc centered at $\frac{1 + r^2}{1 - r^2}$ and of radius $\frac{2r}{1 - r^2}$ (see Figure 2.2). (This can also be seen using a subordination argument.) Thus if $p \in \mathcal{P}$ then

$$\left| p(z) - \frac{1 + r^2}{1 - r^2} \right| \le \frac{2r}{1 - r^2}, \quad |z| = r.$$

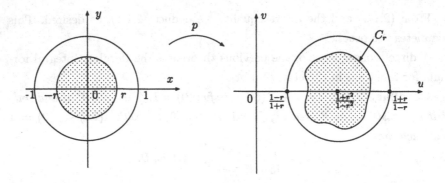

Figure 2.2: C_r = the image of ∂U_r.

The Herglotz formula also gives the following bounds for the coefficients of functions in $\mathcal{P}$. This result is due to Carathéodory [Cara1].

Theorem 2.1.5. *If $p \in \mathcal{P}$ has the power series $p(z) = 1 + \sum\limits_{n=1}^{\infty} p_n z^n$, $z \in U$, then $|p_n| \le 2$ for all $n = 1, 2, \ldots$. These estimates are sharp.*

Proof. Since $p \in \mathcal{P}$, there is a non-decreasing function μ on $[0, 2\pi]$ such that $\mu(2\pi) - \mu(0) = 1$ and

$$p(z) = \int_0^{2\pi} \frac{1 + ze^{-it}}{1 - ze^{-it}} d\mu(t).$$

A binomial expansion of the integrand gives $p_n = 2 \int_0^{2\pi} e^{-int} d\mu(t)$ for all $n = 1, 2, \ldots$. Therefore $|p_n| \le 2$, as claimed. Clearly $|p_n| = 2$, $n = 1, 2, \ldots$, for $p(z) = \dfrac{1 + \lambda z}{1 - \lambda z}$, with $|\lambda| = 1$. This completes the proof.

Finally we state the following compactness result for the class $\mathcal{P}$. We leave the proof for the reader, since it is similar to the proof of Corollary 1.1.7.

Corollary 2.1.6. $\mathcal{P}$ *is a compact set.*

2.1.2   Applications of the subordination principle

Next we obtain simple applications of the subordination principle. These applications, due to Robertson [Robe3], are useful in the study of certain subclasses of univalent functions on the unit disc. We mention that both Theorems 2.1.7 and 2.1.8 can be generalized to several complex variables (see Lemmas 6.1.33 and 6.1.35).

Theorem 2.1.7. *Let* $w(\cdot, t)$ *be a holomorphic function on* U *such that* $w(0, t) = 0$ *for* $t \in [0, 1]$. *Assume that* $|w(z, t)| < 1$ *for* $|z| < 1$, $0 \le t \le 1$, *and* $w(z, 0) = z$, $|z| < 1$. *Let* ρ *be a positive number for which the limit*

$$(2.1.7) \qquad w(z) = \lim_{t \searrow 0} \left[\frac{w(z, t) - z}{zt^\rho} \right]$$

exists, is holomorphic on U, *and is such that* $\operatorname{Re} w(0) \ne 0$. *Then* $\operatorname{Re} w(z) < 0$ *for* $|z| < 1$.

Proof. In view of the Schwarz lemma, $|w(z, t)| \le |z|$ for $z \in U$ and $t \in [0, 1]$. Taking into account (2.1.7), we deduce that $\lim_{t \searrow 0} w(z, t) = z = w(z, 0)$. Then it is obvious that $\operatorname{Re} w(z) \le 0$ and since w is holomorphic on U and $\operatorname{Re} w(0) \ne 0$, we must have $\operatorname{Re} w(z) < 0$ on U by the minimum principle for harmonic functions. This completes the proof.

The above result is a special case of the following. However, the proof of Theorem 2.1.8 depends on Theorem 2.1.7.

Theorem 2.1.8. *Let* $f : U \to \mathbb{C}$ *be a function in the class* S. *Let* $g : U \times [0, 1] \to \mathbb{C}$ *be a function such that* $g(\cdot, t)$ *is holomorphic on* U *for each* $t \in [0, 1]$, $g(z, 0) \equiv f(z)$, $g(0, t) \equiv 0$ *and* $g(z, t) \prec f(z)$ *for* $z \in U$ *and* $t \in [0, 1]$. *Also let* ρ *be a positive number for which the limit*

$$(2.1.8) \qquad G(z) = \lim_{t \searrow 0} \left[\frac{g(z, t) - g(z, 0)}{zt^\rho} \right]$$

exists, is holomorphic on U, *and satisfies the condition* $\operatorname{Re} G(0) \ne 0$. *Then* $\operatorname{Re} [f'(z)/G(z)] < 0$, $|z| < 1$.

Proof. Since $g(z, t) \prec f(z)$, there exists a Schwarz function $w(\cdot, t)$ for each $t \in [0, 1]$, such that $g(z, t) = f(w(z, t))$, $z \in U$ and $t \in [0, 1]$. On the other hand, the relation (2.1.8) yields that $\lim_{t \searrow 0} g(z, t) = f(z)$, and hence $\lim_{t \searrow 0} w(z, t) = z$, $z \in U$.

For $t > 0$, we have

$$(2.1.9) \qquad \frac{g(z, t) - g(z, 0)}{zt^\rho} = \left[\frac{f(w(z, t)) - f(w(z, 0))}{w(z, t) - w(z, 0)} \right] \left[\frac{w(z, t) - w(z, 0)}{zt^\rho} \right].$$

Now let $t \searrow 0$ in (2.1.9). The left side of (2.1.9) has the limit $G(z)$ by (2.1.8). The first factor on the right side of (2.1.9) has the limit $f'(z) \neq 0$. Hence the limit

$$w(z) = \lim_{t \searrow 0} \frac{w(z,t) - w(z,0)}{zt^\rho}$$

exists and in view of (2.1.9), $w(z) = G(z)/f'(z)$. Moreover, since G is holomorphic on U and Re $G(0) \neq 0$, w is also holomorphic on U and Re $w(0) \neq 0$. Using the result of Theorem 2.1.7, we deduce that Re $w(z) < 0$ on U. This completes the proof.

Notes. The material in Section 2.1 is classical and well-known. For additional results see the books [Dur], [Goo1], [Hal-MG], [Pom5] (basic sources used in the preparation of this section), [Gol4], [Hay], [Mill-Moc2], [Neh2], [Scho], and the papers of Rogosinki ([Rog1], [Rog2]), Robertson [Robe3], Robinson [Robi], Herglotz [Her], and MacGregor ([Mac3], [Mac4]).

Problems

2.1.1. Suppose $f \in H(U)$, $g \in S$, and $f \prec g$. Show that

$$|f(z)| \leq \frac{|z|}{(1-|z|)^2} \quad \text{and} \quad |f'(z)| \leq \frac{1+|z|}{(1-|z|)^3} \quad \text{on} \quad U.$$

(Schiffer, 1936 [Sch].)

2.1.2. Let $f, g \in H(U)$ be such that $f \prec g$ and

$$f(z) = \sum_{n=0}^\infty a_n z^n, \quad g(z) = \sum_{n=0}^\infty b_n z^n, \quad z \in U.$$

Show that

$$\sum_{j=0}^n |a_j|^2 \leq \sum_{j=0}^n |b_j|^2, \quad n = 0, 1, 2, \ldots.$$

(Rogosinski, 1943 [Rog2], Robertson, 1970 [Robe4].)

2.1.3. Let $k \in \mathbb{N}$, $p \in \mathcal{P}$ and $|z| = r < 1$. Show that

$$\left| \frac{1}{k!} p^{(k)}(z) \right| \leq \frac{2}{(1-r)^{k+1}},$$

and

$$|\arg p(z)| \le \arcsin\left(\frac{2r}{1+r^2}\right).$$

2.1.4. For $0 \le \alpha < 1$, let

$$\mathcal{P}(\alpha) = \left\{p \in H(U) : p(0) = 1, \operatorname{Re} p(z) > \alpha, z \in U\right\}.$$

Find the growth result and the integral representation formula for functions in $\mathcal{P}(\alpha)$.

2.1.5. Let $f : U \to \mathbb{C}$ be a holomorphic function such that $|f(z)| < 1$, $z \in U$. Show that $|f'(z)| \le (1 - |f(z)|^2)/(1 - |z|^2)$, $z \in U$.

2.1.6. Prove Corollary 2.1.6.

2.1.7. (Generalized Schwarz Lemma) Suppose $\varphi : U \to U$ is a holomorphic function such that $\varphi(0) = \varphi'(0) = \ldots = \varphi^{(k-1)}(0) = 0$, $k \in \mathbb{N}$, $k \ge 2$. Show that $\left|\frac{1}{k!}\varphi^{(k)}(0)\right| \le 1$ and $|\varphi(z)| \le |z|^k$, $z \in U$. Equality holds in each of these relations if and only if $\varphi(z) = \lambda z^k$, where $\lambda \in \mathbb{C}$, $|\lambda| = 1$.

2.1.8. Show that if $f(z) = a_0 + a_1 z + \ldots + a_n z^n + \ldots$ is holomorphic on U and $|f(z)| < 1$ on U, then $|a_n| \le 1 - |a_0|^2$, for $n = 1, 2, \ldots$.

Hint. Apply the inequality $|h'(0)| \le 1 - |h(0)|^2$, to the following bounded holomorphic function on U: $h(z) = g\left(\sqrt[n]{z}\right)$, where

$$g(z) = \frac{1}{n}[f(\omega z) + f(\omega^2 z) + \ldots + f(\omega^n z)]$$

$$= a_0 + a_n z^n + a_{2n} z^{2n} + \ldots + a_{kn} z^{kn} + \ldots, \quad |z| < 1,$$

and $\omega = e^{\frac{2\pi i}{n}}$.

2.1.9. Show that if $p \in \mathcal{P}$ has the power series $p(z) = 1 + p_1 z + p_2 z^2 + \ldots + p_n z^n + \ldots$, $z \in U$, then $|p_2 - \frac{1}{2}p_1^2| \le 2 - \frac{1}{2}|p_1|^2$.

Hint. Use the function $q(z) = \dfrac{p(z) - 1}{z(p(z) + 1)}$, $z \in U$, and apply Problem 2.1.8.

2.1.10. Let $p \in H(U)$ with $p(0) = 1$ and $\alpha \in \mathbb{R}$. Suppose

$$\operatorname{Re}\left[p(z) + \alpha \frac{zp'(z)}{p(z)}\right] > 0, \quad z \in U.$$

Show that $\operatorname{Re} p(z) > 0$, $z \in U$.
(Sakaguchi, 1962 [Sak3].)

2.2   Starlike and convex functions

This section is devoted to the study of two of the most important sub-classes of S, namely the starlike and convex functions. Both classes are defined by geometrical considerations, but both have very useful analytic characterizations. Bounds for the Taylor series coefficients can be obtained much more easily than for the full class S. Starlike functions satisfy the same growth, distortion, and covering theorems as the full class S, but stronger results hold for convex functions and various other subclasses. The analytic characterizations of starlikeness and convexity can be generalized to higher dimensions, but the proofs must be modified considerably.

Definition 2.2.1. Let Ω be a set in $\mathbb{C}$. We say that Ω is *starlike with respect to a fixed point* $w_0 \in \Omega$ if the closed line segment joining w_0 to each point $w \in \Omega$ lies entirely in Ω. Also we say that Ω is *convex* if for all $w_1, w_2 \in \Omega$ the closed line segment between w_1 and w_2 lies entirely in Ω. In other words, Ω is convex if and only if Ω is starlike with respect to each of its points.

Let $r \in (0,1]$, $f \in H(U_r)$, and let $z_0 \in U_r$. We say that f is *starlike* on U_r with respect to z_0 if f is univalent on U_r and the image $f(U_r)$ is a starlike domain with respect to $w_0 = f(z_0)$. The term starlike will mean starlike with respect to zero. Also we say that f is *convex* on U_r if f is univalent on U_r and the image $f(U_r)$ is a convex domain in $\mathbb{C}$. Let $S^*(U_r)$ and $K(U_r)$ denote the subclasses of $S(U_r)$ consisting respectively of the normalized starlike and convex functions on U_r. The classes $S^*(U)$ and $K(U)$ will be denoted by S^* and K. These are the classes of normalized starlike and normalized convex functions in the unit disc U.

We begin with the well known analytical characterization of starlikeness.

Theorem 2.2.2. *Let $f : U \to \mathbb{C}$ be a holomorphic function with $f(0) = 0$. Then f is starlike if and only if $f'(0) \neq 0$ and*

$$\mathrm{Re} \left[\frac{z f'(z)}{f(z)} \right] > 0 \quad on \quad U.$$

Proof. First assume that f is starlike. Then f is univalent and hence $f'(0) \neq 0$. We shall show that $f(U_r)$ is a starlike domain (with respect to zero) for all $r \in (0,1)$. For this purpose fix $r \in (0,1)$, let $t \in (0,1)$ and consider

$g(z) = f^{-1}(tf(z))$ for $z \in U$. This function is well defined and holomorphic on U, since f is starlike. It satisfies $g(0) = 0$ and $|g(z)| < 1$ on U. From the Schwarz lemma we conclude that $|g(z)| \leq |z|$ on U, and hence $tf(z) = f(g(z)) \in f(U_r)$ for all $z \in U_r$. Thus $f(U_r)$ is a starlike domain with respect to zero. Geometric considerations yield that the image of the circle $|z| = r$ is a starlike curve with respect to 0, that is $\arg f(re^{i\theta})$ increases as θ increases on $[0, 2\pi]$. Hence

$$\frac{\partial}{\partial \theta} \arg f(re^{i\theta}) \geq 0, \quad \theta \in [0, 2\pi].$$

Since

$$\frac{\partial}{\partial \theta} \arg f(re^{i\theta}) = \frac{\partial}{\partial \theta} \mathrm{Im} \left[\log f(re^{i\theta}) \right]$$

$$= \mathrm{Im} \left[\frac{iz f'(z)}{f(z)} \right] = \mathrm{Re} \left[\frac{z f'(z)}{f(z)} \right],$$

we conclude that $\mathrm{Re} \left[\frac{z f'(z)}{f(z)} \right] \geq 0$, $|z| = r < 1$. But $f'(0) \neq 0$ and therefore, by applying the minimum principle for harmonic functions, we conclude that

$$\mathrm{Re} \left[\frac{z f'(z)}{f(z)} \right] > 0, \quad z \in U_r.$$

Since r is arbitrary, we obtain

$$\mathrm{Re} \left[\frac{z f'(z)}{f(z)} \right] > 0 \quad \text{on } U,$$

as claimed.

Conversely, assume $f'(0) \neq 0$ and $\mathrm{Re} \left[\frac{z f'(z)}{f(z)} \right] > 0$, $|z| < 1$. Then $f(z) \neq 0$, for $z \in U \setminus \{0\}$, for otherwise the function $\frac{z f'(z)}{f(z)}$ would have a pole in U. Fix $r \in (0, 1)$. A simple computation, as in the first part of the proof, shows that

$$\frac{\partial}{\partial \theta} \arg f(re^{i\theta}) > 0, \quad \theta \in [0, 2\pi].$$

Therefore $\arg f(re^{i\theta})$ is an increasing function of $\theta \in [0, 2\pi]$. Also, since f has only one simple zero in the entire unit disc U, we deduce from the argument principle that the variation of the argument of $f(re^{i\theta})$ for $\theta \in [0, 2\pi]$

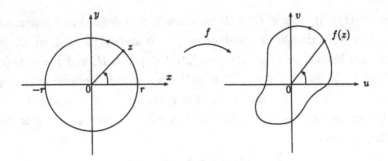

Figure 2.3: Starlikeness of the image of U_r

is equal to 2π. Indeed,

$$\int_0^{2\pi} \frac{\partial}{\partial \theta} \arg f(re^{i\theta})d\theta = \mathrm{Re}\left\{\frac{1}{i}\int\limits_{|z|=r} \frac{f'(z)}{f(z)}dz\right\} = 2\pi.$$

Therefore, the image of $|z| = r$ is a simple starlike curve and $f(U_r)$ is a starlike domain. Moreover, since f is injective on the circle $|z| = r$, the *principle of univalence on the boundary* (see e.g. [Pom5, Lemma 1.1]; see also Problem 1.1.17) implies that f is univalent in the disc U_r too. Since r is arbitrary, we conclude that f is univalent on the whole disc U. Finally since $f(U) = \bigcup\limits_{0<r<1} f(U_r)$, we conclude that $f(U)$ is a starlike domain with respect to zero. This completes the proof.

For convex functions we have the following well known analytical characterization:

Theorem 2.2.3. *Let $f : U \to \mathbb{C}$ be a holomorphic function. Then f is convex if and only if $f'(0) \neq 0$ and*

$$\mathrm{Re}\left[1 + \frac{zf''(z)}{f'(z)}\right] > 0, \quad z \in U.$$

Proof. First assume that f is convex. Then f is univalent and hence $f'(0) \neq 0$. We shall show that $f(U_r)$ is a convex domain for all $r \in (0,1)$. To this end, fix $r \in (0,1)$. Also let $z_1, z_2 \in U_r$ be such that $z_2 \neq 0, |z_1| \leq |z_2|$, and

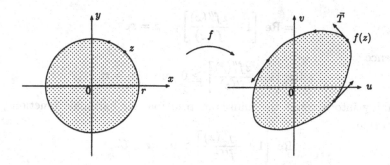

Figure 2.4: Convexity of the image of U_r

let $0 \leq t \leq 1$. Let $g : U \to U$ be defined by

$$g(z) = f^{-1}\left((1-t)f\left(\frac{z_1}{z_2}z\right) + tf(z)\right), \quad z \in U.$$

Since f is convex, g is well defined and holomorphic on U. Moreover, $g(0) = 0$ and hence $|g(z)| \leq |z|$ for $z \in U$, by the Schwarz lemma. For $z = z_2$ we deduce that $|g(z_2)| \leq |z_2| < r$, and hence $(1-t)f(z_1) + tf(z_2) \in f(U_r)$. Therefore we conclude that $f(U_r)$ is a convex domain.

Let Γ_r be the image of the circle $|z| = r$. Then Γ_r is a positively oriented Jordan curve and its interior domain is a convex domain. Moreover, Γ_r is given by the parametric representation $w = f(re^{i\theta})$, $0 \leq \theta \leq 2\pi$.

Since $f(U_r)$ is a convex domain, the argument of the tangent vector to Γ_r is a non-decreasing function of $\theta \in [0, 2\pi]$. That is, letting

$$\psi(\theta) = \arg\left[\frac{\partial}{\partial\theta}f(re^{i\theta})\right] = \arg[ire^{i\theta}f'(re^{i\theta})],$$

we must have $\psi'(\theta) \geq 0$ for $\theta \in [0, 2\pi]$, or

$$\frac{\partial}{\partial\theta}\mathrm{Im}\,[\log(izf'(z))] \geq 0, \quad z = re^{i\theta}.$$

A simple computation shows that

$$\frac{\partial}{\partial\theta}\mathrm{Im}\,[\log(izf'(z))] = \mathrm{Im}\left[i\left(1 + \frac{zf''(z)}{f'(z)}\right)\right]$$

$$= \operatorname{Re}\left[1+\frac{zf''(z)}{f'(z)}\right], \quad z = re^{i\theta},$$

and hence

$$\operatorname{Re}\left[1+\frac{zf''(z)}{f'(z)}\right] \geq 0 \quad \text{on} \quad |z| = r.$$

Taking into account the minimum principle for harmonic functions, we deduce that

$$\operatorname{Re}\left[1+\frac{zf''(z)}{f'(z)}\right] > 0, \quad z \in U_r,$$

since strict inequality holds at $z = 0$. Since r is arbitrary, we obtain

$$\operatorname{Re}\left[1+\frac{zf''(z)}{f'(z)}\right] > 0, \quad z \in U.$$

This completes the first part of the proof.

Conversely, assume that $f'(0) \neq 0$ and

$$\operatorname{Re}\left[1+\frac{zf''(z)}{f'(z)}\right] > 0 \quad \text{on} \quad U.$$

We note that this relation implies that $f'(z) \neq 0$ for $z \in U \setminus \{0\}$. Fix $r \in (0,1)$. Reversing the steps in the above computation, we deduce that

$$\frac{\partial}{\partial\theta}\arg[ire^{i\theta}f'(re^{i\theta})] \geq 0, \quad 0 \leq \theta \leq 2\pi,$$

and hence the argument of the tangent to the curve Γ_r, the image of the circle $|z| = r$, is a non-decreasing function of θ. Moreover, an integration shows that the total increment of $\psi(\theta)$ on $[0, 2\pi]$ is equal to 2π. Indeed,

$$\int_0^{2\pi} \psi'(\theta)d\theta = \int_0^{2\pi} \frac{\partial}{\partial\theta}\arg[ire^{i\theta}f'(re^{i\theta})]d\theta$$

$$= \int_0^{2\pi} \operatorname{Re}\left[1+\frac{re^{i\theta}f''(re^{i\theta})}{f'(re^{i\theta})}\right]d\theta$$

$$= \operatorname{Re}\int_{|z|=r}\left[1+\frac{zf''(z)}{f'(z)}\right]\frac{dz}{iz} = 2\pi.$$

We have therefore shown that Γ_r is a simple convex curve, and thus $f(U_r)$ is a convex domain. Because $f(U) = \bigcup_{0<r<1} f(U_r)$, we deduce that $f(U)$ is a convex domain. Moreover, since f is injective on the circle $|z| = r$, the principle of univalence on the boundary implies that f is also univalent on U_r for each $r \in (0, 1)$. Thus f is univalent on U and hence convex.

We remark that simple proofs of necessity of the conditions in both Theorems 2.2.2 and 2.2.3 follow from Theorem 2.1.8. (See [Robe3]. See also Problems 2.2.12 and 2.2.13.)

There are other necessary and sufficient conditions for convexity which are sometimes useful. The following characterization, which comes from the fact that convexity is equivalent to starlikeness with respect to each interior point, was given by Sheil-Small [She1] and Suffridge [Su1].

Theorem 2.2.4. *Let $f : U \to \mathbb{C}$ be a normalized holomorphic function. Then $f \in K$ if and only if*

$$(2.2.1) \qquad \mathrm{Re} \left[\frac{2zf'(z)}{f(z) - f(\zeta)} - \frac{z + \zeta}{z - \zeta} \right] \geq 0, \quad z, \zeta \in U.$$

Proof. Assume f is convex, in particular univalent. Then the function $g(z, \zeta) = \dfrac{2zf'(z)}{f(z) - f(\zeta)} - \dfrac{z + \zeta}{z - \zeta}$ is holomorphic in the unit polydisc $P = \{(z, \zeta) \in \mathbb{C}^2 : |z| < 1, |\zeta| < 1\}$ of $\mathbb{C}^2$, because $\lim\limits_{\zeta \to z} g(z, \zeta) = 1 + \dfrac{zf''(z)}{f'(z)}$.

Now the image under f of the circle $|z| = r < 1$ is a convex curve, that is a starlike curve with respect to each interior point. Hence letting z vary on the circle $|z| = r$ and fixing ζ with $|\zeta| < r$, we deduce that the argument of the vector between $f(\zeta)$ and $f(z)$ is a non-decreasing function of $\arg z$. The reasoning in the proof of Theorem 2.2.2 shows that the above condition is equivalent to

$$(2.2.2) \qquad \mathrm{Re} \left[\frac{zf'(z)}{f(z) - f(\zeta)} \right] > 0, \quad |\zeta| < |z| = r < 1.$$

If $|\zeta| = |z|$, $\zeta \neq z$, then (2.2.2) implies that

$$\mathrm{Re} \left[\frac{zf'(z)}{f(z) - f(\zeta)} \right] \geq 0,$$

and because $\mathrm{Re} \left[\dfrac{z + \zeta}{z - \zeta} \right] = 0$ in this case, we conclude that $\mathrm{Re}\, g(z, \zeta) \geq 0$. Since $z = \zeta$ is a removable singularity for $g(z, \zeta)$, we obtain

$$\mathrm{Re}\, g(z, \zeta) \geq 0, \quad |z| = |\zeta| = r < 1.$$

Applying the minimum principle for harmonic functions (first by fixing z, $|z| = r$ and varying ζ, $|\zeta| < r$, and then using the same procedure for ζ fixed and varying z), we obtain Re $g(z, \zeta) \geq 0$, $|z| < r$, $|\zeta| < r$. Now letting $r \to 1$, we obtain (2.2.1) as desired.

Conversely, assume (2.2.1) holds. Letting $\zeta \to z$ in (2.2.1), we obtain

$$\operatorname{Re} \left[1 + \frac{zf''(z)}{f'(z)}\right] \geq 0, \quad z \in U,$$

and hence by the minimum principle for harmonic functions and Theorem 2.2.3, we conclude that f is convex. This completes the proof.

Remark 2.2.5. From the proof of Theorem 2.2.4 we observe that for a normalized holomorphic function on the unit disc, we have the following additional characterization of convexity (see [Su1]):

(2.2.3) $\quad f \in K$ if and only if Re $\left[\dfrac{zf'(z)}{f(z) - f(\zeta)}\right] > 0, \quad |\zeta| < |z| < 1.$

Another necessary and sufficient condition for convexity on the unit disc was obtained by Ruscheweyh and Sheil-Small [Rus-She] (see Problem 2.2.14).

The results given in Theorems 2.2.2 and 2.2.3 lead to a very useful connection between the sets S^* and K. This connection was first discovered by Alexander [Al].

Theorem 2.2.6. *Let f be a normalized holomorphic function on U and let $g(z) = zf'(z)$, $z \in U$. Then $f \in K$ if and only if $g \in S^*$.*

The proof is straightforward and we leave it as an exercise for the reader.

Since the Koebe function and its rotations belong to S^*, we conclude that the distortion and growth result for the full class S (Theorem 1.1.6) is also sharp for the class S^*.

Theorem 2.2.7. *Let $f \in S^*$ and $|z| = r < 1$. Then*

(2.2.4) $$\frac{r}{(1+r)^2} \leq |f(z)| \leq \frac{r}{(1-r)^2}$$

and

(2.2.5) $$\frac{1-r}{(1+r)^3} \leq |f'(z)| \leq \frac{1+r}{(1-r)^3}.$$

All of these estimates are sharp. Equality holds in each of these relations for a suitable rotation of the Koebe function.

A direct consequence of this theorem is the fact that the set S^* is a compact subset of $H(U)$. Also the Koebe constant for S^* is $1/4$, as for the class S.

For normalized convex functions we have the following growth and distortion theorem (see [Gro], [Löl]):

Theorem 2.2.8. *Let $f \in K$ and $|z| = r < 1$. Then*

$$(2.2.6) \qquad \frac{r}{1+r} \leq |f(z)| \leq \frac{r}{1-r}$$

and

$$(2.2.7) \qquad \frac{1}{(1+r)^2} \leq |f'(z)| \leq \frac{1}{(1-r)^2}.$$

All of these estimates are sharp. Equality holds at a given point other than 0 for $f(z) = \dfrac{z}{1 - \lambda z}$ for some $\lambda \in \mathbb{C}$, $|\lambda| = 1$.

Proof. The estimate (2.2.7) is a simple consequence of Theorem 2.2.6 and relation (2.2.4). The upper bound in (2.2.6) follows easily by integrating the upper bound in (2.2.7).

In order to obtain the lower bound in (2.2.6), fix z with $|z| = r < 1$. Since $f \in K$, the closed segment Γ between 0 and $f(z)$ is contained in $f(U)$. If γ denotes the inverse image of Γ, then γ is a simple curve from 0 to z. We deduce that

$$|f(z)| = \int_{\Gamma} |dw| = \int_{\gamma} |f'(\zeta)||d\zeta| \geq \int_0^r |f'(\rho e^{i\theta})| d\rho \geq \frac{r}{1+r},$$

making use of the lower estimate in (2.2.7).

It is clear that equality holds for some $z_0 \in U \backslash \{0\}$ in each of the inequalities (2.2.6) and (2.2.7) if and only if $zf'(z)$ is a rotation of the Koebe function, and this implies that $f(z) = z/(1 - \lambda z)$ for some $\lambda \in \mathbb{C}$, $|\lambda| = 1$. For such a function equality holds in the upper or lower estimates along a ray.

The corresponding covering theorem for convex functions may be proved in several different ways:

Theorem 2.2.9. *If $f \in K$ then $f(U)$ contains the disc $U_{1/2}$. This result is sharp.*

Proof. Sharpness follows by considering the function $f(z) = z/(1 - z)$ (a conformal map of U onto a half plane) or its rotations. The statement itself follows from the lower estimate in (2.2.6) on letting $r \to 1$.

A second proof was given by MacGregor in 1964 [Mac1]: We shall show that if $w \notin f(U)$, then $|w| \geq 1/2$. To this end, let $g(z) = [f(z) - w]^2$, $z \in U$. It is easy to see that g is univalent on U, $g(0) = w^2$ and $g'(0) = -2w$. Letting $h(z) = (w^2 - g(z))/(2w)$, $z \in U$, we have $h \in S$ and from Theorem 1.1.5 we conclude that $|w/2| \geq 1/4$. This completes the second proof.

A third proof may be obtained using a subordination argument: Let $w_0 = \rho e^{i\phi} \in \partial f(U)$ be a point at minimum distance from 0. Clearly we may assume that $w_0 = \rho$, for otherwise we can replace f by a suitable rotation of f. Then it is not difficult to see that Re $w < \rho$ for all $w \in f(U)$, and hence $f \prec g$ where $g(z) = 2\rho z/(1 + z)$, $z \in U$. Using the subordination principle, we conclude that $1 = f'(0) \leq g'(0) = 2\rho$, and thus $\rho \geq 1/2$. This completes the proof.

The argument in the third proof actually gives the following (cf. [Gra2], [Min3]):

Theorem 2.2.10. *If f is a normalized holomorphic function on U, then the convex hull $\widehat{f(U)}$ of $f(U)$ contains the disc $U_{1/2}$.*

Further refinements of the growth and covering theorems can be obtained if we assume k-fold symmetry (cf. [Gra-Var2]). We recall that a function $f \in H(U)$ is said to be *k-fold symmetric*, where k is a positive integer, if $e^{-2\pi i/k} f(e^{2\pi i/k}z) = f(z)$, $z \in U$.

Theorem 2.2.11. *(i) Let $f \in S$ and suppose that f is k-fold symmetric. Then $f(U) \supseteq U_{\rho_k}$, where $\rho_k = \dfrac{1}{4^{1/k}}$.*

(ii) Suppose further that $f \in K$. Then $f(U) \supseteq U_{r_k}$, where $r_k = \displaystyle\int_0^1 \dfrac{dt}{(1 + t^k)^{2/k}}$.
Both results are sharp.

Proof. For part (i) it suffices to consider the kth root transform of f, given by $g(z) = \sqrt[k]{f(z^k)}$, $z \in U$. This function belongs to S and using the growth result for the class S (see Theorem 1.1.6), we conclude that

$$(2.2.8) \qquad \frac{|z|}{(1 + |z|^k)^{2/k}} \leq |f(z)| \leq \frac{|z|}{(1 - |z|^k)^{2/k}}, \quad z \in U.$$

Now consider the disc U_r with $0 < r < 1$. The image of U_r is an open set whose boundary is the image of ∂U_r. Using (2.2.8), we see that the image of ∂U_r has distance at least $r/(1 + r^k)^{2/k}$ from the origin, and letting $r \to 1$ we deduce (i).

To prove (ii), apply (2.2.8) to the function $h(z) = zf'(z)$ and then argue as in the proof of (2.2.6) to conclude that

$$\int_0^r \frac{dt}{(1+t^k)^{2/k}} \leq |f(z)| \leq \int_0^r \frac{dt}{(1-t^k)^{2/k}}, \quad |z| = r < 1.$$

Again letting $r \to 1$ in the lower estimate, we deduce our desired conclusion.

Both results are sharp for each k, because for the first we may choose the kth root transform of the Koebe function and for the second we choose the following convex functions:

$$(2.2.9) \qquad g_k(z) = \begin{cases} \dfrac{z}{1-z}, & k = 1 \\[2ex] \dfrac{1}{2} \log\left[\dfrac{1+z}{1-z}\right], & k = 2 \\[2ex] \displaystyle\int_0^z \dfrac{dt}{(1-t^k)^{2/k}}, & k \geq 3. \end{cases}$$

We note that for $k \geq 3$, g_k maps the unit disc conformally onto a regular polygon of order k (see [Neh2, p.196]).

For convex functions, a condition weaker than k-fold symmetry, namely $a_2 = a_3 = \ldots = a_k = 0$, suffices to obtain the same conclusions. In fact we have the following distortion theorem for such functions [Gra-Var2]:

Theorem 2.2.12. *Suppose $f(z) = z + a_{k+1}z^{k+1} + a_{k+2}z^{k+2} + \ldots$ is a convex function on U. Then*

$$\frac{1}{(1+|z^k|)^{2/k}} \leq |f'(z)| \leq \frac{1}{(1-|z|^k)^{2/k}}, \quad z \in U.$$

These estimates are sharp and equality holds for the functions g_k in (2.2.9).

Proof. The method of the proof is based on a subordination argument. If we let

$$g(z) = \frac{zf''(z)}{f'(z)} \quad \text{and} \quad G(z) = \frac{2z}{1-z}, \quad z \in U,$$

then the convexity of f implies that $g \prec G$.

Now $g(0) = 0$ and $g(z) = \dfrac{(k+1)ka_{k+1}z^k + \ldots}{1 + (k+1)a_{k+1}z^k + \ldots}$, $z \in U$, so that $g'(0) = \ldots = g^{(k-1)}(0) = 0$. Using the generalized Schwarz lemma in Problem 2.1.7 and the subordination principle, we conclude that $g(U_r) \subseteq G(U_{r^k})$ for $r \in (0, 1)$.

Next we note that $G(U_{r^k})$ is a disc centered on the real axis at $\dfrac{2r^{2k}}{1 - r^{2k}}$ and of radius $\dfrac{2r^k}{1 - r^{2k}}$. This gives

$$\left| g(z) - \frac{2r^{2k}}{1 - r^{2k}} \right| \leq \frac{2r^k}{1 - r^{2k}}, \quad |z| = r,$$

which is equivalent to

$$\left| \frac{zf''(z)}{f'(z)} - \frac{2r^{2k}}{1 - r^{2k}} \right| \leq \frac{2r^k}{1 - r^{2k}}.$$

Hence

$$\left| \frac{f''(z)}{f'(z)} - \frac{2\bar{z}r^{2k-2}}{1 - r^{2k}} \right| \leq \frac{2r^{k-1}}{1 - r^{2k}}, \quad |z| = r.$$

It suffices to prove our estimates for $z = r$, for otherwise we can use a rotation of f. In this situation we have

$$\left| \frac{f''(r)}{f'(r)} - \frac{2r^{2k-1}}{1 - r^{2k}} \right| \leq \frac{2r^{k-1}}{1 - r^{2k}}.$$

Integrating both sides, we obtain

$$\left| \log f'(r) + \frac{1}{k} \log(1 - r^{2k}) \right| \leq \int_0^r \frac{2t^{k-1}}{1 - t^{2k}} dt = \frac{1}{k} \log \left[\frac{1 + r^k}{1 - r^k} \right],$$

and hence

$$-\frac{1}{k} \log \left[\frac{1 + r^k}{1 - r^k} \right] \leq \log |f'(r)| + \frac{1}{k} \log(1 - r^{2k}) \leq \frac{1}{k} \log \left[\frac{1 + r^k}{1 - r^k} \right]$$

or

$$\frac{1}{k} \log \frac{1}{(1 + r^k)^2} \leq \log |f'(r)| \leq \frac{1}{k} \log \frac{1}{(1 - r^k)^2},$$

which implies the desired relations.

In the case of starlike functions with $a_2 = \ldots = a_k = 0$, we can obtain a corresponding growth result, using Alexander's theorem (Theorem 2.2.6).

Theorem 2.2.13. *Let* $f(z) = z + a_{k+1}z^{k+1} + \ldots$ *be a starlike function on* U. *Then*

$$\frac{|z|}{(1+|z|^k)^{2/k}} \leq |f(z)| \leq \frac{|z|}{(1-|z|^k)^{2/k}}, \quad z \in U.$$

These estimates are sharp and equality holds for the functions $f_k(z) = zg_k'(z)$, *where* g_k *is given by (2.2.9).*

Also from Theorem 2.2.12 we deduce a covering result for normalized convex functions with $a_2 = a_3 = \ldots = a_k = 0$ which generalizes Theorem 2.2.11(ii) [Gra-Var2].

Corollary 2.2.14. *If* $f(z) = z + a_{k+1}z^{k+1} + a_{k+2}z^{k+2} + \ldots$ *is a convex function on* U *then* $f(U)$ *contains the disc* U_{r_k}, *where* $r_k = \int_0^1 \frac{dt}{(1+t^k)^{2/k}}$.

Finally we mention a covering theorem for the convex hull of the image of a function with k-fold symmetry [Gra-Var2]. In this result we do not require univalence; however, it does not generalize to the case of functions with $a_2 = a_3 = \ldots = a_k = 0$. We leave the proof for the reader.

Theorem 2.2.15. *Let* $f : U \to \mathbb{C}$ *be a* k-*fold symmetric function with* $f'(0) = 1$. *Then* $\widehat{f(U)} \supseteq U_{r_k}$, *where* $r_k = \int_0^1 \frac{dt}{(1+t^k)^{2/k}}$.

We now turn to the study of coefficient bounds for functions in S^* and K. In the case of normalized starlike functions Bierberbach's conjecture (now de Branges' theorem) was proved by Nevanlinna [Nev2] as early as 1920:

Theorem 2.2.16. *If* $f(z) = z + a_2z^2 + \ldots + a_nz^n + \ldots$ *is starlike on* U, *then* $|a_n| \leq n$, *for all* $n = 2, 3, \ldots$. *Equality* $|a_n| = n$ *for a given* $n \geq 2$ *holds when* f *is a rotation of the Koebe function.*

Proof. Since $f \in S^*$, the function p defined by $p(z) = \dfrac{zf'(z)}{f(z)}$, $z \in U$, belongs to $\mathcal{P}$. If $p(z) = 1 + p_1z + \ldots + p_nz^n + \ldots$, $z \in U$, then from Theorem 2.1.5 we have $|p_n| \leq 2$, $n \geq 2$. Comparing the coefficients in the power series of $zf'(z)$ and $f(z)p(z)$, we deduce that

$$(n-1)a_n = a_{n-1}p_1 + a_{n-2}p_2 + \ldots + p_{n-1}, \quad n = 2, 3, \ldots.$$

Thus by induction, we obtain

$$(n-1)|a_n| \leq 2(n-1+n-2+\ldots+1) = n(n-1),$$

and hence $|a_n| \le n$, as desired.

It is easy to see that if $|a_n| = n$ for a given n, then the above arguments imply that $|a_2| = 2$, and thus f is a rotation of the Koebe function.

Nevanlinna's result together with Alexander's theorem gives the following bounds for the coefficients of functions in K (see Loewner [Lö1]):

Theorem 2.2.17. *If* $f(z) = z + a_2 z^2 + \ldots + a_n z^n + \ldots$ *is a convex function on* U, *then* $|a_n| \le 1$, *for all* $n = 2, 3, \ldots$. *Equality* $|a_n| = 1$ *for a given* $n \ge 2$ *holds for* $f(z) = \dfrac{z}{1 - \lambda z}$ *for some* $\lambda \in \mathbb{C}$, $|\lambda| = 1$.

Aside from estimates for the kth coefficient alone, there are certain other very useful estimates for the coefficients of functions in K. The following result, first obtained by Hummel [Hum] and then by Trimble [Tri], has some remarkable implications in the theory of univalent functions. We shall give Trimble's proof because it is very simple and elegant. We mention that Hummel's proof uses variational techniques.

Theorem 2.2.18. *Let* $f \in K$ *be defined by* $f(z) = z + a_2 z^2 + a_3 z^3 + \ldots$, $z \in U$. *Then*

(2.2.10) $$|a_3 - a_2^2| \le \frac{1 - |a_2|^2}{3}.$$

Proof. Let $p(z) = 1 + \dfrac{z f''(z)}{f'(z)}$, $z \in U$. Then $p \in \mathcal{P}$ and if we write $q(z) = \dfrac{1 - p(z)}{1 + p(z)}$, $z \in U$, then $q \in \mathcal{V}$ and q has the power series expansion

$$q(z) = -a_2 z + 3(a_2^2 - a_3)z^2 + \ldots, \quad z \in U.$$

If $q(z)/z$ is a unimodular constant then $a_2^2 - a_3 = 0$, and otherwise we can apply the result of Problem 2.1.8. to obtain

$$|3(a_2^2 - a_3)| \le 1 - |a_2|^2,$$

as desired.

We have seen in Problem 1.1.2 that if $f \in S$ has the Taylor series expansion given by (1.1.1), then

$$|a_3 - a_2^2| \le 1.$$

This estimate may be improved for functions in K. The following result was first obtained by Nehari [Neh3] and subsequently by Koepf [Koep2].

Corollary 2.2.19. *If* $f \in K$ *is given by* $f(z) = z + a_2 z^2 + a_3 z^3 + \ldots$, $z \in U$, *then*

(2.2.11)
$$|a_3 - a_2^2| \le \frac{1}{3}.$$

This estimate is sharp. Equality holds if and only if

(2.2.12)
$$f(z) = \frac{1}{2\lambda} \log \left[\frac{1 + \lambda z}{1 - \lambda z} \right], \quad |\lambda| = 1.$$

Proof. It is clear that (2.2.11) follows from (2.2.10). Moreover, from (2.2.10) we conclude that equality in (2.2.11) holds if and only if $a_2 = 0$ and $|a_3| = 1/3$. Thus, f must have the Taylor series expansion

$$f(z) = z + a_3 z^3 + \ldots, \quad z \in U,$$

and hence

$$1 + \frac{z f''(z)}{f'(z)} = 1 + 6 a_3 z^2 + \ldots, \quad z \in U.$$

This function belongs to $\mathcal{P}$, and because $6|a_3| = 2$ and there is no linear term we conclude that

$$1 + \frac{z f''(z)}{f'(z)} = \frac{1 + \lambda^2 z^2}{1 - \lambda^2 z^2}, \quad |\lambda| = 1.$$

(See [Pom5, Corollary 2.3].) This implies (2.2.12) and completes the proof.

The *Schwarzian derivative* of a holomorphic, locally univalent function f on U is defined by

$$\{f; z\} = \left(\frac{f''(z)}{f'(z)} \right)' - \frac{1}{2} \left(\frac{f''(z)}{f'(z)} \right)^2, \quad z \in U.$$

It is important because of its invariance property under linear fractional transformations. Other properties of the Schwarzian derivative will be discussed in the third chapter.

The next result is an estimate for the Schwarzian derivative of a convex function which uses Corollary 2.2.19. It was also obtained by Nehari [Neh3] and was considered in further detail by Koepf [Koep2]. Koepf identified the functions for which equality holds, and gave further characterizations of the functions for which the inequality is strict.

Theorem 2.2.20. *If* $f \in K$ *then*

(2.2.13) $$|\{f; z\}| \leq \frac{2}{(1 - |z|^2)^2}, \quad z \in U.$$

This estimate is sharp and equality holds if and only if $f(U)$ *is a strip domain, i.e.*

(2.2.14) $$f(z) = \frac{1}{\omega + \eta} \log \left[\frac{1 + \omega z}{1 - \eta z}\right], \quad |\omega| = |\eta| = 1, \quad \omega \neq -\eta.$$

Proof. It suffices to prove (2.2.13) for $z \neq 0$. Let g be the Koebe transform of f, given by

(2.2.15) $$g(\zeta) = \frac{f\left(\frac{\zeta + z}{1 + \bar{z}\zeta}\right) - f(z)}{(1 - |z|^2)f'(z)} = \zeta + b_2\zeta^2 + b_3\zeta^3 + \ldots, \quad \zeta \in U.$$

Clearly g is a convex function on U, and thus from Corollary 2.2.19 we have

$$|b_3 - b_2^2| \leq \frac{1}{3}.$$

A simple computation shows that

$$|(1 - |z|^2)^2\{f; z\}| = |\{g; 0\}| = 6|b_3 - b_2^2| \leq 2,$$

hence we obtain (2.2.13).

If we have equality in (2.2.13) for some $z \in U$, then an appropriate Koebe transform g of f has the form (2.2.12). Taking into account (2.2.15), we deduce that f must also map U onto an infinite strip. This completes the proof.

Finally we mention another estimate for the Schwarzian derivative of a normalized convex function on U, which follows from the proof of Theorem 2.2.18 with similar reasoning:

If $f \in K$ then

(2.2.16) $$(1 - |z|^2)^2|\{f; z\}| \leq 2 \left(1 - \left|\frac{1 - |z|^2}{2} \cdot \frac{f''(z)}{f'(z)} - \bar{z}\right|^2\right), \quad z \in U.$$

Remark 2.2.21. For more results concerning the Schwarzian derivative of univalent functions on U, the reader may consult the following references:

[Chu1], [Chu-Os1,2], [Chu-Pom], [Dur], [Har3], [Kra], [Leh], [Min4], [Neh2], [Neh4], [Os1], [Ov], [Pom5].

We end this section with a discussion of the radius of starlikeness and radius of convexity for the full class S. Of course one can also study these notions for subclasses of S.

Let $\mathcal{F}$ be a non-empty subclass of S. Let $r^*(\mathcal{F})$ be the largest positive number such that every function in the class $\mathcal{F}$ is starlike on the disc $U_{r^*(\mathcal{F})}$. This number is called the radius of starlikeness of $\mathcal{F}$. Also let $r_c(\mathcal{F})$ be the largest positive number such that every function in the class $\mathcal{F}$ is convex on $U_{r_c(\mathcal{F})}$. This number is called the radius of convexity of $\mathcal{F}$.

The following result, due to Nevanlinna [Nev1], determines the radius of convexity of S. We remark that Campbell [Cam] showed that if $f \in S$ has radius of convexity $2 - \sqrt{3}$, then f is a rotation of the Koebe function.

Theorem 2.2.22. $r_c(S) = 2 - \sqrt{3}$.

Proof. Let $f \in S$. Then f satisfies inequality (1.1.9) and hence, for $|z| = r < 1$, we have

$$\mathrm{Re}\left[1 + \frac{zf''(z)}{f'(z)}\right] \geq \frac{1 - 4r + r^2}{1 - r^2}.$$

Since $1 - 4r + r^2 \geq 0$ for $0 \leq r \leq 2 - \sqrt{3}$, we conclude that $f \in K(U_r)$ for any such value of r. Therefore $r_c(S) \geq 2 - \sqrt{3}$.

However, if $f(z) = \dfrac{z}{(1-z)^2}$, $z \in U$, and $2 - \sqrt{3} < r < 1$, then f does not map the circle $|z| = r$ onto a convex curve. To see this we note that

$$1 + \frac{zf''(z)}{f'(z)} = \frac{z^2 + 4z + 1}{1 - z^2},$$

and since this expression is negative for z real such that $-1 < z < -(2 - \sqrt{3})$, the conclusion follows. Hence $r_c(S) = 2 - \sqrt{3}$, as desired. This completes the proof.

We state the corresponding result for $r^*(S)$, obtained by Grunsky [Gru2]. We shall give a proof in Chapter 3 as an application of the method of Loewner chains. For other radius problems for univalent functions, the reader may consult [Goo1].

Theorem 2.2.23. $r^*(S) = \tanh \dfrac{\pi}{4}$.

We remark that Brown [Brow] obtained the sharp radius $\rho^* = \rho^*(|\zeta|)$ for which every function $f \in S$ maps the disc $|z - \zeta| < \rho^*$, $|\zeta| < 1$, onto a starlike domain with respect to $f(\zeta)$. We do not give the proof, but we mention that it is of interest for the reader.

Theorem 2.2.24. *Let $r \in [0, 1)$ and ρ^* be the unique root of the equation $A(\rho) = \pi/2$ in $(0, 1 - r)$, where*

$$A(\rho) \equiv \log\left\{\frac{\sqrt{(1-r^2)^2 - r^2\rho^2} + \rho}{\sqrt{(1-r^2)^2 - r^2\rho^2} - \rho}\right\} + \arctan\left\{\frac{r\rho}{\sqrt{(1-r^2)^2 - r^2\rho^2}}\right\}.$$

If $|\zeta| = r$ then every $f \in S$ maps the disc $|z - \zeta| < \rho^*$ onto a starlike domain with respect to $f(\zeta)$. The constant ρ^* is sharp.

Notes. The basic sources used in the preparation of this section are [Dur], [Pom5] and [Goo1]. For additional information about starlike and convex functions, the reader may consult the books [Gol4], [Hal-MG], [Hay], [Neh2], [Rus2], [Scho].

Problems

2.2.1. Let $f(z) = z + az^2$. Find the values of a for which $f \in S^*$, respectively $f \in K$.

2.2.2. Prove Alexander's Theorem (Theorem 2.2.6).

2.2.3. Using the Herglotz formula for functions with positive real part, find the corresponding integral representations for functions in S^* and K.

2.2.4. Show that if $f \in K$, then Re $\sqrt{f'(z)} > 1/2$ and Re $[f(z)/z] > 1/2$ on U. These estimates are sharp.
(Marx, 1932 [Marx] and Strohhäcker, 1933 [Str].)

2.2.5. Let $f \in H(U)$ have the Taylor series expansion given by (1.1.1). Show that if $\sum_{n=2}^{\infty} n|a_n| \leq 1$ then $f \in S^*$, and if $\sum_{n=2}^{\infty} n^2|a_n| \leq 1$ then $f \in K$.

2.2.6. Let $f \in H(U)$ be normalized and let $g \in K$. Show that if $zf'(z) \prec zg'(z)$, then $f(z) \prec g(z)$, $z \in U$.
(Suffridge, 1970 [Su1].)

2.2.7. Show that if $f \in S^*$, then $\mathrm{Re}\,[f(z)/z]^{1/2} > 1/2$, $z \in U$, or equivalently, $f(z)/z \prec 1/(1-z)^2$ on U.

(Robertson, 1936 [Robe1]; Goluzin, 1938 [Gol3].)

Hint. Use Problem 2.2.4 and Theorem 2.2.6.

2.2.8. Show that if $f \in S^*$, then $|\arg[f(z)/z]| \le 2\arcsin r$, for $|z| = r < 1$. Deduce that if $f \in K$, then $|\arg f'(z)| \le 2\arcsin r$, $|z| = r$. All of these bounds are sharp.

2.2.9. Let $f(z) = z + \sum_{n=2}^{\infty} a_{2n-1} z^{2n-1}$, $|z| < 1$, be an odd function such that f is univalent on U. For $t_0 \in [0, 1]$, let

$$g(z, x) = \frac{1}{2}\left\{ f\left(\frac{z+x}{1+\bar{x}z}\right) + f\left(\frac{z-x}{1-\bar{x}z}\right) \right\}, \quad x = te^{i\alpha}, \quad \alpha \in \mathbb{R}, \quad t \in [0, t_0],$$

and suppose $g(z, x) \prec f(z)$, $|z| < 1$. Show that f is convex on U.

(Robertson, 1961 [Robe3].)

2.2.10. Complete the details in the proof of the estimate (2.2.16).

2.2.11. Suppose that $g(z) = \sum_{k=1}^{\infty} g_k z^k$ is convex on U, $f(z) = \sum_{k=1}^{\infty} a_k z^k$ is holomorphic on U and $f \prec g$. Show that $|a_k| \le |g_1|$, $k \ge 1$.

(Rogosinski, 1943 [Rog2].)

2.2.12. Use Theorem 2.1.8 to obtain another proof of necessity of the condition for starlikeness in Theorem 2.2.2.

Hint. Let $\rho = 1$ and $g(z, t) = (1-t)f(z)$ for $z \in U$ and $t \in [0, 1]$. Since f is starlike, it follows that $g(z, t) \prec f(z)$ for $0 \le t \le 1$.

2.2.13. Use Theorem 2.1.8 to obtain another proof of necessity of the condition for convexity in Theorem 2.2.3.

Hint. Let $\rho = 2$ and $g(z, t) = \frac{1}{2}[f(e^{it}z) + f(e^{-it}z)]$ in Theorem 2.1.8. Next, apply a similar argument as in the previous problem.

2.2.14. Let $f : U \to \mathbb{C}$ be a normalized holomorphic function. Show that $f \in K$ if and only if

$$\mathrm{Re}\left[\frac{z(\zeta - w)}{(z - \zeta)(z - w)} \cdot \frac{f(z) - f(w)}{f(\zeta) - f(w)} \right] \ge \frac{1}{2}\mathrm{Re}\left[\frac{z + \zeta}{z - \zeta} \right], \quad z, \zeta, w \in U.$$

(Ruscheweyh and Sheil-Small, 1974 [Rus-She].)

2.2.15. Use the condition in Problem 2.2.14 to derive Theorem 2.2.4.

2.3   Starlikeness and convexity of order α. Alpha convexity

2.3.1   Starlikeness and convexity of order α

We next introduce some subclasses of the normalized starlike and convex functions on the unit disc, and we give some of the basic properties of these classes.

The following notions were introduced by Robertson [Robe1]:

Definition 2.3.1. Let $f : U \to \mathbb{C}$ be a holomorphic function. We say that f is *starlike of order* α, $0 \leq \alpha < 1$, if $f(0) = 0$, $f'(0) \neq 0$ and

$$\mathrm{Re}\left[\frac{zf'(z)}{f(z)}\right] > \alpha, \quad z \in U.$$

Also we say that f is *convex of order* α, $0 \leq \alpha < 1$, if $f'(0) \neq 0$ and

$$\mathrm{Re}\left[1 + \frac{zf''(z)}{f'(z)}\right] > \alpha, \quad z \in U.$$

Let $S^*(\alpha)$ and $K(\alpha)$ denote respectively the classes of normalized starlike and convex functions of order α in the unit disc. There is an Alexander type result relating $S^*(\alpha)$ and $K(\alpha)$: $f \in K(\alpha)$ if and only if $g \in S^*(\alpha)$, where $g(z) = zf'(z)$, $z \in U$.

One of the most important results about these classes is the following theorem of Marx [Marx] and Strohhäcker [Str], which establishes the connection between convexity and starlikeness of order $1/2$. The proof which we give here is due to Suffridge [Su1], and is based on the characterization of convexity given in Theorem 2.2.4.

Theorem 2.3.2. *If* $f \in K$ *then* $f \in S^*(1/2)$. *This result is sharp, i.e. the constant* $1/2$ *cannot be replaced by a larger constant.*

Proof. Since $f \in K$ we have

$$\mathrm{Re}\left[\frac{2zf'(z)}{f(z) - f(\zeta)} - \frac{z+\zeta}{z-\zeta}\right] \geq 0, \quad z, \zeta \in U,$$

by Theorem 2.2.4, and setting $\zeta = 0$ in this relation, we obtain

$$\mathrm{Re}\left[\frac{zf'(z)}{f(z)}\right] \geq \frac{1}{2}, \quad z \in U.$$

Using the minimum principle for harmonic functions, we obtain the desired conclusion.

To see that this result is sharp, it suffices to consider the function $f(z) = z/(1 - z)$, which maps the unit disc conformally onto the half-plane $\{w \in \mathbb{C} : \text{Re } w > -1/2\}$.

Another result due to Marx [Marx] and Strohhäcker [Str] is the following. The proof below is due to Suffridge [Su1].

Corollary 2.3.3. *If $f \in K$ then* $\text{Re } [f(z)/z] > 1/2$, $z \in U$. *This result is sharp.*

Proof. Since $f \in K$, the relation (2.2.1) holds. Let

$$F(z, \zeta) = \frac{zf'(z)}{f(z) - f(\zeta)} - \frac{\zeta}{z - \zeta}, \quad z, \zeta \in U.$$

The relation (2.2.1) is equivalent to $\text{Re } F(z, \zeta) \geq 1/2$ for $z, \zeta \in U$.

Next, fix $\zeta \in U$. Expanding $F(z, \zeta)$ in a power series in the z variable, we deduce that

$$F(z, \zeta) = 1 + \left(\frac{1}{\zeta} - \frac{1}{f(\zeta)}\right) z + \dots.$$

The preceding two relations imply that $2F(\cdot, \zeta) - 1 \in \mathcal{P}$, and hence using Theorem 2.1.5 we conclude that

$$\left| \frac{1}{\zeta} - \frac{1}{f(\zeta)} \right| \leq 1.$$

Multiplying through by ζ gives $|1 - \zeta/f(\zeta)| \leq |\zeta| < 1$, and consequently $\text{Re } [f(\zeta)/\zeta] > 1/2$, as asserted. Again the function $f(z) = z/(1 - z)$ plays an extremal role in the above inequality.

There is a generalization of Theorem 2.3.2 due to Jack [Jac]. However, this result is not sharp. For the sharpness of the order of starlikeness for a given convex function of order α, the reader may consult the book of Miller and Mocanu [Mill-Moc2].

Theorem 2.3.4. *If $f \in K(\alpha)$ with $\alpha \in [0, 1)$ then $f \in S^*(\beta)$, where*

$$\beta = \beta(\alpha) = \frac{2\alpha - 1 + \sqrt{(2\alpha - 1)^2 + 8}}{4}.$$

The following result establishes a duality between the classes $S^*(\alpha)$ and S^*, respectively between $K(\alpha)$ and S^*, $\alpha \in [0, 1)$. The proof is obvious and we leave it for the reader.

Theorem 2.3.5. *Let $\alpha \in [0,1)$. The following assertions hold:*

(i) $f \in S^*(\alpha)$ if and only if $g \in S^*$, where $g(z) = z\left[\dfrac{f(z)}{z}\right]^{\frac{1}{1-\alpha}}$, $z \in U$. The branch of the power function is chosen such that $\left[\dfrac{f(z)}{z}\right]^{\frac{1}{1-\alpha}}\bigg|_{z=0} = 1$.

(ii) $f \in K(\alpha)$ if and only if $h \in S^*$, where $h(z) = z(f'(z))^{1/(1-\alpha)}$, $z \in U$. The branch of the power function is chosen such that $(f'(z))^{1/(1-\alpha)}|_{z=0} = 1$.

Using this result, we may obtain the following growth and distortion theorem for convex functions of order $\alpha \in [0,1)$ due to Robertson [Robe1]:

Theorem 2.3.6. *If $f \in K(\alpha)$, $\alpha \in [0,1)$ and $|z| = r < 1$, then*

$$(2.3.1) \qquad \frac{1}{(1+r)^{2(1-\alpha)}} \leq |f'(z)| \leq \frac{1}{(1-r)^{2(1-\alpha)}}.$$

If $\alpha \neq 1/2$ then

$$(2.3.2) \qquad \frac{(1+r)^{2\alpha-1} - 1}{2\alpha - 1} \leq |f(z)| \leq \frac{1 - (1-r)^{2\alpha-1}}{2\alpha - 1},$$

and if $\alpha = 1/2$ then

$$(2.3.3) \qquad \log(1+r) \leq |f(z)| \leq -\log(1-r).$$

These estimates are sharp. Equality holds in each of the above relations for

$$f(z) = \begin{cases} \dfrac{1 - (1-z)^{2\alpha-1}}{2\alpha - 1}, & \alpha \neq \dfrac{1}{2} \\[3mm] -\log(1-z), & \alpha = \dfrac{1}{2}, \end{cases}$$

where the branches of the power functions $(1-z)^{2\alpha-1}$ and of $\log(1-z)$ are chosen such that $(1-z)^{2\alpha-1}|_{z=0} = 1$ and $\log(1-z)|_{z=0} = 0$.

Proof. The distortion estimate (2.3.1) follows easily from Theorem 2.3.5 (ii) and the growth theorem for starlike functions.

In order to obtain the upper estimates in the growth theorems (2.3.2) and (2.3.3), it suffices to integrate the right-hand inequality in (2.3.1). To prove the lower estimates requires the argument which is given in Theorem 2.2.8.

From Theorem 2.3.5 (i) or from the distortion estimate (2.3.1) we obtain the growth theorem for functions in $S^*(\alpha)$, $\alpha \in [0,1)$.

Theorem 2.3.7. *If* $f \in S^*(\alpha)$, $\alpha \in [0,1)$ *and* $|z| = r < 1$, *then*

$$\frac{r}{(1+r)^{2(1-\alpha)}} \leq |f(z)| \leq \frac{r}{(1-r)^{2(1-\alpha)}}.$$

These estimates are sharp. Equality holds for $f(z) = \dfrac{z}{(1-z)^{2(1-\alpha)}}$, $z \in U$.

Brown [Brow] obtained the following interesting necessary and sufficient condition for convexity of order $\alpha \in [0,1)$. In particular, Brown's result yields that if f is convex on U, then f maps every disc $|z - \zeta| < r < 1 - |\zeta|$, $|\zeta| < 1$, onto a convex domain (see also [Stu] and [Robel]).

Theorem 2.3.8. *Let* $f : U \to \mathbb{C}$ *be a normalized holomorphic function and let* $\alpha \in [0,1)$. *Then* $f \in K(\alpha)$ *if and only if*

$$\text{Re} \left\{ 1 + \frac{(z-\zeta)f''(z)}{f'(z)} \right\} > \alpha, \quad |z - \zeta| < 1 - |\zeta|, \ |\zeta| < 1.$$

Proof. First, assume that

$$\text{Re} \left\{ 1 + \frac{(z-\zeta)f''(z)}{f'(z)} \right\} > \alpha, \quad |z - \zeta| < 1 - |\zeta|, \ |\zeta| < 1.$$

For $\zeta = 0$ in the above we deduce that $f \in K(\alpha)$.

Conversely, suppose $f \in K(\alpha)$. Let ζ be fixed with $|\zeta| = r < 1$ and

$$A \equiv 1 + \left(\frac{z-\zeta}{1-\alpha} \right) \frac{f''(z)}{f'(z)}.$$

In view of the minimum principle for harmonic functions, it suffices to prove that $\text{Re}\, A > 0$ for $|z - \zeta| = \rho < 1 - r$. For this purpose, let $z \in U$ be such that $|z - \zeta| = \rho$. Since $f \in K(\alpha)$ it follows that

$$(2.3.4) \qquad A = \frac{(z-\zeta)(p(z)-1)}{z} + 1,$$

where $p \in \mathcal{P}$. It is well known that the extreme points of $\mathcal{P}$ (p is an extreme point of $\mathcal{P}$ if $p \in \mathcal{P}$ and if $p = tg + (1-t)h$, where $0 < t < 1$, $g \in \mathcal{P}$ and $h \in \mathcal{P}$, implies that $g = h$) are the functions $p_\theta(z) = \dfrac{1 + e^{i\theta}z}{1 - e^{i\theta}z}$, where $\theta \in \mathbb{R}$, and since the quantity on the right side of the equality (2.3.4) is an affine functional on $\mathcal{P}$, it follows that $\min\limits_{p \in \mathcal{P}} \text{Re}\, A$ is attained at such an extreme point of $\mathcal{P}$ (see e.g.

[Hal-MG]). Next, for $\zeta = re^{i\varphi}$ and $z = \zeta + \rho e^{i\phi}$ we obtain after elementary computations that

$$\text{Re } A \geq \min_{0 \leq \theta < 2\pi} \text{Re } \left[\frac{(z - \zeta)(p_\theta(z) - 1)}{z} + 1 \right]$$

$$= \min_{0 \leq \theta < 2\pi} |1 - e^{i\theta} z|^{-2} [-\rho^2 + 1 - 2r\cos(\theta + \varphi) + r^2]$$

$$\geq \min_{0 \leq \theta < 2\pi} |1 - e^{i\theta} z|^{-2} [(1 - r)^2 - \rho^2] > 0.$$

This completes the proof.

Finally we mention that coefficient estimates for functions in $S^*(\alpha)$ and $K(\alpha)$, $\alpha \in [0, 1)$, were obtained by Robertson [Robe1] in his original paper on these classes of functions (see Problem 2.3.10). For more information about starlikeness and convexity of order α, the reader may consult the book [Goo1]. See also [Bern].

2.3.2   Alpha convexity

In this section we introduce another subclass of S, called the class of alpha-convex functions. This notion was introduced by Mocanu in 1969 [Moc1], with the aim of constructing a one-parameter family of subclasses of S which provides a continuous passage from the starlike functions to the convex functions. More detailed treatments may be found in the books of Goodman [Goo1] and Mocanu, Bulboacă, and Sălăgean [Moc-Bu-Să]. Some of the ideas related to alpha-convex functions first occurred in the work of Sakaguchi [Sak3]. We remark that alpha-convex mappings in several complex variables have been considered by Kohr [Koh5]; see also [Koh-Lic2].

Definition 2.3.9. Let f be a normalized holomorphic function on U such that $\frac{f(z)f'(z)}{z} \neq 0$ for $z \in U$. Also let $\alpha \in \mathbb{R}$ and

$$J(\alpha, f; z) = (1 - \alpha)\frac{zf'(z)}{f(z)} + \alpha\left(1 + \frac{zf''(z)}{f'(z)}\right), \quad z \in U.$$

We say that f is α-convex if Re $J(\alpha, f; z) > 0$, $z \in U$.

(It is possible to omit the assumption $\frac{f(z)f'(z)}{z} \neq 0$, $z \in U$, in Definition 2.3.9 since Sakaguchi and Fukui [Sak-Fuk] proved in 1979 that if Re $J(\alpha, f; z) > 0$, $z \in U$, then $\frac{f(z)f'(z)}{z} \neq 0$, $z \in U$.)

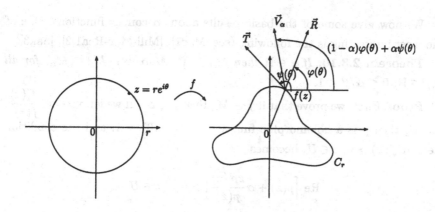

Figure 2.5: α-convexity of C_r

Let M_α denote the class of α-convex functions (or the class of *Mocanu functions*, in honour of P. Mocanu). It is clear that $M_0 = S^*$ and $M_1 = K$.

Mocanu in his original paper [Moc1] defined alpha-convexity geometrically, and deduced the analytic condition in Definition 2.3.9 as a consequence. We shall outline Mocanu's argument, which applies to the case $0 \leq \alpha \leq 1$.

Let $f \in H_u(U)$ with $f(0) = 0$. Also let C_r be the image of the circle $|z| = r$, $r \in (0,1)$. Since f is univalent on U, C_r is a positively oriented Jordan curve given by $w = f(re^{i\theta}) = w(\theta)$, $0 \leq \theta \leq 2\pi$.

We recall that C_r is starlike if $\arg f(z)$ increases as z moves around the circle $|z| = r$ in the positive direction. Moreover, C_r is a convex curve if the argument of the tangent vector to C_r is a non-decreasing function of $\theta \in [0, 2\pi]$.

Let $\varphi(\theta) = \arg f(re^{i\theta})$ be the argument of the position vector and $\psi(\theta)$ be the argument of the tangent vector to C_r at $w = f(re^{i\theta})$. Let $\vec{V}_\alpha = \vec{V}_\alpha(\theta)$ be a vector which starts from $f(re^{i\theta})$ and divides in the ratio α the angle between the tangent vector to C_r at $w = f(re^{i\theta})$ and the position vector $w = f(re^{i\theta})$. Then the angle of inclination of $\vec{V}_\alpha(\theta)$ is $(1 - \alpha)\varphi(\theta) + \alpha\psi(\theta)$.

We say that the curve C_r is α-*convex* if this angle is an increasing function of $\theta \in [0, 2\pi]$. It is easy to see that this relation is equivalent to Re $J(\alpha, f; z) > 0$, $|z| = r$.

We now give some of the basic results about α-convex functions. One of the most interesting is the following (see [Moc1], [Mill-Moc-Rea1,2], [Sak3]):

Theorem 2.3.10. *If $\alpha \in \mathbb{R}$ then $M_\alpha \subseteq S^*$. Moreover, $M_\beta \subset M_\alpha$, for all $\alpha, \beta \in \mathbb{R}$, $0 \leq \alpha/\beta < 1$.*

Proof. First, we prove that if $f \in M_\alpha$ then $f \in S^*$. If we let $p(z) = \dfrac{zf'(z)}{f(z)}$, $z \in U$, then p is a holomorphic function on U, $p(0) = 1$, and the condition $\operatorname{Re} J(\alpha, f; z) > 0$, $z \in U$, becomes

$$\operatorname{Re}\left[p(z) + \alpha\frac{zp'(z)}{p(z)}\right] > 0, \quad z \in U.$$

Taking into account Problem 2.1.10, we deduce that $\operatorname{Re} p(z) > 0$, $z \in U$, and the conclusion follows from Theorem 2.2.2.

Now, let $\alpha, \beta \in \mathbb{R}$ be such that $0 \leq \alpha/\beta < 1$. We shall consider the case $0 \leq \alpha < \beta$; the other case is similar. Suppose that $f \in M_\beta$. As in the proof of the first statement, we have

$$\operatorname{Re}\left[p(z) + \beta\frac{zp'(z)}{p(z)}\right] > 0, \quad z \in U,$$

and $\operatorname{Re} p(z) > 0$, $z \in U$. Next, fix $z \in U$ and let $a = \operatorname{Re} p(z)$ and $b = \operatorname{Re}\left[\dfrac{zp'(z)}{p(z)}\right]$. Let $h(t) = a + tb$, $t \in [0, \beta]$. Since $h(0) = a > 0$ and $h(\beta) > 0$, it is obvious that $h(t) > 0$ for all $t \in [0, \beta]$. This completes the proof.

This theorem implies that the classes M_α increase with α for negative α, and decrease with α when α is positive. The largest of these classes is $M_0 = S^*$. We also note the following consequences of Theorem 2.3.10 (see [Moc1], [Mill-Moc-Rea1,2], [Sak3]):

Corollary 2.3.11. (*i*) *If $\alpha \geq 1$ then $M_\alpha \subseteq K$.*

(*ii*) *Further, if $0 \leq \alpha < 1$ then $K \subseteq M_\alpha$.*

(*iii*) $\displaystyle\bigcap_{\alpha=0}^{\infty} M_\alpha = \{id\}$, *where $id(z) = z$, $z \in U$.*

(Part (*iii*) is an exercise for the reader.)

The following result establishes a duality between the classes S^* and M_α with $\alpha \geq 0$. It can be proved in different ways; see [Moc1], [Mill-Moc-Rea3].

Theorem 2.3.12. *Let $\alpha \geq 0$. Then $f \in M_\alpha$ if and only if the function*

g defined by $g(z) = f(z) \left[\dfrac{zf'(z)}{f(z)}\right]^{\alpha}$, $z \in U$, belongs to S^*. The branch of the power function is chosen such that $\left[\dfrac{zf'(z)}{f(z)}\right]^{\alpha}\Big|_{z=0} = 1$.

For functions in the class M_α, $\alpha > 0$, we have the following growth result, due to Miller [Mill]. For the proof, it suffices to use Theorem 2.3.12 and similar arguments as in the proof of Theorem 2.2.8.

Theorem 2.3.13. *If $\alpha > 0$ and $f \in M_\alpha$, then*

$$-K(-r; \alpha) \le |f(z)| \le K(r; \alpha), \quad |z| = r < 1,$$

where $K(z; \alpha)$ is the α-convex Koebe function,

$$K(z; \alpha) = \left[\frac{1}{\alpha} \int_0^z \frac{\zeta^{\frac{1}{\alpha}} d\zeta}{\zeta(1-\zeta)^{\frac{2}{\alpha}}}\right]^{\alpha}, \quad z \in U.$$

The branch of the power function is chosen such that $K(z; \alpha)$ is normalized holomorphic on U. These estimates are sharp.

Problems

2.3.1. Prove Theorem 2.3.4.

2.3.2. Prove Theorem 2.3.5.

2.3.3. Find integral representations for functions in $S^*(\alpha)$ and $K(\alpha)$, $0 < \alpha < 1$.

2.3.4. Show that the function $f(z) = z/(1-z)^{2(1-\alpha)}$, $z \in U$, is starlike of order $\alpha \in [0, 1)$. (This function is called the *Koebe function of order α*.)

2.3.5. Let $\alpha \in (0, 1)$ and let $f(z) = z + \displaystyle\sum_{n=2}^{\infty} a_n z^n$ be a holomorphic function on U. Show that if $\displaystyle\sum_{n=2}^{\infty}(n - \alpha)|a_n| \le 1 - \alpha$, then $f \in S^*(\alpha)$.

(Merkes, Robertson and Scott, 1962 [Mer-Rob-Sco].)

Hint. Show that $\left|\dfrac{zf'(z)}{f(z)} - 1\right| < 1 - \alpha$, $z \in U$.

2.3.6. State a result similar to that in the preceding problem for functions in $K(\alpha)$.

2.3.7. Show that the function

$$f(z) = \begin{cases} \dfrac{1 - (1-z)^{2\alpha-1}}{2\alpha - 1}, & \alpha \neq \dfrac{1}{2} \\[3mm] -\log(1-z), & \alpha = \dfrac{1}{2} \end{cases}$$

is convex of order $\alpha \in [0,1)$.

2.3.8. Prove part (iii) of Corollary 2.3.11.

2.3.9. Prove Theorems 2.3.12 and 2.3.13.

2.3.10. Let $f(z) = z + \sum\limits_{n=2}^{\infty} a_n z^n$ be a holomorphic function on U and $\alpha \in [0,1)$. Prove the following statements:

(i) If $f \in S^*(\alpha)$ then

$$|a_n| \leq \frac{1}{(n-1)!} \prod_{k=2}^{n} (k - 2\alpha), \quad n = 2, 3, \ldots.$$

These estimates are sharp. Equalities in the above estimates are attained for $f(z) = z/(1-z)^{2(1-\alpha)}$, $z \in U$.

(ii) If $f \in K(\alpha)$ then

$$|a_n| \leq \frac{1}{n!} \prod_{k=2}^{n} (k - 2\alpha), \quad n = 2, 3, \ldots.$$

These estimates are sharp. Equalities are attained for the function f given in Problem 2.3.7.

(Robertson, 1936 [Robe1].)

2.3.11. Goodman [Goo2] introduced the class US^* of uniformly starlike functions on U, consisting of those functions $f \in S^*$ which map each circular arc γ contained in U, with center ζ also in U, onto an arc $f(\gamma)$ which is starlike with respect to $f(\zeta)$. Let $f : U \to \mathbb{C}$ be a normalized holomorphic function on U. Show that $f \in US^*$ if and only if

$$\mathrm{Re}\left[\frac{(z - \zeta)f'(z)}{f(z) - f(\zeta)} \right] > 0, \ z, \zeta \in U.$$

(Goodman, 1991 [Goo2].)

2.3.12. Goodman [Goo3] also introduced the class UK of uniformly convex functions on U, consisting of those functions $f \in K$ which map each circular arc γ contained in U, with center ζ also in U, onto a convex arc $f(\gamma)$. Let $f : U \to \mathbb{C}$ be a normalized holomorphic function on U. Show that $f \in UK$ if and only if

$$\text{Re} \left[1 + \frac{(z - \zeta)f''(z)}{f'(z)} \right] > 0, \ z, \zeta \in U.$$

(Goodman, 1991 [Goo3].) (See also [Ma-Min3], [Røn1,2] and [Kan-Wi].)

2.4 Close-to-convexity, spirallikeness and Φ-likeness in the unit disc

2.4.1 Close-to-convexity in the unit disc

We next introduce two other well known subclasses of univalent functions in the unit disc, namely the close-to-convex and spirallike functions. First we shall study the notion of close-to-convexity, and to this end we begin with the following result, due to Noshiro [Nos], Warschawski [War] and Wolff [Wol]:

Lemma 2.4.1. *Suppose D is a convex domain in $\mathbb{C}$. If $f \in H(D)$ satisfies* $\text{Re } f'(z) > 0$ *on D, then f is univalent on D.*

Proof. Let $z_1, z_2 \in D$ such that $z_1 \neq z_2$. Let γ be the straight line segment joining z_1 to z_2. Let $z(t) = (1 - t)z_1 + tz_2$, $t \in [0, 1]$. Then $z(t) \in D$, $t \in [0, 1]$, and we have

$$f(z_2) - f(z_1) = \int_\gamma f'(\zeta)d\zeta = (z_2 - z_1) \int_0^1 f'(z(t))dt.$$

Taking into account the hypothesis, we deduce that

$$\text{Re} \left[\frac{f(z_2) - f(z_1)}{z_2 - z_1} \right] = \int_0^1 \text{Re } [f'(z(t))]dt > 0,$$

and hence $f(z_2) \neq f(z_1)$. This completes the proof.

Using this lemma we may easily prove the following result, due to Ozaki [Oz] and Kaplan [Kap]:

Lemma 2.4.2. *Let D be a domain in $\mathbb{C}$, and let f, g be holomorphic functions on D such that g is univalent on D and $g(D)$ is a convex domain. If*

$$\operatorname{Re}\left[\frac{f'(z)}{g'(z)}\right] > 0, \quad z \in D,$$

then f is univalent on D.

Proof. Apply Lemma 2.4.1 to the function $h = f \circ g^{-1}$.

The following definition is due to Kaplan [Kap]:

Definition 2.4.3. Let $f \in H(U)$. We say that f is *close-to-convex on U* (or simply *close-to-convex*) if there exists a convex function g on U such that

$$(2.4.1) \qquad \qquad \operatorname{Re}\left[\frac{f'(z)}{g'(z)}\right] > 0, \quad z \in U.$$

Using Theorem 2.2.6, we can replace the condition (2.4.1) by the requirement that

$$(2.4.2) \qquad \qquad \operatorname{Re}\left[\frac{z f'(z)}{h(z)}\right] > 0, \quad z \in U,$$

where h is a starlike function on U. Hence we easily deduce that if f is starlike on U, then f is also close-to-convex.

Let C denote the set of normalized close-to-convex functions on U. Then it is clear that $K \subset S^* \subset C \subset S$.

In 1955 Reade [Rea] proved that the coefficients of close-to-convex functions satisfy the Bieberbach conjecture. The argument is similar to that for starlike functions (Theorem 2.2.16).

Theorem 2.4.4. *If $f(z) = z + a_2 z^2 + \ldots + a_n z^n + \ldots$ is a close-to-convex function then $|a_n| \leq n$, for $n = 2, 3, \ldots$. Equality $|a_n| = n$ for a given $n \geq 2$ holds when f is a rotation of the Koebe function.*

Proof. Since $f \in C$ there exists a convex function h such that

$$\operatorname{Re}\left[\frac{f'(z)}{h'(z)}\right] > 0 \text{ on } U.$$

Therefore $\operatorname{Re}[1/h'(0)] > 0$, and if $\alpha = \arg h'(0)$ we may assume that $|\alpha| < \pi/2$. Also let

$$g(z) = \frac{h(z) - h(0)}{h'(0)}, \quad z \in U.$$

Then $g \in K$ and if we consider the function

$$q(z) = \frac{f'(z)}{e^{i\alpha}g'(z)} = \sum_{n=0}^{\infty} q_n z^n,$$

we deduce that $q_0 = e^{-i\alpha}$ and $\operatorname{Re} q(z) > 0$, $z \in U$. Further, let

$$p(z) = \frac{q(z) + i\sin\alpha}{\cos\alpha} = 1 + p_1 z + \ldots + p_n z^n + \ldots, \quad z \in U.$$

This function belongs to $\mathcal{P}$ and it is easy to see that $q_n = p_n \cos\alpha$, $n \geq 1$. If

$$g(z) = z + c_2 z^2 + \ldots + c_n z^n + \ldots, \quad z \in U,$$

then comparing the coefficients in the power series of $f'(z)$ and $e^{i\alpha}g'(z)q(z)$, we easily deduce that

$$na_n = e^{i\alpha}[nc_n e^{-i\alpha} + (n-1)c_{n-1}q_1 + (n-2)c_{n-2}q_2 + \ldots + 2c_2 q_{n-2} + q_{n-1}].$$

Since $p \in \mathcal{P}$ we have $|p_n| \leq 2$ for $n = 2, 3, \ldots$, and therefore $|q_n| \leq 2\cos\alpha \leq 2$. Also since $g \in K$ we have $|c_n| \leq 1$, $n = 2, 3, \ldots$. Therefore

$$n|a_n| \leq n + 2(n-1) + 2(n-2) + \ldots + 2 \cdot 2 + 2 = n^2,$$

and thus $|a_n| \leq n$ for $n = 2, 3, \ldots$.

It is not difficult to see that equality $|a_n| = n$ for a given $n \geq 2$ holds when f is a rotation of the Koebe function. This completes the proof.

One of the basic results about close-to-convex functions is a geometric characterization due to Kaplan [Kap]. The proof requires the following elementary fact [Kap]:

Lemma 2.4.5. *Let* $v : \mathbb{R} \to \mathbb{R}$ *satisfy the following conditions:*

$$v(t + 2\pi) - v(t) = 2\pi, \quad t \in \mathbb{R},$$

$$v(t_1) - v(t_2) < \pi, \quad t_1 < t_2.$$

Then there is a nondecreasing function $w : \mathbb{R} \to \mathbb{R}$ *such that*

$$w(t + 2\pi) - w(t) = 2\pi, \quad t \in \mathbb{R},$$

and

$$|w(t) - v(t)| \leq \frac{\pi}{2}, \quad t \in \mathbb{R}.$$

Proof. Let

$$w(t) = \sup_{s \leq t} v(s) - \frac{\pi}{2}.$$

It is obvious that w is a nondecreasing function on $\mathbb{R}$, and since

$$v(t + 2\pi) - v(t) = 2\pi, \quad t \in \mathbb{R},$$

we obtain

$$w(t + 2\pi) = \sup_{s \leq t} v(s + 2\pi) - \frac{\pi}{2} = w(t) + 2\pi.$$

On the other hand, since

$$v(s) < v(t) + \pi, \quad s < t,$$

we see that

$$w(t) \leq v(t) + \frac{\pi}{2}.$$

We also have the relation

$$v(t) \leq \sup_{s \leq t} v(s) = w(t) + \frac{\pi}{2}.$$

Combining the above observations, we conclude that

$$|w(t) - v(t)| \leq \frac{\pi}{2},$$

as desired.

Theorem 2.4.6. *Let f be a locally univalent holomorphic function on U such that $f(0) = f'(0) - 1 = 0$. Then f is close-to-convex if and only if*

$$(2.4.3) \qquad \int_{\theta_1}^{\theta_2} \mathrm{Re}\left[1 + \frac{zf''(z)}{f'(z)}\right] d\theta > -\pi, \quad z = re^{i\theta},$$

for each $r \in (0,1)$ and for each pair of real numbers θ_1, θ_2 such that $0 \leq \theta_2 - \theta_1 \leq 2\pi$.

Proof. First assume that f is close-to-convex. Then there is a convex function h such that

$$\mathrm{Re}\left[\frac{f'(z)}{h'(z)}\right] > 0, \quad z \in U,$$

i.e.

$$\left| \arg \left[\frac{f'(z)}{h'(z)} \right] \right| < \frac{\pi}{2}.$$

Since $f'(z) \neq 0$ and $h'(z) \neq 0$, $z \in U$, we may choose branches of the arguments such that

$$| \arg f'(z) - \arg h'(z)| < \frac{\pi}{2}, \quad z \in U.$$

Let $a(z) = \arg f'(z)$ and $b(z) = \arg h'(z)$. For $z = re^{i\theta}$, $0 < r < 1$, $\theta \in \mathbb{R}$, let

$$V(r, \theta) = a(re^{i\theta}) + \theta = \arg[re^{i\theta} f'(re^{i\theta})],$$

and

$$W(r, \theta) = b(re^{i\theta}) + \theta = \arg[re^{i\theta} h'(re^{i\theta})].$$

We then have

$$|V(r, \theta) - W(r, \theta)| < \frac{\pi}{2}.$$

Since h is a convex function, $W(r, \theta)$ is an increasing function of θ. (This may be seen directly or using Alexander's theorem.) By the minimum principle for harmonic functions we must have $\frac{\partial W}{\partial \theta}(r, \theta) > 0$. Combining this observation with the preceding inequality, we deduce that for $0 \leq \theta_2 - \theta_1 \leq 2\pi$,

$$V(r, \theta_1) - V(r, \theta_2) = [V(r, \theta_1) - W(r, \theta_1)]$$

$$+[W(r, \theta_1) - W(r, \theta_2)] + [W(r, \theta_2) - V(r, \theta_2)]$$

$$\leq [V(r, \theta_1) - W(r, \theta_1)] + [W(r, \theta_2) - V(r, \theta_2)] < \pi.$$

We have therefore shown that

$$V(r, \theta_2) - V(r, \theta_1) > -\pi,$$

i.e.

$$\arg[re^{i\theta_2} f'(re^{i\theta_2})] - \arg[re^{i\theta_1} f'(re^{i\theta_1})] > -\pi.$$

This inequality is clearly equivalent to (2.4.3) and this completes the first part of the proof.

Conversely, assume that f is locally univalent on U and satisfies (2.4.3). We must prove that f is close-to-convex. For this purpose, fix $\rho \in (0,1)$ and consider

$$v_\rho(\theta) = V(\rho, \theta) = \arg[\rho e^{i\theta} f'(\rho e^{i\theta})] = \theta + \arg[f'(\rho e^{i\theta})].$$

The condition (2.4.3) is equivalent to

$$v_\rho(\theta_1) - v_\rho(\theta_2) < \pi, \quad \theta_1 < \theta_2.$$

Since $f'(z) \neq 0$, $z \in U$, $\arg[f'(\rho e^{i\theta})]$ must be periodic with period 2π in θ, and therefore

$$v_\rho(\theta + 2\pi) - v_\rho(\theta) = V(\rho, \theta + 2\pi) - V(\rho, \theta) = 2\pi.$$

In view of Lemma 2.4.5 there is a nondecreasing function $w_\rho(\theta)$ such that

$$w_\rho(\theta + 2\pi) - w_\rho(\theta) = 2\pi$$

and

(2.4.4) $$|w_\rho(\theta) - v_\rho(\theta)| \leq \frac{\pi}{2}, \quad \theta \in \mathbb{R}.$$

Now, let $P_\rho(r, \theta)$ be the *Poisson kernel* for the disc U_ρ, i.e.

$$P_\rho(r, \theta) = \frac{\rho^2 - r^2}{\rho^2 - 2\rho r \cos\theta + r^2}, \quad r < \rho,$$

and consider the harmonic function b_ρ on U_ρ given by the *Poisson integral*

$$b_\rho(r, \theta) = \frac{1}{2\pi} \int_{-\pi}^{\pi} P_\rho(r, \theta - s)[w_\rho(s) - s]ds$$

$$= \frac{1}{2\pi} \int_{-\pi}^{\pi} P_\rho(r, t)[w_\rho(t + \theta) - (t + \theta)]dt.$$

Also, let

$$W_\rho(r, \theta) = b_\rho(r, \theta) + \theta.$$

Using the fact that $tP_\rho(r, t)$ is an odd function of t, it is elementary to see that

(2.4.5) $$W_\rho(r, \theta) = \frac{1}{2\pi} \int_{-\pi}^{\pi} P_\rho(r, t)w_\rho(t + \theta)dt.$$

From this we see that $W_\rho(r, \theta)$ is an increasing function of θ, for each fixed $r < \rho$. Indeed, for $\theta_1 < \theta_2$ we have

$$W_\rho(r, \theta_2) - W_\rho(r, \theta_1) = \frac{1}{2\pi} \int_{-\pi}^{\pi} P_\rho(r, t)[w_\rho(t + \theta_2) - w_\rho(t + \theta_1)]dt \geq 0,$$

making use of the fact that $w_\rho(\theta)$ is an increasing function of θ. Therefore

$$\frac{\partial W_\rho}{\partial \theta}(r, \theta) \geq 0,$$

and we may assume that this inequality is strictly satisfied because this derivative is a harmonic function.

On the other hand, from (2.4.4) we have

$$|V(\rho, \theta) - w_\rho(\theta)| \leq \frac{\pi}{2}.$$

Since $V(r, \theta) - \theta = \arg f'(re^{i\theta})$ is a (single-valued) harmonic function on U_ρ, the considerations leading to (2.4.5) show that

$$V(r, \theta) = \frac{1}{2\pi} \int_{-\pi}^{\pi} P_\rho(r, t)V(\rho, t + \theta)dt, \quad r < \rho.$$

Using (2.4.5) and the previous relations we obtain for $r < \rho$ that

$$|V(r, \theta) - W_\rho(r, \theta)|$$

$$\leq \frac{1}{2\pi} \int_{-\pi}^{\pi} P_\rho(r, t)|V(\rho, t + \theta) - w_\rho(t + \theta)|dt \leq \frac{\pi}{2}.$$

Next, let $g_\rho : U_\rho \to \mathbb{C}$ be the holomorphic function which satisfies $\operatorname{Re} g_\rho(0) = 0$ and $\operatorname{Im} g_\rho(z) = b_\rho(r, \theta)$ for $z = re^{i\theta}$. Also let

$$h_\rho(z) = \int_0^z e^{g_\rho(\zeta)}d\zeta, \quad z \in U_\rho.$$

Then h_ρ is holomorphic on U_ρ, $h_\rho(0) = 0$ and $|h'_\rho(0)| = 1$. Moreover, h_ρ is convex on U_ρ. Indeed, for $z = re^{i\theta}$, $r < \rho$, we have

$$\operatorname{Re}\left[1 + \frac{zh''_\rho(z)}{h'_\rho(z)}\right] = 1 + \operatorname{Re}[zg'_\rho(z)] = 1 + \frac{\partial b_\rho}{\partial \theta}(r, \theta) = \frac{\partial W_\rho}{\partial \theta}(r, \theta) > 0.$$

Finally, since

$$V(r, \theta) = \theta + \arg f'(re^{i\theta}),$$

$$W_\rho(r,\theta) = \theta + b_\rho(r,\theta) = \theta + \text{Im } g_\rho(re^{i\theta}) = \theta + \arg h_\rho'(re^{i\theta}),$$

and

$$|V(r,\theta) - W_\rho(r,\theta)| \leq \frac{\pi}{2},$$

we conclude that

$$|\arg f'(z) - \arg h_\rho'(z)| \leq \frac{\pi}{2}, \quad z = re^{i\theta}, r < \rho,$$

or equivalently

$$\text{Re}\left[\frac{f'(z)}{h_\rho'(z)}\right] \geq 0, \quad |z| < \rho.$$

By the minimum principle for harmonic functions, we must have

$$\text{Re}\left[\frac{f'(z)}{h_\rho'(z)}\right] > 0, \quad |z| < \rho.$$

Further, since $h_\rho(0) = 0$, $|h_\rho'(0)| = 1$ and h_ρ is a convex function, it follows that for each $m \in \mathbb{N}$, the family $\{h_{\rho_n}\}_{n \geq m}$ is a normal family of convex functions on the disc U_{ρ_m}, where $\rho_n = 1 - 1/n$. Therefore, by using Montel's theorem and a diagonal sequence argument, we may extract a subsequence $\{h_{\rho_{n_k}}\}_{k \in \mathbb{N}}$ that converges uniformly on each disc U_ρ with $\rho < 1$. The limit function h is convex on U and satisfies

$$\text{Re}\left[\frac{f'(z)}{h'(z)}\right] \geq 0, \quad z \in U.$$

Again using the minimum principle for harmonic functions, we deduce that

$$\text{Re}\left[\frac{f'(z)}{h'(z)}\right] > 0, \quad z \in U,$$

and therefore f is close-to-convex with respect to h. This completes the proof.

We conclude this section with some remarks about close-to-convex functions in the unit disc.

Geometric interpretation of Theorem 2.4.6. Let f be a normalized close-to-convex function and $0 < r < 1$. Also let Γ_r denote the image of the circle $|z| = r$. If $\overrightarrow{T}_1$ and $\overrightarrow{T}_2$ denote the unit tangent vectors to Γ_r at $f(re^{i\theta_1})$ and $f(re^{i\theta_2})$ respectively, then the relation (2.4.3) is equivalent to

$$\arg \overrightarrow{T}_2 - \arg \overrightarrow{T}_1 > -\pi, \quad 0 \leq \theta_2 - \theta_1 \leq 2\pi.$$

In other words the Jordan curve Γ_r does not make a U-turn in the clockwise direction, that is, if θ increases on $[0, 2\pi]$, the tangent vector cannot turn backward through an angle larger than or equal to π.

Linearly accessible domains. Another geometric interpretation of close-to-convexity may be obtained from the work of Lewandowski [Lew1]. Biernacki in 1936 [Bier] introduced the notion of a *linearly accessible* function and Lewandowski in 1958 [Lew1] showed that there is an equivalence between close-to-convexity and linear accessibility. More precisely, we say that a domain $\Omega \subset \mathbb{C}$ is *linearly accessible in the large sense* if the complement of Ω is the union of half-lines or rays. Such a domain is clearly a simply connected domain. If f is a univalent function on U and $f(U)$ is a linearly accessible domain in the large sense, then we say that f *is linearly accessible in the large sense*.

If the complement of the domain Ω is the union of rays which are disjoint (except that the initial point of one ray may lie on another such ray), we say that Ω is *linearly accessible in the strict sense* (or simply *linearly accessible*). A function which is univalent in U and maps U conformally onto such a domain is called a *linearly accessible function in the strict sense* (or simply *linearly accessible*). Biernacki introduced both conditions but worked with the latter one, and Lewandowski's result is that f *is close-to-convex if and only if f is linearly accessible in the strict sense*.

Bielecki and Lewandowski [Biel-Lew] subsequently found a short and elegant proof of Lewandowski's theorem, which we shall give in the next chapter. It relies on another important analytic characterization of close-to-convex functions in terms of Loewner chains. There is also a more recent elementary proof due to Koepf [Koep3].

There is a later study of linearly accessible functions in the large sense due to Sheil-Small [She2], which contains some very interesting results. The following may be viewed as an analog of Kaplan's theorem [She2]:

Theorem 2.4.7. *A function f which is holomorphic and univalent on U is linearly accessible in the large sense if and only if for each r, $0 < r < 1$,*

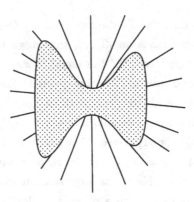

Figure 2.6: A close-to-convex domain is linearly accessible

and each $z_0 \in U$ we have

$$\frac{1}{2}\theta_2 + \arg\frac{f(re^{i\theta_2}) - f(z_0)}{re^{i\theta_2} - z_0} - \frac{1}{2}\theta_1 - \arg\frac{f(re^{i\theta_1}) - f(z_0)}{re^{i\theta_1} - z_0} > -\pi$$

whenever $\theta_2 > \theta_1$.

The next theorem gives a condition which is somewhat analogous to the definition of close-to-convexity, except that the function g below depends on the point $z_0 \in U$ (see [She2]):

Theorem 2.4.8. *A function f holomorphic and univalent on U is linearly accessible in the large sense if and only if corresponding to each fixed point $z_0 \in U$ there is a function $g(z) = g_{z_0}(z)$ starlike of order $1/2$ for which the inequality*

$$\mathrm{Re}\left[\frac{z(f(z) - f(z_0))}{g(z)(z - z_0)}\right] > 0, \quad z \in U,$$

holds.

We remark that the function g in the above is not necessarily normalized. Also note that if we let

(2.4.6) $$h(z) = z_0 g'(0) + (z - z_0)\frac{g(z)}{z}, \quad z \in U,$$

then

(2.4.7) $$z\frac{h(z) - h(z_0)}{z - z_0} = g(z),$$

and the inequality from Theorem 2.4.8 is equivalent to

$$(2.4.8) \qquad \qquad \text{Re} \left[\frac{f(z) - f(z_0)}{h(z) - h(z_0)} \right] > 0.$$

If f is close-to-convex, then it is well known that (2.4.8) holds for some convex function h (see Problem 2.4.11). Also in this case if we define the function $g(z) = g_{z_0}(z)$ by (2.4.7), then g is starlike of order 1/2 [She1], and hence the conditions of Theorem 2.4.8 are satisfied.

We shall leave the proofs of these theorems for the reader.

We remark that Krzyz [Krz3] obtained the following rotation theorem for the set C (compare with Theorem 3.2.6). We leave the proof for the reader.

Theorem 2.4.9. *Let $f \in C$. Then*

$$|\arg f'(z)| \leq 4 \arcsin |z|, \quad z \in U.$$

This estimate is sharp.

Krzyz also found the radius of close-to-convexity r_{cc} for the set S [Krz2]. He showed that r_{cc} is the unique solution of a certain equation such that $0.80 < r_{cc} < 0.81$.

2.4.2 Spirallike functions in the unit disc

In this section we consider another subclass of univalent functions in the unit disc, namely the *spirallike* functions. This class was introduced by Spaček in 1933 [Spa].

Let $\alpha \in (-\pi/2, \pi/2)$. A *logarithmic α-spiral* (or *α-spiral*) is a curve given by

$$w = w_0 \exp(-e^{-i\alpha}t), \quad t \in \mathbb{R},$$

where w_0 is a nonzero complex number. Thus if $w = w(t)$ is a logarithmic α-spiral, then

$$\text{Im} \left[e^{i\alpha} \log w(t) \right] = \text{constant}, \quad t \in \mathbb{R}.$$

If D is a domain in $\mathbb{C}$ such that $0 \in D$, then we say that D is *α-spirallike* (or *spirallike of type α*) if for each $w \in D$, the arc of the α-spiral joining w to the origin lies in D.

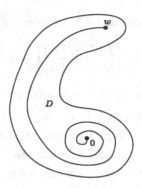

Figure 2.7: Spirallike domain

A closed curve γ is logarithmically spirallike of type α if each α-logarithmic spiral $w = w(t)$ intersects γ at a single point. Such a curve γ must be a Jordan curve and the domain Ω bounded by γ is a spirallike domain of type α.

Let $f : U \to \mathbb{C}$ be a holomorphic function on U such that $f(0) = 0$ and let $\alpha \in (-\pi/2, \pi/2)$. We say that f is *spirallike of type* α (or α-*spirallike*) if f is univalent on U and $f(U)$ is a spirallike domain of type α. We denote the class of normalized spirallike functions of type α by $\widehat{S}_\alpha$. We say simply that f is *spirallike* if f is spirallike of type α for some $\alpha \in (-\pi/2, \pi/2)$. It is obvious that 0-spirallike functions are starlike.

The following theorem, due to Spaček [Spa], provides a necessary and sufficient condition for spirallikeness of type α in the unit disc. In the proof of sufficiency we shall use methods of Al-Amiri and Mocanu [Al-Moc] which apply to the case of nonanalytic functions. In the next chapter we shall give another proof of Theorem 2.4.10 using the method of Loewner chains.

Theorem 2.4.10. *Let* $f : U \to \mathbb{C}$ *be a holomorphic function such that* $f(0) = 0$ *and* $f'(0) \neq 0$. *Also let* $\alpha \in \mathbb{R}$, $|\alpha| < \pi/2$. *Then* f *is spirallike of type* α *if and only if*

$$(2.4.9) \qquad \mathrm{Re}\left[e^{i\alpha}\frac{zf'(z)}{f(z)}\right] > 0, \quad z \in U.$$

Proof. First assume that f satisfies (2.4.9). Then it is elementary to deduce that $f(z) \neq 0$ for $z \in U \setminus \{0\}$ and also that $f'(z) \neq 0$, $z \in U$, i.e. f is locally univalent on U. For $r \in (0,1)$ let $\Gamma_r = f(\partial U_r)$. We shall prove that the curves Γ_r are nonintersecting Jordan curves. This will imply the univalence of f. To this end, let $\{\gamma_\varphi\}$, $\varphi \in [0, 2\pi)$, be the family of spirals defined by

$$(2.4.10) \qquad \gamma_\varphi(t) = e^{i\varphi} \exp(-e^{-i\alpha}t) = e^{-t\cos\alpha} e^{i(\varphi+t\sin\alpha)}, \quad t \in \mathbb{R}.$$

(There is a different choice of sign in [Al-Moc].)

It is clear that each point $w \in \mathbb{C} \setminus \{0\}$ lies on precisely one spiral of the family $\{\gamma_\varphi\}$. Thus for $z = re^{i\theta}$, $0 < r < 1$, $0 \le \theta < 2\pi$, the equation

$$(2.4.11) \qquad f(z) = \gamma_\varphi(t)$$

yields a unique $\varphi = \varphi(r, \theta) \in [0, 2\pi)$.

We first show that Γ_r is a Jordan curve for each $r \in (0,1)$. For this purpose, it suffices to show that

$$(2.4.12) \qquad \frac{\partial\varphi}{\partial\theta}(r,\theta) > 0, \quad \theta \in [0, 2\pi)$$

and

$$(2.4.13) \qquad \mathrm{Var}_{0 \le \theta < 2\pi}\varphi(r,\theta) = 2\pi,$$

where Var stands for the total variation. From (2.4.10) and (2.4.11) we easily obtain that

$$(2.4.14) \qquad -t\cos\alpha = \log|f(z)|$$

and

$$(2.4.15) \qquad \varphi + t\sin\alpha = \arg f(z).$$

Therefore,

$$(2.4.16) \qquad \varphi = \arg f(z) + \tan\alpha \log|f(z)|,$$

and thus

$$\frac{\partial\varphi}{\partial\theta}(r,\theta) = \frac{\partial}{\partial\theta}\arg f(re^{i\theta}) + \tan\alpha\frac{\partial}{\partial\theta}\log|f(re^{i\theta})|$$

$$= \mathrm{Re}\left[\frac{re^{i\theta}f'(re^{i\theta})}{f(re^{i\theta})}\right] - \tan\alpha\,\mathrm{Im}\left[\frac{re^{i\theta}f'(re^{i\theta})}{f(re^{i\theta})}\right]$$

$$= \frac{1}{\cos\alpha}\mathrm{Re}\left[e^{i\alpha}\frac{zf'(z)}{f(z)}\right], \quad z = re^{i\theta}.$$

Taking into account (2.4.9) and the above relation, we obtain (2.4.12). Moreover, since $f(z) \neq 0$, $0 < |z| < 1$, we conclude that the curves Γ_r, $0 < r < 1$, are homotopic in $\mathbb{C} \setminus \{0\}$, and therefore Γ_r has the same index with respect to the origin for each $r \in (0, 1)$. But f is locally univalent and preserves the orientation in a neighbourhood of the origin. Hence, for sufficiently small r the index of Γ_r with respect to zero is 1, and thus the total variation of the argument along Γ_r is 2π. From (2.4.16) and this reasoning, we obtain that for all $r \in (0, 1)$,

$$\text{Var}_{0 \leq \theta < 2\pi} \varphi(r, \theta) = \text{Var}_{0 \leq \theta < 2\pi} \arg f(re^{i\theta}) = 2\pi.$$

We have therefore proved that for each $r \in (0, 1)$, Γ_r is a logarithmically spirallike Jordan curve of type α.

It remains to show that f is univalent on U. To this end, it suffices to show that $\Gamma_{r_1} \cap \Gamma_{r_2} = \emptyset$ for $r_1, r_2 \in (0, 1)$, $r_1 \neq r_2$. For this purpose, fix $\varphi \in [0, 2\pi)$. Considering the system

$$f(z) = \gamma_\varphi(t), \quad |z| = r, \quad 0 < r < 1,$$

we obtain a unique solution $z = re^{i\theta}$, $\theta = \theta(r)$, and a unique $t = t(r, \theta) = t(r)$. We need only show that

$$\frac{dt}{dr} < 0, \quad r \in (0, 1).$$

Differentiating the relations (2.4.14) and (2.4.15) with respect to r, we obtain

$$-\frac{dt}{dr} \cos \alpha = \frac{1}{r} \text{Re} \left[\frac{z f'(z)}{f(z)} \right] - \frac{d\theta}{dr} \text{Im} \left[\frac{z f'(z)}{f(z)} \right],$$

$$\frac{dt}{dr} \sin \alpha = \frac{1}{r} \text{Im} \left[\frac{z f'(z)}{f(z)} \right] + \frac{d\theta}{dr} \text{Re} \left[\frac{z f'(z)}{f(z)} \right].$$

Eliminating $\dfrac{d\theta}{dr}$ in the above equalities, we obtain

$$-|f(z)|^2 \frac{dt}{dr} \text{Re} \left[e^{i\alpha} \frac{z f'(z)}{f(z)} \right] = r |f'(z)|^2.$$

If we now use the hypothesis, we conclude that $\dfrac{dt}{dr} < 0$, as desired.

We have therefore shown that f is univalent on U and each domain $f(U_r)$ is spirallike of type α. Since $f(U) = \bigcup_{0<r<1} f(U_r)$, we conclude that $f(U)$ is a spirallike domain of type α and this completes the first part of the proof.

Conversely, suppose that f is spirallike of type α. Then for each $z \in U$, the range of f is a domain containing the arc of the α-spiral joining $f(z)$ to the origin. That is, $f(z) \exp(-e^{-i\alpha}t) \in f(U)$, for each $t \geq 0$. Let

$$(2.4.17) \qquad \gamma(z,t) = f^{-1}(f(z)\exp(-e^{-i\alpha}t)), \quad 0 \leq t < \infty.$$

Then $\gamma(z,0) = z$ and for each fixed t, $\gamma(z,t)$ is an analytic map of U to itself such that $\gamma(0,t) = 0$. Therefore $|\gamma(z,t)| \leq |z|$, in view of the Schwarz lemma. Moreover, differentiating both sides of (2.4.17), we obtain

$$f'(\gamma(z,t))\frac{\partial \gamma}{\partial t}(z,t) = -e^{-i\alpha}\exp(-e^{-i\alpha}t)f(z),$$

and hence for $t = 0$, we have

$$e^{-i\alpha}\frac{f(z)}{zf'(z)} = -\frac{\frac{\partial \gamma}{\partial t}(z,0)}{z}.$$

Thus, in order to show (2.4.9), it suffices to prove that

$$\text{Re}\left[\frac{1}{z}\frac{\partial \gamma}{\partial t}(z,0)\right] \leq 0.$$

Since $|\gamma(z,t)| \leq |z|$, we obtain

$$\text{Re}\left[\frac{1}{z}\frac{\partial \gamma}{\partial t}(z,0)\right] = \lim_{t\to 0^+}\frac{1}{t}\text{Re}\left[\frac{\gamma(z,t)}{z} - 1\right] \leq 0,$$

which is the desired conclusion. This completes the proof.

We remark that Theorem 2.1.8 can be used to obtain another simple proof of necessity of the condition in Theorem 2.4.10. (See [Robe3]. See also Problem 2.4.14.)

Using the above characterization of spirallikeness, we can give the following correspondence between the classes S^* and $\widehat{S}_\alpha$. This result provides many examples of spirallike functions in the unit disc (see [Goo1]).

Theorem 2.4.11. *Let $\alpha \in (-\pi/2, \pi/2)$ and $\beta = e^{-i\alpha} \cos \alpha$. Then $f \in \widehat{S}_\alpha$ if and only if there is $g \in S^*$ such that*

$$(2.4.18) \qquad\qquad f(z) = z\left(\frac{g(z)}{z}\right)^\beta, \quad z \in U.$$

The branch of the power function is chosen such that $\left(\frac{g(z)}{z}\right)^\beta\Big|_{z=0} = 1.$

Proof. First assume $f \in \widehat{S}_\alpha$. Clearly (2.4.18) is equivalent to

$$g(z) = z\left(\frac{f(z)}{z}\right)^{e^{i\alpha}/\cos\alpha}, \quad z \in U.$$

We choose the branch of the power function such that $\left(\frac{f(z)}{z}\right)^{e^{i\alpha}/\cos\alpha}\Big|_{z=0} = 1.$
A simple computation yields the relation

$$\frac{zg'(z)}{g(z)} = (1 + i\tan\alpha)\frac{zf'(z)}{f(z)} - i\tan\alpha, \quad z \in U.$$

Therefore

$$\operatorname{Re}\frac{zg'(z)}{g(z)} = \frac{1}{\cos\alpha}\operatorname{Re}\left[e^{i\alpha}\frac{zf'(z)}{f(z)}\right] > 0, \quad z \in U,$$

since f is spirallike of type α. Consequently, g is starlike as desired.

Conversely, if $g \in S^*$, then in view of the above relation and the fact that $\alpha \in (-\pi/2, \pi/2)$, one deduces that

$$\operatorname{Re}\left[e^{i\alpha}\frac{zf'(z)}{f(z)}\right] > 0, \quad z \in U.$$

Thus by Theorem 2.4.10, f is spirallike of type α. This completes the proof.

Using (2.4.18), it can be seen that the function

$$f(z) = \frac{z}{(1-z)^{2e^{-i\alpha}\cos\alpha}}, \quad z \in U,$$

is spirallike of type $\alpha \in (-\pi/2, \pi/2)$. This function is called the α-*spiral Koebe function*, and maps the unit disc onto the complement of an arc of a logarithmic α-spiral.

A direct consequence of Theorem 2.4.11 is the following sharp bound for the second order coefficient of functions in $\widehat{S}_\alpha$. For sharp estimates of other coefficients of spirallike functions of type α, the reader may consult [Goo1, vol.1, p. 151].

Corollary 2.4.12. *Let $f(z) = z + a_2 z^2 + \ldots$ be a spirallike function of type $\alpha \in (-\pi/2, \pi/2)$. Then $|a_2| \le 2\cos\alpha$. This bound is sharp. Equality is attained for the α-spiral Koebe function.*

Proof. Since $f \in \widehat{S}_\alpha$ there is a function $g \in S^*$ such that

$$f(z) = z\left(\frac{g(z)}{z}\right)^\beta, \quad z \in U,$$

where $\beta = e^{-i\alpha}\cos\alpha$. Let $g(z) = z + b_2 z^2 + \ldots$. If we identify the coefficients in the power series expansions, we easily deduce that $a_2 = b_2\beta$. Taking into account Theorem 2.2.16, we obtain the desired conclusion.

2.4.3 Φ-like functions on the unit disc

We finish this chapter with a short discussion of Φ-like functions. This concept was introduced and studied by Brickman in 1973 [Bri] as a generalization of starlikeness and spirallikeness. It makes some use of differential equations, though not in as deep a way as the Loewner method, which we shall study in the next chapter. The differential equations which arise are autonomous, and only the most basic existence and uniqueness theorems are needed. A necessary and sufficient condition for univalence is obtained (Corollary 2.4.17). We mention that the sufficient condition for univalence had earlier been established (using a different method) by Kas'yanyuk in 1959 [Kas]. Another proof of Corollary 2.4.17 was given by Avkhadiev and Aksent'ev [Avk-Aks2, Theorem 25]. Related results were obtained by Rahmanov [Rah]. However, we shall use the method of Brickman [Bri].

Definition 2.4.13. Let f be a holomorphic function on U such that $f(0) = 0$ and $f'(0) \ne 0$. Let Φ be a holomorphic function on $f(U)$ such that $\Phi(0) = 0$ and $\mathrm{Re}\,\Phi'(0) > 0$. We say that f is Φ-*like* on U (or simply Φ-like) if

$$(2.4.19) \qquad \mathrm{Re}\left[\frac{zf'(z)}{\Phi(f(z))}\right] > 0, \quad z \in U.$$

We remark that if $\Phi(w) \equiv w$ in the above definition then f is starlike, and if $\Phi(w) = \lambda w$, $\mathrm{Re}\,\lambda > 0$, then f is spirallike of type α, where $\alpha = -\arg\lambda$.

Definition 2.4.14. Let Ω be a region in $\mathbb{C}$ such that $0 \in \Omega$, and let Φ be a holomorphic function on Ω such that $\Phi(0) = 0$ and Re $\Phi'(0) > 0$. We say that Ω is Φ-*like* if for each point $w_0 \in \Omega$ the initial value problem

$$(2.4.20) \qquad \frac{dw}{dt} = -\Phi(w), \quad w(0) = w_0$$

has a solution $w(t)$ defined for $t \geq 0$, such that $w(t) \in \Omega$ for $t \geq 0$, and $w(t) \to 0$ as $t \to \infty$.

Note that if there is a solution of (2.4.20), then this solution is unique. On the other hand, if $\Phi(w) \equiv w$ then the solution of (2.4.20) is $w(t) = w_0 e^{-t}$. Thus in this case Ω is a Φ-like region if and only if Ω is starlike (with respect to the origin). Also if $\Phi(w) \equiv \lambda w$, Re $\lambda > 0$, then $w(t) = w_0 e^{-\lambda t}$, hence Ω is a Φ-like domain if and only if Ω is α-spirallike, where $\alpha = -\arg \lambda$.

The significance of Φ-like functions and Φ-like domains appears in the following results due to Brickman [Bri]. To this end, we need to use the next lemma.

Lemma 2.4.15. *Let p be a holomorphic function on U such that* Re $p(z) > 0$, $z \in U$. *Then for each fixed $z \in U$, the initial value problem*

$$(2.4.21) \qquad \frac{dv}{dt} = -vp(v), \quad v(0) = z,$$

has a unique solution $v(t) = v(z,t)$ defined for all $t \geq 0$, such that $|v(z,t)|$ is a strictly decreasing function of $t \in [0,\infty)$ and $v(z,t) \to 0$ as $t \to \infty$.

Proof. For a detailed proof see the more general existence and uniqueness theorem for the Loewner differential equation (Theorem 3.1.10). (The normalization $p(0) = 1$ required in the Loewner theory plays no role in the existence and uniqueness theorem.) The behaviour of the solutions as $t \to \infty$ can be deduced from the fact that

$$\frac{d}{dt}|v|^2 = -2|v|^2 \text{Re}\, p(v).$$

Theorem 2.4.16. *Let f be a Φ-like function. Then f is univalent on U and $f(U)$ is a Φ-like domain.*

Proof. Let $p(z) = \dfrac{\Phi(f(z))}{zf'(z)}$, $z \in U$. Then from (2.4.19) we deduce that p is holomorphic on U and Re $p(z) > 0$ on U. Fix $z \in U$ and let $v(t) = v(z,t)$,

$t \geq 0$, be the solution of the initial value problem

$$\frac{dv}{dt} = -vp(v), \quad v(0) = z.$$

This solution exists and is unique by Lemma 2.4.15. Since $v(z,t) \in U$, we can define $w_z(t) = w(z,t) = f(v(z,t))$, $t \geq 0$. A short computation shows that w_z is the unique solution of the initial value problem

(2.4.22) $$\frac{dw_z}{dt} = -\Phi(w_z), \quad w_z(0) = f(z).$$

Next, using the result of Lemma 2.4.15, we conclude that

$$\lim_{t \to \infty} w_z(t) = \lim_{t \to \infty} f(v(z,t)) = f(0) = 0.$$

Therefore $f(U)$ is a Φ-like domain.

In order to show that f is univalent, let $a, b \in U$ be such that $f(a) = f(b)$. Then $w_a(t)$ and $w_b(t)$ are solutions of the same initial value problem (2.4.22). Hence $w_a(t) = w_b(t)$, or $f(v(a,t)) = f(v(b,t))$, for all $t \geq 0$, by the uniqueness of solutions. Moreover, since f has a local inverse at the origin, and $v(a,t) \to 0$, $v(b,t) \to 0$ as $t \to \infty$, it follows that there is $t_0 \geq 0$ such that $v(a,t) = v(b,t)$ for $t \geq t_0$. Again using the uniqueness of solutions of (2.4.21), we conclude that $v(a,t) = v(b,t)$ for all $t \geq 0$, because both are solutions of the same initial value problem

$$\frac{dv}{dt} = -vp(v), \quad v(t_0) = v(a,t_0) = v(b,t_0),$$

which has a unique solution for all $t \geq 0$. Hence $a = v(a,0) = v(b,0) = b$. Thus f is univalent, as claimed.

Corollary 2.4.17. *Let f be a holomorphic function on U such that $f(0) = 0$. Then f is univalent on U if and only if f is Φ-like for some Φ.*

Proof. We have already proved in Theorem 2.4.16 that Φ-likeness implies univalence. Thus it suffices to show that if f is univalent then f is Φ-like, for some Φ. For this purpose, let $p \in H(U)$ be such that $\mathrm{Re}\, p(z) > 0$ on U, and consider the equation

$$\Phi(f(z)) = \frac{zf'(z)}{p(z)}.$$

This equation provides a solution Φ such that $\Phi \in H(f(U))$, $\Phi'(0) = 1/p(0)$ and

$$\text{Re}\left[\frac{zf'(z)}{\Phi(f(z))}\right] > 0, \quad z \in U.$$

Thus f is Φ-like.

The converse of Theorem 2.4.16 is given in the next result.

Theorem 2.4.18. *Let f be a univalent function on U such that $f(0) = 0$ and $f(U)$ is a Φ-like domain. Then f is Φ-like.*

Proof. Since $f(U)$ is Φ-like, we may define $w_z(t) = w(z, t)$ (for $z \in U$ and $t \geq 0$) to be the unique solution of the initial value problem

$$(2.4.23) \qquad \frac{dw_z}{dt} = -\Phi(w_z), \quad w_z(0) = f(z).$$

Also let $v_z(t) = v(z, t) = f^{-1}(w_z(t))$ for $z \in U$ and $t \geq 0$. Since f is univalent, $v_z(t)$ is well defined. A simple computation shows that

$$f'(v(z, t))\frac{\partial v}{\partial t}(z, t) = -\Phi(w(z, t)),$$

and setting $t = 0$ we obtain

$$\frac{\partial v}{\partial t}(z, 0) = -\frac{\Phi(f(z))}{f'(z)}.$$

Let

$$p(z) = -\frac{1}{z}\frac{\partial v}{\partial t}(z, 0) = \frac{\Phi(f(z))}{zf'(z)}, \quad z \in U.$$

Then p is holomorphic on U, $\text{Re } p(0) = \text{Re } \Phi'(0) > 0$, and we wish to prove that $\text{Re } p(z) > 0$ on $U \setminus \{0\}$. Since $\Phi(z)$ and $f(z) = w_z(0)$ are holomorphic, a classical theorem in the theory of differential equations implies that $w(z, t)$ is holomorphic in z for fixed $t \geq 0$ (see e.g. [Cart3, Theorem 3.3.1]). Hence $v(z, t)$ is also holomorphic in z for each $t \geq 0$. On the other hand, $|v(z, t)| < 1$ for all $z \in U$ and $t \geq 0$, and also $v(0, 0) = 0$. Moreover, by the uniqueness of solutions to (2.4.23), we know that $w_0(0) = 0$ implies $w_0(t) = 0$ for all $t \geq 0$, and consequently $v_0(t) = 0$ for all $t \geq 0$. In view of the Schwarz lemma we deduce that $|v(z, t)| \leq |z|$ for $z \in U$ and $t \geq 0$. Thus

$$\text{Re } p(z) = -\text{Re}\lim_{t \to 0+}\frac{v(z, t) - z}{tz}$$

$$= - \lim_{t \to 0^+} \text{Re} \left\{ \frac{1}{t} \left(\frac{v(z,t)}{z} - 1 \right) \right\} \geq 0, \quad z \in U.$$

Finally, since $\text{Re } p(0) > 0$, we must have $\text{Re } p(z) > 0$ on U by the minimum principle for harmonic functions. This completes the proof.

Combining the results of Theorems 2.4.16 and 2.4.18 with the observations following Definition 2.4.14, we obtain alternative proofs of Theorems 2.2.2 and 2.4.10.

Corollary 2.4.19. *Let $f : U \to \mathbb{C}$ be a holomorphic function such that $f(0) = 0$ and $f'(0) \neq 0$. Then f is starlike if and only if*

$$\text{Re} \left[\frac{zf'(z)}{f(z)} \right] > 0, \quad z \in U,$$

and f is spirallike of type $\alpha \in (-\pi/2, \pi/2)$ if and only if

$$\text{Re} \left[e^{i\alpha} \frac{zf'(z)}{f(z)} \right] > 0, \quad z \in U.$$

Notes. Much of Section 2.4 is based on [Dur], [Goo1], [Kap], [Pom5]. For further material on close-to-convex, spirallike, and Φ-like functions, the reader may consult the following references: [Al-Moc], [Avk-Aks1,2], [Biel-Lew], [Bier], [Bri], [Clu-Pom], [Gol4], [Hal-MG], [Kas], [Koep3], [Krz2,3], [Lew1], [Moc-Bu-Să], [Rah], [Robe1,3], [Rus-She], [She2], [Spa]. All of these classes of mappings may be generalized to several complex variables.

Problems

2.4.1. Show that if $f \in H(U)$ with $f(0) = f'(0) - 1 = 0$ and if

$$\text{Re} \left[\frac{zf'(z)}{f(z) - f(-z)} \right] > 0, \quad z \in U,$$

then f is close-to-convex on U.
(Sakaguchi, 1959 [Sak2].)

2.4.2. Show that if $f \in H(U)$ is normalized and if

$$\text{Re} \left[\frac{zf'(z)}{f(z) + \overline{f(\overline{z})}} \right] > 0, \quad z \in U,$$

then f is close-to-convex on U.

(Sakaguchi, 1959 [Sak2].)

2.4.3. Show that if $f \in H(U)$ is such that Re $[(1 - z^2)f'(z)] > 0$ on U, then f is univalent on U.

Hint. Use Lemma 2.4.2.

2.4.4. Let D be a convex domain in $\mathbb{C}$, let $\varphi : \mathbb{R} \to \mathbb{R}$ be a continuous function and let $f \in H(D)$ be such that

$$\text{Re } f'(z) + \varphi(\text{Im } f(z))\text{Im } f'(z) > 0, \quad z \in D.$$

Show that f is univalent on D.

(Janiec, 1989 [Jan].)

2.4.5. Show that if $f : U \to \mathbb{C}$ is normalized locally univalent and satisfies

$$\text{Re } \left[1 + \frac{zf''(z)}{f'(z)}\right] > -\frac{1}{2}, \quad z \in U,$$

then f is close-to-convex.

2.4.6. Let $\gamma : [0, 2\pi] \to \mathbb{R}$ be a nonconstant function such that γ is nondecreasing on $[0, \pi]$ and nonincreasing on $[\pi, 2\pi]$. Let

$$f(z) = \int_0^{2\pi} \frac{1 + ze^{-it}}{1 - ze^{-it}}\gamma(t)dt, \quad z \in U.$$

Show that f is close-to-convex.

(Kaplan, 1952 [Kap].)

2.4.7. Prove Theorem 2.4.7.

2.4.8. Prove Theorem 2.4.8.

2.4.9. Show that the α-spiral Koebe function is not close-to-convex when $\alpha \in (-\pi/2, \pi/2) \setminus \{0\}$.

(Duren, 1983 [Dur].)

2.4.10. Show that $f(z) = \dfrac{z(1 - z \cos \varphi)}{(1 - e^{i\varphi}z)^2}$, $\cos \varphi \neq 0$, is close-to-convex, but is not spirallike.

(Duren, 1983 [Dur].)

Hint. The image of the unit disc under this function is the complement of a half-line which is not collinear with the origin.

2.4.11. Show that if f is close-to-convex with respect to the convex function h, then

$$\text{Re}\ \left[\frac{f(z_2) - f(z_1)}{h(z_2) - h(z_1)}\right] > 0, \quad |z_1| < 1, |z_2| < 1.$$

Hint. Use similar reasoning as in the proofs of Lemmas 2.4.1 and 2.4.2.

2.4.12. Prove Theorem 2.4.9.

2.4.13. Let $\lambda, \mu \in \mathbb{C}$ with $|\lambda| = |\mu| = 1$ and

$$f(z) = \frac{z - \frac{1}{2}(\lambda + \mu)z^2}{(1 - \mu z)^2}, \quad z \in U.$$

Show that f is close-to-convex.

2.4.14. Use Theorem 2.1.8 to obtain another proof of necessity of the condition in Theorem 2.4.10.

Hint. Let $g(z, t) = (1 - te^{-i\alpha})f(z)$ and $\rho = 1$ in Theorem 2.1.8.

Chapter 3

The Loewner theory

3.1 Loewner chains and the Loewner differential equation

3.1.1 Kernel convergence

One of the most powerful techniques in the theory of univalent functions is Loewner's theory [Lö2], developed in the early 1920's. This theory enables us to obtain many far-reaching results which are not accessible via the elementary methods we have considered up to this point. Loewner was influenced by the work of S. Lie on semigroups associated to differential equations. His idea was to express an arbitrary function in a dense subclass of S as the limit of a time-dependent flow starting from the identity function. The flow satisfies a differential equation – the Loewner differential equation. From the outset, one of the main applications of Loewner's method was to the study of the Bieberbach conjecture. Not surprisingly, the method plays a central role in de Brange's proof of this famous conjecture [DeB]. However Loewner's method also yields geometric results of considerable interest, such as the radius of starlikeness for the class S and the rotation theorem for the class S. For this reason, we are going to present some of the basic results of Loewner's theory. For further details, the reader may consult [Ahl2], [Bec1-2], [Con], [Gol4], [Gon5], [Hay], [Hen], [Kuf1-4], [Lö2], [Pom3], [Ros-Rov], and especially

[Pom5, Chapter 6] and [Dur, Chapter 3].

We begin this section with some basic ideas about *kernel convergence*.

Definition 3.1.1. Let $\{G_n\}_{n\in\mathbb{N}}$ be a sequence of domains in $\mathbb{C}$ such that $0 \in G_n$, for all $n \in \mathbb{N}$. If 0 is an interior point of $\bigcap_{n=1}^{\infty} G_n$, we define the *kernel G* of $\{G_n\}_{n\in\mathbb{N}}$ to be the largest domain containing 0 such that if K is a compact subset of G, then there exists a positive integer n_0 such that $K \subseteq G_n$ for $n \geq n_0$. If 0 is not an interior point of $\bigcap_{n=1}^{\infty} G_n$, we define the kernel to be $\{0\}$.

We first prove that the kernel is well defined. We may assume that $G \neq \{0\}$. Let $\mathcal{G}$ denote the set of domains D such that $0 \in D$ and every compact subset K of D is contained in all but a finite number of sets G_n. We claim that $G = \bigcup_{D\in\mathcal{G}} D$ belongs to $\mathcal{G}$.

Certainly G is a domain which contains 0. By the definitions of $\mathcal{G}$ and G, we know that for each $z \in G$ there is a domain $D_z \in \mathcal{G}$ such that $z \in D_z$, and there is a disc $U(z, \rho_z)$ centered at z and of radius ρ_z whose closure is contained in D_z. If K is a compact subset of G, we may cover K by a finite number of discs $U(z_j, \rho_{z_j})$, $j = 1, \ldots, k$, such that $\overline{U}(z_j, \rho_{z_j}) \subset D_j$ for some $D_j \in \mathcal{G}$. Again using the definition of $\mathcal{G}$, we have for $j = 1, \ldots, k$, $\overline{U}(z_j, \rho_{z_j}) \subset G_n$, for all but a finite number of integers n. Therefore, $K \subset G_n$ for all but a finite number of sets G_n.

We have thus proved that $G \in \mathcal{G}$, and it is evident that no larger domain can belong to $\mathcal{G}$.

We say that $\{G_n\}_{n\in\mathbb{N}}$ *converges* (*kernel converges*) to G if each subsequence of $\{G_n\}_{n\in\mathbb{N}}$ has the same kernel G. We denote this by $G_n \to G$. The importance of this concept arises from the *Carathéodory convergence theorem* [Cara2].

Theorem 3.1.2. *Let $\{G_n\}_{n\in\mathbb{N}}$ be a sequence of simply connected domains with $0 \in G_n$ and $G_n \subsetneqq \mathbb{C}$, for all $n \in \mathbb{N}$. Let f_n be the conformal mapping of U onto G_n such that $f_n(0) = 0$ and $f_n'(0) > 0$. Let G be the kernel of $\{G_n\}_{n\in\mathbb{N}}$. Then $f_n \to f$ locally uniformly on U if and only if $G_n \to G \neq \mathbb{C}$. In the case of convergence, either $G = \{0\}$ and $f \equiv 0$, or else $G \neq \{0\}$ and then G is a simply connected domain, f gives a conformal map of the unit disc U onto G, and $f_n^{-1} \to f^{-1}$ locally uniformly on G.*

Proof. Necessity. First assume that $f_n \to f$ locally uniformly on U. Then $f \in H(U)$ and it is either constant or a univalent function, by Hurwitz's theorem.

Case (a): If f is constant then it is obvious that $f \equiv 0$ since $f(0) = 0$. To prove that the kernel of $\{G_n\}_{n \in \mathbb{N}}$ is $\{0\}$, we argue by contradiction. If there is a domain H such that $0 \in H$ and each compact subset of H lies in G_n for sufficiently large n, then there exists $\epsilon > 0$ such that $U_\epsilon \subset G_n$ for all n. Let $g_n = f_n^{-1}$, $n \in \mathbb{N}$. Then g_n is a holomorphic function on U_ϵ, $g_n(0) = 0$ and $|g_n(w)| < 1$ for $w \in U_\epsilon$. By the Schwarz lemma, one deduces that $|g_n(w)| \leq |w|/\epsilon$, $w \in U_\epsilon$, and $g_n'(0) \leq 1/\epsilon$. Hence $f_n'(0) \geq \epsilon$, for sufficiently large n. However, this is a contradiction to the fact that $f_n \to 0$ locally uniformly on U as $n \to \infty$. Therefore, the kernel of $\{G_n\}_{n \in \mathbb{N}}$ is $\{0\}$. Similarly, every subsequence of $\{G_n\}_{n \in \mathbb{N}}$ has kernel $\{0\}$, and hence $G_n \to \{0\}$.

Case (b): We now assume $f \not\equiv 0$. Let $\Omega = f(U)$. Then Ω is a simply connected domain in $\mathbb{C}$. We have to prove that $G = \Omega$ and $G_n \to G$ in the sense of kernel convergence. This is carried out in the following three steps:

First step. First, we prove that $\Omega \subseteq G$. To this end, we have to prove that if K is a compact subset of Ω, then $K \subset G_n$ for sufficiently large n. To do so we enclose K within a smooth Jordan curve Γ lying in $\Omega \setminus K$. Let η be the distance between Γ and K, and let $\gamma = f^{-1}(\Gamma)$ (see Figure 3.1). If $v_0 \in K$ then

$$|f(z) - v_0| \geq \eta, \quad z \in \gamma.$$

Now since $f_n \to f$ uniformly on γ, there exists $n_0 \in \mathbb{N}$ such that

$$|f_n(z) - f(z)| < \eta, \quad z \in \gamma, n \geq n_0.$$

Consequently if $n \geq n_0$ and $z \in \gamma$, then

$$|[f(z) - v_0] - [f_n(z) - v_0]| < \eta \leq |f(z) - v_0|.$$

By Rouché's theorem, one deduces that both equations $f(z) - v_0 = 0$ and $f_n(z) - v_0 = 0$ have the same number of zeros inside γ for $n \geq n_0$. Since the function $f(z) - v_0$ has precisely one zero inside γ, we see that $v_0 \in f_n(U) = G_n$ for all $n \geq n_0$. Noting that n_0 is independent of $v_0 \in K$, we conclude that $K \subseteq G_n$ for sufficiently large n, and hence $\Omega \subseteq G$ as claimed.

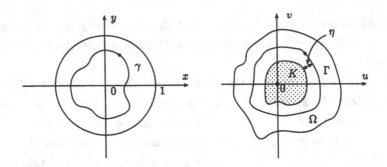

Figure 3.1: A compact subset K of $\Omega = f(U)$

Second step. There exists a subsequence $\{n_k\}_{k\in\mathbb{N}}$ such that $f^{-1} = \lim_{k\to\infty} f_{n_k}^{-1}$ locally uniformly on Ω.

Indeed, the inverse functions $g_n = f_n^{-1}$ are well defined on any fixed compact subset of Ω for n sufficiently large, and also $|g_n(w)| \leq 1$ for such n. By Montel's theorem, there exists a subsequence $\{g_{n_k}\}_{k\in\mathbb{N}}$ of $\{g_n\}_{n\in\mathbb{N}}$ which converges locally uniformly on Ω to a holomorphic function g on Ω. This function satisfies $g(0) = 0$ and $g'(0) > 0$, since

$$0 < \frac{1}{f'(0)} = \lim_{k\to\infty} \frac{1}{f'_{n_k}(0)} = \lim_{k\to\infty} g'_{n_k}(0) = g'(0).$$

Thus g is not constant, and therefore by Hurwitz's theorem g is univalent.

We next prove that $g = f^{-1}$. To this end, let $z_0 \in U$ and $w_0 = f(z_0)$. Let $\delta > 0$ be small enough that the closed disc $\overline{U}(z_0, \delta)$ is contained in U. Also let $\gamma_1 = \partial U(z_0, \delta)$ and $\Gamma_1 = f(\gamma_1)$. If η_1 denotes the distance between w_0 and Γ_1, then clearly $|f(z) - w_0| \geq \eta_1$ for $z \in \gamma_1$, and since $f_n \to f$ locally uniformly on U, there exists a positive integer n'_0 such that $|f_n(z) - f(z)| < \eta_1$ for $n \geq n'_0$ and $z \in \gamma_1$. An argument using Rouché's theorem again implies that both equations $f_n(z) - w_0 = 0$ and $f(z) - w_0 = 0$ have the same number of zeros in $U(z_0, \delta)$ for $n \geq n'_0$, that is one. We then conclude that for each $n \geq n'_0$, there exists a point $z_n \in U(z_0, \delta)$ such that $f_n(z_n) = w_0$, or $z_n = g_n(w_0)$. Thus for sufficiently large k, we have

$$|g_{n_k}(w_0) - g(w_0)| < \delta.$$

Consequently, we obtain that

$$|g(w_0) - z_0| \leq |g(w_0) - g_{n_k}(w_0)| + |z_{n_k} - z_0| < 2\delta.$$

Since δ is arbitrary, we must have $g(w_0) = z_0$, and since z_0 is arbitrary, one deduces that $g = f^{-1}$.

Third step. $f^{-1} = \lim_{n\to\infty} f_n^{-1}$ locally uniformly on Ω and $\Omega = G$.

The argument in the second step implies that each subsequence of $\{g_n\}_{n\in\mathbb{N}}$ contains a further subsequence which converges locally uniformly on Ω to f^{-1}. But this implies that the whole sequence $\{g_n\}_{n\in\mathbb{N}}$ converges to f^{-1}.

On the other hand, by Definition 3.1.1 each compact subset of the kernel G of $\{G_n\}_{n\in\mathbb{N}}$ is contained in all but a finite number of sets G_n. In view of the first step we know that $\Omega \subseteq G$. Since $f^{-1} = \lim_{n\to\infty} f_n^{-1}$ exists locally uniformly on Ω, an application of Vitali's theorem shows that $\phi = \lim_{n\to\infty} f_n^{-1}$ exists locally uniformly on G. Furthermore ϕ is univalent on G and $\phi|_\Omega = f^{-1}$. Now if $\Omega \subsetneqq G$, there exists a point $w_1 \in G \setminus \Omega$ and we must have $\phi(w_1) \in U$. However since f^{-1} is a one-to-one map of Ω onto U, $\phi(w_1) = f^{-1}(w_2) = \phi(w_2)$ for some $w_2 \in \Omega$. But this is a contradiction with the univalence of ϕ on G. Consequently we must have $\Omega = G$, as claimed.

We have therefore proved that the kernel of $\{G_n\}_{n\in\mathbb{N}}$ is $f(U)$, and because each subsequence $\{f_{n_k}\}_{k\in\mathbb{N}}$ of $\{f_n\}_{n\in\mathbb{N}}$ converges locally uniformly to f on U, the corresponding subsequence $\{G_{n_k}\}_{k\in\mathbb{N}}$ has the same kernel $G = f(U)$. Hence $G_n \to G$.

Sufficiency. Conversely, assume that $G_n \to G$ and $G \neq \mathbb{C}$. We shall prove that $f_n \to f$ locally uniformly on U.

Case (a): If $G = \{0\}$ then it is straightforward to prove that $f_n'(0) \to 0$. Otherwise there would exist some $\rho > 0$ and a subsequence $\{f_{n_k}\}_{k\in\mathbb{N}}$ such that $f_{n_k}'(0) \geq \rho$. Then the Koebe $1/4$-theorem would imply that each domain G_{n_k} contains the disc $U_{\rho/4}$. Since this is a contradiction with the assumption that $G_n \to \{0\}$, we must have $f_n'(0) \to 0$. Now, using the growth theorem for the class S (see Theorem 1.1.6), we obtain

$$|f_n(z)| \leq f_n'(0)\frac{|z|}{(1-|z|)^2}, \quad z \in U,$$

and this implies that $f_n \to 0$ locally uniformly on U.

Case (b): If $G \neq \{0\}$ and $G \neq \mathbb{C}$, we first prove that $\{f_n'(0)\}_{n \in \mathbb{N}}$ is a bounded sequence. Otherwise there exists a subsequence $\{n_k\}_{k \in \mathbb{N}}$ such that $f_{n_k}'(0) > k$ for $k \geq 1$. By the Koebe 1/4-theorem, $G_{n_k} \supseteq U_{k/4}$. It follows that the sequence $\{G_{n_k}\}_{k \in \mathbb{N}}$ has the kernel $\mathbb{C}$. However, this is a contradiction with the hypothesis, and hence there exists $M > 0$ such that $f_n'(0) \leq M$ for all $n \in \mathbb{N}$. By the growth theorem for the class S, we have

$$|f_n(z)| \leq M \frac{|z|}{(1 - |z|)^2}, \quad z \in U.$$

Therefore $\{f_n\}_{n \in \mathbb{N}}$ is a locally uniformly bounded sequence, and hence by Montel's theorem $\{f_n\}_{n \in \mathbb{N}}$ has a subsequence $\{f_{n_k}\}_{k \in \mathbb{N}}$ which converges locally uniformly on U to a function f. By the first part of the proof applied to this subsequence, f is a univalent function on U which maps U onto the kernel of $\{G_{n_k}\}_{k \in \mathbb{N}}$. This kernel must be G since $G_n \to G$. Since $f(0) = 0$ and $f'(0) > 0$, f is uniquely determined by the uniqueness assertion of the Riemann mapping theorem. The same reasoning may be applied to subsequences of $\{f_n\}_{n \in \mathbb{N}}$, and shows that any subsequence contains a further subsequence which converges locally uniformly on U to the same function f. Hence $\{f_n\}_{n \in \mathbb{N}}$ converges locally uniformly on U to f, as desired. This completes the proof.

We remark that Theorem 3.1.2 does not hold for conformal mappings of $\mathbb{C}$. Indeed, if $f_n(z) = z/n$, $z \in \mathbb{C}$, then $G_n = f_n(\mathbb{C}) \to \mathbb{C}$, but $f_n \to 0$ which is not a univalent function.

We recall that a function $f \in S$ is called a *slit mapping* if f maps the unit disc conformally onto the complex plane minus a set of Jordan arcs. These arcs must tend to ∞ because $f(U)$ is a simply connected domain. The function f is called a *single-slit mapping* if the complement of $f(U)$ is a single Jordan arc.

Carathéodory's convergence theorem enables us to prove the important result that with respect to uniform convergence on compact sets, the set of single-slit mappings is dense in S.

Theorem 3.1.3. *Let $f \in S$. Then there exists a sequence of single-slit mappings $f_n \in S$ such that $f_n \to f$ locally uniformly on U.*

Proof. By considering dilations of f, i.e. $f_r(z) = f(rz)/r$, $0 < r < 1$, we may reduce to the case in which f is univalent on $\overline{U}$. In this case $f(U)$ is a domain bounded by an analytic Jordan curve $\Gamma: [a, b] \to \mathbb{C}$. Let $w_0 =$

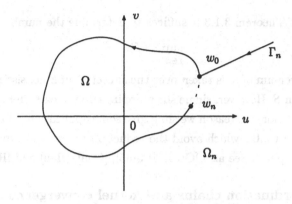

Figure 3.2: Density of single-slit maps

$\Gamma(a) = \Gamma(b)$, and let $\{w_n\}_{n \in \mathbb{N}}$ be a sequence of points on Γ corresponding to an increasing sequence of parameter values $\{t_n\}_{n \in \mathbb{N}} \subset [a, b]$, such that $t_n \to b$ as $n \to \infty$. Also let Γ_n be a Jordan arc consisting of an arc running from ∞ to w_0 in the exterior of Γ, followed by an arc which moves from w_0 along Γ to w_n (see Figure 3.2).

Let Ω_n be the complement of Γ_n. Also let f_n be the univalent function that maps U conformally onto Ω_n such that $f_n(0) = 0$ and $f_n'(0) > 0$. From this construction it is geometrically clear that $\Omega = f(U)$ is the kernel of $\{\Omega_n\}_{n \in \mathbb{N}}$ and that $\Omega_n \to \Omega$ in the sense of kernel convergence. Using the result of Theorem 3.1.2, we conclude that $f_n \to f$ locally uniformly on U. In addition, from the Weierstrass convergence theorem, we deduce that $f_n'(0) \to f'(0) = 1$. Finally, if $g_n(z) = f_n(z)/f_n'(0)$, then $\{g_n\}_{n \in \mathbb{N}}$ is a sequence of single-slit mappings in S such that $g_n \to f$ locally uniformly on U. This completes the proof.

The foregoing result has the following very important consequence which is used frequently in the Loewner theory. Let $\Psi : S \to \mathbb{R}$ be a real-valued continuous functional on S, in the sense that if $\{f_n\}_{n \in \mathbb{N}} \subset S$ and $f_n \to f$ locally uniformly on U, then $\Psi(f_n) \to \Psi(f)$. If we wish to study the problem of finding the number

$$\sup_{f \in S} \Psi(f),$$

then in view of Theorem 3.1.3 it suffices to determine the number

$$\sup_{f \in S'} \Psi(f),$$

where the supremum of Ψ is taken over the subset S' of S consisting of single-slit mappings in S. However, single slit mappings do not exist in higher dimensions, and partly for this reason we shall give some applications of the Loewner method in one variable which avoid the reduction to single-slit mappings.

Notes. See [Dur]. See also [Gol4], [Pom5], [Con], [Hen] and [Ros-Rov].

3.1.2   Subordination chains and kernel convergence

In this section we shall study properties of Loewner chains (univalent subordination chains), and in the next we shall show how they may be described by the Loewner differential equation. Much of this section is based on [Pom5, Chapter 6]. For further discussion about this subject see [Dur], [Gol4], [Ahl2], [Con], [Hay], [Hen], [Ros-Rov].

In the theory of Loewner chains, if h is a function which depends holomorphically on $z \in U$ and is also a function of other (real) variables, it is customary to write $h'(z, \cdot)$ instead of $\frac{\partial h}{\partial z}(z, \cdot)$.

Definition 3.1.4. The function $f : U \times [0, \infty) \to \mathbb{C}$ is called a *subordination chain* if $f(\cdot, t)$ is holomorphic on U, $f'(0, t)$ is a continuous function of $t \geq 0$ with $f'(0, t) \neq 0$ for $t \geq 0$, $|f'(0, t)|$ is strictly increasing, $f'(0, t) \to \infty$ as $t \to \infty$, and

$$(3.1.1) \qquad f(z, s) \prec f(z, t), \quad z \in U, \quad 0 \leq s \leq t < \infty.$$

Hence, if $f(z, t)$ is a subordination chain then there exists a family of Schwarz functions $v = v(z, s, t)$, called the *transition functions*, such that

$$(3.1.2) \qquad f(z, s) = f(v(z, s, t), t), \quad z \in U, \quad 0 \leq s \leq t < \infty.$$

The subordination chain $f(z, t)$ is called a *Loewner chain* (or a *univalent subordination chain*) if $f(\cdot, t)$ is univalent on U for all $t \geq 0$.

There is no loss of generality in assuming that the functions in a subordination chain satisfy the following *normalization*:

$$f(0, t) = 0, \quad f'(0, t) = e^t, \quad t \geq 0.$$

Indeed, if $f(z,t) = a_0 + a_1(t)z + \ldots$ is a general subordination chain, and if $t^* = \log\left|\frac{a_1(t)}{a_1(0)}\right|$ and $\theta(t) = \arg\left[\frac{a_1(t)}{a_1(0)}\right]$, then

$$f_*(z,t^*) = \frac{1}{a_1(0)}\left[f(e^{-i\theta(t)}z,t) - a_0\right]$$

becomes a normalized subordination chain. From now on we shall assume that all Loewner chains under discussion are normalized.

For a Loewner chain $f(z,t)$ let f_t denote the univalent function defined by $f_t(z) = f(z,t)$. We may think of this Loewner chain as a parameterized family of univalent functions $\{f_t\}_{t\in[0,\infty)}$ on U, indexed by time, f_0 being the first element, and as t increases to ∞ the images of the functions expand to fill out the complex plane.

If $f(z,t)$ is a Loewner chain, then there is a unique family of Schwarz functions $v(z,s,t)$ such that $f(z,s) = f(v(z,s,t),t)$, $z \in U$, $0 \le s \le t < \infty$. The normalization implies that $v'(0,s,t) = e^{s-t}$ for $0 \le s \le t$. Moreover, since the functions f_t and f_s are univalent on U, $v(\cdot,s,t)$ is likewise univalent on U. In addition, from (3.1.2) we deduce that the functions $v(z,s,t)$ have the following *semigroup property*:

(3.1.3) $\quad v(z,s,u) = v(v(z,s,t),t,u), \quad z \in U, \quad 0 \le s \le t \le u < \infty.$

To verify this property, it suffices to observe that for $w = v(z,s,t)$, we have

$$f(v(w,t,u),u) = f(w,t) = f(z,s) = f(v(z,s,u),u),$$

hence (3.1.3) follows by the univalence of $f(\cdot,u)$.

Further, using the semigroup property (3.1.3) and the fact that $|v(z,s,t)| \le |z|$, we deduce that

$$|v(z,s,u)| \le |v(z,s,t)|, \quad z \in U, \quad 0 \le s \le t \le u < \infty,$$

and thus $|v(z,s,t)|$ is a decreasing function of $t \in [s,\infty)$.

We now prove that between the notions of a Loewner chain and kernel convergence there is a duality. For this purpose, let $\{G(t)\}_{t\ge 0}$ be a family of simply connected domains such that

(3.1.4) $\qquad\qquad 0 \in G(s) \subsetneq G(t) \subsetneq \mathbb{C}, \quad 0 \le s < t < \infty$

$$(3.1.5) \qquad \begin{cases} G(t_n) \to G(t_0) & \text{if } t_n \to t_0 < \infty \\ G(t_n) \to \mathbb{C} & \text{if } t_n \to \infty. \end{cases}$$

The convergence in question is kernel convergence.

We prove the following result (see [Con], [Pom5]):

Theorem 3.1.5. *(i) Let $\{G(t)\}_{t \geq 0}$ be a family of simply connected domains which satisfy the conditions (3.1.4) and (3.1.5). For each $t \geq 0$ let $g_t(z) = g(z, t)$ be the conformal map of U onto $G(t)$ such that $g_t(0) = 0$ and $g_t'(0) = \alpha(t) > 0$. Let $\alpha_0 = \alpha(0)$.*

(a) Then α is strictly increasing, continuous and $\alpha(t) \to \infty$ as $t \to \infty$.

(b) If $\beta(t) = \log[\alpha(t)/\alpha_0]$ then $f(z, t) = \alpha_0^{-1} g(z, \beta^{-1}(t))$ is a Loewner chain and $f(U, t) = \alpha_0^{-1} G(\beta^{-1}(t))$.

(ii) Conversely, let $f(z, t)$ be a Loewner chain and $G(t) = f(U, t)$. Then the sequence of simply connected domains $\{G(t)\}_{t \geq 0}$ satisfies the conditions (3.1.4) and (3.1.5).

Proof. To prove part (i), let

$$g(z, t) = \alpha(t)[z + b_2(t)z^2 + \cdots], \quad z \in U.$$

Using the relation (3.1.4), we have

$$(3.1.6) \qquad g(z, s) \prec g(z, t), \quad 0 \leq s \leq t < \infty,$$

and therefore there is a family of Schwarz functions $v = v(z, s, t)$ such that

$$g(z, s) = g(v(z, s, t), t), \ z \in U, 0 \leq s \leq t < \infty.$$

Hence

$$\alpha(s) = g'(0, s) = g'(0, t)v'(0, s, t) = \alpha(t)v'(0, s, t).$$

In view of (3.1.4), we must have $0 < v'(0, s, t) < 1$ for $0 \leq s < t < \infty$, and thus $\alpha(\cdot)$ is a strictly increasing function from $[0, \infty)$ into $(0, \infty)$. Moreover, since $G(t_n) \to \mathbb{C}$ as $t_n \to \infty$, we must have $\alpha(t) \to \infty$ as $t \to \infty$. On the other hand, from Theorem 3.1.2 we know that $g_{t_n} \to g_t$ locally uniformly on U as $t_n \to t < \infty$, so the function α is continuous. These arguments prove (a).

We next prove the assertion (b). To this end, it suffices to observe that $\alpha : [0, \infty) \to [\alpha_0, \infty)$ is strictly increasing and continuous, hence one-to-one.

Consequently β is also a strictly increasing function from $[0, \infty)$ onto $[0, \infty)$. Using the relation (3.1.6) and the above argument, we deduce that $f(z, t) = \alpha_0^{-1} g(z, \beta^{-1}(t))$ is a univalent subordination chain. Moreover, if $\tau = \beta^{-1}(t)$ then $t = \beta(\tau)$ and $e^t = \alpha(\tau)/\alpha_0$. Consequently, we obtain

$$f'(0, t) = \alpha_0^{-1} g'(0, \tau) = \alpha_0^{-1} \alpha(\tau) = e^t.$$

We conclude that $f(z, t)$ is normalized, and moreover $f(U, t) = \alpha_0^{-1} G(\beta^{-1}(t))$. This completes the proof of part (i).

We now prove part (ii). Clearly $G(s) \subseteq G(t)$ for $0 \le s \le t < \infty$, and also $G(s) \ne G(t)$ for $s \ne t$, by the uniqueness assertion of the Riemann mapping theorem. This implies (3.1.4). The first part of (3.1.5) follows from Theorem 3.1.2 and the estimate (3.1.9) below. On the other hand, since for each $t \ge 0$, $e^{-t} f(\cdot, t)$ is a function in the class S, the Koebe 1/4-theorem gives $e^{-t} f(U, t) \supseteq U_{1/4}$. From this we can conclude that $G(t_n) = f(U, t_n) \to \mathbb{C}$ as $t_n \to \infty$, and this completes the proof of (ii).

The following estimates are very useful in many situations. Among other things they imply that a (normalized) Loewner chain is locally Lipschitz continuous in t, locally uniformly in z.

Lemma 3.1.6. *If $f(z, t)$ is a Loewner chain then*

$$(3.1.7) \qquad e^t \frac{|z|}{(1 + |z|)^2} \le |f(z, t)| \le e^t \frac{|z|}{(1 - |z|)^2}, \qquad |z| < 1, \quad t \ge 0,$$

$$(3.1.8) \qquad e^t \frac{1 - |z|}{(1 + |z|)^3} \le |f'(z, t)| \le e^t \frac{1 + |z|}{(1 - |z|)^3}, \qquad |z| < 1, \quad t \ge 0,$$

$$(3.1.9) \quad |f(z, t) - f(z, s)| \le \frac{8|z|}{(1 - |z|)^4} (e^t - e^s), \quad z \in U, \ 0 \le s \le t < \infty.$$

Proof. Obviously the relations (3.1.7) and (3.1.8) are simple consequences of the growth and distortion theorems for the class S, because $e^{-t} f(\cdot, t) \in S$ for each $t \ge 0$. In order to prove (3.1.9), let $v(z, s, t)$ be the family of transition functions defined by $f(z, t)$, i.e.

$$f(z, s) = f(v(z, s, t), t), \quad z \in U, \quad 0 \le s \le t < \infty.$$

Also let

$$p(z,s,t) = \frac{1+e^{s-t}}{1-e^{s-t}} \cdot \frac{z - v(z,s,t)}{z + v(z,s,t)}, \quad z \in U, \quad 0 \le s < t < \infty.$$

This function is holomorphic on U and has positive real part. Moreover, $p(0,s,t) = 1$ and hence $p(\cdot, s, t) \in \mathcal{P}$. Therefore, from (2.1.4) we deduce that

$$\frac{1+e^{s-t}}{1-e^{s-t}} \left| \frac{z - v(z,s,t)}{z + v(z,s,t)} \right| = |p(z,s,t)| \le \frac{1+|z|}{1-|z|},$$

and from this it follows that

$$|z - v(z,s,t)| \le 2|z| \frac{1+|z|}{1-|z|}(1 - e^{s-t}).$$

Using (3.1.8) and the above relation we obtain

$$|f(z,t) - f(z,s)| = |f(z,t) - f(v(z,s,t),t)| = \left| \int_{v(z,s,t)}^{z} f'(\zeta,t)d\zeta \right|$$

$$\le |z - v(z,s,t)| \frac{2e^t}{(1-|z|)^3} \le \frac{8|z|}{(1-|z|)^4}(e^t - e^s),$$

proving (3.1.9). This completes the proof.

Example 3.1.7. Let $\gamma \colon [0,\infty) \to \mathbb{C}$ be a Jordan arc that does not contain the origin, such that $\gamma(t) \to \infty$ as $t \to \infty$, and consider the domains $G(t) = \mathbb{C} \setminus \gamma([t,\infty))$, $0 \le t < \infty$. The family $\{G(t)\}_{t\ge0}$ satisfies the conditions (3.1.4) and (3.1.5) and hence, using part (b) of Theorem 3.1.5, we obtain an example of a Loewner chain.

Example 3.1.7 shows that any single-slit mapping is the initial element of a Loewner chain. For example, $f(z,t) = \dfrac{e^t z}{(1-z)^2}$ is a Loewner chain whose initial element is the Koebe function. (It is clear geometrically that $f(z,t) = e^t f(z)$ is a Loewner chain iff f is starlike.) In fact, we have the following deeper result, which asserts that any function in S can be embedded in a Loewner chain.

Theorem 3.1.8. *For each function f in S there is a Loewner chain $f(z,t)$ such that $f(z,0) = f(z)$ on U.*

Proof. We first assume that f is holomorphic on $\overline{U}$. Then the image of the unit circle is a closed Jordan curve C. Let G denote the inner domain of C and let H be the outer domain of C. Let $\hat{\mathbb{C}}$ be the extended complex plane and g be a conformal map of $\hat{\mathbb{C}} \setminus \overline{U}$ onto H such that $g(\infty) = \infty$. For $t \geq 0$ consider the closed Jordan curve $C_t = \{g(e^t e^{i\theta}) : 0 \leq \theta \leq 2\pi\}$ and its inner domain $G(t)$. Then $G(0) = G$ and the family $\{G(t)\}_{t \geq 0}$ satisfies (3.1.4) and (3.1.5). If $h(z,t)$ is the conformal map of U onto $G(t)$ with $h(0,t) = 0$, $h'(0,t) = \alpha(t) > 0$, then the uniqueness assertion of the Riemann mapping theorem implies that $h(z,0) = f(z)$. Hence letting $\beta(t) = \log[\alpha(t)]$, we deduce from Theorem 3.1.5 (i) (b) that $f(z,t) = h(z, \beta^{-1}(t))$ is a Loewner chain satisfying $f(z,0) = f(z)$, $z \in U$.

In the general case let $f \in S$ and let $f_n(z) = f(r_n z)/r_n$, where $r_n = 1 - 1/n$ and $n \in \mathbb{N}\setminus\{1\}$. Then $f_n \in S$ and is analytic on $\overline{U}$. Using the above arguments we conclude that f_n can be embedded as the initial element of a Loewner chain $f_n(z,t)$. The result is now a consequence of the following theorem:

Theorem 3.1.9. *Any sequence of Loewner chains $\{f_n(z,t)\}_{n \in \mathbb{N}}$ contains a subsequence which converges locally uniformly on U for each fixed $t \geq 0$ to a Loewner chain.*

Proof. Using the relation (3.1.9), we deduce that for each $r \in (0,1)$,

$$|f_n(z,t) - f_n(z,s)| \leq \frac{8r}{(1-r)^4}(e^t - e^s), \quad n \in \mathbb{N}, |z| \leq r, 0 \leq s \leq t < \infty.$$

The upper bound in (3.1.8) implies that

$$|f_n(z,t) - f_n(w,t)| \leq |z-w| \int_0^1 |f'((1-\tau)z + \tau w, t)| d\tau$$

$$\leq e^t \frac{1+r}{(1-r)^3}|z-w|, \quad |z| \leq r, |w| \leq r, t \geq 0, n \in \mathbb{N}.$$

The above estimates imply that the functions $f_n(z,t)$, $n \in \mathbb{N}$, are equicontinuous on $\{(z,t) : |z| \leq 1 - 1/k, 0 \leq t \leq k\}$, $k = 2,3,\ldots$. Applying the Arzela-Ascoli theorem on the compact space $\overline{U}_{1-1/k} \times [0,k]$ (see e.g. [Roy, p. 167-169]), we conclude that for k fixed, there is a subsequence $\{f_{n_p}(z,t)\}_{p \in \mathbb{N}}$ which converges pointwise on $\{(z,t) : |z| \leq 1 - 1/k, 0 \leq t \leq k\}$ to a limit $\tilde{f}_k(z,t)$, and furthermore the convergence is uniform. A diagonal sequence argument then shows that there exists a subsequence, which we denote again by $\{f_{n_p}(z,t)\}_{p \in \mathbb{N}}$, which converges pointwise in $\{|z| < 1, 0 \leq t < \infty\}$, and furthermore the convergence is locally uniform on this space. In particular, the convergence is locally uniform on U for each fixed $t \geq 0$. Since the limit function $f(z, t)$ satisfies the conditions $f(0, t) = 0$ and $f'(0, t) = e^t$, it follows that $f(z, t)$ cannot be identically constant, and hence is univalent on U, for each $t \geq 0$. Also since $f_{n_p}(z, s) \prec f_{n_p}(z, t)$ for $p = 1, 2, \ldots, 0 \leq s \leq t < \infty$, we deduce that

$$f_{n_p}(z, s) = f_{n_p}(v_{n_p}(z, s, t), t), \quad p = 1, 2, \ldots,$$

where $v_{n_p}(z, s, t)$ are univalent Schwarz functions. Since the set $\{v_{n_p}(\cdot, s, t)\}_{p \in \mathbb{N}}$ is locally uniformly bounded on U, there is a subsequence of $\{n_p\}_{p \in \mathbb{N}}$, which we denote again by $\{n_p\}_{p \in \mathbb{N}}$, such that $\{v_{n_p}(z, s, t)\}_{p \in \mathbb{N}}$ converges locally uniformly on U to an analytic function $v(z, s, t)$ which is easily seen to be a Schwarz function. Moreover, taking limits for this subsequence, we deduce that $f(z, s) = f(v(z, s, t), t)$, that is $f(z, s) \prec f(z, t)$. We have therefore proved that $f(z, t)$ is a Loewner chain. This completes the proof.

3.1.3   Loewner's differential equation

We now derive one of the basic results in Loewner's theory, the Loewner differential equation. There are two versions of the equation, one for the Loewner chain itself and one for the transition functions. A (time-dependent) function of positive real part appears in both of them, and this corresponds to the fact that a Loewner chain represents an expanding flow. There is a more specialized version of the differential equation for single-slit mappings, first considered by Loewner [Lö2] (see (3.1.29)) and later by Kufarev [Kuf1-4]. However, we shall work mainly with the general version, which was first studied by Kufarev [Kuf1,3]. (Loewner did, however, write down equation (3.1.10).) Important contributions were made by Pommerenke [Pom3] and also Becker [Bec1-3]. Our treatment is similar in spirit to that of [Pom5, Chapter 6].

Theorem 3.1.10. *Assume $p(z, t)$ belongs to $\mathcal{P}$ for each $t \geq 0$ and is a measurable function of $t \in [0, \infty)$ for each $z \in U$. Then for each $z \in U$ and $s \geq 0$, the initial value problem*

$$(3.1.10) \qquad \frac{\partial v}{\partial t} = -vp(v, t), \quad a.e. \quad t \geq s, \quad v(z, s, s) = z,$$

*has a unique locally absolutely continuous solution $v(t) = v(z, s, t) = e^{s-t}z +$
.... Furthermore, for fixed $s \geq 0$ and $z \in U$, $v(z, s, t)$ is Lipschitz continuous in
$t \geq s$ locally uniformly with respect to z. The functions $v(\cdot, s, t)$ are univalent
Schwarz functions for $0 \leq s \leq t < \infty$, and for each $s \geq 0$, the limit*

$$(3.1.11) \qquad f(z, s) = \lim_{t \to \infty} e^t v(z, s, t)$$

*exists locally uniformly on U. Moreover $f(\cdot, s)$ is univalent and $f(v(z, s, t), t) =
f(z, s)$ for all $z \in U$ and $s \leq t < \infty$. Thus the function $f : U \times [0, \infty) \to \mathbb{C}$
given by (3.1.11) is a Loewner chain, and in addition it satisfies the differential
equation*

$$(3.1.12) \qquad \frac{\partial f}{\partial t}(z, t) = zp(z, t)f'(z, t) \quad a.e. \quad t \geq 0,$$

for all $z \in U$. (The exceptional set of measure 0 is independent of z.)

Proof. The proof combines the classical method of successive approximations with properties of the set $\mathcal{P}$, which allow us to conclude that global
rather than local solutions (for a.e. $t \geq s$) are obtained.

Fix $s \geq 0$. Also fix $r \in (0, 1)$ and let $z \in \overline{U}_r$. We shall prove that (3.1.10)
has a unique solution on each interval $s \leq t \leq t_1$ with $t_1 > s$. We define the
successive approximations $v_n(t) = v_n(z, t)$ such that $v_0(t) \equiv 0$ and

$$(3.1.13) \quad v_{n+1}(t) = z \exp\left[-\int_s^t p(v_n(\tau), \tau)d\tau\right], \quad n = 0, 1, 2, \ldots,$$

for $t \geq s$. These approximations are well defined, since $|v_n(t)| \leq r$ by an
inductive argument using the fact that $p(z, t) \in \mathcal{P}$. For each $t \geq s$, $v_n(z, t)$
is holomorphic in z, $v_n(0, t) = 0$, $v'_n(0, t) = e^{s-t}$, and $v_n(z, s) = z$, for $n =
1, 2, \ldots$.

Now, the fact that $\text{Re } p(z, t) > 0$ implies by (2.1.6) that

$$|p'(z, t)| \leq \frac{2}{(1 - r)^2}, \quad |z| \leq r, \quad t \geq 0.$$

Since for $a, b \in U$, we have

$$p(b, t) - p(a, t) = \int_a^b p'(\zeta, t)d\zeta,$$

where the integration is over the straight line segment [a,b], we easily deduce
that

(3.1.14) $|p(a,t) - p(b,t)| \leq \dfrac{2|a-b|}{(1-r)^2}$,

for $|a| \leq r$, $|b| \leq r$, $t \geq 0$.

Using the above argument and the elementary relation

(3.1.15) $|e^{-z_1} - e^{-z_2}| \leq |z_1 - z_2|$,

for all z_1, $z_2 \in \mathbb{C}$ with $\mathrm{Re}\, z_1 \geq 0$, $\mathrm{Re}\, z_2 \geq 0$, we deduce that

$$|v_{n+1}(t) - v_n(t)| \leq r \int_s^t |p(v_n(\tau), \tau) - p(v_{n-1}(\tau), \tau)| d\tau$$

$$\leq \frac{2r}{(1-r)^2} \int_s^t |v_n(\tau) - v_{n-1}(\tau)| d\tau.$$

From this we obtain by induction that

$$|v_{n+1}(t) - v_n(t)| \leq \frac{2^n r^n (t-s)^n}{(1-r)^{2n} n!}, \quad |z| \leq r, \quad n = 0, 1, \ldots.$$

Since

$$v_n(t) = \sum_{j=1}^{n} \{v_j(t) - v_{j-1}(t)\},$$

the above estimates now imply that the limit

$$v(t) = v(z, s, t) = \lim_{n \to \infty} v_n(t) = \lim_{n \to \infty} \sum_{j=1}^{n} \{v_j(t) - v_{j-1}(t)\}$$

exists uniformly on $|z| \leq r$, $s \leq t \leq t_1$, for each t_1. This limit is a continuous
function of t, and is analytic as a function of z. Moreover, using (3.1.13), we
have

$$|v(z, s, t)| \leq |z|.$$

Taking into account the fact that $v'_n(0, t) = e^{s-t}$, $n = 1, 2, \ldots$, we deduce
that $v'(0, s, t) = e^{s-t}$.

Using Lebesgue's dominated convergence theorem, we conclude that
$v(z, s, t)$ satisfies the integral equation

$$v(z, s, t) = z \exp \left[-\int_s^t p(v(z, s, \tau), \tau) d\tau \right].$$

Hence $v(z, s, t)$ is a locally absolutely continuous function of $t \in [s, \infty)$ and satisfies the differential equation

$$\frac{\partial v}{\partial t} = -vp(v, t), \quad a.e. \quad t \geq s, \quad v(s) = z.$$

Moreover, $v(z, s, t)$ is a Lipschitz continuous function of $t \geq s$ locally uniformly with respect to z. Indeed, using the relations (3.1.15) and (2.1.4), we obtain

$$|v(z, s, t_1) - v(z, s, t_2)|$$

$$= |z| \cdot \left| \exp\left[-\int_s^{t_1} p(v(z, s, \tau), \tau)d\tau \right] - \exp\left[-\int_s^{t_2} p(v(z, s, \tau), \tau)d\tau \right] \right|$$

$$\leq r \int_{t_1}^{t_2} |p(v(z, s, \tau), \tau)|d\tau \leq \frac{r(1+r)}{1-r}(t_2 - t_1),$$

for $|z| < r$ and $s \leq t_1 \leq t_2 < \infty$.

We next prove that the solution of (3.1.10) is unique. To this end, suppose $u(t)$ is another locally absolutely continuous solution of (3.1.10) such that $u(s) = z$.

Using the relations (3.1.14) and (3.1.15), we obtain

$$|u(t) - v(t)|$$

$$= \left| z \exp\left[-\int_s^t p(u(\tau), \tau)d\tau \right] - z \exp\left[-\int_s^t p(v(\tau), \tau)d\tau \right] \right|$$

$$\leq \frac{2r}{(1-r)^2} \int_s^t |u(\tau) - v(\tau)|d\tau,$$

for $t \geq s$ and $|z| \leq r$. Consider a subinterval $[s, t_1]$ of $[s, \infty)$, and let $M > 0$ be such that $|u(t) - v(t)| \leq M$ on $[s, t_1]$. Then for $t \in [s, t_1]$, we obtain the estimate

$$|u(t) - v(t)| \leq \frac{2rM}{(1-r)^2} \int_s^t d\tau = \frac{2rM}{(1-r)^2}(t - s).$$

Again using the method of mathematical induction, we conclude that

$$|u(t) - v(t)| \leq \frac{M}{n!} \left[\frac{2r}{(1-r)^2} \right]^n (t - s)^n, \quad t \in [s, t_1],$$

for all $n \in \mathbb{N}$. Therefore $u(t) = v(t)$ on $[s, t_1]$ and this assures us that the solution of (3.1.10) is unique, as claimed.

The semigroup property

(3.1.16) $v(z, s, t) = v(v(z, s, \tau), \tau, t), \quad 0 \leq s \leq \tau \leq t < \infty,$

also follows from the uniqueness of solutions to initial value problems of the form (3.1.10). Indeed, both sides of (3.1.16) satisfy the differential equation and have the same value when $t = \tau$.

We next prove that the solution $v(z, s, t)$ of (3.1.10) is univalent for each $t \geq s$. Indeed, suppose that $t_0 \geq s$ and $z_1, z_2 \in U$ are such that $v(z_1, s, t_0) = v(z_2, s, t_0)$. Also let $v_j(t) = v(z_j, s, t)$ for $j = 1, 2$, and let $w(t) = v_1(t) - v_2(t)$. Then $w(t_0) = 0$.

Since Re $p(z, t) > 0$ and $p(0, t) = 1$, the Herglotz integral formula (2.1.3) gives the estimate

$$|z_1 p(z_1, t) - z_2 p(z_2, t)| \leq \frac{1 + |z_1|}{1 - |z_1|} \cdot \frac{1 + |z_2|}{1 - |z_2|} |z_1 - z_2|,$$

for $z_1, z_2 \in U$ and $t \geq 0$ (see Problem 3.1.3). Consequently, we obtain

$$\left| \frac{\partial w}{\partial t}(t) \right| = |v_1(t) p(v_1(t), t) - v_2(t) p(v_2(t), t)|$$

$$\leq \frac{1 + |z_1|}{1 - |z_1|} \cdot \frac{1 + |z_2|}{1 - |z_2|} |w(t)|, \quad s \leq t \leq t_0.$$

Now choose $K > 0$ such that $|w(t)| \leq K$ for $s \leq t \leq t_0$. Then

$$|w(t)| = \left| \int_t^{t_0} \frac{\partial w}{\partial \tau}(\tau) d\tau \right|$$

$$\leq \int_t^{t_0} K \frac{1 + |z_1|}{1 - |z_1|} \cdot \frac{1 + |z_2|}{1 - |z_2|} d\tau$$

$$\leq K \frac{1 + |z_1|}{1 - |z_1|} \cdot \frac{1 + |z_2|}{1 - |z_2|} (t_0 - t).$$

Substituting back into the estimate for $\left| \dfrac{\partial w}{\partial t}(t) \right|$ gives

$$|w(t)| = \left| \int_t^{t_0} \frac{\partial w}{\partial \tau}(\tau) d\tau \right|$$

$$\leq \int_t^{t_0} \frac{1 + |z_1|}{1 - |z_1|} \cdot \frac{1 + |z_2|}{1 - |z_2|} K \frac{1 + |z_1|}{1 - |z_1|} \cdot \frac{1 + |z_2|}{1 - |z_2|} (t_0 - \tau) d\tau$$

$$= K \left[\frac{1 + |z_1|}{1 - |z_1|} \cdot \frac{1 + |z_2|}{1 - |z_2|} \right]^2 \frac{(t_0 - t)^2}{2}.$$

By induction, we may obtain a sequence of estimates which lead to the conclusion that $w(t) = 0$ on $[s, t_0]$. Then the relation $w(s) = 0$ and the fact that $v(z_j, s, s) = z_j$ for $j = 1, 2$, imply that $z_1 = z_2$, as claimed.

Now, since the function $e^{t-s} v(\cdot, s, t)$, $t \geq s$, belongs to S, the growth theorem gives

(3.1.17) $$|v(z, s, t)| \leq \frac{|z|}{(1 - |z|)^2} e^{s-t}, \quad z \in U, \quad t \geq s.$$

Further, using (3.1.14) and (3.1.17), we obtain for $|z| \leq r$ and $t \geq s$ that

$$|p(v(z, s, t), t) - 1| \leq \frac{2|v(z, s, t)|}{(1 - r)^2} \leq \frac{2e^{s-t} r}{(1 - r)^4}.$$

Combining this relation with the equality

$$e^{t-s} v(z, s, t) = z \exp \left[\int_s^t (1 - p(v(z, s, \tau), \tau)) d\tau \right],$$

we conclude that the family $\{e^{t-s} v(z, s, t)\}_{t \geq s}$ is Cauchy (locally uniformly in z), and hence the limit

$$f(z, s) = \lim_{t \to \infty} e^t v(z, s, t)$$

exists locally uniformly in z. From (3.1.16) we deduce that

$$f(z, s) = \lim_{t \to \infty} e^t v(z, s, t)$$

$$= \lim_{t \to \infty} e^t v(v(z, s, \tau), \tau, t) = f(v(z, s, \tau), \tau),$$

for $z \in U$ and $0 \leq s \leq \tau < \infty$. Thus $f(z, t)$ is a subordination chain. Also $f(0, t) = 0$ and $f'(0, t) = e^t$ for $t \geq 0$. Then by Hurwitz's theorem, $f(\cdot, t)$ is univalent on U for each $t \geq 0$. Therefore $f(z, t)$ is a Loewner chain.

The proof that equation (3.1.12) is satisfied is contained in Theorem 3.1.12.

We are now able to prove the following characterization of Loewner chains, which is one of the main results of the theory. There is also a version of this theorem for non-normalized Loewner chains [Pom3, Folgerung3] (see Problem 3.1.6). For further discussion of equation (3.1.19), the reader may consult [Bec1] and [Pom3].

Theorem 3.1.11. *The function* $f : U \times [0, \infty) \to \mathbb{C}$ *with* $f(0, t) = 0$, $f'(0, t) = e^t$, $t \geq 0$, *is a Loewner chain if and only if the following conditions hold:*

(i) There exist $r \in (0, 1)$ *and a constant* $M > 0$ *such that* $f(z, t)$ *is holomorphic on* U_r *for each* $t \geq 0$, *locally absolutely continuous in* $t \geq 0$ *locally uniformly with respect to* $z \in U_r$, *and*

$$(3.1.18) \qquad |f(z, t)| \leq Me^t, \quad |z| < r, \quad t \geq 0.$$

(ii) There exists a function $p(z, t)$ *such that* $p(\cdot, t) \in \mathcal{P}$ *for each* $t \geq 0$, $p(z, \cdot)$ *is measurable on* $[0, \infty)$ *for each* $z \in U$, *and for all* $z \in U_r$,

$$(3.1.19) \qquad \frac{\partial f}{\partial t}(z, t) = zf'(z, t)p(z, t), \quad a.e. \quad t \geq 0.$$

(iii) For each $t \geq 0$, $f(\cdot, t)$ *is the analytic continuation of* $f(\cdot, t)|_{U_r}$ *to* U, *and furthermore this analytic continuation exists under the assumptions (i) and (ii).*

Proof. First assume that $f(z, t)$ is a Loewner chain. We shall show that the conditions (i) and (ii) hold for each $r \in (0, 1)$.

Since $e^{-t}f(\cdot, t)$ is a function in S for each $t \geq 0$, the growth theorem implies that for each $r \in (0, 1)$ there is a positive number $M = M(r)$ such that $|f(z, t| \leq Me^t$ for $|z| < r$, $t \geq 0$. This proves (i).

Let $v(z, s, t)$ denote the transition functions defined by the chain $f(z, t)$, i.e.

$$(3.1.20) \qquad f(z, s) = f(v(z, s, t), t), \quad z \in U, \quad 0 \leq s \leq t < \infty.$$

By applying the mean value theorem to the real and imaginary parts of f, we obtain

$$(3.1.21) \qquad \frac{1}{h}\Big[f(z, t + h) - f(z, t)\Big]$$

$$= \frac{1}{h}\Big[f(z, t + h) - f(v(z, t, t + h), t + h)\Big]$$

$$= A(z, t, h)\Big(\frac{1}{h}\big[z - v(z, t, t + h)\big]\Big), \quad z \in U, t \geq 0, h > 0,$$

where $A(z, t, h)$ is a real-linear operator which tends to the complex-linear operator $f'(z, t)$ as $h \to 0^+$. Now (3.1.9) implies that $f(z, t)$ is locally Lipschitz in t, locally uniformly in z. Hence from Vitali's theorem, except for a set of measure 0 in t, the left-hand side of (3.1.21) has a limit as $h \to 0$ for all z. For such t, the one-sided limit as $h \to 0^+$ of the right-hand side must also exist, and we obtain the function $p(\cdot, t)$ in (3.1.19) using Theorem 2.1.7 and the normalization of the transition functions. For t in the exceptional set of measure 0, we define $p(z, t) = 1$, $z \in U$.

The measurability of $p(z, \cdot)$ on $[0, \infty)$ follows from the measurability of $\frac{\partial f}{\partial t}(z, \cdot)$ and $f'(z, \cdot)$ on $[0, \infty)$. (From the Cauchy integral formula and the local Lipschitz continuity of $f(z, \cdot)$, it follows that $f'(z, \cdot)$ is also locally Lipschitz continuous on $[0, \infty)$.) This completes the proof of (ii).

We now prove the converse statement. For this purpose, let $r \in (0, 1)$, $M > 0$, $f(z, t)$ and $p(z, t)$ satisfy the assumptions (i) and (ii). We show that $f(z, t)$ is locally Lipschitz continuous in t locally uniformly with respect to $z \in U_r$. To accomplish this, let $\rho \in (0, r)$ and let $T > 0$. Using the Cauchy integral formula and (3.1.18), we deduce that there exists an $L = L(\rho, T) > 0$ such that

$$(3.1.22) \qquad |f'(z, t)| \leq L(\rho, T), \quad |z| \leq \rho, \quad t \in [0, T].$$

Moreover, in view of (3.1.19), (3.1.22), and the fact that

$$|p(z, t)| \leq \frac{1 + \rho}{1 - \rho}, \quad |z| \leq \rho, \quad t \geq 0,$$

one deduces that there is an $N = N(\rho, T) > 0$ such that

$$\left| \frac{\partial f}{\partial t}(z, t) \right| \leq N(\rho, T), \quad |z| \leq \rho, \quad a.e. \quad t \in [0, T].$$

Further, since

$$f(z, t_2) - f(z, t_1) = \int_{t_1}^{t_2} \frac{\partial f}{\partial t}(z, t) dt, \quad 0 \leq t_1 < t_2 \leq T,$$

we deduce that

$$(3.1.23) \qquad |f(z, t_1) - f(z, t_2)| \leq N(\rho, T)(t_2 - t_1)$$

for $|z| \leq \rho$ and $0 \leq t_1 < t_2 \leq T$. Since $\rho \in (0, r)$ and $T > 0$ were arbitrarily chosen, the conclusion follows.

Taking into account Theorem 3.1.10, we conclude that the initial value problem

$$\frac{\partial v}{\partial t} = -vp(v,t), \quad a.e. \quad t \geq s, \quad v(z,s,s) = z,$$

has a unique locally absolutely continuous solution $v(t) = v(z,s,t)$. Moreover, for fixed s and t, $v(\cdot,s,t)$ is a univalent Schwarz function. For $z \in U_r$, $s \geq 0$ and $t \geq s$, let $f(z,s,t) = f(v(z,s,t),t)$. Since $v = v(z,s,t)$ is Lipschitz continuous in $t \in [s,\infty)$ locally uniformly with respect to $z \in U$ by the proof of Theorem 3.1.10, and $f(z,t)$ is also locally Lipschitz in t, we easily deduce that $f(z,s,t)$ is locally Lipschitz continuous in t for $t \in [s,\infty)$ locally uniformly with respect to $z \in U_r$ as well. Indeed, from (3.1.22) we obtain

$$|f(z,t) - f(w,t)| \leq \int_0^1 |f'((1-\tau)z + \tau w, t)|d\tau \leq L(\rho,T)|z - w|,$$

for all $t \in [0,T]$, $T > 0$, $|z| \leq \rho$, $|w| \leq \rho$, and $\rho \in (0,r)$. Hence, if $s \geq 0$ and $T > s$ we obtain in view of (3.1.23) and the above relation that

$$|f(z,s,t_1) - f(z,s,t_2)|$$

$$\leq |f(v(z,s,t_1),t_1) - f(v(z,s,t_1),t_2)| + |f(v(z,s,t_1),t_2) - f(v(z,s,t_2),t_2)|$$

$$\leq N(\rho,T)(t_2 - t_1) + L(\rho,T)|v(z,s,t_1) - v(z,s,t_2)|$$

$$\leq N(\rho,T)(t_2 - t_1) + R(\rho,T)(t_2 - t_1),$$

for $|z| \leq \rho$ and $s \leq t_1 \leq t_2 \leq T$.

It follows that for all $z \in U_r$, $\frac{\partial}{\partial t}f(z,s,t)$ exists for almost all $t \geq s$, and moreover

$$\frac{\partial}{\partial t}f(z,s,t) = f'(v(z,s,t),t)\frac{\partial v}{\partial t}(z,s,t) + \frac{\partial f}{\partial t}(v(z,s,t),t)$$

$$= f'(v(z,s,t),t)\left(\frac{\partial v}{\partial t}(z,s,t) + v(z,s,t)p(v(z,s,t),t)\right) = 0, \quad a.e. \quad t \geq s.$$

Because $v(z,s,s) = z$ and $f(z,s,t)$ is a locally absolutely continuous function of t, we deduce that $f(z,s,t)$ is identically constant as a function of t, i.e. $f(z,s,t) = f(z,s,s)$, and hence

(3.1.24) $f(v(z,s,t),t) = f(z,s), \quad |z| < r, \quad 0 \leq s \leq t < \infty.$

We next extend the function $f(z,t)$ univalently to the whole disc U. Using (3.1.18), we have

$$|e^{-t}f(z,t) - z| \le 1 + M, \quad |z| < r, \quad t \ge 0,$$

and since $e^{-t}f(z,t) - z = a_2(t)z^2 + \cdots$, we can apply the Schwarz lemma to obtain

$$|e^{-t}f(z,t) - z| \le (1 + M)\frac{|z|^2}{r^2}, \quad |z| < r, \quad t \ge 0.$$

Also since $e^{t-s}v(\cdot, s, t)$ belongs to S, the growth theorem gives

$$|v(z,s,t)| \le \frac{|z|}{(1-|z|)^2}e^{s-t}, \quad z \in U, \quad 0 \le s \le t < \infty.$$

Hence from (3.1.24) we obtain

$$|f(z,s) - e^t v(z,s,t)| = e^t|e^{-t}f(v(z,s,t),t) - v(z,s,t)|$$

$$\le \frac{e^t(1+M)}{r^2}|v(z,s,t)|^2 \le \frac{e^{2s-t}}{(1-r)^4}(1+M), \quad |z| < r.$$

From this we see that

$$(3.1.25) \qquad e^t v(z,s,t) \to f(z,s) \text{ as } t \to \infty,$$

uniformly on $|z| < r$.

On the other hand, if $g = g(z,s)$ is the function given by

$$g(z,s) = \lim_{t \to \infty} e^t v(z,s,t),$$

then this limit exists locally uniformly on U for each $s \ge 0$, and $g : U \times [0, \infty) \to \mathbb{C}$ is a Loewner chain by Theorem 3.1.10. Moreover, $g(z,s) = f(z,s)$ for $|z| < r$ and $s \ge 0$, by (3.1.25). Finally, using the condition (iii) and the identity theorem for holomorphic functions, we deduce that $g(z,s) = f(z,s)$ for $z \in U$ and $s \ge 0$. This completes the proof.

Theorem 3.1.12. *Let $f(z,t)$ be a Loewner chain and let $v(z,s,t)$ be the transition functions associated to $f(z,t)$. Then there exists a function $p(z,t)$ such that $p(\cdot,t) \in \mathcal{P}$, $t \ge 0$, $p(z,t)$ is measurable in $t \in [0,\infty)$ for each $z \in U$, and*

$$(3.1.26) \qquad \frac{\partial f}{\partial t}(z,t) = zf'(z,t)p(z,t), \quad a.e. \quad t \ge 0, \quad \forall z \in U.$$

Moreover, for each $s \geq 0$ and $z \in U$, $v = v(z, s, t)$ is the unique locally absolutely continuous solution of the initial value problem

$$(3.1.27) \qquad \frac{\partial v}{\partial t} = -vp(v, t), \quad a.e. \quad t \geq s, \quad v(z, s, s) = z,$$

and the limit

$$(3.1.28) \qquad \lim_{t \to \infty} e^t v(z, s, t) = f(z, s)$$

exists locally uniformly on U.

Proof. The existence of the function $p(z, t)$ such that (3.1.26) holds follows from the first part of the proof of Theorem 3.1.11.

Let $u = u(z, s, t)$ be the locally absolutely continuous solution of the initial value problem

$$\frac{\partial u}{\partial t} = -up(u, t), \quad a.e. \quad t \geq s, \quad u(z, s, s) = z,$$

for each fixed $s \geq 0$ and $z \in U$. Then $u(\cdot, s, t)$ is a univalent Schwarz function for $t \geq s$. Since $f(z, t)$ is a locally absolutely continuous function of t, it is differentiable a.e on $[0, \infty)$, and with similar reasoning as in the proof of Theorem 3.1.11 we deduce that $f(u(z, s, t), t)$ is also a locally absolutely continuous function of t and hence differentiable a.e. on $[s, \infty)$. Therefore in view of (3.1.26) we obtain that

$$\frac{\partial}{\partial t} f(u(z, s, t), t)$$

$$= f'(u, t) \frac{\partial u}{\partial t} + uf'(u, t)p(u, t) = 0, \quad a.e. \quad t \geq s.$$

Hence

$$f(u(z, s, t), t) = f(u(z, s, s), s) = f(z, s),$$

and thus

$$f(v(z, s, t), t) = f(u(z, s, t), t), \quad z \in U, \quad t \geq s.$$

Since $f(\cdot, t)$ is univalent on U, we must have $u(z, s, t) = v(z, s, t)$ for all $z \in U$ and $0 \leq s \leq t < \infty$. Consequently, $v(z, s, t)$ satisfies the initial value problem (3.1.27). Moreover, from Theorem 3.1.10, (3.1.28) follows. This completes the proof.

We remark that both equations (3.1.26) and (3.1.27) are called the *Loewner differential equations*. Also we remark that Theorems 3.1.10, 3.1.11 and 3.1.12 imply the following uniqueness result for Loewner chains which satisfy the differential equation (3.1.26) (compare with [Bec3-4]). The corresponding uniqueness result in several variables is false, as we shall see in Chapter 8.

Theorem 3.1.13. *Let* $p : U \times [0, \infty) \to \mathbb{C}$ *be a function such that* $p(\cdot, t) \in \mathcal{P}$, $t \geq 0$, *and* $p(z, \cdot)$ *is measurable on* $[0, \infty)$, $z \in U$. *Then there exists a unique Loewner chain* $f(z, t)$ *which satisfies the Loewner differential equation (3.1.26).*

Proof.

Existence. In view of Theorem 3.1.10, the initial value problem

$$\frac{\partial v}{\partial t} = -vp(v, t), \quad a.e. \quad t \geq s, \quad v(s) = z,$$

has a unique locally absolutely continuous solution $v(t) = v(z, s, t)$ for all $z \in U$ and $s \geq 0$, and the limit

$$\lim_{t \to \infty} e^t v(z, s, t) = f(z, s)$$

exists and gives a Loewner chain which satisfies (3.1.26).

Uniqueness. Let $g(z, t)$ be a Loewner chain which satisfies (3.1.26). Also let $w(z, s, t)$ be the transition functions associated to $g(z, t)$. From Theorem 3.1.12,

$$g(z, s) = \lim_{t \to \infty} e^t w(z, s, t)$$

locally uniformly on U for each $s \geq 0$, and $w(t) = w(z, s, t)$ satisfies the initial value problem (3.1.27). Thus $g(z, s) = f(z, s)$ for all $z \in U$ and $s \geq 0$. This completes the proof.

Loewner realized that the single-slit mappings in S could play a special role in his theory because of Theorem 3.1.3. In this case the associated Loewner chains are of the type discussed in Example 3.1.7, and the Loewner differential equation takes the form

$$\frac{\partial f}{\partial t}(z, t) = zf'(z, t)\frac{1 + k(t)z}{1 - k(t)z},$$

where k is a continuous function from $[0, \infty)$ into $\mathbb{C}$ such that $|k(t)| = 1$ for $0 \leq t < \infty$. The basic result for single-slit mappings is the following [Lö2] (cf. [Ahl2], [Con], [Dur], [Gol4], [Hay], [Hen]):

Theorem 3.1.14. *(i) Let $c : [0, \infty) \to \mathbb{C}$ be a continuous function such that $|c(t)| = 1$, $t \in [0, \infty)$. Then for each $z \in U$, there exists a unique function $w = w(z, t)$ such that $w(z, \cdot)$ is of class C^1 on $[0, \infty)$ and satisfies the differential equation*

$$(3.1.29) \qquad \frac{\partial w}{\partial t} = -w \frac{1 + c(t)w}{1 - c(t)w}, \quad t \geq 0, \quad w(z, 0) = z.$$

Moreover, $w(\cdot, t)$ is a univalent Schwarz function and $w'(0, t) = e^{-t}$ for each $t \geq 0$.

(ii) If $f \in S$ is a single-slit mapping, then there exists a continuous function $k : [0, \infty) \to \mathbb{C}$ with $|k(t)| = 1$, $t \geq 0$, such that $\lim_{t \to \infty} e^t v(z, t) = f(z)$ locally uniformly on U, where $v = v(z, t)$ is the solution of the differential equation

$$\frac{\partial v}{\partial t} = -v \frac{1 + k(t)v}{1 - k(t)v}, \quad t \geq 0, \quad v(z, 0) = z,$$

for all $z \in U$.

(iii) Further, in case (ii) there exists a Loewner chain $f(z, t)$ such that $f(z) = f(z, 0)$, $z \in U$, $f(z, \cdot) \in C^1([0, \infty))$ for each $z \in U$, and

$$\frac{\partial f}{\partial t}(z, t) = \frac{1 + k(t)z}{1 - k(t)z} z f'(z, t), \quad z \in U, \quad t \geq 0.$$

The equation (3.1.29) was also studied in detail by Kufarev in the 1940's [Kuf1-4]. In particular he observed that given a continuous function $c : [0, \infty) \to \mathbb{C}$ with $|c(t)| = 1$, $t \geq 0$, the solution $w = w(z, t)$ of the Loewner differential equation (3.1.29) need not give rise to a single-slit mapping [Kuf2].

3.1.4 Remarks on Bieberbach's conjecture

Perhaps the most famous problem associated with the class S is the Bieberbach conjecture, formulated in 1916 [Bie1] and solved by de Branges [DeB] in 1985. This conjecture and two related conjectures due to Robertson and Milin influenced the development of univalent function theory for more than sixty years. The Loewner method played an important role in the solution, but there are other ideas involved which do not have generalizations to several variables, and we shall therefore not give a proof of de Branges' theorem here. As well as the paper of de Branges, the reader may consult the references [Fit-Pom], [Con], [Gon5], [Hay], [Hen], [Ros-Rov] for the complete proof.

The Bieberbach conjecture is the following:

Conjecture 3.1.15. If $f \in S$ has the power series expansion $f(z) = z + a_2 z^2 + \ldots + a_n z^n + \ldots$, then

$$(3.1.30) \qquad |a_n| \le n, \quad n \ge 2.$$

Moreover, if there is an integer $n \ge 2$ such that equality holds in (3.1.30), then f is a rotation of the Koebe function.

Robertson [Robe2] observed in 1936 that the following statement about odd functions in S implies the Bieberbach conjecture:

Conjecture 3.1.16. For each odd function $h \in S$, $h(z) = z + b_3 z^3 + b_5 z^5 + \ldots + b_{2n-1} z^{2n-1} + \ldots$, the inequality

$$(3.1.31) \qquad 1 + |b_3|^2 + \ldots + |b_{2n-1}|^2 \le n,$$

holds for each $n \ge 2$. If there is an integer $n \ge 2$ such that equality holds in (3.1.31), then $[h(z)]^2 = f(z^2)$, $z \in U$, and f is a rotation of the Koebe function.

Next we mention Milin's conjecture (see [Mili2]) which implies Robertson's conjecture, and thus the Bieberbach conjecture.

Conjecture 3.1.17. Let $f \in S$ and $g(z) = \log[z^{-1} f(z)] = \sum_{k=1}^{\infty} c_k z^k$, $z \in U$, where the branch of the logarithm is chosen such that $g(0) = 0$. Then the inequality

$$(3.1.32) \qquad \sum_{l=1}^{n} \sum_{k=1}^{l} \left[k|c_k|^2 - \frac{4}{k} \right] \le 0$$

holds for each $n \ge 1$. If there is an integer n such that equality holds in (3.1.32), then f is a rotation of the Koebe function.

In a remarkable paper in 1985, de Branges [DeB] proved

Theorem 3.1.18. *The Milin conjecture is true.*

De Branges' proof of the Milin conjecture uses the reduction to single-slit mappings and some ingenious ideas about systems of special functions. A simplified proof was given by FitzGerald and Pommerenke [Fit-Pom]. Other versions of the proof are presented in the books cited above.

Loewner in his original paper [Lö2] applied the differential equation (3.1.29) to prove the Bieberbach conjecture in the case $n = 3$. Here we show how the general form of the Loewner differential equation (3.1.26) can be used to prove the same result (see also [Pom5]).

Theorem 3.1.19. *If $f \in S$ is given by $f(z) = z + a_2 z^2 + a_3 z^3 + \ldots, z \in U$, then $|a_3| \leq 3$. This bound is sharp.*

Proof. Since $f \in S$, it follows from Theorem 3.1.8 that there is a Loewner chain $f(z, t)$ such that $f(z) = f(z, 0)$, $z \in U$. Let

$$f(z, t) = e^t z + a_2(t) z^2 + \ldots + a_n(t) z^n + \ldots, \quad z \in U, t \geq 0,$$

where the coefficients $a_n(t)$ are locally absolutely continuous functions of $t \in [0, \infty)$ (see Problem 3.1.2) and $a_n = a_n(0)$ for $n = 2, 3, \ldots$. Now by Theorem 3.1.12, there is a function $p(z, t)$ such that $p(\cdot, t) \in \mathcal{P}$, $p(z, t)$ is measurable in $t \in [0, \infty)$ for all $z \in U$, and for almost all $t \geq 0$,

$$(3.1.33) \qquad \frac{\partial f}{\partial t}(z, t) = z f'(z, t) p(z, t), \quad \forall z \in U.$$

Let

$$p(z, t) = 1 + p_1(t) z + p_2(t) z^2 + \ldots + p_n(t) z^n + \ldots, \quad z \in U, t \geq 0.$$

Identifying the coefficients in both sides of (3.1.33), we deduce that

$$\frac{d}{dt} a_n(t) = \sum_{j=1}^{n-1} j a_j(t) p_{n-j}(t) + n a_n(t), \quad a.e. \quad t \geq 0,$$

for $n = 2, 3, \ldots$. Multiplying both sides of this equality by e^{-nt} and integrating, we obtain for $0 \leq s < t < \infty$ that

$$\int_s^t \frac{d}{d\tau} [e^{-n\tau} a_n(\tau)] d\tau = \sum_{j=1}^{n-1} j \int_s^t e^{-n\tau} a_j(\tau) p_{n-j}(\tau) d\tau.$$

Letting $t \to \infty$ in the above, we obtain

(3.1.34) $$a_n(s) = -e^{ns} \sum_{j=1}^{n-1} j \int_s^\infty e^{-n\tau} a_j(\tau) p_{n-j}(\tau) d\tau,$$

for $n = 2, 3, \ldots$, and $s \geq 0$. Here we have used the fact that $\lim_{t\to\infty} e^{-nt} a_n(t) = 0$, $n = 2, 3, \ldots$, since $e^{-t} f(\cdot, t) \in S$, $t \geq 0$ (compare with Chapter 1). Consequently, for $n = 2$ in (3.1.34), one deduces that

(3.1.35) $$a_2(s) = -e^{2s} \int_s^\infty e^{-\tau} p_1(\tau) d\tau.$$

Moreover, for $n = 3$ and $s = 0$ in (3.1.34), we obtain by an elementary computation using (3.1.35) the equality

$$a_3 = -\int_0^\infty e^{-2\tau} p_2(\tau) d\tau - 2 \int_0^\infty e^{-3\tau} a_2(\tau) p_1(\tau) d\tau$$

$$= -\int_0^\infty e^{-2\tau} p_2(\tau) d\tau + 2 \int_0^\infty e^{-\tau} p_1(\tau) \left(\int_\tau^\infty e^{-t} p_1(t) dt \right) d\tau$$

$$= -\int_0^\infty e^{-2\tau} p_2(\tau) d\tau + \left(\int_0^\infty e^{-\tau} p_1(\tau) d\tau \right)^2.$$

It suffices to show that Re $a_3 \leq 3$, since there is a rotation of f whose third coefficient is $|a_3|$. We have

(3.1.36)   Re $a_3 = -\int_0^\infty e^{-2\tau} \mathrm{Re}\, p_2(\tau) d\tau + \left(\int_0^\infty e^{-\tau} \mathrm{Re}\, p_1(\tau) d\tau \right)^2$

$$-\left(\int_0^\infty e^{-\tau} \mathrm{Im}\, p_1(\tau) d\tau \right)^2.$$

Using Problem 2.1.9, we have

$$\left| p_2(\tau) - \frac{1}{2} p_1^2(\tau) \right| \leq 2 - \frac{1}{2} |p_1(\tau)|^2, \quad \tau \geq 0.$$

Taking the negative real part on the left-hand side, we obtain

$$2 + \mathrm{Re}\, p_2(\tau) \geq \frac{1}{2} \mathrm{Re}\, p_1^2(\tau) + \frac{1}{2} |p_1(\tau)|^2 = [\mathrm{Re}\, p_1(\tau)]^2, \quad \tau \geq 0.$$

We use this estimate in the first integral on the right side of (3.1.36). We apply the Schwarz inequality to the second integral, breaking the integrand into the factors $e^{-\tau/2}$ and $e^{-\tau/2}\operatorname{Re} p_1(\tau)$, and we neglect the third term. This gives

$$\operatorname{Re} a_3 \le \int_0^\infty [2e^{-2\tau} + (e^{-\tau} - e^{-2\tau})(\operatorname{Re} p_1(\tau))^2]d\tau \le 3,$$

using the fact that $|\operatorname{Re} p_1(\tau)| \le 2$, $\tau \ge 0$ (see Theorem 2.1.5).

Equality $|a_3| = 3$ occurs for the Koebe function and its rotations. This completes the proof.

Notes. The results of Section 3.1 concerning Loewner chains and the Loewner differential equation may be found along with further material in [Pom3], [Pom5], [Dur], [Ros-Rov], [Ahl2], [Con], [Gol4], [Hay], [Hen], [Gon5], [Bec1-4], [Bran], [Lö2], [Kuf1-5], [Biel-Lew]. See also [Bae-Dra-Dur-Mar].

Problems

3.1.1. (i) The transition functions $v(z, s, t)$ of a Loewner chain satisfy estimates which correspond to (3.1.9), namely

$$|v(z, s_1, t) - v(z, s_2, t)| \le \frac{2|z|}{(1 - |z|)^2}(1 - e^{s_1 - s_2}), \quad 0 \le s_1 \le s_2 \le t < \infty,$$

and

$$|v(z, s, t_1) - v(z, s, t_2)| \le 2|z|\frac{1 + |z|}{1 - |z|}(1 - e^{t_1 - t_2}), \quad 0 \le s \le t_1 \le t_2 < \infty.$$

Prove these estimates.

(ii) Prove that on any interval $(0, t]$,

$$\frac{\partial v}{\partial s}(z, s, t) = zp(z, s)v'(z, s, t), \quad a.e. \quad s \in (0, t], \quad v(z, t, t) = z,$$

where $p(z, s)$ is the function given by Theorem 3.1.12.

3.1.2. Show that the Taylor series coefficients of the functions f_t in a Loewner chain $f(z, t)$ are locally absolutely continuous functions of t.

3.1.3. Show that if $p \in \mathcal{P}$ and z_1, $z_2 \in U$, then

$$|\tilde{z_1}p(z_1) - z_2p(z_2)| \le \frac{1 + |z_1|}{1 - |z_1|} \cdot \frac{1 + |z_2|}{1 - |z_2|} \cdot |z_1 - z_2|.$$

3.1.4. If f is the Koebe function and $f(z,t)$ is the Loewner chain associated to f, i.e.

$$f(z,t) = \frac{e^t z}{(1-z)^2}, \quad z \in U, \quad t \geq 0,$$

find the function k with $|k(t)| = 1$ that appears in Loewner's equation

$$\frac{\partial f}{\partial t}(z,t) = zf'(z,t)\frac{1+k(t)z}{1-k(t)z}.$$

Do the same for a rotation of the Koebe function.

3.1.5. Prove Theorem 3.1.14.

3.1.6. Deduce the following from Theorem 3.1.11 by making a change of variable:

Let $r \in (0,1)$. The function $f(z,t) = a_1(t)z + \cdots$ on $U_r \times [0,\infty)$ is the restriction of a (non-normalized) Loewner chain defined on $U \times [0,\infty)$ if the following conditions hold:

(i) $f(\cdot,t)$ is holomorphic on U_r for each $t \geq 0$, and $f(z,\cdot)$ is locally absolutely continuous on $[0,\infty)$ locally uniformly with respect to $z \in U_r$;

(ii) $a_1(\cdot) \in C^1([0,\infty))$, $a_1(t) \neq 0$, $t \geq 0$, $|a_1(\cdot)|$ is strictly increasing on $[0,\infty)$, $|a_1(t)| \to \infty$ as $t \to \infty$, and there exists a constant $M > 0$ such that

$$|f(z,t)| \leq M|a_1(t)|, \quad |z| < r, \quad t \geq 0.$$

(iii) There exists a function $p(z,t)$ on $U \times [0,\infty)$ such that $p(\cdot,t)$ is holomorphic and has positive real part on U (but is not normalized), $p(z,\cdot)$ is measurable on $[0,\infty)$ for each $z \in U$, and for all $z \in U_r$,

$$\frac{\partial f}{\partial t}(z,t) = zf'(z,t)p(z,t), \quad a.e. \quad t \geq 0.$$

(Pommerenke, 1965 [Pom3].)

3.2 Applications of Loewner's differential equation to the study of univalent functions

In this section we shall investigate certain applications of Loewner's method to geometric problems such as the radius of starlikeness and the rotation theorem for the class S. We shall also show that for certain subclasses of S there exist alternative characterizations in terms of Loewner chains.

3.2.1 The radius of starlikeness for the class S and the rotation theorem

We first apply Loewner's method to determine the radius of starlikeness of S. This result was obtained by Grunsky [Gru2]; Goluzin [Gol2] applied Loewner's theory to the related Theorems 3.2.1 and 3.2.4 and Corollary 3.2.2.

We begin with the following bound for $\left|\arg\left[\dfrac{f(z)}{z}\right]\right|$ for $f \in S$ (see [Gru1]), where the branch is chosen such that $\arg\left[\dfrac{f(z)}{z}\right]\Big|_{z=0} = 0$.

Theorem 3.2.1. *If $f \in S$ then*

$$\left|\arg\left[\frac{f(z)}{z}\right]\right| \le \log\left[\frac{1+|z|}{1-|z|}\right], \quad z \in U.$$

This bound is sharp for each $z \in U$.

Proof.

Step 1. Using Theorem 3.1.3, we may suppose that f is a single-slit map. Then from Theorem 3.1.14 there exists a continuous function $k : [0,\infty) \to \mathbb{C}$ with $|k(t)| = 1$, $t \ge 0$, such that

$$(3.2.1) \qquad\qquad \lim_{t\to\infty} e^t v(z,t) = f(z)$$

locally uniformly on U, where $v = v(z,t)$ is the unique solution of the initial value problem

$$(3.2.2) \qquad\qquad \frac{\partial v}{\partial t} = -v\frac{1+k(t)v}{1-k(t)v}, \quad t \ge 0, \quad v(z,0) = z.$$

For fixed $z \in U \setminus \{0\}$, $\log v(z,t)$ can be defined as a continuous function of t beginning with a particular value of $\log z$ when $t = 0$, and (3.2.2) yields

$$\frac{\partial}{\partial t}\log v(z,t) = -\frac{1+k(t)v(z,t)}{1-k(t)v(z,t)}.$$

Taking real and imaginary parts, we obtain

$$(3.2.3) \qquad\qquad \frac{\partial}{\partial t}|v(z,t)| = -|v(z,t)|\frac{1-|v(z,t)|^2}{|1-k(t)v(z,t)|^2}$$

and

$$(3.2.4) \qquad\qquad \frac{\partial}{\partial t}\arg v(z,t) = -2\frac{\mathrm{Im}\,[k(t)v(z,t)]}{|1-k(t)v(z,t)|^2}.$$

Hence from (3.2.3) we deduce that $|v(z,t)|$ decreases monotonically from $|z|$ to 0 as t increases from 0 to ∞, and therefore there is a one-to-one correspondence between t and $|v|$, which enables us to consider $|v|$ as the independent variable. Moreover, from (3.2.3) we obtain

$$(3.2.5) \qquad dt = -\frac{d_t|v|}{|v|} \cdot \frac{|1 - kv|^2}{1 - |v|^2}.$$

Taking into account the relations (3.2.3), (3.2.4) and (3.2.5) we easily deduce that

$$|d_t \arg v(z,t)| \leq \frac{2|v(z,t)|dt}{|1 - k(t)v(z,t)|^2} = -\frac{2d_t|v(z,t)|}{1 - |v(z,t)|^2}.$$

On the other hand, since

$$\left| \arg\left[\frac{v(z,t)}{z} \right] \right| = \left| \int_0^t \frac{\partial}{\partial \tau} \arg v(z,\tau)d\tau \right|,$$

we obtain

$$\left| \arg\left[\frac{v(z,t)}{z} \right] \right| \leq 2 \int_0^t \frac{|v(z,\tau)|}{|1 - k(\tau)v(z,\tau)|^2}d\tau$$

$$\leq 2 \int_0^\infty \frac{|v(z,\tau)|}{|1 - k(\tau)v(z,\tau)|^2}d\tau = 2 \int_0^{|z|} \frac{du}{1 - u^2} = \log\left[\frac{1 + |z|}{1 - |z|} \right].$$

Letting $t \to \infty$ in this inequality and using (3.2.1), we obtain the desired bound.

Step 2. To prove the statement about sharpness, it is enough to show that for each $z \in U \setminus \{0\}$, we can find a continuous function k with $|k(t)| = 1$ on $[0, \infty)$, and for which the function $v = v(z,t)$ given by (3.2.2) satisfies the relation

$$(3.2.6) \qquad \text{Im} \, [k(t)v(z,t)] = -|v(z,t)|, \quad t \in [0, \infty).$$

Indeed, if the relation (3.2.6) holds with $|k(t)| = 1$, then

$$\text{Re} \, [k(t)v(z,t)] = 0, \quad t \in [0, \infty),$$

and in view of (3.2.3) and (3.2.4) we deduce that

$$(3.2.7) \qquad \frac{\partial}{\partial t}|v(z,t)| = -|v(z,t)|\frac{1 - |v(z,t)|^2}{1 + |v(z,t)|^2}$$

and

(3.2.8)
$$\frac{\partial}{\partial t} \arg v(z,t) = \frac{2|v(z,t)|}{1+|v(z,t)|^2}.$$

Separating variables in (3.2.7), integrating from 0 to t, and using the fact that $v(z,0) = z$, we deduce that

$$\frac{|v(z,t)|}{1-|v(z,t)|^2} = \frac{e^{-t}|z|}{1-|z|^2}.$$

Solving this equation, we obtain $|v(z,t)|$ as a function of $t \in [0,\infty)$ which decreases monotonically from $|z|$ to 0 as t increases from 0 to ∞. Next, we may use this function and the relation (3.2.8) to deduce that

$$d_t \arg v(z,t) = -\frac{2d_t|v(z,t)|}{1-|v(z,t)|^2}.$$

Hence, integrating both sides of this equality from 0 to t, we obtain

(3.2.9)
$$\arg \left[\frac{v(z,t)}{z}\right] = \log\left[\frac{1+|z|}{1-|z|}\right] - \log\left[\frac{1+|v(z,t)|}{1-|v(z,t)|}\right],$$

where again we have used the equality $v(z,0) = z$. Therefore we have defined the function $v = v(z,t)$, and taking into account (3.2.6) the function k is determined by the equality

$$k(t)v(z,t) = -i|v(z,t)|, \quad t \geq 0.$$

This function is continuous in $t \in [0,\infty)$ and satisfies the condition $|k(t)| = 1$, $t \geq 0$. Also it is clear that the Loewner differential equation (3.2.2), corresponding to the function k, has exactly the solution $v(z,t)$ determined above. Moreover, if we let $f(z) = \lim_{t\to\infty} e^t v(z,t)$, we deduce in view of (3.2.9) that

$$\arg\left[\frac{f(z)}{z}\right] = \log\left[\frac{1+|z|}{1-|z|}\right].$$

This proves the sharpness of the upper bound in Theorem 3.2.1. In a similar manner, we can prove the sharpness of the lower bound.

A direct consequence of the above result is the following bound for $\left|\arg\left[\frac{zf'(z)}{f(z)}\right]\right|$ when $f \in S$ (see [Gru1]). The branch is chosen such that

$\arg\left[\dfrac{zf'(z)}{f(z)}\right]\Bigg|_{z=0} = 0$. This result can be used to obtain the radius of starlike-ness of S.

Corollary 3.2.2. *If $f \in S$ then*

$$(3.2.10) \qquad \left|\arg\left[\frac{zf'(z)}{f(z)}\right]\right| \le \log\left[\frac{1+|z|}{1-|z|}\right], \quad z \in U.$$

This estimate is sharp for each $z \in U$.

Proof. Fix $z_0 \in U \setminus \{0\}$ and consider the Koebe transform

$$g(z) = \frac{f\left(\dfrac{z+z_0}{1+\overline{z}_0 z}\right) - f(z_0)}{(1-|z_0|^2)f'(z_0)}, \quad z \in U.$$

Then $g \in S$ and in view of Theorem 3.2.1 we obtain

$$\left|\arg\left[\frac{g(z)}{z}\right]\right| \le \log\left[\frac{1+|z|}{1-|z|}\right], \quad z \in U.$$

On the other hand, it is obvious that

$$\frac{g(-z_0)}{-z_0} = \frac{f(z_0)}{z_0 f'(z_0)} \cdot \frac{1}{1-|z_0|^2},$$

and hence the above relations imply that

$$\left|\arg\left[\frac{z_0 f'(z_0)}{f(z_0)}\right]\right| \le \log\left[\frac{1+|z_0|}{1-|z_0|}\right],$$

as desired.

The sharpness of (3.2.10) follows from the sharpness of the estimate in Theorem 3.2.1. This completes the proof.

The following result, due to Grunsky [Gru2], establishes the radius of star-likeness of S. For comparison, we recall that the radius of convexity of S is $2 - \sqrt{3} \approx 0.27$ (see Theorem 2.2.22), while the radius of close-to-convexity is given in Theorem 3.2.5.

Corollary 3.2.3. *The radius of starlikeness of S is* $\tanh\dfrac{\pi}{4}$, *i.e.*

$$r^*(S) = \tanh\frac{\pi}{4} = \frac{e^{\frac{\pi}{2}} - 1}{e^{\frac{\pi}{2}} + 1} \approx 0.66.$$

Proof. Taking into account Theorem 2.2.2 and Corollary 3.2.2, we conclude that $f \in S^*(U_r)$, where r is determined by solving the equation

$$\log \left[\frac{1+r}{1-r} \right] = \frac{\pi}{2},$$

that is, $r = \tanh \dfrac{\pi}{4}$. Thus

$$\mathrm{Re} \left[\frac{zf'(z)}{f(z)} \right] > 0, \quad |z| < r,$$

and this relation may fail for $|z| \geq r$, in view of the sharpness of the estimate (3.2.10). Hence $r^*(S) = \tanh \dfrac{\pi}{4}$.

There is a related estimate for $\left| \log \left[\dfrac{zf'(z)}{f(z)} \right] \right|$ when $f \in S$, where the branch is chosen such that $\log \left[\dfrac{zf'(z)}{f(z)} \right] \bigg|_{z=0} = 0$. This result was also obtained by Grunsky [Gru1]. We leave the proof for the reader, since it suffices to use similar arguments as in the proof of Theorem 3.2.1.

Theorem 3.2.4. *If $f \in S$ then*

$$\left| \log \left[\frac{zf'(z)}{f(z)} \right] \right| \leq \log \left[\frac{1+|z|}{1-|z|} \right], \quad z \in U.$$

This estimate is sharp for each $z \in U$.

In 1962 Krzyz [Krz2] obtained the radius of close-to-convexity for the class S. This is the largest number r_{cc} such that every function in the class S is close-to-convex on $U_{r_{cc}}$. We omit the proof of this result; however we mention that it is of interest for the reader.

Theorem 3.2.5. *For $r \in (r^*(S), 1)$, let*

$$A = A(r) = (1 + r^2)(1 - r^2)^{-1},$$

and let $X_0(r)$ be the unique root of the polynomial

$$P(X) \equiv X^3 - AX^2 + A^2 X - A = 0.$$

Then the radius of close-to-convexity of the class S is the unique real root r_{cc} of the equation

$$2 \operatorname{arccot} \left[\frac{1 - r^2}{1 + r^2} X_0(r) \right] + \log(1 + X_0^2(r)) - 2 \log \left[\frac{2r}{1 - r^2} \right] = 0,$$

contained in the interval $(r^*(S), 1)$. *In particular,* $0.80 < r_{cc} < 0.81$.

Another application of Loewner's theory is the following estimate for $|\arg f'(z)|$ due to Goluzin [Gol2]. (In fact the sharpness of the bound for $|z| > 1/\sqrt{2}$ was first established by Bazilevich [Baz1].) Because of the geometric interpretation of $\arg f'(z)$ as the local rotation factor under the mapping $f \in S$, Theorem 3.2.6 is called the *rotation theorem* for the class S. Again the branch is chosen such that $\arg f'(z)|_{z=0} = 0$. We note that with an elementary proof, based on inequality (1.1.9), it is possible to show that if $f \in S$, then

$$|\arg f'(z)| \le 2 \log \left[\frac{1 + |z|}{1 - |z|} \right], \quad z \in U.$$

However, this estimate is not sharp when $z \ne 0$.

Theorem 3.2.6. *Let* $f \in S$. *Then*

$$|\arg f'(z)| \le \begin{cases} 4 \arcsin |z|, & |z| \le \dfrac{1}{\sqrt{2}} \\[2mm] \pi + \log \left[\dfrac{|z|^2}{1 - |z|^2} \right], & 1 > |z| \ge \dfrac{1}{\sqrt{2}}. \end{cases}$$

These estimates are sharp for each $z \in U$.

Proof.

Step 1. Again using Theorem 3.1.3 we may suppose that f is a single-slit mapping. We may also assume that $z \ne 0$. Then using Theorem 3.1.14 we may express f as the limit

$$f(z) = \lim_{t \to \infty} e^t v(z, t)$$

locally uniformly on U, where $v = v(z, t)$ is the solution of the initial value problem

(3.2.11) $$\frac{\partial v}{\partial t} = -v \frac{1 + kv}{1 - kv}, \quad v(z, 0) = z,$$

and $k : [0, \infty) \to \mathbb{C}$ is a continuous function such that $|k(t)| = 1$, $t \ge 0$. By Weierstrass' theorem, we conclude that

$$f'(z) = \lim_{t \to \infty} e^t v'(z, t).$$

Differentiating both sides of equation (3.2.11) with respect to z (this is possible since $v(z, \cdot)$ is of class C^1 on $[0, \infty)$ for each $z \in U$), one deduces that

$$\frac{\partial}{\partial z}\left(\frac{\partial v}{\partial t}\right) = \frac{\partial v'}{\partial t} = -v'\frac{1 + 2kv - k^2v^2}{(1 - kv)^2}$$

and hence

$$\frac{\partial}{\partial t}\log v' = 1 - \frac{2}{(1 - kv)^2}.$$

Taking imaginary parts in this equality, we have

(3.2.12)
$$\frac{\partial}{\partial t}\arg v' = 2\frac{\text{Im}\,(1 - kv)^2}{|1 - kv|^4}.$$

On the other hand, since

$$\frac{\partial}{\partial t}\log|v| = \text{Re}\left[\frac{\partial v}{\partial t}\frac{1}{v}\right],$$

we obtain from (3.2.11) the equality

(3.2.13)
$$\frac{\partial}{\partial t}\log|v| = -\frac{1 - |v|^2}{|1 - kv|^2}.$$

This equation shows that $|v(z, t)|$ decreases from $|z|$ to 0 as t increases from 0 to ∞. Hence there is a one-to-one correspondence between $|v|$ and t, and we can regard $|v|$ as the independent variable.

Eliminating t between (3.2.12) and (3.2.13), one concludes that

(3.2.14)
$$d_t \arg v'(z, t) = -2\frac{\text{Im}\,(1 - kv)^2 d_t|v|}{|1 - kv|^2|v|(1 - |v|^2)}.$$

Since $|k(t)| = 1$ we see that

(3.2.15)
$$\frac{|\text{Im}\,(1 - kv)^2|}{|1 - kv|^2} = |\sin\{2\arg(1 - kv)\}|$$

$$\leq \begin{cases} \sin(2\arcsin|v|) = 2|v|\sqrt{1 - |v|^2}, & |v| \leq \dfrac{1}{\sqrt{2}} \\[2ex] 1, & |v| > \dfrac{1}{\sqrt{2}}. \end{cases}$$

Now, because $|v(z,t)|$ decreases as t increases we have $d_t|v| < 0$, and from (3.2.14) we obtain

$$|d_t \arg v'(z,t)| \leq \begin{cases} -\dfrac{4d_t|v|}{\sqrt{1-|v|^2}}, & |v| \leq \dfrac{1}{\sqrt{2}} \\[3mm] -\dfrac{2d_t|v|}{|v|(1-|v|^2)}, & |v| > \dfrac{1}{\sqrt{2}}. \end{cases}$$

Integrating the above inequality from $t = 0$ to $t = \infty$, and noting that $\arg v'(z,t) \to \arg f'(z)$ as $t \to \infty$, we obtain

$$|\arg f'(z)| \leq \int_0^{|z|} \frac{4dx}{\sqrt{1-x^2}} = 4\arcsin|z|, \quad |z| \leq \frac{1}{\sqrt{2}},$$

and

$$|\arg f'(z)| \leq \int_0^{1/\sqrt{2}} \frac{4dx}{\sqrt{1-x^2}} + \int_{\frac{1}{\sqrt{2}}}^{|z|} \frac{2dx}{x(1-x^2)}$$

$$= \pi + \log\left[\frac{|z|^2}{1-|z|^2}\right], \quad 1 > |z| > \frac{1}{\sqrt{2}}.$$

This is the desired relation.

Step 2. In order to prove that these estimates are sharp for each $z \in U$ we show that we can find a continuous function k with $|k(t)| = 1$ on $[0,\infty)$, such that if $v(z,t)$ is the solution of the differential equation (3.2.11), then for this function equality holds for all t in (3.2.15). Fix $z \in U \setminus \{0\}$ and consider the resulting equation (3.2.15). An equivalent form of this equation is the following:

$$|\sin\{\arg(1 - kv)\}| = \begin{cases} |v|, & |v| \leq \dfrac{1}{\sqrt{2}} \\[3mm] \dfrac{1}{\sqrt{2}}, & |v| > \dfrac{1}{\sqrt{2}}. \end{cases}$$

The above equation enables us to determine $k(t)v(z,t)$ as a function of $|v(z,t)|$. (The function $k(t)v(z,t)$ is uniquely determined for $|v| \leq 1/2$, while for $|v| > 1/2$ there are two possible choices, and we can make a continuous choice.) If we substitute this function kv into (3.2.13), we may compute $|v(z,t)|$ as a function of t.

On the other hand, from (3.2.11) we have

$$\frac{\partial}{\partial t} \log v(z,t) = -\frac{1 + k(t)v(z,t)}{1 - k(t)v(z,t)}.$$

Hence, taking real and imaginary parts we obtain (3.2.13) and also

(3.2.16) $$\frac{\partial}{\partial t} \arg v(z,t) = -\frac{2\mathrm{Im}\,[k(t)v(z,t)]}{|1 - k(t)v(z,t)|^2}.$$

Therefore, from (3.2.16) we may determine $\arg v(z,t)$ in terms of t. Thus we have sufficient information to determine $v(z,t)$ and $k(t)$ in such a way that the Loewner equation (3.2.11) and (3.2.15) are satisfied for all t. This proves the sharpness of Theorem 3.2.6.

Remark 3.2.7. There are refinements of the rotation theorem for certain subclasses of S: for normalized close-to-convex functions, we have $|\arg f'(z)| \leq 4 \arcsin|z|$ (Theorem 2.4.9), while for normalized convex functions we have $|\arg f'(z)| \leq 2\arcsin|z|$ (see Problem 2.2.8).

3.2.2   Applications of the method of Loewner chains to characterize some subclasses of S

We begin by using Theorem 3.1.11 to give an alternative characterization of spirallike functions of type α with $|\alpha| < \pi/2$. At the same time, we give another proof of Theorem 2.4.10 by using the method of subordination chains (see Spaček [Spa], Robertson [Robe3], [Pom5, Theorem 6.6]).

Theorem 3.2.8. *Let $f : U \to \mathbb{C}$ be a normalized holomorphic function and let $\alpha \in \mathbb{R}$ with $|\alpha| < \pi/2$. Then f is spirallike of type α if and only if*

(3.2.17) $$\mathrm{Re}\,\left[e^{i\alpha} \frac{z f'(z)}{f(z)} \right] > 0 \text{ on } U.$$

Proof. First assume the condition (3.2.17) holds.

Let $f: U \times [0,\infty) \to \mathbb{C}$ be defined by

(3.2.18) $$f(z,t) = e^{(1-ia)t} f(e^{iat}z), \quad z \in U, \quad t \geq 0,$$

where $a = \tan\alpha$. Then $f(z,t)$ is holomorphic on $|z| < 1$, $f(0,t) = 0$ and $f'(0,t) = e^t$ for $t \geq 0$. For $r \in (0,1)$ let $M = \max_{|z| \leq r} |f(z)|$. Then $|f(z,t)| \leq Me^t$ for $|z| \leq r$ and $t \geq 0$, and hence the condition (3.1.18) is satisfied.

A simple computation yields that

(3.2.19) $$\frac{\partial f}{\partial t}(z,t) = zf'(z,t)p(z,t), \quad z \in U, \quad t \geq 0,$$

where

(3.2.20) $$p(z,t) = ia + (1 - ia)\frac{f(e^{iat}z)}{e^{iat}zf'(e^{iat}z)}.$$

Obviously $p(z,t)$ is holomorphic on $|z| < 1$, $p(0,t) = 1$ and from (3.2.17), Re $p(z,t) > 0$, $|z| < 1$, $t \geq 0$. The measurability condition from Theorem 3.1.11 is also satisfied, and hence $f(z,t)$ is a Loewner chain. In particular f is univalent on U and

$$f(z) \prec f(z,t) \prec e^{(1-ia)t}f(z), \quad \text{for } z \in U \text{ and } t \geq 0.$$

This implies that

$$\exp\left(-e^{-ia}\frac{t}{\cos\alpha}\right)f(z) \in f(U), \quad z \in U, \quad t \geq 0,$$

and we conclude that f is a spirallike function of type α.

Conversely, assume that f is spirallike of type α. Then f is univalent on U and $f(z) \prec e^{(1-ia)t}f(z)$ for $z \in U$ and $t \geq 0$. Let

$$v(z,s,t) = e^{-iat}f^{-1}\left(\exp\left(-\frac{e^{-ia}}{\cos\alpha}(t-s)\right)f(e^{ias}z)\right),$$

for $z \in U$ and $0 \leq s \leq t < \infty$. This function is well defined on U since f is spirallike. Also $v(0,s,t) = 0$ and $|v(z,s,t)| < 1$ on U, hence v is a Schwarz function. Moreover, $v(\cdot,s,t)$ is univalent on U for $0 \leq s \leq t < \infty$.

Now, if $f(z,t)$ is the map defined by (3.2.18) then it is easy to see that

$$f(z,s) = f(v(z,s,t),t), \quad z \in U, \quad 0 \leq s \leq t < \infty,$$

and therefore $f(z,t)$ is a subordination chain. Moreover, since f is univalent on U it follows that $f(\cdot,t)$ is also univalent on U, and $f(z,t)$ is normalized, so that $f(z,t)$ is a Loewner chain.

Using the relations (3.2.19), (3.2.20) and Theorem 3.1.12, we must have Re $p(z,t) > 0$ for $|z| < 1$ and $t \geq 0$. Setting $t = 0$ in (3.2.20), we deduce that

$$\text{Re}\left[e^{ia}\frac{zf'(z)}{f(z)}\right] > 0, \quad z \in U.$$

This completes the proof.

Using the above arguments we obtain the following (cf. [Pom5]):

Corollary 3.2.9. *Let f be a normalized holomorphic function on U and $\alpha \in \mathbb{R}$ with $|\alpha| < \pi/2$. Also let $a = \tan \alpha$. Then f is spirallike of type α if and only if*

$$f(z,t) = e^{(1-ia)t} f(e^{iat}z)$$

is a Loewner chain. In particular, f is starlike if and only if $f(z,t) = e^t f(z)$ is a Loewner chain.

With similar reasoning as in the proof of Theorem 3.2.8 we may prove the following alternative characterization of close-to-convexity (see [Biel-Lew]):

Theorem 3.2.10. *Let $f : U \to \mathbb{C}$ be a normalized holomorphic function and let g be starlike on U. Then f is close-to-convex with respect to the convex function $h(z) = \int_0^z \dfrac{g(t)}{t} dt$ if and only if*

$$f(z,t) = f(z) + (e^t - 1)g(z), \quad z \in U, \quad t \geq 0,$$

is a Loewner chain. (The Loewner chain $f(z,t)$ is not normalized if g is not normalized.)

Proof. First we suppose that f is close-to-convex with respect to h. Then $f(\cdot, t)$ is also close-to-convex with respect to h for any fixed $t \geq 0$, and hence univalent. It is also easily checked that

$$(3.2.21) \qquad \mathrm{Re}\left[\frac{\partial f}{\partial t}(z,t)/\{zf'(z,t)\}\right] > 0, \quad t \geq 0, z \in U.$$

This implies that $f(z,t)$ is a subordination chain.

Conversely, the condition (3.2.21) when $t = 0$ implies that f is close-to-convex with respect to h. This completes the proof.

We remark that the above characterization of close-to-convexity in terms of Loewner chains may be used to prove the linear accessibility property of image domains for close-to-convex functions.

Corollary 3.2.11. *Let f be a normalized close-to-convex function on U. Then for each r, $0 < r < 1$, the complement of $f(U_r)$ is a union of nonintersecting rays.*

Proof. Since f is close-to-convex on U there exists a starlike function g such that $f(z,t) = f(z) + (e^t - 1)g(z)$ is a Loewner chain. Hence

$$L(z,r) = \left\{ f(z) + tg(z) : \ t \geq 0, \ z \text{ fixed}, \ |z| = r \right\}$$

are disjoint rays which fill up the complement of $f(U_r)$.

It is also possible to give a characterization of convexity in terms of Loewner chains, using Theorems 2.2.6 and 3.2.10.

Theorem 3.2.12. *Let* $f : U \to \mathbb{C}$ *be a normalized holomorphic function. Then* $f \in K$ *if and only if*

$$(3.2.22) \qquad f(z,t) = f(z) + (e^t - 1)zf'(z), \quad z \in U, \quad t \geq 0,$$

is a Loewner chain.

Proof. First assume $f \in K$. Then $g(z) = zf'(z)$ is normalized starlike on U, by Theorem 2.2.6. Using Theorem 3.2.10, it is obvious that $f(z,t)$ given by (3.2.22) is a Loewner chain.

Conversely, assume $f(z,t)$ given by (3.2.22) is a Loewner chain. In view of Theorem 3.1.12 we obtain

$$\text{Re}\left[\frac{\frac{\partial f}{\partial t}(z,t)}{zf'(z,t)}\right] > 0, \quad z \in U, \quad t \geq 0.$$

A straightforward computation yields

$$\text{Re}\left[1 + (1 - e^{-t})\frac{zf''(z)}{f'(z)}\right] > 0, \quad z \in U, \quad t \geq 0.$$

Letting $t \to \infty$ in the above, we deduce that

$$\text{Re}\left[1 + \frac{zf''(z)}{f'(z)}\right] > 0, \quad z \in U,$$

as desired. This completes the proof.

Notes. In this section we have proved some of the most well known and basic applications of Loewner's differential equation, apart from coefficient estimates. For most of the results in this section, we have used the following references: [Dur], [Gol4], [Hay], [Pom3], [Pom5]. For other applications of Loewner's differential equation the reader may consult [Ahl2], [Avk-Aks2], [Baz3], [Bec1-4], [Bran], [Con], [Gon5], [Hen], [Leu], [Lew1,2], [Ros], [Ros-Rov], [Rot], [Rov].

Problems

3.2.1. Prove Theorem 3.2.4.

3.2.2. Prove Theorem 3.2.5.

3.2.3. Complete the details in the proof of Theorem 3.2.10.

3.3   Univalence criteria

3.3.1   Becker's univalence criteria

We finish this chapter with some applications of the method of Loewner chains to univalence criteria involving higher derivatives of the function.

First we prove the following result due to Becker, 1972 [Bec1] (an improvement of an earlier result of Duren, Shapiro, and Shields [Dur-Sh-Sh]):

Theorem 3.3.1. *Let* $f : U \to \mathbb{C}$ *be a holomorphic function such that* $f'(0) \neq 0$. *If*

$$(3.3.1) \qquad (1 - |z|^2)\left|\frac{z f''(z)}{f'(z)}\right| \leq 1, \quad z \in U,$$

then f *is univalent on* U.

Proof. Clearly we may suppose that f is normalized. Using the relation (3.3.1), one deduces that $f'(z) \neq 0$ for $z \in U$. Let

$$f(z,t) = f(e^{-t}z) + (e^t - e^{-t})z f'(e^{-t}z), \quad z \in U, \quad t \geq 0.$$

Then $f(\cdot, t)$ is holomorphic on U, $f(0,t) = 0$, $f'(0,t) = e^t$ for $t \geq 0$, and $f(z, \cdot) \in C^\infty([0,\infty))$ for each $z \in U$. It is clear that as $t \to \infty$,

$$e^{-t}f(z,t) = z + O(e^{-t})$$

locally uniformly in z, and hence

$$\lim_{t \to \infty} e^{-t}f(z,t) = z$$

locally uniformly in z. Consequently $\{e^{-t}f(z,t)\}_{t \geq 0}$ is a normal family, and for each $r \in (0,1)$ there is a constant $M = M(r) > 0$ such that $|f(z,t)| \leq Me^t$

for $|z| \leq r$ and $t \geq 0$. Hence the condition (i) from Theorem 3.1.11 is satisfied. To check that condition (ii) is satisfied, we first note that

$$\frac{\partial f}{\partial t}(z,t) = e^t z f'(e^{-t}z) - (1 - e^{-2t})z^2 f''(e^{-t}z),$$

and

$$zf'(z,t) = e^t z f'(e^{-t}z) + (1 - e^{-2t})z^2 f''(e^{-t}z)$$

$$= e^t z f'(e^{-t}z)(1 - E(z,t)),$$

where

$$E(z,t) = -(1 - e^{-2t})e^{-t}\frac{zf''(e^{-t}z)}{f'(e^{-t}z)}.$$

Using the relation (3.3.1) and the fact that $1 - e^{-2t} < 1 - |e^{-t}z|^2$, $z \in U$, we easily deduce that

$$|E(z,t)| = (1 - e^{-2t})\left|\frac{e^{-t}zf''(e^{-t}z)}{f'(e^{-t}z)}\right| < 1,$$

and hence $f'(z,t) \neq 0$ for $z \in U$ and $t \geq 0$. If we define $p(z,t)$ by

$$p(z,t) = \frac{\frac{\partial f}{\partial t}(z,t)}{zf'(z,t)}, \quad z \in U, \quad t \geq 0,$$

then straightforward computation yields that

$$p(z,t) = \frac{1 + E(z,t)}{1 - E(z,t)}, \quad z \in U, \quad t \geq 0,$$

and also $p(0,t) = 1$. The holomorphy and measurability conditions of Theorem 3.1.11 are clearly satisfied and we have Re $p(z,t) > 0$ for $|z| < 1$ and $t \geq 0$.

Since all conditions of Theorem 3.1.11 are satisfied, we conclude that $f(z,t)$ is a Loewner chain. Hence f, as the first element of this chain, is univalent on U. This completes the proof.

It is of interest to see whether there is any room for improvement of the constant on the right-hand side of (3.3.1). Consider the function $f(z) = e^{\beta z}$, $z \in U$, where $\beta > \pi$. Then it is clear that f is not univalent on U. On the other hand, since

$$\frac{zf''(z)}{f'(z)} = \beta z$$

and

$$\sup\left\{(1-|z|^2)\left|\frac{zf''(z)}{f'(z)}\right| : |z| \le 1\right\}$$

$$= \sup\left\{\beta r(1-r^2) : 0 \le r \le 1\right\} = \frac{2}{9}\sqrt{3}\beta,$$

we see that the constant 1 in (3.3.1) cannot be replaced by any number strictly greater than $\frac{2}{9}\sqrt{3}\pi \approx 1.21$.

Pommerenke [Pom8] gave an example which shows that the constant 1 in (3.3.1) cannot be replaced by 1.121. This improves the above bound obtained by Becker [Bec1].

We remark that an interesting extension of Becker's result was obtained by Ruscheweyh [Rus1].

With similar reasoning as in the proof of Theorem 3.3.1, one may prove the following criterion for univalence, due to Ahlfors [Ahl3] and Becker [Bec1]:

Theorem 3.3.2. *Let* $f : U \to \mathbb{C}$ *be a holomorphic function with* $f'(0) \ne 0$. *Let* $c \in \mathbb{C}$ *with* $|c| \le 1$, $c \ne -1$, *and assume that*

$$\left|(1-|z|^2)\frac{zf''(z)}{f'(z)} + c|z|^2\right| \le 1, \quad z \in U.$$

Then f *is univalent on* U.

In this case it suffices to prove that

$$f(z,t) = f(e^{-t}z) + \frac{1}{1+c}(e^t - e^{-t})zf'(e^{-t}z)$$

is a (non-normalized) univalent subordination chain. We leave the details for the reader.

3.3.2   Univalence criteria involving the Schwarzian derivative

Another very important class of univalence criteria are those involving the Schwarzian derivative. The basic result was discovered by Nehari [Neh1] and is a sufficient condition for univalence. A more elementary necessary condition was found earlier by Kraus [Krau] and rediscovered by Nehari. The aim of this section is to present the results of Nehari and Kraus and a general criterion

for univalence due to Epstein. We shall see that there are some connections
with the theory of Loewner chains.

Recall that the Schwarzian derivative of a locally univalent function f is
defined by

$$\{f; z\} = \left(\frac{f''(z)}{f'(z)}\right)' - \frac{1}{2}\left(\frac{f''(z)}{f'(z)}\right)^2, \quad z \in U.$$

It is noteworthy because of its invariance properties under linear fractional
transformations. If $h = f \circ g$ is the composition of locally univalent functions,
then

$$\{h; z\} = \{f; g(z)\}(g'(z))^2 + \{g; z\}.$$

If T is a linear fractional transformation then $\{T; z\} = 0$, and hence we
obtain $\{T \circ g; z\} = \{g; z\}$ and $\{f \circ T; z\} = \{f; T(z)\}(T'(z))^2$.

Theorem 3.3.3. *(i) If $f \in S$ then*

$$|\{f; z\}| \leq \frac{6}{(1 - |z|^2)^2}, \quad z \in U.$$

This result is sharp.

(ii) Conversely, if $f : U \to \mathbb{C}$ is a holomorphic function such that

$$(3.3.2) \qquad |\{f; z\}| \leq \frac{2}{(1 - |z|^2)^2}, \quad z \in U,$$

then f is univalent on U.

Proof. First we prove the necessary condition for univalence given in part
(i). To this end, fix $z \in U$ and consider the Koebe transform of f:

$$g(t) = \frac{f\left(\frac{t+z}{1+\bar{z}t}\right) - f(z)}{(1 - |z|^2)f'(z)} = t + b_2 t^2 + b_3 t^3 + \ldots, \quad t \in U.$$

Then $g \in S$ and simple computations yield that

$$b_2 = \frac{1}{2}\left[(1 - |z|^2)\frac{f''(z)}{f'(z)} - 2\bar{z}\right]$$

and

$$b_3 = \frac{1}{6}\left[(1 - |z|^2)^2\frac{f'''(z)}{f'(z)} - 6\bar{z}(1 - |z|^2)\frac{f''(z)}{f'(z)} + 6\bar{z}^2\right].$$

We now consider the function φ defined by

$$\varphi(\zeta) = \frac{1}{g(1/\zeta)} = \zeta + \alpha_0 + \frac{\alpha_1}{\zeta} + \dots, \quad \zeta \in \Delta.$$

Since $g \in S$ we have $\varphi \in \Sigma$ and it is easy to see that $\alpha_0 = -b_2$ and

$$\alpha_1 = b_2^2 - b_3 = -\frac{1}{6}(1 - |z|^2)^2\{f; z\}.$$

From Corollary 1.1.3 we have $|\alpha_1| \leq 1$, and thus

$$|\{f; z\}| \leq \frac{6}{(1 - |z|^2)^2}, \quad z \in U.$$

The sharpness of the above estimate follows from the sharpness of the estimate $|\alpha_1| \leq 1$. In fact, equality occurs for the Koebe function.

The sufficient condition for univalence in part (ii) is usually deduced from a comparison theorem for ordinary differential equations. A different proof, using the Loewner method, was given by Becker [Bec1]. We shall present Becker's proof.

It is known that if p is a holomorphic function on U, then any (meromorphic) function f on U such that $\{f; z\} = p(z)$ has the form $f = w_1/w_2$, where w_1 and w_2 are linearly independent solutions of the differential equation

$$(3.3.3) \qquad\qquad w''(z) + \frac{p(z)}{2}w(z) = 0.$$

The converse is also true. (See [Neh1], [Dur].) It follows that if f and g are holomorphic functions on U and $\{f; z\} = \{g; z\}$, then there exists a linear fractional transformation T such that $g = T \circ f$. Now, assume that f is holomorphic on U and satisfies (3.3.2). Then $f'(z) \neq 0$ for $z \in U$ (Problem 3.3.1), and if $p(z) = \{f; z\}$, $z \in U$, then p is also a holomorphic function on the unit disc.

With this choice of p, let w_1 and w_2 be the solutions of equation (3.3.3) such that

$$(3.3.4) \qquad\qquad w_1(0) = 0, \; w_1'(0) = 1,$$

$$(3.3.5) \qquad\qquad w_2(0) = 1, \; w_2'(0) = 0.$$

The Wronskian of these solutions is constant, and taking into account the conditions (3.3.4) and (3.3.5), we see that

$$(3.3.6) \qquad w_1'(z)w_2(z) - w_1(z)w_2'(z) = 1, \quad z \in U.$$

Let $u(z) = w_1(z)/w_2(z)$, for $z \in U$. Then $\{u; z\} = p(z)$, and using (3.3.3), (3.3.4), and (3.3.5), we obtain the expansion

$$u(z) = z + \frac{p(0)}{6}z^3 + \dots, \quad z \in U.$$

As already noted, there exists a linear fractional transformation T such that $u = T \circ f$. Consequently, in order to prove that f is univalent, it suffices to show this property for the function u. To this end, let

$$f(z,t) = \frac{w_1(e^{-t}z) + (e^t - e^{-t})zw_1'(e^{-t}z)}{w_2(e^{-t}z) + (e^t - e^{-t})zw_2'(e^{-t}z)}, \quad z \in U.$$

Then $f(0,t) = 0$ and $f'(0,t) = e^t$ for $t \geq 0$. Also $f(z,0) = w_1(z)/w_2(z) = u(z)$. We shall show that $f(z,t)$ is a Loewner chain.

Let

$$g(z,t) = w_2(e^{-t}z) + (e^t - e^{-t})zw_2'(e^{-t}z)$$

$$= w_2(e^{-t}z) + (1 - e^{-2t})z^2 \left[\frac{w_2'(e^{-t}z)}{e^{-t}z} \right], \quad z \in U, t \geq 0.$$

Then $g(\cdot, t)$ is holomorphic on U and using (3.3.3) and (3.3.5), we obtain

$$(3.3.7) \qquad \lim_{t \to \infty} g(z,t) = w_2(0) + w_2''(0)z^2 = 1 - \frac{p(0)}{2}z^2,$$

locally uniformly on U. In view of (3.3.2), we have $|p(0)| \leq 2$, and thus for each $r \in (0,1)$ we have $g(z,t) \neq 0$ for $|z| \leq r$ and t sufficiently large. Together with the fact that $g(0,t) = w_2(0) = 1$ for $t \geq 0$, this implies that there exists a number $r \in (0,1)$ such that $g(z,t) \neq 0$ for $|z| \leq r$ and $t \geq 0$. For this r, the function $f(\cdot, t)$ is holomorphic on U_r for each $t \geq 0$, and $f(z, \cdot)$ is of class C^∞ on $[0, \infty)$ for each $z \in U_r$. Further, using the relations (3.3.4), (3.3.5) and (3.3.7), we deduce that

$$\lim_{t \to \infty} e^{-t} f(z,t) = \frac{z}{1 - \dfrac{p(0)}{2}z^2}$$

locally uniformly on U_r.

Taking into account (3.3.3) and (3.3.6), by differentiation we easily obtain the relations

$$f'(z,t) = e^t \frac{1 + \frac{1}{2}z^2(1 - e^{-2t})^2 p(e^{-t}z)}{[w_2(e^{-t}z) + z(e^t - e^{-t})w_2'(e^{-t}z)]^2}$$

and

$$\frac{\partial f}{\partial t}(z,t) = e^t z \frac{1 - \frac{1}{2}z^2(1 - e^{-2t})^2 p(e^{-t}z)}{[w_2(e^{-t}z) + z(e^t - e^{-t})w_2'(e^{-t}z)]^2},$$

for $|z| < r$ and $t \geq 0$. Let

$$\lambda(z,t) = \frac{\dfrac{\partial f}{\partial t}(z,t) - zf'(z,t)}{\dfrac{\partial f}{\partial t}(z,t) + zf'(z,t)}$$

$$= -\frac{1}{2}z^2(1 - e^{-2t})^2 p(e^{-t}z), \quad |z| < r, \quad t \geq 0.$$

Then $\lambda(z,t)$ is holomorphic on $|z| < r$ for $t \geq 0$, and clearly we may extend this function holomorphically to the whole disc U. Moreover, using the assumption (3.3.2), one deduces that

$$|\lambda(z,t)| = \frac{1}{2}|z|^2(1 - e^{-2t})^2|p(e^{-t}z)|$$

$$\leq \frac{1}{2}|z|^2(1 - |e^{-t}z|^2)^2|p(e^{-t}z)| \leq |z|^2,$$

for $|z| < 1$ and $t \geq 0$. Hence, if $h(z,t) = [1 + \lambda(z,t)]/[1 - \lambda(z,t)]$, then in view of the above inequality we conclude that $h(\cdot,t)$ is a holomorphic function on U for $t \geq 0$, and $\operatorname{Re} h(z,t) > 0$ for $z \in U$ and $t \geq 0$. Moreover, it is obvious that $h(0,t) = 1$, $h(z,\cdot)$ is a measurable function on $[0,\infty)$ for each $z \in U$, and

$$\frac{\partial f}{\partial t}(z,t) = zf'(z,t)h(z,t), \quad |z| < r, \quad t \geq 0.$$

Since all conditions of Theorem 3.1.11 are satisfied, we conclude that $f(z,t)$ is a Loewner chain, and therefore $f(z,0) = u(z)$ is univalent on U, as desired. This completes the proof.

Remark 3.3.4. The constant 2 in the estimate (3.3.2) is best possible. To see this, let

$$f(z) = \left(\frac{1-z}{1+z}\right)^\gamma, \quad \gamma \in \mathbb{C}.$$

In this case

$$\{f; z\} = \frac{2(1-\gamma^2)}{(1-z^2)^2}, \quad z \in U.$$

It is easy to see that if $\gamma = i\beta$ with $\beta \in \mathbb{R}$ and $\beta \neq 0$, then f is not univalent on U. In this case

$$|\{f; z\}| \leq \frac{2(1+\beta^2)}{(1-|z|^2)^2}, \quad z \in U.$$

This example is due to Hille [Hil1].

In addition to the univalence condition (3.3.2), Nehari [Neh1] obtained another sufficient condition for univalence, as follows. We leave the details of the proof for the reader.

Theorem 3.3.5. *If f is a holomorphic function on U and*

(3.3.8)
$$|\{f; z\}| \leq \frac{\pi^2}{2}, \quad z \in U,$$

then f is univalent on U.

Again the constant $\pi^2/2$ is best possible in (3.3.8), because if $f(z) = e^{\lambda z}$, $\lambda > \pi$, then f is not univalent on U, and in this case $|\{f; z\}| = \lambda^2/2$.

Nehari gave further univalence criteria in [Neh4]. Osgood and Stowe [Os-St1-2] showed that the criteria of Nehari in Theorems 3.3.3 and 3.3.5 could be viewed as particular cases of a single theorem, in which the Schwarzian derivative is defined relative to conformal metrics on the domain and range of f.

Epstein [Ep1,2] obtained a general criterion for univalence, which includes most previously known criteria. His proof uses hyperbolic geometry and is quite different from classical approaches.

Pommerenke [Pom9] gave a very interesting proof of the analytic function case of Epstein's result, using the method of Loewner chains. He also showed that an additional assumption on g made by Epstein was not necessary. Here we shall present Pommerenke's proof. (In the proofs of Pommerenke and Epstein the function f is allowed to be meromorphic.)

Theorem 3.3.6. *Let f and g be holomorphic and locally univalent functions on U. If*

$$\left| \frac{1}{2}(1 - |z|^2)^2 \left[\{f; z\} - \{g; z\} \right] + (1 - |z|^2) \frac{\overline{z} g''(z)}{g'(z)} \right| \leq 1, \ z \in U,$$

then f is univalent on U.

Proof. Suppose

$$f(z) = a_0 + a_1 z + \ldots, \quad z \in U,$$

and

$$g(z) = b_0 + b_1 z + \ldots, \quad z \in U.$$

Since f and g are locally univalent, it follows that $a_1 \neq 0$ and $b_1 \neq 0$. Also since the Schwarzian is invariant under linear fractional transformations, we may assume without loss of generality that f and g are normalized.

We next introduce the functions

$$w_1(z) = \sqrt{\frac{g'(z)}{f'(z)}} = 1 + cz^2 + O(z^3)$$

and

$$w_2(z) = f(z) w_1(z) = z + dz^2 + O(z^3).$$

These functions are holomorphic on U since f' and g' cannot have zeros on U.

Now, let $f_t(z) = f(z, t)$ be the chain given by

$$f(z, t) = \frac{w_2(e^{-t}z) + (e^t - e^{-t})z w_2'(e^{-t}z)}{w_1(e^{-t}z) + (e^t - e^{-t})z w_1'(e^{-t}z)}, \ z \in U, \ t \geq 0.$$

(Compare with the proof of Theorem 3.3.3.)

Then $f(\cdot, t)$ is meromorphic on U. Since $w_1(z) = 1 + cz^2 + \ldots$ it follows that the denominator in the expression of $f(z, t)$ has the form $1 + O(z^2)$ as $z \to 0$, uniformly in t. Hence there exist constants $r \in (0, 1)$ and $M > 0$ such that $f(\cdot, t)$ is holomorphic on U_r for each $t \geq 0$, and

$$|f(z, t)| \leq M e^t, \quad |z| < r, \ t \geq 0.$$

Moreover, it is easy to see that

$$f_t(z) = e^t z + O(z^2) \quad \text{as } z \to 0,$$

since the numerator is $e^t z + O(z^2)$ as $z \to 0$. Thus $f(z,t)$ is a normalized chain. Letting $\alpha(t) = 1 - e^{-2t}$, we obtain after elementary computations that

$$\lambda(z,t) \equiv \frac{\frac{\partial f}{\partial t}(z,t) - z f'(z,t)}{\frac{\partial f}{\partial t}(z,t) + z f'(z,t)}$$

$$= -\frac{e^{-t}\alpha z(w_2'' w_1 - w_1'' w_2) + \alpha^2 z^2(w_2'' w_1' - w_1'' w_2')}{w_2' w_1 - w_1' w_2},$$

where w_1, w_2, w_1', w_2', w_1'', w_2'' are computed at $e^{-t}z$. Using the definitions of w_1 and w_2, and writing $S_f(z) = \{f; z\}$ and $S_g(z) = \{g; z\}$, we deduce that

$$w_2' w_1 - w_1' w_2 = f' w_1^2,$$

$$w_2'' w_1 - w_1'' w_2 = f'' w_1^2 + 2 f' w_1' w_1 = f' w_1^2 g''/g',$$

$$w_2'' w_1' - w_1'' w_2' = f'' w_1' w_1 - f' w_1'' w_1 + 2 f' w_1'^2$$

$$= \frac{1}{2} f' w_1^2 (S_f - S_g).$$

Therefore, we obtain

$$\lambda(z,t) = \frac{1}{2}\alpha^2(t) z^2 [S_g(e^{-t}z) - S_f(e^{-t}z)] - \alpha(t)\frac{e^{-t}z g''(e^{-t}z)}{g'(e^{-t}z)},$$

for all $z \in U$ and $t \geq 0$.

It can be seen that $\lambda(\cdot, t)$ is zero for $t = 0$ and is holomorphic on $\overline{U}$ if $t > 0$. In view of the maximum modulus theorem, we obtain for $t > 0$ that

$$|\lambda(z,t)| \leq \max_{|w|=1} |\lambda(w,t)| = |\lambda(w_0,t)| = |\overline{w}_0^2 \lambda(w_0,t)|,$$

for some $w_0 \in \mathbb{C}$, $|w_0| = 1$. Next, using the inequality from the hypothesis for $z = e^{-t}w_0$, we have

$$|\lambda(w_0,t)| = \left| \frac{1}{2}(1 - |e^{-t}w_0|^2)^2 [S_f(e^{-t}w_0) - S_g(e^{-t}w_0)] \right.$$

$$+(1 - |e^{-t}w_0|^2)\frac{e^{-t}\overline{w}_0 g''(e^{-t}w_0)}{g'(e^{-t}w_0)}\Bigg| \leq 1.$$

This estimate together with the fact that $\lambda(0,t) = 0$, $t \geq 0$, implies that if we define

$$p(z,t) = \frac{\frac{\partial f}{\partial t}(z,t)}{zf'(z,t)} = \frac{1 + \lambda(z,t)}{1 - \lambda(z,t)},$$

then $p(\cdot,t) \in \mathcal{P}$, $t \geq 0$. Also it is obvious that $p(z,\cdot)$ is measurable on $[0,\infty)$ for each $z \in U$.

Taking into account Theorem 3.1.11 we deduce that $f(z,t)$ is a Loewner chain, and thus $f(z,0) = w_2(z)/w_1(z) = f(z)$ is univalent on U. This completes the proof.

Remark 3.3.7. If $g(z) \equiv z$ in Theorem 3.3.6, we obtain the sufficient condition for univalence due to Nehari (see Theorem 3.3.3). On the other hand, if $f \equiv g$ in Theorem 3.3.6, we obtain Becker's univalence result given in Theorem 3.3.1.

Recently Schippers [Schip] derived a differential equation for the Loewner flow of the Schwarzian derivative of a univalent function on U, and used this result to obtain sharp bounds for higher order analogs of the Schwarzian derivative. Introducing the operators $\sigma_n(f)$, defined by

$$\sigma_3(f) = \{f; z\} \quad \text{and} \quad \sigma_{n+1}(f) = (\sigma_n(f))' - (n-1)\frac{f''}{f'}\sigma_n(f),$$

he obtained the following results via the Loewner method:

Theorem 3.3.8. $\displaystyle\sup_{f \in H_u(U)} [\lambda(z)]^{n-1}|\sigma_n(f)(z)| = 4^{n-3}(n-2)!6,$

where $\lambda(z) = 1/(1 - |z|^2)$ is the hyperbolic metric on U.

This estimate is sharp for a suitable rotation of the Koebe function.

(Higher order Schwarzians have been considered by other authors, e.g. [Aha], [Har1-2], [Tam].)

3.3.3   A generalization of Becker's and Nehari's univalence criteria

We finish this chapter with a result which generalizes Theorems 3.3.1, 3.3.2 and 3.3.3. This generalization was obtained by Pascu [Pas], using the method of Loewner chains.

Theorem 3.3.9. *Let* $F = F(u,v) : U \times \mathbb{C} \to \mathbb{C}$ *satisfy the following assumptions:*

(i) The function $L(z,t) = F(e^{-t}z, e^t z)$ *is holomorphic on* U *for all* $t \in [0,\infty)$ *and is locally absolutely continuous in* $t \in [0,\infty)$ *locally uniformly with respect to* $z \in U$.

(ii) The function $\dfrac{\partial L(z,t)}{\partial t}/[zL'(z,t)]$ *is holomorphic on* $\overline{U}$ *for* $t > 0$ *and is holomorphic on* U *for* $t = 0$.

(iii) $\dfrac{\partial F}{\partial v}(0,0) \neq 0$, $\dfrac{\partial F}{\partial u}(0,0)/\dfrac{\partial F}{\partial v}(0,0) \notin (-\infty,-1]$, *and there exists* $a_1 :$ $[0,\infty) \to \mathbb{C}$ *such that* $|a_1(t)|$ *is strictly increasing to* ∞ *as* $t \to \infty$, *and*

$$a_1(t) = e^{-t}\frac{\partial F}{\partial u}(0,0) + e^t\frac{\partial F}{\partial v}(0,0), \quad t \in [0,\infty).$$

(iv) The family of functions $\left\{\dfrac{F(e^{-t}z, e^t z)}{a_1(t)}\right\}_{t\geq 0}$ *is a normal family on* U.
Let

$$G(u,v) = u\frac{\partial F}{\partial u}(u,v)/[v\frac{\partial F}{\partial v}(u,v)].$$

If

(3.3.9) $$|G(z,z)| < 1, \quad z \in U,$$

and

(3.3.10) $$|G(z, 1/\bar{z})| \leq 1, \quad z \in U \setminus \{0\},$$

then $F(z,z)$ *is a univalent function on* U.

Remark 3.3.10. From the assumption (iii) it follows that $a_1(t) \neq 0$ for $t \in [0,\infty)$.

Proof of Theorem 3.3.9. Let

$$p(z,t) = \frac{\partial L}{\partial t}(z,t)/[zL'(z,t)], \quad z \in U, \quad t \in [0,\infty).$$

From the condition (ii) one deduces that $p(\cdot,t)$ is holomorphic on U for each $t \geq 0$, and $p(z,\cdot)$ is a measurable function on $[0,\infty)$ for each $z \in U$. We prove that $\operatorname{Re} p(z,t) > 0$ for $z \in U$ and $t \geq 0$. To this end, let

$$w(z,t) = \frac{1 - p(z,t)}{1 + p(z,t)}, \quad z \in U, \quad t \geq 0.$$

After short computations, it is easy to obtain that $w(z,t) = G(e^{-t}z, e^t z)$ for $z \in U$ and $t \geq 0$. Hence $w(\cdot, t)$ is a holomorphic function of U for each $t \geq 0$. From the condition (3.3.9) it follows that $|w(z,0)| = |G(z,z)| < 1$, $z \in U$.

If $t > 0$ is fixed, then $p(\cdot, t)$ is holomorphic on $\overline{U}$, by assumption (ii). Also, from (3.3.9) we have

(3.3.11) $|w(0,t)| = |G(0,0)| < 1$.

Further, from (3.3.10) we obtain

(3.3.12) $|w(z,t)| \leq \max_{|z|=1} |w(z,t)| = \max_{|z|=1} |G(e^{-t}z, e^t z)|$

$$= |G(e^{-t}\lambda, e^t \lambda)| = |G(z_0, 1/\overline{z}_0)| \leq 1,$$

where λ is a complex number with $|\lambda| = 1$ and $z_0 = e^{-t}\lambda \in U$. Consequently, from (3.3.11), (3.3.12) and using the maximum principle for holomorphic functions, one obtains that $|w(z,t)| < 1$ for $z \in U$ and $t \geq 0$. Thus $\operatorname{Re} p(z,t) > 0$, as claimed.

Taking into account Theorem 3.1.11 and Problem 3.1.6, one concludes that $L(z,t)$ is a univalent subordination chain. Therefore $L(z,0) = F(z,z)$ is univalent on U. This completes the proof.

Remark 3.3.11. Let $f : U \to \mathbb{C}$ be a normalized holomorphic function. Also let $F_j : U \times \mathbb{C} \to \mathbb{C}$, $j = 1, 2, 3$, be the functions given by

$$F_1(u,v) = f(u) + (v-u)f'(u),$$

$$F_2(u,v) = f(u) + \frac{v-u}{1+c}f'(u),$$

where $c \in \mathbb{C}$ with $|c| \leq 1$, $c \neq -1$, and

$$F_3(u,v) = f(u) + \frac{(v-u)f'(u)}{1 - \frac{v-u}{2} \cdot \frac{f''(u)}{f'(u)}}.$$

Setting successively $F = F_j$, $j = 1, 2, 3$, in Theorem 3.3.9 yields the results of Theorems 3.3.1, 3.3.2 and the sufficient condition in Theorem 3.3.3. On the other hand, if

$$F(u,v) = \frac{a(u) + (v-u)b(u)}{c(u) + (v-u)d(u)}$$

in Theorem 3.3.9, where a, b, c, d are analytic functions on U such that $a(z)d(z) - c(z)b(z) \neq 0$ for $z \in U$, then we obtain a sufficient condition of univalence due to Betker [Bet2].

Notes. We remark that in the literature there are many generalizations of Nehari's univalence criteria, given in Theorems 3.3.3 (ii) and 3.3.5. The reader may consult the following papers: [Ahl3], [Bec1], [And-Hin], [Chu1], [Ep1,2], [Geh-Pom], [Lew-St], [Neh4], [Os-St1], [Pfa4], [Schw].

Problems

3.3.1. Let f be a holomorphic function on U and suppose that $f'(a) = 0$ for some point $a \in U$. Show that $\{f; z\}$ cannot have a removable singularity at a.

3.3.2. Prove Theorem 3.3.2.

3.3.3. Prove Theorem 3.3.5.

3.3.4. Let $f : U \to \mathbb{C}$ be an analytic and locally univalent function such that
$$|\{f; z\}| \leq \frac{4}{1 - |z|^2}, \quad z \in U.$$
Show that f is univalent on U.
(Pokornyi, 1951 [Pok], Nehari, 1979 [Neh4].)

3.3.5. Let $f : U \to \mathbb{C}$ be an analytic and locally univalent function such that
$$\left|\{f; z\} + \frac{4\bar{z}^2}{(1 - |z|^2)^2}\right| \leq \frac{4}{(1 - |z|^2)^2}, \quad z \in U.$$
Show that f is univalent on U.
(Chuaqui, 1995 [Chu1].)

3.3.6. Complete the details in the proof of Theorem 3.3.6.

Chapter 4

Bloch functions and the Bloch constant

In this chapter we shall discuss some of the basic properties of Bloch functions, and their role in the study of the Bloch constant problem and other problems of the same type. We also point out some other connections between Bloch functions and univalent functions, and between the related but larger class of normal functions and univalent functions. We show how the classical lower estimate of Ahlfors for the Bloch constant was obtained by Bonk [Bon1,2] as a consequence of a function theoretic distortion theorem. We also discuss analogous results of Liu and Minda [Liu-Min] for locally univalent Bloch functions. Finally we consider the analog of the Bloch constant problem for normalized convex functions. This problem can be solved exactly, as shown by Szegö [Sze].

4.1 Preliminaries concerning Bloch functions

Definition 4.1.1. Let f be a holomorphic function on U and let

$$(4.1.1) \qquad \|f\| = \sup_{|z|<1} (1 - |z|^2)|f'(z)|.$$

The function f is called a *Bloch function* if $\|f\| < \infty$, and in this case $\|f\|$ is called the *Bloch seminorm* of f.

145

The Bloch seminorm is invariant under pre-composition with automorphisms of the unit disc and post-composition with (orientation preserving) Euclidean motions of the complex plane.

Let $\mathcal{B}$ denote the set of Bloch functions on U. This set is a Banach space with respect to the norm

$$\|f\|_{\mathcal{B}} = \|f\| + |f(0)|.$$

It is known that $\mathcal{B}$ is isomorphic to ℓ^{∞} (the space of bounded sequences with the supremum norm) (see [Shi-Wil]) and also $\mathcal{B}$ is isomorphic to the second dual of the subspace of $\mathcal{B}$ spanned by the polynomial functions (see [And-Cl-Pom], [Shi-Wil]). On the other hand, $\mathcal{B}$ is not separable (see [Cam-Cim-Pfa]; [And-Cl-Pom]).

An alternative characterization of Bloch functions is given in the following theorem (see [Pom4]). First we note that the group of holomorphic automorphisms of the unit disc U is given by

$$\text{Aut}(U) = \left\{ \varphi \in H(U) : \varphi(z) = e^{i\theta} \frac{z+a}{1+\bar{a}z}, \, \theta \in \mathbb{R}, \, |a| < 1 \right\}.$$

Now if $f \in H(U)$, we consider the family $\mathcal{F}_f$ defined by

$$\mathcal{F}_f = \left\{ g \in H(U) : g(z) = f(\varphi(z)) - f(\varphi(0)), \, \varphi \in \text{Aut}(U) \right\}.$$

Then we have

Theorem 4.1.2. *The function $f \in H(U)$ is a Bloch function if and only if the family $\mathcal{F}_f$ is a (finitely) normal family.*

Proof. If $\mathcal{F}_f$ is a normal family, then $\{(f \circ \varphi)'(0) : \varphi \in \text{Aut}(U)\}$ is a bounded set. Taking $\varphi(z) = (z+a)/(1+\bar{a}z)$ yields that $\{f'(a)(1-|a|^2) : a \in U\}$ is a bounded set, and thus f is a Bloch function.

Conversely, if f is a Bloch function and $g(z) = f(\varphi(z)) - f(\varphi(0))$ where φ is a disc automorphism, then

$$g(z) = \int_0^z g'(\zeta)d\zeta = \int_0^z f'(\varphi(\zeta))\varphi'(\zeta)d\zeta, \quad z \in U,$$

where the integration is along the straight line segment from 0 to z. It is elementary to see that the integrand is bounded by

$$\frac{C}{1-|\varphi(\zeta)|^2} \cdot \frac{1-|a|^2}{|1+\bar{a}\zeta|^2} = \frac{C}{1-|\zeta|^2},$$

where $C = \|f\|$, and hence

$$|g(z)| \le \frac{C}{2} \log \left[\frac{1+|z|}{1-|z|}\right], \quad z \in U.$$

This shows that the family $\mathcal{F}_f$ is locally uniformly bounded on U and completes the proof.

There is a variety of other useful characterizations of Bloch functions (see [And], [And-Cl-Pom], [Cim], [Pom4], [Tim1]). We shall give some of them in this chapter (Theorem 4.1.7, Problem 4.1.4, Problem 4.2.2).

The related notion of normal function was introduced by Lehto and Virtanen [Leh-Vir1].

Definition 4.1.3. Let f be a meromorphic function on U and let

$$f^*(z) = \frac{|f'(z)|}{1+|f(z)|^2}, \quad z \in U.$$

f^* is called the *spherical derivative* of f. Also let

$$\nu(f) = \sup_{|z|<1} (1 - |z|^2) f^*(z).$$

The function f is called *normal* if $\nu(f) < \infty$. The quantity $\nu(f)$ is called the *order of the normal function* f and is invariant under pre-composition with disc automorphisms and post-composition with rotations of the Riemann sphere. (See the proof of [Min2, Theorem 3].)

Again there is a characterization in terms of normal families, whose proof we omit. If f is a meromorphic function on U, we let

$$\widetilde{\mathcal{F}}_f = \Big\{g: \; g = f(\varphi(z)), \; \varphi \in \mathrm{Aut}(U)\Big\}.$$

Then we have [Leh-Vir1]

Theorem 4.1.4. *The meromorphic function f on U is normal if and only if the family $\widetilde{\mathcal{F}}_f$ is a normal family (viewed as a family of mappings into the Riemann sphere with the spherical metric $\dfrac{|dw|}{1+|w|^2}$).*

It is clear that a Bloch function is normal, and that a bounded holomorphic function on U is Bloch. In fact, for bounded holomorphic functions on U we have the inequality [And-Cl-Pom]

$$(4.1.2) \qquad\qquad \|f\|_B \le 2\|f\|_\infty,$$

where $\|f\|_\infty = \sup_{|z|<1} |f(z)|$.

We also note the following examples:

Example 4.1.5. (i) Let $f \in H(U)$ be given by

$$f(z) = \frac{1}{2} \log \left[\frac{1+z}{1-z} \right], \quad z \in U,$$

where the branch of the logarithm is chosen so that $\log 1 = 0$. Then f is an unbounded Bloch function and $\|f\| = \|f\|_B = 1$.

(ii) If $\alpha > 0$ and

$$f(z) = \left(\frac{1+z}{1-z} \right)^\alpha, \quad z \in U,$$

then f is not a Bloch function.

In fact if f is a Bloch function of Bloch seminorm 1 such that $f(0) = 0$, it follows from the proof of Theorem 4.1.2 that

$$|f(z)| \leq \frac{1}{2} \log \left[\frac{1+|z|}{1-|z|} \right], \quad z \in U.$$

Hence the function $f(z) = \frac{1}{2} \log \left[\frac{1+z}{1-z} \right]$, $z \in U$, is extremal for the growth of such functions.

We now consider some connections between Bloch functions and univalent functions. First we have the following theorem of Pommerenke [Pom5]:

Theorem 4.1.6. *If $f : U \to \mathbb{C}$ is a univalent function then $f(z)$ and $f'(z)$ are normal functions.*

Proof. Since $f \in H_u(U)$, f may be written in the form

$$f(z) = \alpha g(z) + \beta, \quad z \in U,$$

where $g \in S$ and $\alpha, \beta \in \mathbb{C}$, $\alpha \neq 0$. Hence we have

$$(1 - |z|^2) f^*(z) = \frac{(1 - |z|^2)|f'(z)|}{1 + |f(z)|^2}$$

$$= \frac{|\alpha g(z)|}{1 + |\alpha g(z) + \beta|^2} (1 - |z|^2) \left| \frac{g'(z)}{g(z)} \right|.$$

Using the fact that $g \in S$ and the estimate (1.1.7), we conclude that

$$\sup_{|z|<1} (1 - |z|^2) f^*(z) < \infty.$$

Thus f is normal.

Next we consider f'. In this case we obtain

$$\frac{(1-|z|^2)|f''(z)|}{1+|f'(z)|^2} \le \frac{1}{2}(1-|z|^2)\left|\frac{f''(z)}{f'(z)}\right| = \frac{1}{2}(1-|z|^2)\left|\frac{g''(z)}{g'(z)}\right| \le 3,$$

making use of the fact that the function $g \in S$ satisfies (see (1.1.9))

$$\left|\frac{1-|z|^2}{2}\frac{g''(z)}{g'(z)} - \bar{z}\right| \le 2, \quad z \in U.$$

Therefore f' is also a normal function.

The following beautiful result of Pommerenke [Pom4,5] (see also [Dur-Sh-Sh]) exhibits a surprising connection between the set S and the set of Bloch functions in U.

Theorem 4.1.7. *Let $f \in H(U)$. Then f is a Bloch function if and only if there exist a function $g \in S$ and a complex number α such that*

(4.1.3) $$f(z) = \alpha \log g'(z) + f(0), \quad z \in U.$$

The branch of the logarithm is chosen so that $\log g'(0) = 0$.

Proof. First assume there exist $\alpha \in \mathbb{C}$ and $g \in S$ such that f can be written as in (4.1.3). Then

$$(1-|z|^2)|f'(z)| \le |\alpha|(1-|z|^2)\left|\frac{g''(z)}{g'(z)}\right| \le 6|\alpha|, \quad z \in U,$$

where again we have used (1.1.9). Thus f is a Bloch function and $\|f\| \le 6|\alpha|$.

Conversely, suppose that f is a Bloch function. Without loss of generality, we may assume that $\|f\| \ne 0$ (otherwise, if $\|f\| = 0$, we can take $\alpha = 0$ and $g(z) \equiv z$) and let $g : U \to \mathbb{C}$ be given by

$$g(z) = \int_0^z \exp\left[\frac{f(\zeta)-f(0)}{\|f\|}\right] d\zeta, \quad z \in U,$$

where the integration is along the straight line segment from 0 to z. Then $g \in H(U)$, $g(0) = 0$ and $g'(0) = 1$. Also let $\alpha = \|f\|$. Obviously (4.1.3) holds, and taking into account (4.1.1) we obtain

$$(1-|z|^2)\left|\frac{g''(z)}{g'(z)}\right| = \frac{1}{\|f\|}(1-|z|^2)|f'(z)| \le 1, \quad z \in U.$$

Hence, in view of the Becker univalence criterion (see [Bec1] or Theorem 3.3.1), we conclude that $g \in S$, as desired.

Remark 4.1.8. More precisely, if $\mathcal{B}_S$ denotes the set of functions

$$\mathcal{B}_S = \{\log g'(z) : g \in S\},$$

then we have ([And], [Bec1])

$$\{f \in \mathcal{B} : \|f\| \leq 1 \text{ and } f(0) = 0\} \subsetneqq \mathcal{B}_S \subsetneqq \{f \in \mathcal{B} : \|f\| \leq 6 \text{ and } f(0) = 0\}.$$

The inclusions follow from arguments in the above proof. (Showing that the inclusions are strict is an exercise.) The constant 6 is sharp, as one can see by considering the Koebe function.

An elementary result about the coefficients of Bloch functions is given in the following theorem (see [And-Cl-Pom], [Pom4]). We leave the proof for the reader.

Theorem 4.1.9. Let $f : U \to \mathbb{C}$ be a Bloch function such that

$$f(z) = \sum_{n=0}^{\infty} a_n z^n, \ z \in U.$$

Then $|a_n| \leq 2\|f\|_{\mathcal{B}}$, $n = 0, 1, \ldots$.

In particular, this theorem states that the coefficients of a Bloch function are bounded. Of course, not every holomorphic function on U with bounded coefficients is a Bloch function. A counterexample is given by the function $f(z) = 1/(1-z)$, $z \in U$.

Other theorems about the coefficients of Bloch functions are proved in [And-Cl-Pom]. The Banach space structure of $\mathcal{B}$ is also studied in that paper; in addition see the survey papers [And] and [Cim].

There are no inclusions between the set of Bloch functions and any of the classical H^p spaces ($p \neq \infty$) or the Nevanlinna class. In one direction, this is shown by Example 4.1.5 (ii); in the other, there exist Bloch functions constructed using gap series which do not belong to the Nevanlinna class (see the discussion following Theorem 2 in [And]).

Coifman, Rochberg, and Weiss [Coi-Roc-Wei] showed that Bloch functions could be characterized by a bounded mean oscillation condition on U. Relations between univalent functions, analytic functions on U whose boundary values have bounded mean oscillation, and Bloch functions have been studied by Baernstein [Bae] and Pommerenke [Pom6,7]. In particular, Pommerenke showed that a univalent Bloch function belongs to the space BMOA of analytic functions whose boundary values on the unit circle have bounded mean oscillation.

For further information about the material in this section, see [And], [And-Cl-Pom], [Cim], [Pom4,5], [Tim1].

Problems

4.1.1. Prove the estimate (4.1.2).

4.1.2. Give examples which show that the inclusions in Remark 4.1.8 are proper.

4.1.3. Prove Theorem 4.1.9.

4.1.4. Show that if f is a Bloch function and d_h is the hyperbolic distance on U, then

$$\|f\| = \sup_{\substack{a,b \in U \\ a \neq b}} \frac{|f(a) - f(b)|}{d_h(a,b)}.$$

Hence f is a Bloch function if and only if there is a constant $M > 0$ such that

$$|f(a) - f(b)| \leq M d_h(a,b), \quad a, b \in U.$$

It follows that any Bloch function is uniformly continuous with respect to the hyperbolic metric on U and the Euclidean metric on its range.

4.1.5. Let $f : U \to \mathbb{C}$ be a holomorphic function and $n \geq 2$. Show that $f \in \mathcal{B}$ if and only if $(1 - |z|^2)^n f^{(n)}(z)$ is bounded in U.
(K. Zhu, 1990 [Zhu].)

4.2 The Bloch constant problem and Bonk's distortion theorem

Bloch functions arose originally in the study of the Bloch constant problem. In this section we discuss this problem and related problems, and we show how the classical lower estimate of Ahlfors for the Bloch constant was obtained from a function-theoretic distortion theorem by Bonk [Bon1,2].

In 1924 Bloch [Blo1,2] formulated a rather intricate covering theorem for non-univalent holomorphic functions on the unit disc, requiring only the condition $f'(0) = 1$. To state it, we need

Definition 4.2.1. Let $f \in H(U)$ and $a \in U$.

(i) A *schlicht disc* of f centered at $f(a)$ is a disc with center $f(a)$ such that f maps conformally a subdomain of U containing a onto this disc.

(ii) Let $r(a, f)$ be the radius of the largest schlicht disc of f centered at $f(a)$.

(iii) Let $r(f) = \sup \left\{ r(a, f) : a \in U \right\}$.

(iv) Let $\mathbf{B} = \inf \left\{ r(f) : f \in H(U),\ f'(0) = 1 \right\}$.

$\mathbf{B}$ is called the *Bloch constant*.

Bloch showed that $\mathbf{B}$ is positive. In 1929 Landau [Lan] gave numerical estimates for $\mathbf{B}$ and certain related constants, making use of an important reduction which accounts for the name "Bloch function". (However, this terminology was not introduced until much later by Pommerenke [Pom4].) To state it, we introduce a subclass $\mathcal{B}_1$ of $\mathcal{B}$ defined by

$$(4.2.1) \qquad \mathcal{B}_1 = \left\{ f \in \mathcal{B} : f(0) = 0,\ f'(0) = 1,\ \|f\| = 1 \right\}.$$

Landau's reduction is the following:

Theorem 4.2.2. $\mathbf{B} = \inf \left\{ r(f) : f \in \mathcal{B}_1 \right\}$.

Proof. In computing $\mathbf{B}$ it is clear that we can take the infimum over functions which are holomorphic in $\overline{U}$. For any such function f, the quantity $(1 - |z|^2)|f'(z)|$ has a maximum at an interior point z_0 of U. Suppose that the maximum value C is larger than 1. Then $z_0 \neq 0$.

Let $\varphi(z) = \dfrac{z + z_0}{1 + \overline{z}_0 z}$ and let $g(z) = \dfrac{e^{-i\alpha}}{C}[(f \circ \varphi)(z) - (f \circ \varphi)(0)]$, $z \in U$, where $\alpha = \arg\{(1 - |z_0|^2)f'(z_0)\}$. Then use of the relation

$$(1 - |z|^2)|(f \circ \varphi)'(z)| = (1 - |\varphi(z)|^2)|f'(\varphi(z))|$$

shows that $g \in \mathcal{B}_1$ and $r(g) \leq r(f)$. This implies that

$$\inf \left\{ r(f) : f \in \mathcal{B}_1 \right\} \leq \inf \left\{ r(f) : f \in H(U),\ f'(0) = 1 \right\}.$$

Since the opposite inequality is obvious, we are done.

Remark 4.2.3. Functions $f \in \mathcal{B}_1$, $f(z) = z + \sum\limits_{n=2}^{\infty} a_n z^n$, satisfy the growth estimate

$$|f(z)| \leq \frac{1}{2} \log \left[\frac{1+|z|}{1-|z|} \right] = |z| + \frac{|z|^3}{3} + \ldots, \ z \in U.$$

Together with the condition $f'(0) = 1$, this implies $a_2 = 0$ and $|a_3| \leq 1/3$. Bonk [Bon2] proved that $|a_4| \leq 5/4$ and Chen and Gauthier [Che-Gau1] improved this bound by showing that $|a_4| \leq 4.2/4 = 1.05$.

Landau also introduced two closely-related constants - the *Landau constant* L and the *univalent Bloch constant* A. To define L one omits the requirement that the discs in the range of f be schlicht, i.e. we set

$$L_f = \sup \left\{ r : f(U) \text{ contains a disc of radius } r \right\}$$

and

$$L = \inf \left\{ L_f : f \in H(U), \ f'(0) = 1 \right\}.$$

For A one restricts to univalent functions, so that the distinction between schlicht discs and arbitrary discs in the range of f disappears. Thus we define

$$A = \inf \left\{ r(f) : f \in S \right\}.$$

Then $\mathbf{B} \leq L \leq A$.

The locally univalent Bloch constant $\mathbf{B}_0$ was introduced much later in the 1960's (Peschl [Pes1], Chern [Cher], Pommerenke [Pom4]). It is defined by

$$(4.2.2) \quad \mathbf{B}_0 = \inf \left\{ r(f) : f \in H(U), \ f'(z) \neq 0, \ z \in U, \ f'(0) = 1 \right\}.$$

None of these constants is known precisely, but the classical upper bounds for $\mathbf{B}, \mathbf{B}_0$, and L, which are based on constructions of Ahlfors and Grunsky [Ahl-Gru] in the case of $\mathbf{B}$ and Rademacher [Rad] in the case of L (and $\mathbf{B}_0$) are conjectured to be sharp. Proving this is a long-standing open problem in geometric function theory. These bounds are

$$\mathbf{B} \leq \frac{\Gamma(1/3)\Gamma(11/12)}{\sqrt{1+\sqrt{3}}\,\Gamma(1/4)} \approx 0.4719,$$

and

$$\mathbf{B}_0 \le L \le \frac{\Gamma(1/3)\Gamma(5/6)}{\Gamma(1/6)} \approx 0.5433.$$

For further details see [Ahl-Gru], [Gol4], [Hil2], [Min1]. (The upper bound for L is incorrectly stated in [Gol4] and [Hil2].)

There is no conjecture for the precise value of A, but the estimates

$$0.57088 < A < 0.6564155$$

are known ([Bel-Hum], [ZhaS], [Jen3]).

It should also be mentioned that in 1923, prior to Bloch's work, Szegö [Sze] had solved the univalent Bloch constant problem exactly for the case of convex univalent functions. We shall consider this case below.

In a remarkable paper in 1938 demonstrating a connection between the Schwarz lemma and curvature, Ahlfors [Ahl1] obtained the lower bound $\mathbf{B} \ge \sqrt{3}/4$, using a differential geometric argument and the maximum principle for subharmonic functions. Extending Ahlfors'methods, Heins [Hei] showed in 1962 that $\mathbf{B} > \sqrt{3}/4$.

It was not until the 1988 thesis and a related publication in 1990 by Bonk [Bon1,2], that it was finally shown that Ahlfors' lower estimate for $\mathbf{B}$ could be obtained by function-theoretic methods. Bonk also obtained a small numerical improvement in the lower bound for $\mathbf{B}$: $\mathbf{B} \ge \dfrac{\sqrt{3}}{4} + 10^{-14}$. Subsequently Chen and Gauthier [Che-Gau1] showed that $\mathbf{B} \ge \dfrac{\sqrt{3}}{4} + 2 \times 10^{-4}$. See also [CheH].

Here, however, we are mainly interested in Bonk's proof of the classical lower estimate of Ahlfors, since the methods are similar to those encountered in covering theorems in univalent function theory. We shall consider lower bounds for $\mathbf{B}_0$ in the next section.

The following distortion result is due to Bonk [Bon2]:

Theorem 4.2.4. *Suppose* $f \in \mathcal{B}_1$. *Then*

$$(4.2.3) \qquad \operatorname{Re} f'(z) \ge \frac{1 - \sqrt{3}|z|}{\left(1 - \dfrac{1}{\sqrt{3}}|z|\right)^3}, \quad |z| \le \frac{1}{\sqrt{3}}.$$

Proof. It suffices to prove (4.2.3) for $z \in [0, 1/\sqrt{3}]$, since we may introduce rotations of f.

Let

$$g(w) = \frac{1}{\sqrt{3}} \cdot \frac{1 - w}{1 - \dfrac{w}{3}}$$

and

$$h(w) = \frac{9}{4} w \left(1 - \frac{1}{3}w\right)^2.$$

An easy computation yields

(4.2.4)
$$|h(w)|(1 - |g(w)|^2) = 1, \quad |w| = 1.$$

We now define the function

$$q(w) = \left\{ \frac{f'(g(w))}{h(w)} - 1 \right\} \frac{w}{(1 - w)^2}.$$

We note that $f'(g(w))$ and $h(w)$ are holomorphic on U, and q has removable singularities at 0 and 1. Indeed, it is easy to see that $w = 0$ is a removable singularity, and $w = 1$ is also removable since $f''(0) = 0$ by Remark 4.2.3. Therefore q is holomorphic on $\overline{U}$.

From the fact that $f \in \mathcal{B}_1$ and (4.2.4) we deduce that

$$\left| \frac{f'(g(w))}{h(w)} \right| \leq \frac{1}{(1 - |g(w)|^2)|h(w)|} = 1, \quad |w| = 1.$$

We also note that in the expression for $q(w)$, the factor $\dfrac{w}{(1 - w)^2}$ is the Koebe function and has negative real part on $\partial U \setminus \{1\}$. These two observations imply that Re $q(w) \geq 0$ for $w \in \partial U \setminus \{1\}$. Using the fact that q has a removable singularity at 1 and the minimum principle for harmonic functions, we therefore obtain

$$\text{Re } q(w) \geq 0, \quad |w| \leq 1.$$

In particular for $0 < w < 1$, where the Koebe function is positive, we deduce that

$$\text{Re } \left[\frac{f'(g(w))}{h(w)} \right] \geq 1.$$

This is equivalent to

(4.2.5) $\operatorname{Re} f'(g(w)) \geq h(w), \quad 0 < w < 1,$

since $h(w)$ is positive for $0 < w < 1$.

Setting $z = g(w)$ in (4.2.5) and noting that z ranges from $1/\sqrt{3}$ to 0 as w ranges from 0 to 1 gives the desired result.

Corollary 4.2.5. $B \geq \sqrt{3}/4$.

Proof. By Landau's reduction it suffices to show that $r(f) \geq \sqrt{3}/4$ for all $f \in \mathcal{B}_1$. In fact we shall show that $r(0, f) \geq \sqrt{3}/4$ for such f. If $f \in \mathcal{B}_1$, then Bonk's distortion theorem implies that

$$\operatorname{Re} f'(z) > 0, \quad |z| < \frac{1}{\sqrt{3}}.$$

Using the Wolff-Noshiro-Warschawski theorem (see Lemma 2.4.1), we deduce that f is univalent on the disc $U_{1/\sqrt{3}}$. Thus f maps the disc $U_{1/\sqrt{3}}$ conformally onto a simply connected domain D, and clearly $0 \in D$.

Moreover, the boundary of D is the image of the circle $|z| = 1/\sqrt{3}$ and if $w = f(e^{i\theta}/\sqrt{3})$, $\theta \in [0, 2\pi]$, is a point on this image, then

$$\left| f\left(\frac{1}{\sqrt{3}} e^{i\theta}\right) \right| = \left| \int_0^{\frac{1}{\sqrt{3}}} f'(\rho e^{i\theta}) d\rho \right| \geq \int_0^{\frac{1}{\sqrt{3}}} \operatorname{Re} f'(\rho e^{i\theta}) d\rho$$

$$\geq \int_0^{\frac{1}{\sqrt{3}}} \frac{1 - \sqrt{3}\rho}{\left(1 - \frac{1}{\sqrt{3}}\rho\right)^3} d\rho = \frac{\sqrt{3}}{4}.$$

From this we conclude that the domain D contains the disc $U_{\sqrt{3}/4}$ and hence $r(0, f) \geq \sqrt{3}/4$. Since $f \in \mathcal{B}_1$ is arbitrary, we have $B \geq \sqrt{3}/4$, as desired.

In order to understand fully the roles of the functions g and h in the proof of Bonk's distortion theorem, one needs to study the theory of extremal problems for Bloch functions, as developed by Cima and Wogen [Cim-Wog], Ruscheweyh and Wirths [Rus-Wir1,2], and Bonk himself in his thesis [Bon1]. Bonk's thesis contains another proof of his distortion theorem.

A more geometric proof of Bonk's distortion theorem based on Julia's lemma was given by Minda [Min5]. Minda's proof is essentially the one-variable case of Theorem 9.1.6. Bonk [Bon1] and Minda [Min5] also determined the extremal functions. They are given by the function

$$f(z) = \frac{\sqrt{3}}{4}\left(1 - 3\left(\frac{z - 1/\sqrt{3}}{1 - z/\sqrt{3}}\right)^2\right)$$

and its rotations. This function is a two-sheeted branched analytic covering of U onto the disc $U\left(\frac{\sqrt{3}}{4}, \frac{3\sqrt{3}}{4}\right)$.

Problems

4.2.1. Show that $\inf\left\{r(0, f) :\ f \in H(U),\ f'(0) = 1\right\} = 0$. (This is why one must allow the schlicht discs to have arbitrary centers in Bloch's theorem.)

4.2.2. Show that a holomorphic function f on U is Bloch if and only if there exists $M > 0$ such that $r(a, f) \leq M$ for all $a \in U$.

4.2.3. Let f be a holomorphic function on U such that $f(0) = 0$, $f'(0) = \alpha > 0$ and $|f(z)| < M$, $z \in U$. Show that

(i) f is univalent on the disc U_{ρ_0}, where

$$\rho_0 = \frac{\alpha}{M + \sqrt{M^2 - \alpha^2}} > \frac{\alpha}{2M};$$

(ii) For any positive number $\rho \leq \rho_0$, $f(U_\rho)$ contains the disc U_R, where

$$R = M\frac{\rho(\alpha - M\rho)}{M - \alpha\rho} \geq M\rho_0\rho.$$

(Landau, 1929 [Lan], Dieudonné, 1931, [Die2].)

4.2.4. Let $\mathcal{F}$ be the set of functions f which satisfy the conditions in Problem 4.2.3. Show that ρ_0 is the radius of univalence of $\mathcal{F}$ (i.e. ρ_0 is the largest value of ρ such that each function in $\mathcal{F}$ is univalent on U_ρ). (Dieudonné, 1931 [Die2].)

4.3   Locally univalent Bloch functions

4.3.1   Distortion results for locally univalent Bloch functions

It is also possible to obtain a distortion theorem for locally univalent Bloch functions using a version of Julia's lemma. This was carried out by Liu and

Minda [Liu-Min] who deduced the known estimate $\mathbf{B}_0 > 1/2$. (Peschl [Pes1], Chern [Cher], and Pommerenke [Pom4] all showed that $\mathbf{B}_0 \geq 1/2$, and Pommerenke showed that the inequality is strict.) In this section we shall present their proofs. Part of the distortion theorem was obtained much earlier by Peschl [Pes1,2] using different methods. We also note that Yanagihara [Yan2] obtained a very small numerical improvement in the lower estimate for the locally univalent Bloch constant $(\mathbf{B}_0 > 0.5 + 10^{-335})$. A further improvement $\mathbf{B}_0 > 0.5 + 2 \times 10^{-8}$ was announced by Chen in [CheH].

We introduce the subclass $\mathcal{B}_0$ of $\mathcal{B}$ defined by

$$\mathcal{B}_0 = \left\{ f \in \mathcal{B} : f'(z) \neq 0, \ z \in U, \ \|f\| = 1, \ f(0) = f'(0) - 1 = 0 \right\}.$$

Because of Landau's reduction it suffices to consider the class $\mathcal{B}_0$ when estimating the locally univalent Bloch constant, i.e.

(4.3.1) $\mathbf{B}_0 = \inf \left\{ r(f) : \ f \in \mathcal{B}_0 \right\}.$

For $r > 0$, let

$$\Delta(1, r) = \left\{ z \in U : \frac{|1 - z|^2}{1 - |z|^2} < r \right\} = \left\{ z \in U : \left| z - \frac{1}{1 + r} \right| < \frac{r}{1 + r} \right\}.$$

Then $\Delta(1, r)$ is a horodisc in U, i.e. a disc in U that is internally tangent to ∂U at 1. Also let $\overline{\Delta}(1, r)$ denote the closure of $\Delta(1, r)$ relative to U (which means that $1 \notin \overline{\Delta}(1, r)$).

We begin with Julia's lemma on the unit disc (in less than full generality, since we do not need the version which considers the angular derivative).

Lemma 4.3.1. *If f is a holomorphic function on $U \cup \{1\}$, f maps U into U and $f(1) = 1$, then $f'(1) = \alpha > 0$ and for each $r > 0$, the function f maps the closed horodisc $\overline{\Delta}(1, r)$ into the closed horodisc $\overline{\Delta}(1, \alpha r)$. Moreover, a point on the boundary of $\Delta(1, r)$ (relative to U) is mapped into the boundary of $\Delta(1, \alpha r)$ if and only if f is a conformal automorphism of the unit disc U such that $f(1) = 1$.*

Julia's lemma is proved in full generality in [Ahl2, Section 1-4]; the restricted version given above appears in [Poly-Sze, Problem 292]. By composing with a suitable linear fractional transformation, we obtain the following version of Julia's lemma for maps from the unit disc into the right half plane.

Lemma 4.3.2. *Suppose f is a holomorphic function on $U \cup \{1\}$, f maps U into the right half plane $\Pi = \{z \in \mathbb{C} : \text{Re } z > 0\}$ and $f(1) = 0$. Then for each $r > 0$, f maps the closed horodisc $\overline{\Delta}(1, r)$ into the closed disc $\overline{U}(\beta r, \beta r)$, where $\beta = -f'(1) > 0$. Moreover, a point on the boundary of $\Delta(1, r)$ is mapped into the boundary of $U(\beta r, \beta r)$ if and only if f is a conformal mapping of U onto Π such that $f(1) = 0$.*

A direct consequence of Lemma 4.3.2 is the following (see [Liu-Min]):

Corollary 4.3.3. *Suppose f is a holomorphic function on $U \cup \{1\}$ satisfying the assumptions of Lemma 4.3.2. Then for any $x \in (-1, 1)$,*

$$\text{Re } f(x) \leq 2\beta \frac{1-x}{1+x},$$

with equality for some $x \in (-1, 1)$ if and only if

$$f(z) = 2\beta \frac{1-z}{1+z}, \quad z \in U.$$

Proof. Let $x \in (-1, 1)$ be fixed and $r = (1-x)/(1+x)$. Then x lies on the boundary of $\overline{\Delta}(1, r)$, and in view of Lemma 4.3.2 we deduce that $f(x)$ lies inside the circle which passes through 0 and $2\beta(1-x)/(1+x)$ and intersects the real axis orthogonally. The statement regarding equality is clear. This completes the proof.

Before proceeding to the distortion theorem of Liu and Minda [Liu-Min], we consider a certain function which is extremal for this theorem.

Example 4.3.4. Let

$$(4.3.2) \qquad F(z) = -\frac{e}{2} \exp\left\{-\frac{1+z}{1-z}\right\} + \frac{1}{2}, \quad z \in U.$$

It is easy to see that F is a universal covering projection of U onto the punctured disc $\{w : 0 < |w - 1/2| < e/2\}$. Moreover,

$$(1 - |z|^2)|F'(z)| = \frac{e(1 - |z|^2)}{|1 - z|^2} \exp\left\{-\frac{1 - |z|^2}{|1 - z|^2}\right\} \leq 1, \quad z \in U,$$

and equality holds if and only if z belongs to the circle

$$\left\{z : \frac{|1 - z|^2}{1 - |z|^2} = 1\right\} = \left\{z : \left|z - \frac{1}{2}\right| = \frac{1}{2}\right\}.$$

(To see this it suffices to note that for $t > 0$ the function te^{1-t} assumes its maximum value 1 when $t = 1$.) Therefore F is a Bloch function and $\|F\| \le 1$.

It is easy to see that $r(0, F) = 1/2$ and $r(F) = e/4$. One may also verify (Problem 4.3.1) that

$$(4.3.3) \qquad |F'(z)| \ge F'(|z|) = \frac{1}{(1-|z|)^2} \exp\left\{-\frac{2|z|}{1-|z|}\right\}, \; z \in U.$$

We now present the distortion result of Liu and Minda [Liu-Min]. We remark that the inequality $|f'(z)| \ge F'(|z|)$, $z \in U$, was first obtained by Peschl [Pes1,2], though Peschl did not determine all the extremal functions.

Theorem 4.3.5. *Suppose $f \in \mathcal{B}_0$ and let F denote the function given by (4.3.2). Then the following inequalities hold:*

(i) For all $z \in U$,

$$|f'(z)| \ge F'(|z|) = \frac{1}{(1-|z|)^2} \exp\left\{-\frac{2|z|}{1-|z|}\right\},$$

and equality holds for some $z = re^{i\theta} \ne 0$ if and only if $f(z) = e^{i\theta} F(e^{-i\theta}z)$ for some $\theta \in \mathbb{R}$.

(ii) For $|z| \le 1/2$,

$$\operatorname{Re} f'(z) \ge F'(|z|) = \frac{1}{(1-|z|)^2} \exp\left\{-\frac{2|z|}{1-|z|}\right\},$$

and equality holds for some $z = re^{i\theta} \ne 0$ if and only if $f(z) = e^{i\theta} F(e^{-i\theta}z)$ for some $\theta \in \mathbb{R}$.

(iii) $\operatorname{Re} f'(z) > 0$ for $|z| < \sqrt{\dfrac{\pi}{4+\pi}} \approx 0.6633$. In particular f is univalent in this disc.

Proof. (i) Let θ be a real number and let

$$g(z) = \left(\frac{1+z}{2}\right)^2 f'\left(e^{i\theta}\frac{1-z}{2}\right), \quad z \in U.$$

Clearly g is holomorphic on $\overline{U}$ with the possible exception of the point $z = -1$. We also observe that $g(1) = 1$ and $g(z) \ne 0$, $z \in U$. Further, since $f'(0) = 1$ and $\|f\| = 1$, we deduce from Remark 4.2.3 that $f''(0) = 0$, and from this it follows that $g'(1) = 1$. Using the fact that $\|f\| = 1$, we deduce that

$$|g(z)| = \left|\frac{1+z}{2}\right|^2 \cdot \left|f'\left(e^{i\theta}\frac{1-z}{2}\right)\right|$$

$$\leq \left(1 - \left|\frac{1-z}{2}\right|^2\right)\left|f'\left(e^{i\theta}\frac{1-z}{2}\right)\right| \leq 1, \quad z \in U.$$

Moreover, since $g'(1) = 1$, g cannot be constant and thus g maps U into $U \setminus \{0\}$. We can therefore find a holomorphic function h on U which maps U into Π such that $g(z) = \exp(-h(z))$, $z \in U$, and $h(1) = 0$. A simple computation shows that $h'(1) = -g'(1) = -1$, and in view of Corollary 4.3.3 we deduce that

$$|g(x)| = \exp\{-\mathrm{Re}\, h(x)\} \geq \exp\left\{-2\frac{1-x}{1+x}\right\}, \quad x \in (-1,1).$$

Furthermore, equality holds for some $x \in (-1,1)$ if and only if

$$h(z) = 2\frac{1-z}{1+z}, \quad z \in U.$$

Replacing $(1-x)/2$ by t in the above inequality, we conclude that

$$|f'(e^{i\theta}t)| \geq \frac{1}{(1-t)^2}\exp\left\{-\frac{2t}{1-t}\right\}, \quad t \in [0,1).$$

Moreover, equality holds for some $t \in (0,1)$ if and only if $f(z) = e^{i\theta}F(e^{-i\theta}z)$ for some real θ.

(ii) Keeping the same notation as in (i), we deduce that for all $x \in (-1,1)$, $h(x)$ lies in the disc centered at $(1-x)/(1+x)$ and of radius $(1-x)/(1+x)$. Since

$$(4.3.4) \min\left\{\mathrm{Re}\,\exp(-w) : \left|w - \frac{1-x}{1+x}\right| \leq \frac{1-x}{1+x}\right\} = \exp\left\{-2\frac{1-x}{1+x}\right\},$$

for all $x \in [0,1]$ (see Problem 4.3.2), we obtain

$$\mathrm{Re}\, g(x) = \mathrm{Re}\,\exp(-h(x)) \geq \exp\left\{-2\frac{1-x}{1+x}\right\}, \quad x \in [0,1].$$

The above relation is equivalent to

$$\mathrm{Re}\, f'(e^{i\theta}t) \geq \frac{1}{(1-t)^2}\exp\left\{-\frac{2t}{1-t}\right\}, \quad t \in [0,1/2].$$

The sharpness of this result follows as in (i).

(iii) Let $\rho = \sqrt{\dfrac{\pi}{4+\pi}}$ and let

$$p(z) = \left(\frac{1-\rho^2}{1-\rho^2 z}\right)^2 f'\left(e^{i\theta}\frac{\rho(1-z)}{1-\rho^2 z}\right), \quad z \in U.$$

Then p is holomorphic on $\overline{U}$ and, as in the proof of (i), p maps U into $U \setminus \{0\}$. Moreover, $p(1) = 1$ and $p'(1) = \pi/2$. (The value of ρ is chosen so that the latter condition is satisfied.) Hence we can find a function q which is holomorphic on U, maps U into Π, and satisfies $p(z) = \exp(-q(z))$, $q(1) = 0$ and $q'(1) = -\pi/2$. Again applying Corollary 4.3.3, we conclude that $q(x)$ lies in the disc with center $\pi(1-x)/[2(1+x)]$ and radius $\pi(1-x)/[2(1+x)]$.

On the other hand, since

$$\min\left\{\text{Re }\exp(-\zeta) : \left|\zeta - \frac{\pi}{2}\frac{1-x}{1+x}\right| \le \frac{\pi}{2}\frac{1-x}{1+x}\right\} > 0,$$

for all $x \in (0,1]$, we conclude that $\text{Re } p(x) > 0$ for all $x \in (0,1]$, and hence $\text{Re } f'(e^{i\theta}t) > 0$ for all $t \in [0,\rho]$. This completes the proof.

(An improvement of the constant in part (iii) of Theorem 4.3.5 (to $\approx$ 0.6654) was recently obtained by Chen [CheH].)

As shown by Liu and Minda [Liu-Min], the lower bound $\mathbf{B}_0 \ge 1/2$ follows directly from the distortion theorem, and strict inequality follows from a knowledge of the extremal functions for the distortion theorem.

Corollary 4.3.6. *Assume $f \in \mathcal{B}_0$. Then $r(0,f) \ge 1/2$ with equality if and only if $f(z) = e^{i\theta} F(e^{-i\theta}z)$ for some real θ, where F is given by (4.3.2). Moreover, $\mathbf{B}_0 > 1/2$.*

Proof. Let $\Omega \subseteq U$ be the domain containing 0 which is mapped conformally onto $U_{r(0,f)}$ by f. Since f is locally univalent, the boundary of $U_{r(0,f)}$ must meet the boundary of the Riemann surface $f(U)$. Thus there is a line segment Γ in $U_{r(0,f)}$ joining 0 to a boundary point of $f(U)$. Then $\gamma = (f|_\Omega)^{-1} \circ \Gamma$ is an arc beginning at 0 which tends to the boundary of U. Using part (i) of Theorem 4.3.5, we deduce that

$$r(0,f) = \int_\Gamma |dw| = \int_\gamma |f'(\zeta)||d\zeta| \ge \int_0^1 F'(|\zeta|)d|\zeta| = \frac{1}{2}.$$

It is clear that equality occurs if and only if f is a rotation of F.

We now prove that $\mathbf{B_0} > 1/2$. Landau's reduction together with a normal family argument shows that we can find a function $f \in \mathcal{B}_0$ such that $r(f) = \mathbf{B_0}$. The first part of the proof implies that $r(0, f) \geq 1/2$. If $r(0, f) > 1/2$ then we are done. On the other hand, if $r(0, f) = 1/2$ then f is a rotation of F, and from the discussion in Example 4.3.4 we have $r(f) = r(F) = e/4 > 1/2$. This completes the proof.

4.3.2   The case of convex functions

We end this chapter with a classical theorem of Szegö [Sze] which solves the Bloch constant problem for the case of normalized convex functions. We also mention a related theorem of Graham and Varolin [Gra-Var2].

Theorem 4.3.7. *If $f \in K$ then $f(U)$ contains a disc of radius $\pi/4$ and this result is sharp. The extremal functions are given by*

$$f(z) = \frac{1}{2} \log \left[\frac{1+z}{1-z} \right], \quad z \in U,$$

and its rotations.

Proof. There are at least three different ways of proving this result. Szegö's proof used a subordination argument to reduce to the case of conformal maps onto infinite strips or triangles. Computations show that a strip mapping is the extremal case.

A differential geometric proof using Ahlfors' method was given by Zhang [ZhaM] and rediscovered by Minda [Min3].

The third proof is based on Landau's reduction, which shows that it is sufficient to consider the case of normalized convex functions with vanishing second coefficient (cf. Remark 4.2.3). The covering theorem for such functions (Corollary 2.2.14) states precisely that the image contains a disc of radius $\pi/4$ centered at 0.

One can prove an analogous theorem for normalized convex functions with $a_2 = \ldots = a_k = 0$. This condition holds in particular for convex functions with k-fold symmetry. The following result was obtained by Graham and Varolin [Gra-Var2] and uses Theorem 2.2.12.

Theorem 4.3.8. *For the class of normalized convex functions of the form* $f(z) = z + a_{k+1}z^{k+1} + a_{k+2}z^{k+2} + \ldots, z \in U,$ *the Bloch constant coincides with the Koebe constant and has the value*

$$r_k = \int_0^1 \frac{dt}{(1+t^k)^{2/k}}, \quad k \geq 2.$$

Extremal functions are given by

$$f_k(z) = \begin{cases} \dfrac{1}{2} \log \left[\dfrac{1+z}{1-z}\right], & k = 2 \\[3mm] \displaystyle\int_0^z \dfrac{dt}{(1-t^k)^{2/k}}, & k \geq 3. \end{cases}$$

Problems

4.3.1. Prove the inequality in (4.3.3).

4.3.2. Prove (4.3.4). Why is the restriction $x \in [0, 1]$ needed ?

Chapter 5

Linear invariance in the unit disc

5.1 General ideas concerning linear-invariant families

In this chapter we shall investigate some growth and distortion results for locally univalent functions in the unit disc. The setting of this work is Pommerenke's theory of linear-invariant families of locally univalent functions

$$f(z) = z + a_2 z^2 + a_3 z^3 + \ldots, \quad z \in U.$$

(See [Pom1], [Pom2].)

The basic notion in this theory is the order of a given linear-invariant family, introduced by Pommerenke [Pom1]. We shall give a number of applications of this notion, including generalizations of some classical results for the set S which have been studied in the previous chapters.

Let $\mathcal{L}S$ denote the set of normalized locally univalent functions on the unit disc U. If $f \in \mathcal{L}S$ and ϕ is an automorphism of U, we let $\Lambda_\phi(f)$ denote the Koebe transform of f with respect to ϕ, i.e.

$$\Lambda_\phi(f)(z) = \frac{(f \circ \phi)(z) - (f \circ \phi)(0)}{(f \circ \phi)'(0)}, \quad z \in U.$$

We note that the Koebe transform has the following group property:

$$\Lambda_{\phi\circ\psi} = \Lambda_\psi \circ \Lambda_\phi.$$

We recall that $\mathrm{Aut}(U)$ denotes the set of holomorphic automorphisms of the unit disc.

Definition 5.1.1. The family $\mathcal{F}$ is called a *linear-invariant family* (L.I.F.) if

(i) $\mathcal{F} \subseteq \mathcal{LS}$,

(ii) $\Lambda_\phi(f) \in \mathcal{F}$, for all $f \in \mathcal{F}$ and $\phi \in \mathrm{Aut}(U)$.

Definition 5.1.2. For a linear-invariant family $\mathcal{F}$, let

$$\mathrm{ord}\,\mathcal{F} = \sup\left\{\left|\frac{f''(0)}{2}\right| : f \in \mathcal{F}\right\}$$

denote the *order* of the L.I.F. $\mathcal{F}$.

Evidently, if $\mathcal{F}$ is a compact family, then $\mathrm{ord}\,\mathcal{F} < \infty$.

The first result of this chapter is due to Pommerenke [Pom1] and it is useful in many applications.

Lemma 5.1.3. *Let $\mathcal{F}$ be a linear-invariant family and $\alpha = \mathrm{ord}\,\mathcal{F}$. Then*

$$(5.1.1) \qquad \alpha = \sup_{f\in\mathcal{F}}\sup_{|\zeta|<1}\left|-\bar\zeta + \frac{1}{2}(1-|\zeta|^2)\frac{f''(\zeta)}{f'(\zeta)}\right|.$$

Proof. Fix $\zeta \in U$. Given $f \in \mathcal{F}$, we consider the Koebe transform $f(z;\zeta)$ of f with respect to the disc automorphism $\phi(z) = (z+\zeta)/(1+\bar\zeta z)$, i.e.

$$(5.1.2) \qquad f(z;\zeta) = \frac{f\left(\dfrac{z+\zeta}{1+\bar\zeta z}\right) - f(\zeta)}{(1-|\zeta|^2)f'(\zeta)}.$$

Letting $z^* = \dfrac{z+\zeta}{1+\bar\zeta z}$, it is clear that

$$(5.1.3) \qquad \frac{\partial f}{\partial z}(z;\zeta) = f'(z;\zeta) = \frac{f'(z^*)}{(1+\bar\zeta z)^2 f'(\zeta)}.$$

Hence

$$(5.1.4) \qquad \frac{\partial^2 f}{\partial z^2}(z;\zeta) = f''(z;\zeta) = -\frac{2\bar\zeta}{(1+\bar\zeta z)^3}\frac{f'(z^*)}{f'(\zeta)} + \frac{1-|\zeta|^2}{(1+\bar\zeta z)^4}\frac{f''(z^*)}{f'(\zeta)},$$

so that

$$\left| \frac{1}{2} f''(0; \zeta) \right| = \left| -\bar{\zeta} + \frac{1}{2}(1 - |\zeta|^2) \frac{f''(\zeta)}{f'(\zeta)} \right| \leq \alpha,$$

and thus

(5.1.5)
$$\sup_{f \in \mathcal{F}} \sup_{|\zeta| < 1} \left| -\bar{\zeta} + \frac{1}{2}(1 - |\zeta|^2) \frac{f''(\zeta)}{f'(\zeta)} \right| \leq \alpha.$$

On the other hand, since $\mathcal{F} \subset \mathcal{LS}$ we clearly have

$$\sup_{f \in \mathcal{F}} \sup_{|\zeta| < 1} \left| -\bar{\zeta} + \frac{1}{2}(1 - |\zeta|^2) \frac{f''(\zeta)}{f'(\zeta)} \right| \geq \sup_{f \in \mathcal{F}} \left| \frac{f''(0)}{2} \right| = \alpha,$$

and together with (5.1.5), this implies (5.1.1) as desired.

We now give some examples of L.I.F.'s in the unit disc (for further examples see [Pom1]).

Example 5.1.4. The class S of normalized univalent functions on U is a L.I.F.

Example 5.1.5. The class K of normalized convex functions on U is a L.I.F. The class C of normalized close-to-convex functions on U is also a L.I.F. However, the class S^* of normalized starlike functions on U is not a L.I.F.

Example 5.1.6. $\mathcal{LS}$ is a L.I.F. of infinite order. To see this, let $k > 0$ and $f(z) = [e^{kz} - 1]/k$, $z \in U$. Then it is obvious that $f \in \mathcal{LS}$ and $|f''(0)/2| = k/2 \to \infty$ as $k \to \infty$.

The universal L.I.F. $\mathcal{U}(\alpha)$, consisting of all L.I.F.'s contained in $\mathcal{LS}$ with order not greater than α, is also a L.I.F.

Example 5.1.7. Let $\mathcal{G}$ be a non-empty subset of $\mathcal{LS}$ and let

$$\Lambda[\mathcal{G}] = \left\{ \Lambda_\phi(f) : f \in \mathcal{G}, \ \phi \in \mathrm{Aut}(U) \right\}$$

denote the L.I.F. generated by $\mathcal{G}$. Clearly $\mathcal{G}$ is a L.I.F. if and only if $\Lambda[\mathcal{G}] = \mathcal{G}$.

(This construction gives a number of interesting examples of L.I.F.'s on U.)

We now consider a basic distortion and growth theorem for a L.I.F. of order α which generalizes many classical results from the theory of univalent functions on the unit disc. This result was obtained in 1964 by Pommerenke, together with other estimates for L.I.F.'s [Pom1, Satz 1.1]. We give only an upper bound in the growth result (5.1.7), since the general L.I.F. may include functions that have zeros in $0 < |z| < 1$.

Theorem 5.1.8. *Let $\mathcal{F}$ be a L.I.F. with $\alpha = \operatorname{ord}\mathcal{F} < \infty$. Also let $f \in \mathcal{F}$ and $|z| = r < 1$. Then*

$$(5.1.6) \qquad \left| \log\left[(1 - |z|^2)f'(z)\right] \right| \le \alpha \log\left[\frac{1+r}{1-r}\right]$$

and

$$(5.1.7) \qquad |f(z)| \le \frac{1}{2\alpha}\left[\left(\frac{1+r}{1-r}\right)^{\alpha} - 1\right].$$

Equality in (5.1.6) and (5.1.7) is achieved by the function

$$f(z) = \frac{1}{2\alpha}\left[\left(\frac{1+z}{1-z}\right)^{\alpha} - 1\right], \quad z \in U.$$

Remark 5.1.9. Taking real parts in (5.1.6) and exponentiating, we deduce the following distortion theorem for the L.I.F. $\mathcal{F}$ with $\operatorname{ord}\mathcal{F} = \alpha < \infty$:

$$(5.1.8) \qquad \frac{(1-r)^{\alpha-1}}{(1+r)^{\alpha+1}} \le |f'(z)| \le \frac{(1+r)^{\alpha-1}}{(1-r)^{\alpha+1}}, \quad |z| = r, \quad f \in \mathcal{F}.$$

Also, taking imaginary parts in (5.1.6), we obtain

$$|\arg f'(z)| \le \alpha \log\left[\frac{1+r}{1-r}\right], \quad |z| = r, \quad f \in \mathcal{F}.$$

We mention that Campbell [Cam] showed that equality in (5.1.8) holds if and only if f is a rotation of the generalized Koebe function, i.e.

$$(5.1.9) \qquad f(z) = \frac{e^{-i\theta}}{2\alpha}\left[\left(\frac{1 + ze^{i\theta}}{1 - ze^{i\theta}}\right)^{\alpha} - 1\right], \quad z \in U, \quad \theta \in \mathbb{R}.$$

We note that for $\alpha = 2$ in (5.1.8), the distortion result is the same as for the class S (cf. (1.1.6)).

The case $\alpha = 1$ is of special interest because, as we shall see, this is the minimum possible order and any L.I.F. with order 1 must be a subfamily of K.

We also mention that if $\mathcal{F}$ is a L.I.F. with $\operatorname{ord}\mathcal{F} = \alpha < \infty$, then the estimate (5.1.7) implies that $\mathcal{F}$ is a normal family.

Proof of Theorem 5.1.8. Taking into account the relation (5.1.1), we deduce that

$$\left|\frac{\partial}{\partial r}\log\left[(1-r^2)f'(re^{i\theta})\right]\right| = \left|-\frac{2r}{1-r^2} + e^{i\theta}\frac{f''(re^{i\theta})}{f'(re^{i\theta})}\right| \le \frac{2\alpha}{1-r^2}.$$

Integrating both sides of this inequality, we obtain (5.1.6).

Furthermore, since for $z = re^{i\theta}$ we have

$$f(z) = \int_0^r f'(\rho e^{i\theta})e^{i\theta}d\rho,$$

we obtain in view of the upper bound in (5.1.8) that

$$|f(z)| \leq \int_0^r |f'(\rho e^{i\theta})|d\rho$$

$$\leq \int_0^r \frac{(1+\rho)^{\alpha-1}}{(1-\rho)^{\alpha+1}}d\rho = \frac{1}{2\alpha}\left[\left(\frac{1+r}{1-r}\right)^\alpha - 1\right].$$

Hence we deduce the estimate (5.1.7).

Remark 5.1.10. Let $\mathcal{F}$ be a L.I.F. with $\operatorname{ord}\mathcal{F} = \alpha < \infty$, such that $\mathcal{F}$ is a subset of S. If $f \in \mathcal{F}$ and $|z| = r \in [0,1)$, then we also have the estimates (see [Pom1], [Cam], [Gon5])

$$(5.1.10) \qquad |f(z)| \geq \frac{1}{2\alpha}\left[1 - \left(\frac{1-r}{1+r}\right)^\alpha\right]$$

and

$$(5.1.11)\, \frac{1}{2\alpha}\left[1 - \left(\frac{1-r}{1+r}\right)^\alpha\right] \leq \frac{|f(z)|}{(1-r^2)|f'(z)|} \leq \frac{1}{2\alpha}\left[\left(\frac{1+r}{1-r}\right)^\alpha - 1\right].$$

Proof. In order to derive the lower estimate in (5.1.10), let

$$\rho(r) = \min\left\{|f(z)| : |z| = r\right\}.$$

Obviously, the closed disc $\overline{U}_{\rho(r)}$ is contained in the image of the closed disc $\overline{U}_r$. Next, let $z_1 \in U$, $|z_1| = r$, so that $|f(z_1)| = \rho(r)$. It is clear that the closed segment between 0 and $f(z_1)$ lies in the closed disc $\overline{U}_{\rho(r)}$. Denote by Γ this segment and let γ be the inverse image of Γ. Then γ is contained in $\overline{U}_r$ and moreover, using the lower bound in (5.1.8), we have

$$\rho(r) = |f(z_1)| = \int_\gamma |f'(z)||dz| \geq \int_\gamma \frac{(1-|z|)^{\alpha-1}}{(1+|z|)^{\alpha+1}}|dz|$$

$$\geq \int_0^r \frac{(1-t)^{\alpha-1}}{(1+t)^{\alpha+1}}dt = \frac{1}{2\alpha}\left[1 - \left(\frac{1-r}{1+r}\right)^\alpha\right].$$

Next, we prove (5.1.11). For this purpose, let $\zeta \in U$ and $f(z; \zeta)$ be given by (5.1.2). Then $f(z; \zeta) \in \mathcal{F}$ and from (5.1.7) and (5.1.10) we obtain

$$\frac{1}{2\alpha}\left[1 - \left(\frac{1 - |z|}{1 + |z|}\right)^{\alpha}\right] \le |f(z; \zeta)| \le \frac{1}{2\alpha}\left[\left(\frac{1 + |z|}{1 - |z|}\right)^{\alpha} - 1\right], \quad z \in U.$$

Letting $z = -\zeta$ in the above, we deduce (5.1.11), as desired. This completes the proof.

We remark that Campbell [Cam] extended the results of Theorems 1.1.8 and 1.1.9 to the theory of the universal L.I.F. $\mathcal{U}(\alpha)$.

The foregoing proof contains the following covering result for L.I.F.'s, due to Pommerenke [Pom1]:

Corollary 5.1.11. *If $\mathcal{F}$ is a L.I.F. with* $\mathrm{ord}\,\mathcal{F} = \alpha < \infty$ *and $f \in \mathcal{F}$, then $f(U_r)$ contains the schlicht disc U_ρ, where*

$$\rho = \rho(\alpha, r) = \frac{1}{2\alpha}\left[1 - \left(\frac{1 - r}{1 + r}\right)^{\alpha}\right],$$

for all $r \in (0, 1)$. In particular, $f(U)$ contains the schlicht disc $U_{1/(2\alpha)}$.

We now prove that for each L.I.F. $\mathcal{F}$ the order is always at least 1 (cf. [Pom1, Folgerung 1.1]).

Theorem 5.1.12. *Let $\mathcal{F}$ be a L.I.F. and let $\alpha = \mathrm{ord}\,\mathcal{F}$. Then $\alpha \ge 1$.*

Proof. Suppose $\alpha < 1$. Then from (5.1.8) we conclude that if $f \in \mathcal{F}$ then $|f'(z)| \to \infty$ as $|z| \to 1$. This gives a contradiction to the maximum principle for the holomorphic function $1/f'$. Therefore we must have $\alpha \ge 1$, as stated.

The following result provides another way to generate L.I.F.'s in the unit disc (see [Pom1, Satz 1.2]). To this end, let $\mathcal{W}$ denote the set of univalent functions φ such that $|\varphi(z)| < 1$, $z \in U$, i.e.

$$\mathcal{W} = \left\{\varphi : \ \varphi \in H_u(U), \ |\varphi(z)| < 1, \ z \in U\right\}.$$

Theorem 5.1.13. *Let $\mathcal{F}$ be a L.I.F. with* $\mathrm{ord}\,\mathcal{F} = \alpha < \infty$ *and let $\mathcal{M}$ denote the set of functions of the form*

$$\Lambda_\varphi(f)(z) = \frac{(f \circ \varphi)(z) - (f \circ \varphi)(0)}{(f \circ \varphi)'(0)}, \quad z \in U,$$

where $f \in \mathcal{F}$ and $\varphi \in \mathcal{W}$. Then $\mathcal{M}$ is a L.I.F. and $\mathrm{ord}\,\mathcal{M} = \max\{\alpha, 2\}$.

Proof. First we show that if $g \in \mathcal{M}$ and $\psi \in \mathcal{W}$, then $\Lambda_\psi(g) \in \mathcal{M}$. Now g must be of the form $\Lambda_\varphi(f)$, where $f \in \mathcal{F}$ and $\varphi \in \mathcal{W}$, so in view of the group property of the Koebe transform, we have

$$\Lambda_\psi(g) = \Lambda_\psi(\Lambda_\varphi(f)) = \Lambda_{\varphi \circ \psi}(f) \in \mathcal{M},$$

because $\varphi \circ \psi \in \mathcal{W}$. In particular, this is true if $\psi \in \mathrm{Aut}(U)$. It is clear that $\mathcal{M} \subset \mathcal{LS}$, and hence $\mathcal{M}$ is a L.I.F.

Now let $\beta = \mathrm{ord}\,\mathcal{M}$ and with g as above, let $\varphi(0) = \zeta$. Also let

$$T(z) = \frac{z - \zeta}{1 - \bar{\zeta}z}, \quad z \in U,$$

and let $\lambda = T \circ \varphi$ and $h = \Lambda_{T^{-1}}(f)$.

Then $\lambda(0) = 0$ and again using the group property of the Koebe transform, we obtain

$$g(z) = \Lambda_\varphi(f)(z) = \Lambda_\lambda(h)(z) = \frac{h(\lambda(z))}{\lambda'(0)}.$$

Hence

$$g'(z) = \frac{\lambda'(z)}{\lambda'(0)}h'(\lambda(z))$$

and

$$(5.1.12) \qquad \frac{g''(0)}{2} = \frac{1}{2}\frac{\lambda''(0)}{\lambda'(0)} + \frac{1}{2}h''(0)\lambda'(0).$$

Since $\dfrac{\lambda(z)}{\lambda'(0)}$ is a function in S and $\left|\dfrac{\lambda(z)}{\lambda'(0)}\right| < \dfrac{1}{|\lambda'(0)|}$, $z \in U$, it follows from Problem 1.1.1 that

$$\frac{1}{2}\left|\frac{\lambda''(0)}{\lambda'(0)}\right| \le 2(1 - |\lambda'(0)|).$$

Using (5.1.12), we conclude that

$$\frac{1}{2}|g''(0)| \le 2(1 - |\lambda'(0)|) + \alpha|\lambda'(0)| \le \max\{\alpha, 2\},$$

and therefore $\beta \le \max\{\alpha, 2\}$.

Now let $F(z) = z + a_2 z^2 + \ldots$ belong to $\mathcal{M}$ and choose ϕ to be

$$\phi(z) = \frac{z(1 - r)^2}{(1 - rz)^2}, \quad z \in U, \quad 0 < r < 1.$$

Then $\phi \in \mathcal{W}$ and

$$\phi(z) = (1-r)^2 z + 2r(1-r)^2 z^2 + \ldots, \quad z \in U.$$

(Compare with the Koebe function.)

After simple computations, we obtain

$$\Lambda_\phi(F)(z) = z + (2r + a_2(1-r)^2)z^2 + \ldots, \quad z \in U,$$

and letting $r \nearrow 1$, we observe that

$$\left| \frac{1}{2}[\Lambda_\phi(F)]''(0) \right| \to 2.$$

Hence $\beta \geq 2$. On the other hand, because $\mathcal{M} \supset \mathcal{F}$, we have $\beta \geq \alpha$, so $\beta \geq \max\{\alpha, 2\}$. This completes the proof.

5.2 Extremal problems and radius of univalence

5.2.1 Bounds for coefficients of functions in linear-invariant families

In Section 5.1 we have seen that the notion of linear invariance plays an important role in several problems of function theory of one complex variable. In this section we are going to study coefficient bounds as well as the radius of univalence, starlikeness, and convexity for L.I.F.'s of finite order.

We begin with the following estimate involving the second and third coefficients for functions in compact L.I.F.'s on the unit disc. This result was obtained by Pommerenke in 1964 [Pom1].

Theorem 5.2.1. Let $\mathcal{F}$ be a compact L.I.F. with $\alpha = \operatorname{ord} \mathcal{F}$. If $f(z) = z + a_2 z^2 + a_3 z^3 + \ldots$ is a function in $\mathcal{F}$ such that $a_2 = \alpha$, then

(5.2.1) $$a_3 = \frac{1}{3}(2\alpha^2 + 1).$$

Proof. Let $z, \zeta \in U$ and consider the Koebe transform $f(z; \zeta)$ given by (5.1.2). Expanding $f(z; \zeta)$ in powers of ζ and $\overline{\zeta}$ and taking into account (5.1.3) and (5.1.4), we deduce that

$$f(z; \zeta) = f(z) + (f'(z) - 1 - 2a_2 f(z))\zeta - z^2 f'(z)\overline{\zeta} + O(|\zeta|^2)$$

for $|\zeta|$ small. Now expanding in z, we obtain

$$f(z; \zeta) = z + [a_2 + (3a_3 - 2a_2^2)\zeta - \overline{\zeta} + O(|\zeta|^2)]z^2 + \dots.$$

Since f is such that $a_2 = \alpha$, the preceding expansion gives

$$|\alpha + (3a_3 - 2\alpha^2)\zeta - \overline{\zeta} + O(|\zeta|^2)| \le \alpha,$$

and hence

$$\alpha + \operatorname{Re}\,[(3a_3 - 2\alpha^2 - 1)\zeta] + O(|\zeta|^2) \le \alpha.$$

Considering small values of ζ, we see that this is only possible if (5.2.1) holds. This completes the proof.

Another bound for the coefficients of functions that belong to L.I.F.'s of finite order is given in the following [Pom1, Satz 2.4]. (Recall that $|a_3 - a_2^2| \le 1$ for functions in S and $|a_3 - a_2^2| \le 1/3$ for functions in K.)

Theorem 5.2.2. *Let $\mathcal{F}$ be a L.I.F. with* $\operatorname{ord} \mathcal{F} = \alpha < \infty$ *and let* $\lambda \in \mathbb{R}$. *Then*

(5.2.2)
$$\left| \left(\frac{2}{3} - \lambda \right) \alpha^2 + \frac{1}{3} \right|$$

$$\le \sup \left\{ |a_3 - \lambda a_2^2| : f \in \mathcal{F}, \ f(z) = z + a_2 z^2 + a_3 z^3 + \dots \right\}$$

$$\le \left| \frac{2}{3} - \lambda \right| \alpha^2 + \sqrt{3}\alpha + 1.$$

Proof. Since $\operatorname{ord} \mathcal{F} = \alpha < \infty$ we may assume that $\mathcal{F}$ is a compact family (see Problem 5.2.4). Choose $h \in \mathcal{F}$ such that $h(z) = z + b_2 z^2 + b_3 z^3 + \dots$ and $b_2 = \alpha$. Then from Theorem 5.2.1 we deduce that

$$b_3 - \lambda \alpha^2 = \left(\frac{2}{3} - \lambda \right) \alpha^2 + \frac{1}{3},$$

and hence we obtain the lower bound in (5.2.2).

We now let $f(z) = z + a_2 z^2 + a_3 z^3 + \dots \in \mathcal{F}$, and consider

$$\frac{f''(z)}{f'(z)} = 2a_2 + (3a_3 - 2a_2^2)z + \dots, \quad z \in U.$$

Applying Lemma 5.1.3, we conclude that

$$|3a_3 - 2a_2^2|r = \frac{1}{2\pi} \left| \int_0^{2\pi} \left[e^{-i\theta} \frac{f''(re^{i\theta})}{f'(re^{i\theta})} \right] d\theta \right| \le 2\frac{\alpha + r}{1 - r^2},$$

for $0 < r < 1$. The optimal bound on $|3a_3 - 2a_2^2|$ is obtained when $r = 1/\sqrt{3}$, namely

$$|3a_3 - 2a_2^2| \leq 3\sqrt{3}\left(\alpha + \frac{1}{\sqrt{3}}\right).$$

The triangle inequality now gives

$$|a_3 - \lambda a_2^2| \leq \left|\frac{2}{3} - \lambda\right||a_2|^2 + \left|a_3 - \frac{2}{3}a_2^2\right| \leq \left|\frac{2}{3} - \lambda\right|\alpha^2 + \sqrt{3}\alpha + 1,$$

which completes the proof.

5.2.2 Radius problems for linear-invariant families

We now give some results involving linear invariance and radius problems. In 1964 Pommerenke [Pom1] obtained the following remarkable result concerning the radius of convexity $r_c(\mathcal{F})$ of a linear-invariant family $\mathcal{F}$. Recall that $r_c(\mathcal{F})$ is the largest number such that every function in the set $\mathcal{F}$ is convex on $U_{r_c(\mathcal{F})}$.

Theorem 5.2.3. *Let $\mathcal{F}$ be a linear-invariant family with $\alpha = \text{ord}\,\mathcal{F} < \infty$. Then*

$$(5.2.3) \qquad\qquad r_c(\mathcal{F}) = \alpha - \sqrt{\alpha^2 - 1}.$$

Remark 5.2.4. For $\alpha = 2$ we obtain the familiar radius of convexity for the class S, i.e. $r_c(S) = 2 - \sqrt{3}$ (see Theorem 2.2.22). We remark that in [Cam-Zi] the authors obtained the radius of close-to-convexity of various linear-invariant families.

Proof of Theorem 5.2.3. Let $f \in \mathcal{F}$. Using the relation (5.1.1), we obtain

$$\text{Re}\left[1 + \frac{zf''(z)}{f'(z)}\right] \geq \frac{1 - 2\alpha|z| + |z|^2}{1 - |z|^2}.$$

For $|z| < \alpha - \sqrt{\alpha^2 - 1}$ one deduces that

$$(5.2.4) \qquad\qquad \text{Re}\left[1 + \frac{zf''(z)}{f'(z)}\right] > 0,$$

and thus, from Theorem 2.2.3 we conclude that f is convex on U_ρ, where $\rho = \rho(\alpha) = \alpha - \sqrt{\alpha^2 - 1}$. Therefore $r_c(\mathcal{F}) \geq \rho(\alpha)$.

We now show that $r_c(\mathcal{F}) \leq \rho(\alpha)$. To this end, let $f \in \mathcal{F}$ and let ζ with $|\zeta| < r = r_c(\mathcal{F})$. Consider the map $f(z;\zeta)$ given by (5.1.2) and set $z = -\zeta$. Using (5.1.3), (5.1.4) and (5.2.4), we conclude that

$$\text{Re} \left[\frac{2|\zeta|^2}{1 - |\zeta|^2} - \frac{\zeta}{1 - |\zeta|^2} f''(0) \right] = \text{Re} \left[-\frac{\zeta f''(-\zeta;\zeta)}{f'(-\zeta;\zeta)} \right] > -1.$$

Therefore $\text{Re}\, [\zeta f''(0)] < 1 + |\zeta|^2$, and since ζ is arbitrary chosen on the disc U_r, we deduce by letting $|\zeta| \to r$ that

$$r|f''(0)| \leq 1 + r^2.$$

Consequently,

$$\alpha = \sup_{f \in \mathcal{F}} \left| \frac{1}{2} f''(0) \right| \leq \frac{1 + r^2}{2r},$$

that is $r \leq \alpha - \sqrt{\alpha^2 - 1}$. Thus we obtain $r_c(\mathcal{F}) = \rho(\alpha)$, as claimed. This completes the proof.

Corollary 5.2.5. *Let $\mathcal{F}$ be a L.I.F. with $\alpha = \text{ord}\,\mathcal{F}$. Then $\mathcal{F}$ is a subfamily of K if and only if $\alpha = 1$.*

This result was obtained in [Pom1].

Another important consequence of Theorem 5.2.3 is the following (see [Pom1, Folgerung 2.5]):

Corollary 5.2.6. *Let $\mathcal{F}$ be a L.I.F. with $\alpha = \text{ord}\,\mathcal{F} < \infty$. Then each $f \in \mathcal{F}$ maps the disc $U_{1/\alpha}$ conformally onto a starlike domain with respect to zero.*

Proof. Let z_0 be given with $|z_0| < 1/\alpha$ and choose ζ such that $z_0 = 2\zeta/(1 + |\zeta|^2)$. Then it is clear that $|\zeta| < \alpha - \sqrt{\alpha^2 - 1}$. If $f(z;\zeta)$ denotes the function given by (5.1.2), then from Theorem 5.2.3 we know that $f(z;\zeta)$ is convex for $|z| < \alpha - \sqrt{\alpha^2 - 1}$. Therefore $f(z;\zeta) - f(-\zeta;\zeta)$ is starlike with respect to $-\zeta$ for $|z| < \alpha - \sqrt{\alpha^2 - 1}$, and hence from Theorem 2.2.2 we deduce that

$$\text{Re} \left[\frac{z f'(z;\zeta)}{f(z;\zeta) - f(-\zeta;\zeta)} \right] > 0$$

for $|\zeta| \leq |z| < \alpha - \sqrt{\alpha^2 - 1}$. Moreover, since $z_0 = \dfrac{2\zeta}{1 + |\zeta|^2}$, we obtain

$$\text{Re} \left[z_0 \frac{f'(z_0)}{f(z_0)} \right] = 2 \frac{1 + |\zeta|^2}{1 - |\zeta|^2} \text{Re} \left[\frac{\zeta f'(\zeta;\zeta)}{f(\zeta;\zeta) - f(-\zeta;\zeta)} \right] > 0.$$

Because z_0 is arbitrary, we deduce that f is starlike on $U_{1/\alpha}$.

We finish this section by obtaining a relation between the radius of univalence of a L.I.F. $\mathcal{F}$ and its radius of nonvanishing.

Let r_0 denote the largest positive number such that no function in the L.I.F. $\mathcal{F}$ has zeros on $U_{r_0} \setminus \{0\}$. Also let r_1 denote the largest number such that every function in the L.I.F. $\mathcal{F}$ is univalent on U_{r_1}. Then we have the following result, due to Pommerenke [Pom1]:

Theorem 5.2.7. *If $\mathcal{F}$ is a L.I.F. of finite order, then* $r_1 = \dfrac{r_0}{1 + \sqrt{1 - r_0^2}}$.

Proof. Let $f \in \mathcal{F}$ and $r \leq \dfrac{r_0}{1 + \sqrt{1 - r_0^2}}$. Also let z_1, z_2 be such that $|z_1| < r$, $|z_2| < r$, $z_1 \neq z_2$. Then

$$0 < \left| \frac{z_1 - z_2}{1 - \bar{z}_1 z_2} \right| < \frac{2r}{1 + r^2} \leq r_0.$$

Since $f(z; z_1) \in \mathcal{F}$, where $f(z; \zeta)$ is defined by (5.1.2), we have

$$\frac{f(z_2) - f(z_1)}{(1 - |z_1|^2)f'(z_1)} = f\left(\frac{z_2 - z_1}{1 - \bar{z}_1 z_2}; z_1 \right) \neq 0.$$

This implies $f(z_1) \neq f(z_2)$, and thus f is univalent on U_r. Hence

$$r_1 \geq \frac{r_0}{1 + \sqrt{1 - r_0^2}}.$$

In order to prove the converse, let $z_0, \zeta \in U$ be such that $0 < |z_0| < \dfrac{2r_1}{1 + r_1^2}$ and $z_0 = \dfrac{2\zeta}{1 + |\zeta|^2}$. Then $0 < |\zeta| < r_1$ and

$$f(z_0) = f\left(\frac{2\zeta}{1 + |\zeta|^2} \right) = (1 - |\zeta|^2)f'(\zeta)\left[f(\zeta; \zeta) - f(-\zeta; \zeta) \right] \neq 0.$$

This implies that $r_0 \geq \dfrac{2r_1}{1 + r_1^2}$, i.e. $r_1 \leq \dfrac{r_0}{1 + \sqrt{1 + r_0^2}}$. This completes the proof.

Remark 5.2.8. Linear invariance is closely related to the concepts of uniform local univalence and uniform local convexity. These latter notions are defined with respect to hyperbolic geometry on the unit disc U. If $d_h(a, b)$ is the hyperbolic distance between $a, b \in U$ (see (1.1.12)), then $U_h(a, \rho) = \{z \in$

$U : d_h(a, z) < \rho\}$ denotes the hyperbolic disc in U with center a and radius ρ, $0 < \rho \leq \infty$.

For $f \in H(U)$, let $\rho(z, f)$ denote the hyperbolic radius of the largest hyperbolic disc in U centered at z in which f is univalent. Define

$$\rho(f) = \inf\{\rho(z, f) : z \in U\}.$$

The function f is called *uniformly locally univalent* (in the hyperbolic sense) provided $\rho(f) > 0$.

For a locally univalent function f in U, let

$$\rho_c(f) = \sup\left\{\rho : f \text{ is univalent on } U_h(a, \rho) \text{ and } f(U_h(a, \rho))\right.$$

$$\left. \text{is a convex domain, for all } a \in U\right\}.$$

Then $\rho_c(f) \leq \rho(f)$. The locally univalent function f is called *uniformly locally convex* (in the hyperbolic sense) provided $\rho_c(f) > 0$.

We remark that some of the radii problems we have just studied can be formulated in a more invariant way. Ma and Minda [Ma-Min2] proved the following beautiful result:

Theorem 5.2.9. *Given $\rho > 0$, the L.I.F. $\mathcal{K}(\rho)$ consisting of functions f in $\mathcal{LS}$ such that $\rho_c(f) \geq \rho$ coincides with the universal L.I.F. $\mathcal{U}(\coth(2\rho))$.*

In other words, linear invariance of finite order can be characterized geometrically.

Similarly the family $\mathcal{S}(\rho)$ of functions f in $\mathcal{LS}$ such that $\rho(f) \geq \rho$ is linear-invariant. Ma and Minda [Ma-Min2] (see also [Pom1]) gave upper and lower bounds for its order.

Notes. In addition to the basic papers of Pommerenke [Pom1], [Pom2], the reader may consult [Cam], [Cam-Cim-Pfa], [Cam-Pf], [Cam-Zi], [Gon5], [Koep1], [Ma-Min1], and [Ma-Min2] for material on L.I.F.'s.

Problems

5.2.1. Let f be a normalized locally univalent function on U. Also let

$$\sigma = \sup_{z \in U}(1 - |z|^2)^2|\{f; z\}|,$$

where $\{f; z\}$ denotes the Schwarzian derivative of f at z. Show that $f \in \mathcal{U}(\beta)$, where

$$\beta = \left(1 + \frac{1}{2}\sigma\right)^{1/2}$$

(Pommerenke, 1964 [Pom1].) (For other relations of this type see [Har3].)

5.2.2. Let $\mathcal{F}$ be a family of normalized locally univalent functions on U. For $0 \leq r < 1$, let

$$G(r, \mathcal{F}) = G(r) = \sup_{f \in \mathcal{F}} \max_{|z|=r} \arg f'(z),$$

where the argument varies continuously from the initial value of $\arg f'(0) = 0$. Show that if $\mathcal{F}$ is a linear-invariant family, then

$$G(r) = - \inf_{f \in \mathcal{F}} \min_{|z|=r} \arg f'(z).$$

(Campbell and Ziegler, 1974 [Cam-Zi].)

5.2.3. a) Let $\mathcal{F}$ be a linear-invariant family of finite order. Let $cl(\mathcal{F})$ denote the closure of $\mathcal{F}$ in the topology of uniform convergence on compact sets. Let $\mathcal{F}(*) = \{f(tz)/t : f \in \mathcal{F}, \ 0 < t \leq 1\}$. Show that

$$G(r, \mathcal{F}) = G(r, \mathcal{F}(*)) = G(r, cl(\mathcal{F})),$$

where $G(r, \mathcal{F})$ is defined in the previous problem. Moreover, show that $G(r, \mathcal{F})$ is an increasing continuous function of r, satisfying $2\arcsin r \leq G(r)$ for $0 \leq r < 1$.

b) If $\mathcal{F}$ is any linear-invariant family of normalized convex functions, show that $G(r) \equiv 2\arcsin r$.

(Campbell and Ziegler, 1974 [Cam-Zi].)

5.2.4. Let $\mathcal{F}$ be a L.I.F. such that $\text{ord}\,\mathcal{F} = \alpha$. Show that $cl(\mathcal{F})$ is a L.I.F. and $\text{ord}\,[cl(\mathcal{F})] = \alpha$.

5.2.5. Show that the class C of normalized close-to-convex functions on U is a L.I.F.

5.2.6. Let $\mathcal{F}$ be a L.I.F. such that $\text{ord}\,\mathcal{F} = \alpha < \infty$. Also let ρ be the radius of univalence of $\mathcal{F}$. Show that if $f \in \mathcal{F}$, then

$$\frac{1}{2\alpha}\left[1 - \left(\frac{1-|z|}{1+|z|}\right)^{\alpha}\right] \leq \min\left\{|f(z)|, \frac{|f(z)|}{(1-|z|^2)|f'(z)|}\right\}, \qquad |z| < \rho.$$

(Campbell, 1974 [Cam].)

5.2.7. Show that equality in (5.1.8) holds if and only if f is given by (5.1.9).
(Campbell, 1974 [Cam].)

5.2.8. Let $\mathcal{F}$ be a L.I.F. of finite order and $f \in \mathcal{F}$. Show that $f'(z)$ is a normal function.
(Pommerenke, 1964 [Pom1].)

Part II

Univalent mappings in several complex variables and complex Banach spaces

Chapter 6

Univalence in several complex variables

The purpose of the second part of this book is to study the theory of biholomorphic mappings of the unit ball and certain bounded domains in $\mathbb{C}^n$ and in complex Banach spaces. The domains will be such that precise results can be given for the solutions of extremal problems for holomorphic mappings – growth, distortion, and covering theorems, coefficient estimates, etc. This means that we shall be primarily interested in the Euclidean unit ball and the unit polydisc in $\mathbb{C}^n$, more generally in the unit ball with respect to an arbitrary norm, and in some cases bounded balanced pseudoconvex domains (see Section 6.1.2), where the growth of holomorphic functions may be measured relative to the Minkowski function.

For reasons which are explained in Section 6.1.7, we are obliged to focus on proper subclasses of the normalized biholomorphic mappings of such domains into the ambient space (which we denote by $S(B)$ in the case of the unit ball). As in one variable, such subclasses will frequently be defined by geometrical conditions which admit reformulations as analytic conditions. The most basic conditions of this type are of course starlikeness and convexity, and we shall study criteria for starlikeness and convexity in this chapter and extremal problems for these classes in the next. We shall also discuss a number of other classes which are generalizations of familiar subclasses of S to several variables – close-to-starlikeness (a generalization of close-to-convexity), spiral-likeness, etc. It is to be expected, however, that new subclasses of $S(B)$ which are not direct generalizations of previously-studied classes in one variable will also be of interest. For example, we consider the quasi-convex maps introduced by Roper and Suffridge [Rop-Su2] in Section 6.3 and Chapter 7.

A powerful tool for studying extremal problems for the class S in one variable is the Loewner method. We shall give a detailed account of the extension of this method to several variables, including recent improvements of the known existence theorems and new applications. In several variables the set of normalized biholomorphic maps which can be embedded in a nice way as the initial element of a Loewner chain (i.e. such that the initial map has a parametric representation) is a proper subclass of $S(B)$. It is obviously a subclass of special interest and it contains many of the other subclasses considered here.

We shall study Bloch mappings in several variables and also linear-invariant families. Finally we shall study extension operators which map particular subclasses of S to particular subclasses of $S(B)$, for the case that B is the Euclidean unit ball in $\mathbb{C}^n$. The most important such operator is the Roper-Suffridge operator, which extends a convex univalent function of one variable to a convex mapping of B and has many other nice properties.

6.1 Preliminaries concerning holomorphic mappings in $\mathbb{C}^n$ and complex Banach spaces

6.1.1 Holomorphic functions in $\mathbb{C}^n$

We shall begin with a brief discussion of holomorphic functions of several variables. We shall present some of the most important results about such functions, in some cases without complete proofs.

Let $\mathbb{C}^n$ denote the space of n-complex variables $z = (z_1, \ldots, z_n)$ where $z_j \in \mathbb{C}$, $1 \leq j \leq n$. Thus $\mathbb{C}^n$ is the Cartesian product of n copies of $\mathbb{C}$, and is a complex vector space: addition of two elements as well as the multiplication of an element of $\mathbb{C}^n$ by a complex scalar is defined componentwise. As a complex vector space $\mathbb{C}^n$ is n-dimensional, but as a real vector space it is $2n$-dimensional. Naturally, $\mathbb{C}^n$ may be identified with $\mathbb{R}^{2n}$ via the map

$$(x_1 + iy_1, \ldots, x_n + iy_n) \leftrightarrow (x_1, y_1, \ldots, x_n, y_n).$$

This identification gives the topology of $\mathbb{C}^n$: all the usual concepts from topology and analysis of the Euclidean space $\mathbb{R}^{2n}$ may be transferred to $\mathbb{C}^n$. The Euclidean inner product

$$\langle z, w \rangle = \sum_{j=1}^{n} z_j \overline{w}_j, \quad z, w \in \mathbb{C}^n,$$

and the associated Euclidean norm $\|z\| = \langle z, z \rangle^{1/2}$ make $\mathbb{C}^n$ into an n-dimensional complex Hilbert space. If $a \in \mathbb{C}^n$ and $r > 0$, we let $B(a, r) = \{z \in \mathbb{C}^n : \|z - a\| < r\}$ be the open ball of center a and radius r. The closure of $B(a, r)$ will be denoted by $\overline{B}(a, r)$ and its boundary by $\partial B(a, r)$. Instead of $B(0, r)$ we shall write B_r, and the open unit ball B_1 will be denoted simply by B and will be called the open Euclidean unit ball of $\mathbb{C}^n$. When it is necessary to mention the dimension explicitly, the ball B_r (resp. B) will be denoted by B_r^n (resp. B^n). In the case $n = 1$ we shall write U instead of B^1.

We shall also make use of other norms, retaining the notation B for the corresponding unit ball (except in certain special cases).

An open polydisc of center $a = (a_1, \ldots, a_n)$ and polyradius $R = (r_1, \ldots, r_n)$ is a subset of $\mathbb{C}^n$ of the form

$$P(a, R) = P(a_1, \ldots, a_n, r_1, \ldots, r_n) = \left\{ z \in \mathbb{C}^n : |z_j - a_j| < r_j, \ 1 \le j \le n \right\}.$$

Thus $P(a, R)$ is the Cartesian product of n discs. When $r_1 = \ldots = r_n = r$ we denote the polydisc $P(a, R)$ by $P(a, r)$. Instead of $P(0, r)$ we shall write P_r. The unit polydisc $P(0, 1)$ will be denoted by P and is the unit ball with respect to the maximum norm $\| \cdot \|_\infty$, $\|z\|_\infty = \max_{1 \le j \le n} |z_j|$.

We shall usually write vectors in components as row vectors; however, it is understood that the action of a matrix on a vector is given by writing the vector as a column vector. For vectors and matrices A, we shall denote the transpose of A by A^t, the complex conjugate of A by $\overline{A}$, and the conjugate transpose of A by A^*.

The set of bounded complex-linear operators from $\mathbb{C}^n$ into $\mathbb{C}^m$ will be denoted by $L(\mathbb{C}^n, \mathbb{C}^m)$; the identity in $L(\mathbb{C}^n, \mathbb{C}^n)$ is denoted by I.

If $\alpha = (\alpha_1, \ldots, \alpha_n)$ is an ordered n-tuple of nonnegative integers α_i, we shall use the standard multi-index notations

$$\alpha_1 + \cdots + \alpha_n = |\alpha|$$

$$(\alpha_1!) \cdots (\alpha_n!) = \alpha!$$

$$z_1^{\alpha_1} \cdots z_n^{\alpha_n} = z^\alpha, \quad z = (z_1, \ldots, z_n) \in \mathbb{C}^n.$$

Writing the coordinates z_j of a point $z \in \mathbb{C}^n$ in the form $z_j = x_j + iy_j$, $1 \le j \le n$, we introduce the differential operators

$$\frac{\partial}{\partial z_j} = \frac{1}{2}\left(\frac{\partial}{\partial x_j} - i\frac{\partial}{\partial y_j}\right) \quad \text{and} \quad \frac{\partial}{\partial \bar{z}_j} = \frac{1}{2}\left(\frac{\partial}{\partial x_j} + i\frac{\partial}{\partial y_j}\right),$$

which act on complex-valued functions of class C^1.

There are several equivalent ways of defining a holomorphic function on a domain in $\mathbb{C}^n$.

Definition 6.1.1. Let Ω be a domain in $\mathbb{C}^n$ and $f : \Omega \to \mathbb{C}$. We say that f is holomorphic if f is continuous on Ω and holomorphic in each variable separately, i.e. for each $z^0 = (z_1^0, \ldots, z_n^0) \in \Omega$ and $j \in \{1, \ldots, n\}$, the function

$$f(z_1^0, \ldots, z_{j-1}^0, \cdot, z_{j+1}^0, \ldots, z_n^0)$$

is holomorphic on the open set $\{\zeta \in \mathbb{C} : (z_1^0, \ldots, z_{j-1}^0, \zeta, z_{j+1}^0, \ldots, z_n^0) \in \Omega\}$. Such a function obviously satisfies the Cauchy-Riemann equations

$$\frac{\partial f}{\partial \bar{z}_j}(z) = 0, \quad z \in \Omega, 1 \le j \le n.$$

(The assumption of continuity in Definition 6.1.1 can be omitted, but it simplifies matters to make this assumption. See [Kran].)

Definition 6.1.2. The function $f : \Omega \to \mathbb{C}$ is holomorphic if for each point $a \in \Omega$ there is a mapping in $L(\mathbb{C}^n, \mathbb{C})$, denoted by $Df(a)$, such that

$$f(z) = f(a) + Df(a)(z - a) + o(z - a)$$

near a, where $\lim_{z \to a} \frac{o(z - a)}{\|z - a\|} = 0$.

Definition 6.1.3. A function $f : \Omega \to \mathbb{C}$ is holomorphic if for each point $a \in \Omega$ there is an open neighbourhood V of a such that f can be represented as a multiple power series expansion

$$(6.1.1) \qquad f(z) = \sum_{\alpha_1, \ldots, \alpha_n = 0}^{\infty} c_{\alpha_1 \ldots \alpha_n} (z_1 - a_1)^{\alpha_1} \ldots (z_n - a_n)^{\alpha_n}$$

that converges absolutely and uniformly on each compact subset of V. (One can choose V to be any polydisc centered at a that lies in Ω.)

Using multi-index notation, (6.1.1) can be written as

$$f(z) = \sum_{\alpha} c_{\alpha}(z - a)^{\alpha}.$$

We denote the set of holomorphic functions on a domain Ω in $\mathbb{C}^n$ by $H(\Omega, \mathbb{C})$.

If f is a holomorphic function in a neighbourhood of a closed polydisc $\overline{P}(a, R)$, then Cauchy's integral formula can be generalized as follows:

$$(6.1.2) \quad f(z) = \frac{1}{(2\pi i)^n} \int_{|\zeta_n - a_n| = r_n} \cdots \int_{|\zeta_1 - a_1| = r_1} \frac{f(\zeta_1, \ldots, \zeta_n) d\zeta_1 \ldots d\zeta_n}{(\zeta_1 - z_1) \ldots (\zeta_n - z_n)}$$

$$= \frac{1}{(2\pi i)^n} \int_{\partial_0 P(a, R)} \frac{f(\zeta) d\zeta}{(\zeta_1 - z_1) \ldots (\zeta_n - z_n)}, \quad z \in P(a, R).$$

This formula is known as *Cauchy's integral formula on the polydisc*. Note that if $n > 1$, the integration in (6.1.2) does not extend over the full boundary of $P(a, R)$, but only over a subset $\partial_0 P(a, R)$ of real dimension n called the *distinguished boundary*. (There are other versions of Cauchy's integral formula for domains in $\mathbb{C}^n$ with smooth boundary in which the integration is taken over the full boundary, but we shall not need them in this book. See [Hör2], [Kran], [Ran].)

By expanding the kernel in (6.1.2) as a multiple geometric series, one sees that the coefficients in (6.1.1) are given by

$$(6.1.3) \quad c_{\alpha} = c_{\alpha_1 \ldots \alpha_n} = \frac{1}{(2\pi i)^n} \int_{\partial_0 P(a, R)} \frac{f(\zeta_1, \ldots, \zeta_n) d\zeta_1 \ldots d\zeta_n}{(\zeta_1 - a_1)^{\alpha_1 + 1} \ldots (\zeta_n - a_n)^{\alpha_n + 1}}.$$

On the other hand, by differentiating under the integral sign in (6.1.2) we obtain, using multi-index notation,

$$(6.1.4) \qquad \frac{\partial^{|\alpha|} f}{\partial z^{\alpha}}(z) = \frac{\alpha!}{(2\pi i)^n} \int_{\partial_0 P(a,R)} \frac{f(\zeta_1, \ldots, \zeta_n) d\zeta_1 \ldots d\zeta_n}{(\zeta_1 - z_1)^{\alpha_1+1} \ldots (\zeta_n - z_n)^{\alpha_n+1}},$$

for any ordered n-tuple $\alpha = (\alpha_1, \ldots, \alpha_n)$ of nonnegative integers α_i and $z \in P(a, R)$.

Comparing this formula with (6.1.3), one deduces that

$$\alpha! c_\alpha = \frac{\partial^{|\alpha|} f}{\partial z^{\alpha}}(a).$$

Therefore, the expansion (6.1.1) is unique. Moreover, holomorphic functions have derivatives of all orders.

Some additional consequences of the Cauchy integral formula on the polydisc are given in the following theorems:

Theorem 6.1.4. (Cauchy estimates on the polydisc) *Let Ω be an open neighbourhood of a closed polydisc $\overline{P}(a, R)$ in $\mathbb{C}^n$ and let $f : \Omega \to \mathbb{C}$ be a holomorphic function. If $|f(z)| \le M$ for all $z \in \overline{P}(a, R)$ and $\alpha = (\alpha_1, \ldots, \alpha_n) \in \mathbb{N}_0^n$ (where $\mathbb{N}_0 = \mathbb{N} \cup \{0\}$), then*

$$\left| \frac{\partial^{\alpha_1 + \cdots + \alpha_n} f}{\partial z_1^{\alpha_1} \ldots \partial z_n^{\alpha_n}}(a) \right| \le \frac{M(\alpha_1!) \cdots (\alpha_n!)}{r_1^{\alpha_1} \ldots r_n^{\alpha_n}},$$

or in multi-index notation,

$$(6.1.5) \qquad \left| \frac{\partial^{|\alpha|} f}{\partial z^{\alpha}}(a) \right| \le M(\alpha!) R^{-\alpha},$$

where $R^{-\alpha} = r_1^{-\alpha_1} \cdots r_n^{-\alpha_n}$.

Theorem 6.1.5. (Maximum modulus theorem) *Let Ω be a domain in $\mathbb{C}^n$ and $f : \Omega \to \mathbb{C}$ be a holomorphic function. If $|f|$ achieves its maximum at an interior point of Ω, then f is constant.*

6.1.2   Classes of domains in $\mathbb{C}^n$. Pseudoconvexity

We begin with domains which have certain types of rotational symmetry.

Definition 6.1.6. Let Ω be a domain in $\mathbb{C}^n$.

(i) We say that Ω is *circular* if $e^{i\theta}z \in \Omega$ whenever $z \in \Omega$ and $\theta \in \mathbb{R}$.

(ii) We say that Ω is *complete circular* or *balanced* if $\lambda z \in \Omega$ whenever $z \in \Omega$ and $\lambda \in \overline{U}$.

(iii) We say that Ω is a *Reinhardt domain* if, whenever $(z_1, \ldots, z_n) \in \Omega$ and $\theta_j \in \mathbb{R}$, $j = 1, \ldots, n$, we have $(e^{i\theta_1}z_1, \ldots, e^{i\theta_n}z_n) \in \Omega$.

(iv) Finally, we say that Ω is a *complete Reinhardt domain* if, whenever $(z_1, \ldots, z_n) \in \Omega$ and $\lambda_j \in \overline{U}$, $j = 1, \ldots, n$, we have $(\lambda_1 z_1, \ldots, \lambda_n z_n) \in \Omega$.

Obviously the complete Reinhardt domains in $\mathbb{C}^n$ are those domains which can be written as the union of polydiscs centered at 0. Such domains arise in the study of power series expansions: the largest domain in which a power series $\sum a_\alpha z^\alpha$ converges absolutely is a complete Reinhardt domain.

For balanced domains in $\mathbb{C}^n$, there is an important auxiliary function known as the Minkowski function:

Definition 6.1.7. Suppose that Ω is a balanced domain in $\mathbb{C}^n$. The Minkowski function of Ω is defined for $z \in \mathbb{C}^n$ by

$$h(z) = \inf\left\{t > 0 : \frac{z}{t} \in \Omega\right\}.$$

Clearly $\Omega = \{z \in \mathbb{C}^n : h(z) < 1\}$. It can be shown that $h : \mathbb{C}^n \to [0, \infty)$ is upper semicontinuous, and it is clear that $h(\lambda z) = |\lambda|h(z)$ for $\lambda \in \mathbb{C}$, $z \in \mathbb{C}^n$, and $h \equiv 0$ iff $\Omega = \mathbb{C}^n$. When Ω is a bounded balanced convex domain in $\mathbb{C}^n$, it is well known that h is a norm.

We shall also consider certain pseudoconvex domains in this book. Pseudoconvexity is a basic concept in several complex variables and is defined in terms of plurisubharmonic functions.

Definition 6.1.8. Let D be a domain in $\mathbb{C}$ and let $\varphi : D \to [-\infty, \infty)$ be upper semicontinuous. We say that φ is *subharmonic* if for every $z \in D$ and $\rho > 0$ such that $\overline{U}(z, \rho) \subset D$ and for every function g, continuous on $\overline{U}(z, \rho)$ and harmonic on $U(z, \rho)$ such that $\varphi \leq g$ on $\partial U(z, \rho)$, we have $\varphi \leq g$ on $U(z, \rho)$.

Definition 6.1.9. Let Ω be a domain in $\mathbb{C}^n$ and let $\varphi : \Omega \to [-\infty, \infty)$ be an upper semicontinuous function on Ω. We say that φ is *plurisubharmonic* (psh) on Ω if for each $z \in \Omega$ and $w \in \mathbb{C}^n$ the function of one complex variable

$$\zeta \mapsto \varphi(z + \zeta w) \in [-\infty, \infty)$$

is subharmonic where it is defined.

We note the following properties of psh functions:

Theorem 6.1.10. *Let $\Omega \subset \mathbb{C}^n$ and $\Omega' \subset \mathbb{C}^m$ be domains.*

(i) If $f \in H(\Omega, \mathbb{C})$ then $|f|^\alpha$, where $\alpha > 0$, and $\log |f|$ are psh on Ω.

(ii) Let φ be a real-valued function of class C^2 on Ω. Then φ is psh on Ω iff the complex Hessian

$$\sum_{j,k=1}^n \frac{\partial^2 \varphi}{\partial z_j \partial \overline{z}_k}(z) w_j \overline{w}_k$$

is positive semi-definite at each point z of Ω.

(iii) If ψ is psh on Ω' and $f : \Omega \to \Omega'$ is holomorphic then $\psi \circ f$ is psh on Ω.

Definition 6.1.11. Let Ω be a domain in $\mathbb{C}^n$ and for $z \in \Omega$ let $\delta_\Omega(z)$ denote the Euclidean distance from z to $\partial\Omega$. We say that Ω is *pseudoconvex* if the function $-\log \delta_\Omega(\cdot)$ is psh on Ω.

The importance of pseudoconvexity stems from the study of the Levi problem (one of the main lines of development of the early theory of several complex variables) and its solution: a domain $\Omega \subset \mathbb{C}^n$ is pseudoconvex iff it is a domain of holomorphy, i.e. the natural domain of definition of some holomorphic function. For further details, see [Gun-Ros], [Hör2], [Kran], or [Ran].

For balanced domains, pseudoconvexity can be characterized in terms of the Minkowski function:

Theorem 6.1.12. *Let Ω be a balanced domain in $\mathbb{C}^n$ with Minkowski function h. Then Ω is pseudoconvex iff h is psh.*

For a proof, see [Din1, Lemma 2].

As shown by Hamada and Kohr and as we shall see in this book, quite a number of results in geometric function theory on the unit ball in $\mathbb{C}^n$ can be generalized to balanced pseudoconvex domains in $\mathbb{C}^n$ whose Minkowski function is C^1 on $\mathbb{C}^n \backslash \{0\}$. (Caveat: A balanced domain with C^1 boundary need not have a Minkowski function which is C^1 on $\mathbb{C}^n \backslash \{0\}$, or even continuous. See [Ham4].)

6.1.3 Holomorphic mappings

Let Ω be a domain in $\mathbb{C}^n$ and let $f : \Omega \to \mathbb{C}^m$ be a mapping. By writing $f = (f_1, \ldots, f_m)$ and $f_k = u_k + iv_k$, $1 \leq k \leq m$, where $u_k, v_k : \Omega \to \mathbb{R}$, we may identify f with the mapping from $\Omega \subset \mathbb{R}^{2n}$ into $\mathbb{R}^{2m}$ given by

$$(x_1, y_1, \ldots, x_n, y_n) \mapsto (u_1, v_1, \ldots, u_m, v_m).$$

The mapping f is called *holomorphic* on Ω if each of its components $f_1, \ldots, f_m$ is holomorphic on Ω. In this case the differential $Df(z)$ at $z \in \Omega$ is a complex linear mapping from $\mathbb{C}^n$ into $\mathbb{C}^m$, which can be identified with the (complex) Jacobian matrix

$$(6.1.6) \qquad Df(z) = \begin{bmatrix} \dfrac{\partial f_1}{\partial z_1}(z) & \cdots & \dfrac{\partial f_1}{\partial z_n}(z) \\ \cdots & \cdots & \cdots \\ \dfrac{\partial f_m}{\partial z_1}(z) & \cdots & \dfrac{\partial f_m}{\partial z_n}(z) \end{bmatrix},$$

and the relation

$$f(z + h) = f(z) + Df(z)h + o(\|h\|)$$

holds for h in a neighbourhood of the origin of $\mathbb{C}^n$.

We denote the set of holomorphic mappings from Ω into a domain $\Omega' \subset \mathbb{C}^m$ by $H(\Omega, \Omega')$. We shall write $H(\Omega)$ in place of $H(\Omega, \mathbb{C}^n)$ (the equidimensional case). If $0 \in \Omega$, a mapping $f \in H(\Omega)$ will be said to be *normalized* if $f(0) = 0$ and $Df(0) = I$.

Since our objective is to study univalent mappings, we begin by recalling the inverse mapping theorem.

If $f : \Omega \subset \mathbb{C}^n \to \mathbb{C}^n$ is holomorphic, we say that f *is nonsingular* at $z_0 \in \Omega$ if $Df(z_0)$ is invertible. The mapping f is nonsingular on Ω if it is nonsingular at each $z \in \Omega$.

Theorem 6.1.13. (Inverse mapping theorem) *If Ω is a domain in $\mathbb{C}^n$ and $f : \Omega \to \mathbb{C}^n$ is a holomorphic mapping such that $Df(z_0)$ is invertible for some point z_0 of Ω, then there are neighbourhoods V of z_0 and W of $f(z_0)$ such that f is a one-to-one map of V onto W whose inverse is holomorphic on W.*

If $f : \Omega \subset \mathbb{C}^n \to \mathbb{C}^n$ is a holomorphic mapping on the domain Ω, let $J_f(z) = \det Df(z)$, $z \in \Omega$, denote the (complex) Jacobian determinant of f at z. Also let $D_r f(z)$ be the real Jacobian matrix of f at $z \in \Omega$, that is

$$(6.1.7) \qquad D_r f(z) = \begin{bmatrix} \dfrac{\partial u_1}{\partial x_1} & \dfrac{\partial u_1}{\partial y_1} & \cdots & \dfrac{\partial u_1}{\partial y_n} \\ \dfrac{\partial v_1}{\partial x_1} & \dfrac{\partial v_1}{\partial y_1} & \cdots & \dfrac{\partial v_1}{\partial y_n} \\ \cdots & \cdots & \cdots & \cdots \\ \dfrac{\partial v_n}{\partial x_1} & \dfrac{\partial v_n}{\partial y_1} & \cdots & \dfrac{\partial v_n}{\partial y_n} \end{bmatrix}$$

and $J_r f(z) = \det D_r f(z)$. Then elementary computations using matrix theory in (6.1.6) and (6.1.7) give the relation

$$(6.1.8) \qquad J_r f(z) = |J_f(z)|^2 \geq 0, \quad z \in \Omega.$$

We say that f is *locally biholomorphic* on Ω if $J_f(z) \neq 0$, $z \in \Omega$. In this case, f *preserves orientation*.

We say that $f : \Omega \subset \mathbb{C}^n \to \mathbb{C}^n$ is *biholomorphic* on Ω if f is a holomorphic map from Ω onto a domain $\Omega' \subset \mathbb{C}^n$ and has a holomorphic inverse defined on Ω'. In this case the domains Ω and Ω' are called *biholomorphically equivalent*.

Let $S(B)$ be the set of normalized biholomorphic mappings from the unit ball B of $\mathbb{C}^n$ into $\mathbb{C}^n$.

If Ω is a domain in $\mathbb{C}^n$ and $f : \Omega \to \mathbb{C}^n$ is an injective holomorphic mapping, we say that f is *univalent* on Ω. Such a mapping is necessarily nonsingular on Ω, and is thus biholomorphic on Ω. For this reason, there is no confusion in using the term univalent mapping instead of biholomorphic mapping. (In infinite dimensions we must be more careful.)

The chain rule of multivariable calculus can be expressed in complex form as follows: if $\Omega \subset \mathbb{C}^n$ and $\Omega' \subset \mathbb{C}^m$ are domains, if $g \colon \Omega \to \Omega'$ and $f : \Omega' \to \mathbb{C}^p$ are C^1 maps, and if $\zeta = g(z)$, $w = f(\zeta)$, then

$$\frac{\partial w_\ell}{\partial z_j} = \sum_{k=1}^m \left(\frac{\partial w_\ell}{\partial \zeta_k} \frac{\partial \zeta_k}{\partial z_j} + \frac{\partial w_\ell}{\partial \overline{\zeta}_k} \frac{\partial \overline{\zeta}_k}{\partial z_j} \right), \quad j = 1, \ldots, n, \ \ell = 1, \ldots, p.$$

and

$$\frac{\partial w_\ell}{\partial \overline{z}_j} = \sum_{k=1}^m \left(\frac{\partial w_\ell}{\partial \zeta_k} \frac{\partial \zeta_k}{\partial \overline{z}_j} + \frac{\partial w_\ell}{\partial \overline{\zeta}_k} \frac{\partial \overline{\zeta}_k}{\partial \overline{z}_j} \right), \quad j = 1, \ldots, n, \ \ell = 1, \ldots, p.$$

In particular, if f and g are holomorphic then so is $f \circ g$, and $D(f \circ g)(z) = Df(g(z))Dg(z)$.

If φ is a C^1 function on Ω we shall sometimes write

$$\frac{\partial \varphi}{\partial z} = \left(\frac{\partial \varphi}{\partial z_1}, \ldots, \frac{\partial \varphi}{\partial z_n} \right) \quad \text{and} \quad \frac{\partial \varphi}{\partial \bar{z}} = \left(\frac{\partial \varphi}{\partial \bar{z}_1}, \ldots, \frac{\partial \varphi}{\partial \bar{z}_n} \right).$$

Moreover, if φ is a C^2 function on Ω, let

$$\frac{\partial^2 \varphi}{\partial z^2}(z) = \left[\frac{\partial^2 \varphi}{\partial z_i \partial z_j}(z) \right]_{1 \leq i,j \leq n} \quad \text{and} \quad \frac{\partial^2 \varphi}{\partial z \partial \bar{z}}(z) = \left[\frac{\partial^2 \varphi}{\partial z_i \partial \bar{z}_j}(z) \right]_{1 \leq i,j \leq n}.$$

The theory of normal families generalizes to several variables, and we have the following basic theorem:

Theorem 6.1.14. *If $\mathcal{F}$ is a locally uniformly bounded family of holomorphic mappings from a domain $\Omega \subset \mathbb{C}^n$ into $\mathbb{C}^m$ then $\mathcal{F}$ is normal.*

A related result which is very useful in the study of Loewner chains is Vitali's theorem. To state it we need the following definition:

Definition 6.1.15. Let Ω be a domain in $\mathbb{C}^n$. A subset A of Ω is called a *set of uniqueness* for the holomorphic functions on Ω if whenever $f \in H(\Omega, \mathbb{C})$ and $f|_A = 0$ then $f \equiv 0$.

We note that there exist countable sets of uniqueness, for example any countable dense subset of Ω is a set of uniqueness. Of course, if A is a nonempty open subset of Ω, then A is a set of uniqueness for the holomorphic functions on Ω. The reader may show that if $J = \{1/k : k = 2, 3, \ldots\}$ and $A = \underbrace{(J \times J \times \ldots \times J)}_{n-times} \cap B$, then A is a set of uniqueness for the holomorphic functions on the unit ball B of $\mathbb{C}^n$.

Theorem 6.1.16. (Vitali's theorem in several complex variables) *Suppose that Ω is a domain in $\mathbb{C}^n$ and $A \subset \Omega$ is a set of uniqueness for the holomorphic functions on Ω. Let $\{f_k\}_{k \in \mathbb{N}}$ be a sequence of functions in $H(\Omega, \mathbb{C})$ which is locally uniformly bounded and which is such that $\{f_k(z)\}_{k \in \mathbb{N}}$ converges for all $z \in A$. Then there exists $f \in H(\Omega, \mathbb{C})$ such that $f_k \to f$ locally uniformly on Ω as $k \to \infty$.*

We note that this theorem also applies to holomorphic mappings whose target space is $\mathbb{C}^n$.

There is also an n-dimensional version of Hurwitz's theorem:

Theorem 6.1.17. *Suppose that Ω is a domain in $\mathbb{C}^n$ and that $\{f_k\}_{k\in\mathbb{N}} \subset H(\Omega)$ is a sequence of biholomorphic maps such that $f_k \to f$ locally uniformly on Ω. Then either f is biholomorphic on Ω or $J_f \equiv 0$.*

This is complemented by the following convergence result:

Theorem 6.1.18. *Suppose that Ω is a domain in $\mathbb{C}^n$ and that $f \in H(\Omega)$ is a biholomorphic mapping. Suppose $\{f_k\}_{k\in\mathbb{N}} \subset H(\Omega)$ is a sequence of mappings such that $f_k \to f$ locally uniformly on Ω. Let K be a compact subset of Ω. Then there exists $k_0 \geq 1$ such that $f_k|_K$ is injective for $k \geq k_0$.*

Proof. Suppose the contrary. Then there exists a subsequence of $\{f_k\}_{k\in\mathbb{N}}$, which we denote again by $\{f_k\}_{k\in\mathbb{N}}$, and there exist sequences $\{a_k\}_{k\in\mathbb{N}}$ and $\{b_k\}_{k\in\mathbb{N}}$ of points in K such that $a_k \neq b_k$ and $f_k(a_k) = f_k(b_k)$. By passing to subsequences we may assume that $a_k \to a$ and $b_k \to b$ as $k \to \infty$, where a and b are points in K. If $a \neq b$ then we obtain $f(a) = f(b)$, which is a contradiction. Hence $a = b$. We have

$$0 = \frac{f_k(a_k) - f_k(b_k)}{\|a_k - b_k\|} = Df(a)\left(\frac{a_k - b_k}{\|a_k - b_k\|}\right)$$

$$+ \left(Df_k(b_k) - Df(b_k)\right)\left(\frac{a_k - b_k}{\|a_k - b_k\|}\right)$$

$$+ \left(Df(b_k) - Df(a)\right)\left(\frac{a_k - b_k}{\|a_k - b_k\|}\right) + O(a_k - b_k).$$

Passing to subsequences again, we may assume that there is a unit vector $v \in \mathbb{C}^n$ such that $\dfrac{a_k - b_k}{\|a_k - b_k\|} \to v$ as $k \to \infty$. But then $Df(a)(v) = 0$, which is a contradiction.

There are two basic theorems on holomorphic mappings in several variables due to H. Cartan [Cart1]. The first is his theorem on fixed points:

Theorem 6.1.19. *Suppose that Ω is a bounded domain in $\mathbb{C}^n$, and that $f \in H(\Omega, \Omega)$ has a fixed point z_0 at which $Df(z_0) = I$. Then f is the identity map.*

This theorem may be viewed as a generalization of the Schwarz lemma. From it we obtain

Theorem 6.1.20. *Suppose that Ω_1 and Ω_2 are bounded circular domains in $\mathbb{C}^n$ containing 0, and that f is a biholomorphic map from Ω_1 onto Ω_2 such that $f(0) = 0$. Then f is linear.*

This theorem shows that biholomorphic mappings in several variables are very rigid. In particular we obtain the following result, which was first proved by Poincaré [Poi]:

Theorem 6.1.21. *When $n \geq 2$ the Euclidean unit ball B and the unit polydisc P are not biholomorphically equivalent.*

Proof. Theorem 6.1.20 shows that there is no biholomorphism of these domains which takes 0 to 0. We may reduce to this case using the automorphisms of P or of B. (See Section 6.1.4.)

Consequently, the Riemann mapping theorem fails in several complex variables.

There is another generalization of the Schwarz lemma which will be useful [Ham1] (see also [Din1]):

Lemma 6.1.22. *Let Ω be a balanced pseudoconvex domain in $\mathbb{C}^n$ with the Minkowski function h. Suppose that $f : \Omega \to \Omega$ is a holomorphic mapping and that $f(0) = 0$. Then $h(f(z)) \leq h(z)$ for all $z \in \Omega$.*

Proof. Let $v \in \mathbb{C}^n$ be such that $h(v) = 1$ and consider the mapping $g \in H(U, \Omega)$ given by $g(\zeta) = f(\zeta v)$. Then $g(\zeta)/\zeta$ has a removable singularity at $\zeta = 0$ and $h(g(\zeta)/\zeta) \leq 1/r$ when $|\zeta| = r < 1$. Letting $r \nearrow 1$ and using the maximum principle for plurisubharmonic functions (see Theorem 6.1.12) gives $h(g(\zeta)/\zeta) \leq 1$, or $h(f(\zeta v)) \leq |\zeta| = h(\zeta v)$.

6.1.4 Automorphisms of the Euclidean unit ball and the unit polydisc

Theorem 6.1.20 may be used to determine the biholomorphic automorphisms of the unit ball and the unit polydisc in $\mathbb{C}^n$. For more details, the reader may consult [Rud1-2].

To this end, let $\mathcal{U}$ denote the set of unitary transformations of $\mathbb{C}^n$. If Ω is a domain in $\mathbb{C}^n$, let

$$\mathrm{Aut}(\Omega) = \Big\{ f : \Omega \to \Omega : f \text{ is a biholomorphic mapping of } \Omega \text{ onto itself} \Big\}$$

denote *the group of automorphisms of* Ω. We say that $\mathrm{Aut}(\Omega)$ *acts transitively on* Ω if for any $z, w \in \Omega$ there exists an automorphism $f \in \mathrm{Aut}(\Omega)$ such that $f(z) = w$. A domain Ω is *homogeneous* if $\mathrm{Aut}(\Omega)$ acts transitively on Ω.

Let B be the unit ball of $\mathbb{C}^n$ with respect to the Euclidean norm. We have

Theorem 6.1.23. *Up to multiplication by a unitary transformation,* $\text{Aut}(B)$ *consists of mappings*

$$(6.1.9) \qquad \phi_a(z) = T_a\left(\frac{z-a}{1-\langle z,a\rangle}\right), \quad z \in B, \quad a \in B,$$

where $T_0 = I$ *and for* $a \neq 0$, T_a *is the linear operator given by*

$$(6.1.10) \qquad T_a(z) = \frac{\langle z,a\rangle}{1+s_a}a + s_a z, \quad z \in \mathbb{C}^n,$$

and

$$(6.1.11) \qquad s_a = \sqrt{1-\|a\|^2}.$$

That is,

$$\text{Aut}(B) = \Big\{V\phi_a : a \in B, V \in \mathcal{U}\Big\} = \Big\{\phi_b W : b \in B, W \in \mathcal{U}\Big\}.$$

Proof. The fact that ϕ_a is an automorphism of B follows from parts (ii) and (iii) of the following lemma. The fact that all automorphisms of B have the indicated form follows from Theorem 6.1.20.

As we shall remark in the next section, Theorem 6.1.23 also gives the automorphisms of the unit ball in a complex Hilbert space.

Some of the basic properties of the mappings in $\text{Aut}(B)$ are as follows (see [Rud2, p.25-30]):

Lemma 6.1.24. *Let* $a \in B$ *and let* ϕ_a *be given by (6.1.9). Then*

(i) $\phi_a(a) = 0$.

(ii) $1 - \|\phi_a(z)\|^2 = \dfrac{(1-\|a\|^2)(1-\|z\|^2)}{|1-\langle z,a\rangle|^2}, \ z \in B$.

(iii) $(\phi_a)^{-1}(z) = \phi_{-a}(z), \ z \in B$.

(iv) ϕ_a *extends to a homeomorphism of* $\overline{B}$ *onto* $\overline{B}$.

(v) *If* $\psi \in \text{Aut}(B)$ *and* $a = \psi^{-1}(0)$, *then*

$$(6.1.12) \qquad |J_\psi(z)| = \left[\frac{s_a}{|1-\langle z,a\rangle|}\right]^{n+1}, \quad z \in B.$$

Moreover,

$$(6.1.13) \qquad |J_\psi(z)| = \left[\frac{1-\|\psi(z)\|^2}{1-\|z\|^2}\right]^{\frac{n+1}{2}}, \quad z \in B.$$

We note that $\text{Aut}(B)$ acts transitively on B, and thus B is a homogeneous domain in $\mathbb{C}^n$. In fact the action is transitive on directions at a given point.

If the unit ball B is replaced by the unit polydisc P of $\mathbb{C}^n$, then we have

Theorem 6.1.25. *Up to multiplication by a diagonal unitary transformation and a permutation of the coordinates, the set* $\text{Aut}(P)$ *consists of mappings*

$$(6.1.14) \quad \psi(z) = \psi_a(z) = (\psi_{a_1}(z_1), \ldots, \psi_{a_n}(z_n)), \quad z = (z_1, \ldots, z_n) \in P,$$

where $a = (a_1, \ldots, a_n) \in P$ *and*

$$(6.1.15) \qquad\qquad \psi_{a_j}(z_j) = \frac{z_j - a_j}{1 - \overline{a}_j z_j}, \quad 1 \le j \le n.$$

That is,

$$\text{Aut}(P) = \Big\{ (z_1, \ldots, z_n) \mapsto (e^{i\theta_1}\psi_{a_1}(z_{\sigma(1)}), \ldots, e^{i\theta_n}\psi_{a_n}(z_{\sigma(n)})) :$$

$$\theta_1, \ldots, \theta_n \in \mathbb{R}, \ |a_j| < 1, \ 1 \le j \le n, \ \sigma \text{ is an arbitrary}$$

$$\text{permutation of } (1, \ldots, n) \Big\}.$$

Again we note that P is a homogeneous domain; however this time the action of $\text{Aut}(P)$ is not transitive on directions at a given point.

6.1.5 Holomorphic mappings in complex Banach spaces

Let X be a complex Banach space with respect to the norm $\|\cdot\|$. Again we shall denote the open ball of radius r centered at $z_0 \in X$ by $B(z_0, r)$. Instead of $B(0, r)$ we shall write B_r, and for the unit ball B_1 we shall write simply B. As always, U denotes the open unit disc in $\mathbb{C}$. The closure of the ball $B(z_0, r)$ will be indicated by $\overline{B}(z_0, r)$ and its boundary by $\partial B(z_0, r)$.

Now let X and Y be two complex Banach spaces with respect to the norms $\|\cdot\|_1$ and $\|\cdot\|_2$ respectively. For simplicity, we denote both norms by $\|\cdot\|$, when there is no possibility of confusion. Let $L(X, Y)$ denote the Banach space of all continuous complex-linear operators from X into Y with the standard operator norm

$$\|A\| = \sup\{\|A(z)\| : \|z\| = 1\}, \quad A \in L(X, Y).$$

The identity mapping in $L(X, X)$ will be denoted by I.

Let Ω be a domain in X and let $f : \Omega \to Y$ be a mapping. Then f is called *holomorphic* on Ω if for any $z \in \Omega$ there is a mapping $Df(z) \in L(X, Y)$, called the Fréchet derivative of f at z, such that

$$\lim_{h \to 0} \frac{\|f(z + h) - f(z) - Df(z)h\|}{\|h\|} = 0,$$

i.e.

$$f(z + h) = f(z) + Df(z)(h) + o(\|h\|).$$

Let $H(\Omega, \Omega')$ be the set of holomorphic mappings from a domain $\Omega \subseteq X$ into a domain $\Omega' \subseteq Y$, and let $H(\Omega) = H(\Omega, X)$. If $f \in H(\Omega, Y)$ and $z \in \Omega$, then for each $k = 1, 2, \ldots$, there is a bounded symmetric k-linear mapping $D^k f(z) : \prod_{j=1}^{k} X \to Y$, called the kth-order Fréchet derivative of f at z such that

(6.1.16)
$$f(w) = \sum_{k=0}^{\infty} \frac{1}{k!} D^k f(z)((w - z)^k),$$

for all w in some neighbourhood of z. If for some $u \in X$ the disc $\{z + \zeta u : |\zeta| < r\}$ is contained in Ω then the expansion is valid for all w in this disc. If $\Omega - z$ is balanced, the series converges uniformly to f in some neighbourhood of each compact subset of Ω. If f is bounded on the ball $B(z, r) \subseteq \Omega$, then we have uniform convergence on $B(z, s)$ for $0 < s < r$ (see [Muj, Theorems 7.11 and 7.13]).

It is understood that for $h \in X$, $D^0 f(z)(h^0) = f(z)$ and for $k \geq 1$,

$$D^k f(z)(h^k) = D^k f(z)(\underbrace{h, h, \ldots, h}_{k-times}).$$

Moreover, if $f \in H(\Omega, Y)$ and the closed segment $[a, a + h]$ is contained in Ω, then the *Taylor formula with remainder*

(6.1.17)
$$f(a + h) = f(a) + Df(a)(h) + \ldots + \frac{1}{k!} D^k f(a)(h^k)$$

$$+ \int_0^1 \frac{(1 - t)^k}{k!} D^{k+1} f(a + th)(h^{k+1}) dt$$

holds for all $k \in \mathbb{N}$.

The terms in the expansion (6.1.16) can be expressed in terms of Cauchy's integral formula. We begin with the case of a holomorphic map from a domain in the complex plane into X.

If D is a domain in $\mathbb{C}$ and $\gamma : [a, b] \to D$ is a piecewise C^1 curve, and if $q : D \to X$ is a continuous function, then the complex line integral of q along γ is defined by

$$\int_\gamma q(\zeta)d\zeta = \int_a^b q(\gamma(t))\gamma'(t)dt.$$

We then have the following Cauchy integral theorem and Cauchy integral formulas for mappings in $H(D, X)$:

Theorem 6.1.26. *If D is a domain in $\mathbb{C}$ and $g \in H(D, X)$, then*

$$\int_\gamma g(\zeta)d\zeta = 0$$

for each simple closed piecewise C^1 contour γ in D such that the interior of γ is contained in D. If $\zeta_0 \in D$ and $r > 0$ are such that $\overline{U}(\zeta_0, r)$ is contained in D, then

$$g^{(k)}(\zeta_0) = \frac{k!}{2\pi i} \int_{|\zeta-\zeta_0|=r} \frac{g(\zeta)d\zeta}{(\zeta - \zeta_0)^{k+1}}, \quad k = 0, 1, 2 \ldots.$$

These formulas lead to the Cauchy estimates:

$$\|g^{(k)}(\zeta_0)\| \le M r^{-k} k!, \quad k = 0, 1, 2, \ldots,$$

whenever g is a holomorphic mapping from a domain $D \supset \overline{U}(\zeta_0, r)$ into X such that $\|g(\zeta)\| \le M$ for $\zeta \in \overline{U}(\zeta_0, r)$.

Going back to the general case, if Ω is a domain in X, $z \in \Omega$ and $w \in X$, the set

$$\Omega_{z,w} = \{\zeta \in \mathbb{C} : z + \zeta w \in \Omega\}$$

is an open subset of $\mathbb{C}$, and the function $f_{z,w} : \Omega_{z,w} \to Y$, given by $f_{z,w}(\zeta) = f(z + \zeta w)$, is holomorphic on $\Omega_{z,w}$. We may therefore apply the foregoing discussion to $f_{z,w}$ on each connected component of $\Omega_{z,w}$.

Let r be sufficiently small that the Taylor series expansion

$$f(z+h) = \sum_{k=0}^{\infty} \frac{1}{k!} D^k f(z)(h^k)$$

is valid in a neighbourhood of $\overline{B}(z,r)$. Then the term which is homogeneous of degree k may be expressed as

(6.1.18) $$\frac{1}{k!} D^k f(z)(h^k) = \frac{1}{2\pi i} \int_{|\zeta|=r} \frac{f(z+\zeta h)}{\zeta^{k+1}} d\zeta,$$

for all $h \in X$ with $\|h\| \leq 1$ and $k = 0, 1, 2, \dots$. Moreover, if $\|f(w)\| \leq M$ for $w \in \overline{B}(z,r)$, then the above formulas lead to the analog of the classical *Cauchy estimates*:

(6.1.19) $$\|D^k f(z)(h^k)\| \leq M(k!) r^{-k},$$

whenever $h \in X$ with $\|h\| \leq 1$ and $k = 0, 1, 2, \dots$.

In fact when h is fixed, $\|h\| = 1$, we may take r in (6.1.18) and (6.1.19) to be any positive number such that $f_{z,h}$ is holomorphic in a neighbourhood of $\overline{U}_r$ (adjusting M as needed).

Cauchy's integral formula allows us to extend other results in complex function theory to infinite dimensions, for example the maximum modulus theorem:

Theorem 6.1.27. *Let $D \subset X$ be a domain and $f : D \to Y$ be a holomorphic mapping. Then $\|f(z)\|$ can have no maximum in D unless $\|f(z)\|$ is of constant value throughout D.*

We also have the following version of the Schwarz lemma (see [Din1], [Harr1], [Hil-Phi], [Fra-Ve]):

Lemma 6.1.28. *Let $M > 0$ and $f : B \to Y$ be a holomorphic mapping such that $f(0) = 0$ and $\|f(z)\| < M$ for $z \in B$. Then $\|f(z)\| \leq M\|z\|$, $z \in B$. Further, if there is a point $z_0 \in B \setminus \{0\}$ such that $\|f(z_0)\| = M\|z_0\|$, then $\|f(\zeta z_0)\| = M\|\zeta z_0\|$, for all $\zeta \in \mathbb{C}$, $|\zeta| < 1/\|z_0\|$.*

Moreover, if $Df(0) = 0, \dots, D^{k-1}f(0) = 0$, $k \in \mathbb{N}$, $k \geq 2$, then $\|f(z)\| \leq M\|z\|^k$ for $\|z\| < 1$ and $\left\| \frac{1}{k!} D^k f(0)(w^k) \right\| \leq M$ for $\|w\| = 1$.

Proof. Fix $z \in B \setminus \{0\}$ and define $g(\zeta) = f(\zeta z)/\zeta$, $|\zeta| < 1/\|z\|$. Then g is a holomorphic mapping from $|\zeta| < 1/\|z\|$ into Y (it is easy to see that

$Dg(0)(h) = 1/2D^2f(0)(hz, z)$ for $h \in \mathbb{C}$). For $0 < r < 1/\|z\|$ and $|\zeta| = r$, we have $\|g(\zeta)\| \leq M/r$, i.e. $\|f(\zeta z)/\zeta\| \leq M/r$. In view of Theorem 6.1.27, we deduce that $\|f(\zeta z)\| \leq M|\zeta|/r$ for $|\zeta| \leq r$. We now let $r \nearrow 1/\|z\|$, and this yields that $\|f(\zeta z)\| \leq M|\zeta|\|z\|$ for $|\zeta| < 1/\|z\|$. For $\zeta = 1$, the conclusion follows.

If $\|f(z_0)\| = M\|z_0\|$, then from Theorem 6.1.27 and the above reasoning, one deduces that $\|f(\zeta z_0)/\zeta\| = M\|z_0\|$, for all ζ, $|\zeta| \leq r$, $0 < r < 1/\|z_0\|$. Letting $r \nearrow 1/\|z_0\|$, one obtains the claimed result.

For the last statement, it suffices to use similar reasoning as above for the map $h(\zeta) = f(\zeta z)/\zeta^k$, $|\zeta| < 1/\|z\|$, where $z \in B \setminus \{0\}$.

As in one complex variable, a mapping $v \in H(B)$ is a *Schwarz mapping* if $v(0) = 0$ and $\|v(z)\| < 1$, $z \in B$.

If $0 \in \Omega$ and $f \in H(\Omega)$, we say that f is *normalized* if $f(0) = 0$ and $Df(0) = I$. A mapping $f \in H(\Omega, Y)$ is called *locally biholomorphic* on the domain Ω if for each $z \in \Omega$ there are neighbourhoods V of z and W of $f(z)$ such that f is a one-to-one map of V onto W whose inverse is holomorphic on W. It is known that f is locally biholomorphic on Ω if and only if the Fréchet derivative $Df(z)$ has a bounded inverse at each $z \in \Omega$. If $X = Y = \mathbb{C}^n$, the condition that $Df(z)$ has a bounded inverse is just the condition that $J_f(z) \neq 0$ for $z \in \Omega$.

A mapping $f \in H(\Omega, Y)$ is said to be *biholomorphic* on the domain Ω if $f(\Omega)$ is a domain in Y and the inverse f^{-1} exists and is holomorphic on $f(\Omega)$. As in the finite dimensional case, let $S(B)$ denote the set of normalized biholomorphic mappings from B into X.

If $X = \mathbb{C}^n$ and f is a univalent mapping from B into $\mathbb{C}^n$, then f is a biholomorphic mapping from B onto the domain $f(B)$ in $\mathbb{C}^n$. However, in the case of complex Banach spaces, this result is not necessarily true. Heath and Suffridge [Hea-Su] gave an example of a univalent (holomorphic and injective) mapping of the unit ball B of a complex Banach space such that f^{-1} is not holomorphic on $f(B)$, $f(B)$ contains an open set, but $f(B)$ is not open. Hence we must be careful to specify that we are studying biholomorphic maps rather than univalent maps.

As observed by Franzoni and Vesentini [Fra-Ve], Cartan's theorem on fixed

points (Theorem 6.1.19) may be generalized to complex Banach spaces. There
is a generalization of Theorem 6.1.20 as well [Fra-Ve]: If Ω is a bounded circular
domain in a complex Banach space containing 0, and if $f \in \mathrm{Aut}(\Omega)$ is such
that $f(0) = 0$, then f is (the restriction of) a linear isomorphism of X.

It follows that the automorphism group of the unit ball in a complex Hilbert
space is given by Theorem 6.1.23 (with the appropriate notion of unitary
transformation).

Notes. The main sources that have been used in this section for holomorphic
mappings in Banach spaces are [Fra-Ve], [Hil-Phi], [Muj], [Su3-4]. For further
results, the reader may consult [Boc-Sic], [Din2], or [Herv].

6.1.6   Generalizations of functions with positive real part

We wish to present some basic results concerning a class of mappings in
higher dimensions related to the Carathéodory class. The definition makes use
of linear functionals. For further discussion see [Gur], [Su3-4], [Rop-Su2].

For the complex Banach space X, let X^* denote the dual of X (i.e. $X^* = L(X, \mathbb{C})$). For each $z \in X \setminus \{0\}$, we define

$$(6.1.20) \qquad T(z) = \left\{ l_z \in X^* : \ l_z(z) = \|z\|, \ \|l_z\| = 1 \right\},$$

where

$$\|l_z\| = \sup \left\{ |l_z(w)| : \ \|w\| = 1 \right\}.$$

By the Hahn-Banach theorem, $T(z)$ is nonempty. This set plays a central
role in the study of biholomorphic mappings of the unit ball in a complex
Banach space. A case of particular interest is that of the p-norm in $\mathbb{C}^n$, i.e.

$$(6.1.21) \qquad \|z\|_p = \begin{cases} \left[\displaystyle\sum_{j=1}^{n} |z_j|^p \right]^{1/p}, & 1 \le p < \infty \\[2mm] \displaystyle\max_{1 \le j \le n} |z_j|, & p = \infty. \end{cases}$$

Let $B(p)$ denote the unit ball with respect to $\| \cdot \|_p$. When $p = \infty$, $B(p)$ reduces
to the unit polydisc P of $\mathbb{C}^n$. It is known that if $X = \mathbb{C}^n$ with respect to a

p-norm, $1 \leq p \leq \infty$, and $l_z \in T(z)$, then

$$(6.1.22) \qquad l_z(w) = \frac{\sum\limits_{\{j:z_j \neq 0\}} |z_j|^p w_j / z_j}{\|z\|_p^{p-1}} \qquad \text{for} \quad 1 < p < \infty,$$

$$(6.1.23) \qquad l_z(w) = \sum_{\{j:z_j \neq 0\}} |z_j| w_j / z_j + \sum_{\{j:z_j = 0\}} \gamma_j w_j \qquad \text{for} \quad p = 1,$$

where $\gamma_j \in \mathbb{C}$, $|\gamma_j| \leq 1$, for all $j \in \{1, \ldots, n\}$,

$$(6.1.24) \qquad l_z(w) = \sum_{\{j:\|z\|_\infty = |z_j|\}} t_j \frac{w_j \bar{z}_j}{\|z\|_\infty} \qquad \text{for} \quad p = \infty,$$

where each $t_j \geq 0$ and $\sum\limits_{\{j:\|z\|_\infty = |z_j|\}} t_j = 1$. Of course, when $p = 2$, (6.1.22) reduces to $l_z(w) = \left\langle w, \dfrac{z}{\|z\|} \right\rangle$.

The following families play a key role in our discussion:

$$\mathcal{N}_0 = \{f : B \to X : \ f \in H(B), \ f(0) = 0, \ \text{Re} \ [l_z(f(z))] \geq 0,$$

$$z \in B \setminus \{0\}, \ l_z \in T(z)\},$$

$$\mathcal{N} = \{f \in \mathcal{N}_0 : \ \text{Re} \ [l_z(f(z))] > 0, \ z \in B \setminus \{0\}, \ l_z \in T(z)\},$$

$$\mathcal{M} = \{f \in \mathcal{N} : \ Df(0) = I\}.$$

When $X = \mathbb{C}^n$ with respect to a p-norm, the set $\mathcal{M}$ is sometimes denoted by $\mathcal{M}(p)$, and consists of those normalized holomorphic mappings $w : B(p) \to \mathbb{C}^n$ such that for $z = (z_1, \ldots, z_n) \in B(p) \setminus \{0\}$,

$$(6.1.25) \ \begin{cases} \text{Re} \ \sum\limits_{\{j:z_j \neq 0\}} w_j(z) \dfrac{|z_j|^p}{z_j} > 0, \quad 1 < p < \infty \\[2ex] \text{Re} \ \sum\limits_{\{j:z_j \neq 0\}} w_j(z) \dfrac{|z_j|}{z_j} - \sum\limits_{\{j:z_j = 0\}} |w_j(z)| > 0, \quad p = 1 \\[2ex] \text{Re} \ \left[\dfrac{w_j(z)}{z_j} \right] > 0, \ \|z\|_\infty = |z_j| > 0 \ (1 \leq j \leq n), \quad p = \infty. \end{cases}$$

Again we note the special case $p = 2$, for which we have

$$\text{Re} \ \langle w(z), z \rangle > 0, \qquad z \in B \setminus \{0\}.$$

In the case $X = \mathbb{C}$, if $f \in \mathcal{N}_0$ then Re $[\bar{z}f(z)] \geq 0$ on U, or equivalently Re $[f(z)/z] \geq 0$ on U. Thus, by the minimum principle for harmonic functions, either Re $[f(z)/z] \equiv 0$ or Re $[f(z)/z] > 0$. Hence a function f has the property that $f \in \mathcal{N}_0 \setminus \mathcal{N}$ if and only if $f(z) = iaz$ for some real a. However, in dimensions greater than 1, the set $\mathcal{N}_0 \setminus \mathcal{N}$ can be larger. The following example is due to Suffridge [Su4].

Example 6.1.29. Let $X = \mathbb{C}^2$ be the Euclidean space of two complex variables. Let $w : B \subset \mathbb{C}^2 \to \mathbb{C}^2$ be given by $w(z) = (-z_2, z_1)$. Then $w \in H(B)$, $w(0) = 0$ and

$$\text{Re } [l_z(w(z))] = \text{Re } \left[\frac{-\bar{z}_1 z_2 + z_1 \bar{z}_2}{\|z\|} \right] = 0, \quad z \in B \setminus \{0\}.$$

Thus $w \in \mathcal{N}_0 \setminus \mathcal{N}$.

For $X = \mathbb{C}$, the above characterization of $\mathcal{N}_0$ implies that if $w \in \mathcal{N}_0$ and $\alpha \in \mathbb{C}$, $|\alpha| < 1$, then $w(\alpha z)/\alpha \in \mathcal{N}_0$. (It is understood that $w(\alpha z)/\alpha$ denotes the limiting value $w'(0)z$ when $\alpha = 0$.) Moreover, $w \in \mathcal{N}$ unless Re $w'(0) = 0$ and in this case $w(z)/z \equiv$ constant.

In the case of normed spaces the analogous result is the following (see [Su3]):

Lemma 6.1.30. *If $w \in \mathcal{N}_0$ and $|\alpha| < 1$ then $w(\alpha z)/\alpha \in \mathcal{N}_0$ ($w(\alpha z)/\alpha$ is understood to have the limiting value $Dw(0)(z)$ when $\alpha = 0$). Further, if $l_z \in T(z)$, $0 < \|z\| < 1$, then* Re $[l_z(w(z))] = 0$ *if and only if* Re $[l_z(Dw(0)(z))] = 0$, *and in this case* Re $[l_z(w(\alpha z)/\alpha)] \equiv 0$ *when $|\alpha| < 1/\|z\|$.*

Proof. For $0 < |\alpha| < 1$, $z \in B \setminus \{0\}$, and $l_z \in T(z)$, let $A_\alpha(y) = l_z(|\alpha|y/\alpha)$ for $y \in X$. Then $A_\alpha(\alpha z) = \|\alpha z\|$ and $\|A_\alpha(y)\| \leq \|y\|$ for $y \in X$, i.e. $A_\alpha \in T(\alpha z)$. Since $w \in \mathcal{N}_0$ we have

$$0 \leq \frac{1}{|\alpha|} \text{Re } [A_\alpha(w(\alpha z))] = \text{Re } \left[l_z \left(\frac{w(\alpha z)}{\alpha} \right) \right].$$

Using the continuity of l_z, one deduces that Re $[l_z(Dw(0)(z))] \geq 0$. Thus $w(\alpha z)/\alpha \in \mathcal{N}_0$ for $|\alpha| < 1$. Further, in view of the fact that Re $[l_z(w(\alpha z)/\alpha)]$ is a nonnegative harmonic function of α for fixed z with $|\alpha| < 1/\|z\|$, we conclude by the minimum principle that either Re $[l_z(w(\alpha z)/\alpha)] > 0$ or Re $[l_z(w(\alpha z)/\alpha)] \equiv 0$. This completes the proof.

The following lemma, due to Gurganus [Gur], will be applied in several situations in forthcoming chapters.

Lemma 6.1.31. *Let X be a complex Banach space and $h \in \mathcal{M}$. Then for all $z \in B \setminus \{0\}$ and $l_z \in T(z)$,*

$$(6.1.26) \qquad \frac{1 - \|z\|}{1 + \|z\|} \|z\| \leq \operatorname{Re} \left[l_z(h(z)) \right] \leq \frac{1 + \|z\|}{1 - \|z\|} \|z\|.$$

Proof. Let $z \in B \setminus \{0\}$ and $l_z \in T(z)$ be fixed. Also let

$$p(\zeta) = \frac{1}{\zeta} l_z \left(h \left(\frac{z}{\|z\|} \zeta \right) \right), \quad \zeta \in U.$$

Since $h(0) = 0$, p is a holomorphic function in the unit disc U, and since $h \in \mathcal{M}$ it is clear that $\operatorname{Re} p(\zeta) > 0$ on U. Indeed, since there is a one-to-one correspondence between $T(\alpha z)$ and $T(z)$ given by $l_{\alpha z}(\cdot) = \dfrac{|\alpha|}{\alpha} l_z(\cdot)$, for each $\alpha \in \mathbb{C} \setminus \{0\}$, one deduces for $0 < |\zeta| < 1$ that

$$\operatorname{Re} p(\zeta) = \operatorname{Re} \left[\frac{1}{\zeta} l_z \left(h \left(\frac{z}{\|z\|} \zeta \right) \right) \right] = \operatorname{Re} \left[\frac{1}{\zeta} l_{z/\|z\|} \left(h \left(\zeta \frac{z}{\|z\|} \right) \right) \right] =$$

$$= \frac{1}{|\zeta|} \operatorname{Re} \left[l_{\zeta z / \|z\|} \left(h \left(\zeta \frac{z}{\|z\|} \right) \right) \right] > 0.$$

Therefore $\operatorname{Re} p(\zeta) > 0$ on U, and $p(0) = 1$ from the normalization of h. From Theorem 2.1.3 we therefore deduce that

$$\frac{1 - |\zeta|}{1 + |\zeta|} < \operatorname{Re} p(\zeta) < \frac{1 + |\zeta|}{1 - |\zeta|}, \quad \zeta \in U.$$

Substituting $\zeta = \|z\|$ in the above, we obtain the relation (6.1.26), as desired. This completes the proof.

We shall see in Theorem 6.1.39 that a stronger result holds, namely the set $\mathcal{M}$ is compact when $X = \mathbb{C}^n$ with respect to an arbitrary norm. We note the following refinement of Lemma 6.1.31 (originally proved by Pfaltzgraff [Pfa1]) in the case $X = \mathbb{C}^n$ with the Euclidean structure.

Lemma 6.1.32. *If $h \in \mathcal{M}$ then*

$$(6.1.27) \qquad \|z\|^2 \frac{1 - \|z\|}{1 + \|z\|} \leq \operatorname{Re} \langle h(z), z \rangle \leq \|z\|^2 \frac{1 + \|z\|}{1 - \|z\|}, \quad z \in B.$$

We next consider generalizations by Suffridge [Su3-4] of two theorems of Robertson [Robe3] to complex Banach spaces (see Theorems 2.1.7 and 2.1.8). In fact Lemmas 6.1.33 and 6.1.35 are slight modifications of Suffridge's results.

Lemma 6.1.33. *Let $v : B \times [0,1] \to B$ be a mapping such that $v(\cdot, t)$ is holomorphic on B for each fixed $t \in [0,1]$, $v(0,t) = 0$ and $v(z,0) = z$. Suppose that*

$$(6.1.28) \qquad \lim_{t \to 0^+} \frac{z - v(z,t)}{t^\rho} = w(z)$$

exists and is holomorphic on B for some $\rho > 0$. Then $w \in \mathcal{N}_0$.

Proof. Fix $z \in B \setminus \{0\}$ and $l_z \in T(z)$. In view of (6.1.28), we deduce that $\lim_{t \to 0^+} v(z,t) = z$. From Lemma 6.1.28 we obtain that $\|v(z,t)\| \leq \|z\|$, and hence

$$\text{Re}\left[l_z\left(\frac{z - v(z,t)}{t^\rho}\right)\right] = \frac{\|z\| - \text{Re}\,[l_z(v(z,t))]}{t^\rho}$$

$$\geq \frac{1}{t^\rho}[\|z\| - \|v(z,t)\|] \geq 0, \quad t \in (0,1].$$

Using the continuity of l_z and this relation, one deduces by letting $t \to 0^+$ that $\text{Re}\,[l_z(w(z))] \geq 0$, as desired.

The following example of Suffridge [Su3] shows that there are situations in which the mapping w obtained in Lemma 6.1.33 belongs to $\mathcal{N}_0 \setminus \mathcal{N}$.

Example 6.1.34. Let $X = \mathbb{C}^2$ be the Euclidean space of two complex variables $z = (z_1, z_2)$. Let $v : B \times [0,1] \to B$ be given by

$$v(z,t) = \left(\sqrt{1 - t^2}\,z_1 + tz_2, -tz_1 + \sqrt{1 - t^2}\,z_2\right).$$

Obviously v satisfies the requirements of the above lemma and

$$\lim_{t \to 0^+} \frac{z - v(z,t)}{t} = (-z_2, z_1).$$

Hence $w(z) = (-z_2, z_1)$ and from Example 6.1.29 we have $w \in \mathcal{N}_0 \setminus \mathcal{N}$.

Lemma 6.1.35. *Let $f : B \subset X \to Y$ be a biholomorphic mapping of B onto a domain $f(B)$ of Y such that $f(0) = 0$. Assume $F : B \times [0,1] \to Y$ is holomorphic on B for each $t \in [0,1]$, $F(z,0) = f(z)$, $F(0,t) = 0$, and $F(B,t) \subset f(B)$ for each $t \in [0,1]$. Moreover, let $\rho > 0$ be such that*

$$(6.1.29) \qquad \lim_{t \to 0^+} \frac{F(z,0) - F(z,t)}{t^\rho} = g(z)$$

exists and is holomorphic on B. Then $g(z) = Df(z)w(\tilde{z})$ on B, where $w \in \mathcal{N}_0$.

Proof. Since $F(B, t) \subset f(B)$ for $t \in [0, 1]$, there is a mapping $v :$ $B \times [0, 1] \to B$ such that $F(z, t) = f(v(z, t))$ for all $z \in B$. Then $v(z, t) = f^{-1}(F(z, t))$ is holomorphic on B for fixed t, $v(0, t) = 0$ for $t \in [0, 1]$ and $v(z, 0) = z$, $z \in B$. From the Schwarz lemma (Lemma 6.1.28) we obtain that $\|v(z, t)\| \leq \|z\|$ for $z \in B$ and $t \in [0, 1]$. On the other hand, from (6.1.29) we deduce that $\lim_{t \to 0^+} F(z, t) = f(z)$ and hence, $\lim_{t \to 0^+} v(z, t) = z$, $z \in B$.

Next, fix $z \in B \setminus \{0\}$. Then expanding f in a Taylor series, we have

$$f(v(z, t)) = f(z) + Df(z)(v(z, t) - z) + R(v(z, t), z),$$

where $\dfrac{\|R(y, z)\|}{\|y - z\|} \to 0$ as $\|y - z\| \to 0$. Therefore for $t \in (0, 1]$, we obtain

$$(6.1.30) \qquad \frac{F(z, 0) - F(z, t)}{t^\rho} = Df(z) \left(\frac{z - v(z, t)}{t^\rho} \right) - \frac{R(v(z, t), z)}{t^\rho}.$$

We next prove that $\|z - v(z, t)\|/t^\rho$ is bounded as $t \to 0^+$. Assuming the contrary, we may suppose for a certain sequence $\{t_m\}_{m \in \mathbb{N}}$, $t_m \to 0^+$, that

$$\frac{\|z - v(z, t_m)\|}{t_m^\rho} \to \infty \text{ as } m \to \infty.$$

Obviously, from (6.1.30) we have

$$g(z) = \lim_{m \to \infty} \left\{ \left[Df(z) \left(\frac{z - v(z, t_m)}{\|z - v(z, t_m)\|} \right) - \frac{R(v(z, t_m), z)}{\|z - v(z, t_m)\|} \right] \frac{\|z - v(z, t_m)\|}{t_m^\rho} \right\}.$$

On the other hand, since $v(z, t_m) \to z$ and $\dfrac{\|R(v(z, t_m), z)\|}{\|z - v(z, t_m)\|} \to 0$ as $m \to \infty$, the preceding equality yields that

$$\lim_{m \to \infty} Df(z) \left(\frac{z - v(z, t_m)}{\|z - v(z, t_m)\|} \right) = 0.$$

However, this relation is impossible since f is biholomorphic (thus $Df(z)$ is invertible). Consequently, we deduce that $\|z - v(z, t)\|/t^\rho$ must be bounded as $t \to 0^+$. Hence, taking into account the fact that the limit (6.1.29) exists and is holomorphic, using (6.1.30) and the invertibility of $Df(z)$, we conclude that the limit

$$\lim_{t \to 0^+} \frac{z - v(z, t)}{t^\rho} = w(z)$$

exists and is holomorphic on B. It now follows from Lemma 6.1.33 that $w \in \mathcal{N}_0$. The equality $g(z) = Df(z)w(z)$ is clear from (6.1.30), and this completes the proof.

We conclude this section with an important result, which establishes the compactness of $\mathcal{M}$ on the unit ball of $\mathbb{C}^n$ with an arbitrary norm $\|\cdot\|$. This result has recently been obtained in [Gra-Ham-Koh] (see also [Ham-Koh10] for the infinite dimensional case). In the next chapter we shall give another proof of this result, using coefficient estimates for mappings in $\mathcal{M}$.

We need to introduce some notions about the numerical radius and numerical range of holomorphic functions in complex Banach spaces [Harr2].

Definition 6.1.36. (i) Let $h : B \to X$ be a holomorphic mapping such that h has a continuous extension to $\overline{B}$, again denoted by h. Let

$$V(h) = \left\{ l_z(h(z)) : l_z \in T(z), \|z\| = 1 \right\}$$

be the *numerical range of h* (taken with respect to T).

(ii) Let $h : B \to X$ be a holomorphic mapping. Define $h_s(z) = h(sz)$ for $0 < s < 1$ and $z \in \overline{B}$. The *numerical radius of h* is the number

$$|V(h)| = \limsup_{s \to 1^-} \left\{ |\lambda| : \lambda \in V(h_s) \right\}.$$

Clearly if h has a uniformly continuous extension to $\overline{B}$, then $|V(h)| = \sup\{|\lambda| : \lambda \in V(h)\}$. Harris [Harr2] proved the following result, which provides an estimate of the norm of a homogeneous polynomial in terms of the numerical radius. (Abstract homogeneous polynomials are discussed in Section 7.2.3.)

Lemma 6.1.37. *Let $P_m : X \to X$ be a continuous homogeneous polynomial of degree $m \geq 1$. Then*

$$(6.1.31) \qquad\qquad \|P_m\| \leq k_m |V(P_m)|,$$

where

$$\|P_m\| = \sup\{\|P_m(z)\| : \|z\| \leq 1\},$$

$k_m = m^{m/(m-1)}$ *for $m \geq 2$, and $k_1 = e$.*

Further, let

$$L(h) = \limsup_{s \to 1^-} \left\{ \text{Re} \, [l_z(h_s(z))] : l_z \in T(z), \|z\| = 1 \right\},$$

where $h : B \to X$ is a holomorphic mapping. Using Lemma 6.1.37, Harris, Reich, and Shoikhet [Harr-Re-Sh] recently obtained an interesting distortion result for holomorphic mappings with restricted numerical range.

Lemma 6.1.38. *Let $h : B \to X$ be a holomorphic mapping such that $h(0) = 0$ and $L(h)$ is finite. Then*

$$(6.1.32) \qquad \|h(z) - z\| \leq \frac{8\|z\|^2}{(1 - \|z\|)^2}(L(h) - 1), \quad z \in B.$$

We are now able to prove the following theorem on the unit ball of $\mathbb{C}^n$ with an arbitrary norm (cf. [Ham-Koh10], [Gra-Ham-Koh]). In Chapter 7 we shall give another proof which uses Lemma 6.1.37 directly.

Theorem 6.1.39. *The set $\mathcal{M}$ is compact.*

Proof. First we prove that $\mathcal{M}$ is locally uniformly bounded on B. For this purpose, let $h \in \mathcal{M}$. In view of (6.1.26) we have

$$\mathrm{Re}\,[l_z(h(tz))] \leq \frac{1+t}{1-t}t, \quad l_z \in T(z), \; \|z\| = 1, \quad 0 < t < 1.$$

Let $\widetilde{h}_t(z) = h(tz)/t$ for $0 < t < 1$. Then it is obvious that $\widetilde{h}_t \in \mathcal{M}$. Let

$$L(\widetilde{h}_t) = \limsup_{s \to 1^-} \left\{ \mathrm{Re}\,[l_z(\widetilde{h}_t(sz))] : l_z \in T(z), \|z\| = 1 \right\}.$$

It is clear that

$$L(\widetilde{h}_t) \leq \frac{1+t}{1-t} < \infty, \quad 0 < t < 1,$$

by the above arguments. Further, using the relation (6.1.32), one deduces that

$$\|\widetilde{h}_t(z) - z\| \leq \frac{8\|z\|^2}{(1 - \|z\|)^2}(L(\widetilde{h}_t) - 1) \leq \frac{8\|z\|^2}{(1 - \|z\|)^2} \cdot \frac{2t}{1-t}, \quad z \in B.$$

Next, for any r with $0 < r < 1$, let t be such that $r < t < 1$. Also let $s = r/t$. Then $0 < s < 1$ and

$$\|h(tz) - tz\| \leq \frac{8s^2}{(1 - s)^2} \cdot \frac{2t^2}{1-t}, \quad \|z\| \leq s.$$

Consequently, we have proved that for each $r \in (0, 1)$, there exists a constant $M = M(r) > 0$ such that $\|h(z)\| \leq M(r)$ for all $h \in \mathcal{M}$ and $z \in \overline{B}_r$, as claimed.

We next prove that $\mathcal{M}$ is closed. For this purpose, let $\{p_k\}_{k\in\mathbb{N}}$ be a sequence in $\mathcal{M}$ such that $p_k \to p$ locally uniformly on B. Then $p \in H(B)$, $p(0) = 0$ and $Dp(0) = I$, by the normalization of p_k, $k \in \mathbb{N}$. Next, fix $z \in B \setminus \{0\}$ and $l_z \in T(z)$. Since Re $[l_z(p_k(z))] > 0$ for $k \in \mathbb{N}$, we deduce by taking limits and using the continuity of l_z that Re $[l_z(p(z))] \geq 0$. Since p is normalized, we must have Re $[l_z(p(z))] > 0$, by Lemma 6.1.30. Thus $p \in \mathcal{M}$ and this completes the proof.

6.1.7   Examples and counterexamples

There are many results in univalent function theory in one complex variable which cannot be extended (at least not without restrictions) to higher dimensions. We have already noted the failure of the Riemann mapping theorem.

The analog of the class S on the unit ball in $\mathbb{C}^n$ is

$$S(B) = \Big\{ f \in H(B) : f \text{ is biholomorphic on } B,\ f(0) = 0,\ Df(0) = I \Big\}.$$

It is easily seen that $S(B)$ is not a normal family when the dimension is greater than one, and that there can be no growth or covering theorems or coefficient bounds for the full class. Indeed, if g is an arbitrary holomorphic function on the unit disc such that $g(0) = g'(0) = 0$, and B is the Euclidean unit ball in $\mathbb{C}^2$, then the map f defined by

$$(6.1.33) \qquad f(z_1, z_2) = (z_1, z_2 + g(z_1)), \quad (z_1, z_2) \in B,$$

belongs to $S(B)$. Examples of this type were given by H. Cartan [Cart2] in his appendix to Montel's book on univalent functions, published in 1933.

Partly on the basis of such examples, Cartan conjectured that the Jacobian determinant of a mapping in $S(B)$ should satisfy bounds depending only on $\|z\|$. (We recall that $|J_f(z)|^2$ gives the infinitesimal magnification factor for the volume.) However, if h is an arbitrary holomorphic function on U such that $h(0) = 0$, the mapping

$$(6.1.34) \qquad f(z) = (z_1, z_2 e^{h(z_1)}), \quad z = (z_1, z_2) \in B,$$

belongs to $S(B)$ and since $J_f(z) = e^{h(z_1)}$, Cartan's conjecture is false. (This example was given by Duren and Rudin [Dur-Rud]; the special cases

$$f(z) = \left(z_1, \frac{z_2}{(1 - z_1)^k} \right), \quad k \in \mathbb{N}, \ z = (z_1, z_2) \in B,$$

were considered by Barnard, FitzGerald, and Gong [Bar-Fit-Gon2].)

Cartan also suggested that particular subclasses of $S(B)$, such as the starlike and convex mappings, should be singled out for further study. Indeed, many of the results of univalent function theory do have extensions to higher dimensions for these classes.

The failure of Bloch's theorem for normalized holomorphic maps $f : B \to \mathbb{C}^n$, $n \geq 2$, is a consequence of elementary examples of Harris [Harr3], Graham and Wu [Gra-Wu], and Duren and Rudin [Dur-Rud]. Graham and Wu showed that a normalized mapping $f \in H(B)$ such that $J_f \equiv 1$ could have an image of arbitrarily small volume. In particular, there can be no lower bound, independent of f, for the supremum of the radii of balls contained in the image of f.

Example 6.1.40. Let $n = 2$ and let Q be the square of side-length equal to 2 centered at the origin in $\mathbb{C}$, i.e. $Q = \{z \in \mathbb{C} : |\operatorname{Re} z| < 1, |\operatorname{Im} z| < 1\}$. Fix a positive integer k and let $f : Q \times Q \to \mathbb{C}^2$ be given by

$$f(z_1, z_2) = \left(\frac{1}{2\pi k} \exp(2\pi k z_1), z_2 \exp(-2\pi k z_1) \right).$$

Then $J_f(z_1, z_2) = 1$ and f is a covering map onto its image. Using the periodicity of the exponential function, it is easy to see that f is a 2k-to-1 map (except for a set of measure 0), and hence

$$\operatorname{volume}(f(Q \times Q)) = \frac{1}{2k} \operatorname{volume}(Q \times Q) = \frac{8}{k}.$$

Restricting f to the unit ball of $\mathbb{C}^2$, we see that by taking k large we can make $\operatorname{volume}(f(B))$ arbitrarily small.

The example of Harris [Harr3] and Duren and Rudin [Dur-Rud] shows that even the univalent Bloch theorem fails in $\mathbb{C}^n$, $n \geq 2$. The example is a simple polynomial map of degree 2.

Example 6.1.41. If $\delta > 0$, then the map $f : B \subset \mathbb{C}^2 \to \mathbb{C}^2$ given by

$$f(z) = (z_1, z_2 + (z_1/\delta)^2), \quad z = (z_1, z_2) \in B,$$

belongs to $S(B)$, but $f(B)$ contains no closed ball of radius δ.

Proof. We show that there is no $(w_1, w_2) \in \mathbb{C}^2$ such that $f(B)$ contains the circle $\{(w_1 + \delta e^{i\theta}, w_2) : -\pi \leq \theta \leq \pi\}$. To see this, fix $(w_1, w_2) \in \mathbb{C}^2$. If $(w_1 + \lambda, w_2) \in f(B)$, then we must have

$$\left| w_2 - \frac{1}{\delta^2}(w_1 + \lambda)^2 \right| < 1.$$

So if $\{(w_1 + \delta e^{i\theta}, w_2) : -\pi \leq \theta \leq \pi\} \subset f(B)$, then the inequality

$$|\delta^2 w_2 - w_1^2 - 2w_1 \delta e^{i\theta} - \delta^2 e^{2i\theta}| < \delta^2$$

holds for all θ. However, the L^2-norm of the function on the left-hand side of this inequality is

$$\left[2\pi \left(|\delta^2 w_2 - w_1^2|^2 + |2w_1\delta|^2 + \delta^4 \right) \right]^{1/2} \geq (2\pi)^{1/2}\delta^2,$$

and this is a contradiction.

Other interesting examples and counterexamples are given in the recent monographs of Gong [Gon4-5].

Although it is too large a subject to include in this book, we mention that the theory of entire biholomorphic mappings is much different in higher dimensions then in one variable. In one variable an entire univalent function must be linear, and in particular any automorphism of $\mathbb{C}$ must be linear. Both statements are completely false in higher dimensions. If the functions g and h in (6.1.33) and (6.1.34) are entire, the resulting maps f give automorphisms of $\mathbb{C}^n$. Moreover, in higher dimensions $\mathbb{C}^n$ can be biholomorphic to a proper subset of itself. This has been known since the 1920's when examples were given by Fatou and Bieberbach (see [Ros-Rud]), and there has been much study of Fatou-Bieberbach maps in recent years.

Recent years have also seen many developments in the theory of univalent mappings of bounded domains, and it is to these that we now turn.

6.2   Criteria for starlikeness

6.2.1   Criteria for starlikeness on the unit ball in $\mathbb{C}^n$ or in a complex Banach space

In the following we are going to present some generalizations to higher dimensions of the well-known criterion for starlikeness on the unit disc (Theorem 2.2.2). We begin with the Euclidean unit ball in $\mathbb{C}^n$, and then give generalizations to the unit ball with respect to an arbitrary norm in $\mathbb{C}^n$ and the unit ball of a Banach space. We also consider bounded balanced pseudoconvex domains in $\mathbb{C}^n$, and we briefly mention a class of domains introduced by Kikuchi. Finally we obtain starlikeness criteria for mappings of class C^1 (which need not be holomorphic). Before going further let us define the notion of a starlike mapping, which will recur.

Definition 6.2.1. Let X and Y be two complex Banach spaces and let Ω be a domain in X. Also let $f \in H(\Omega, Y)$ and $z_0 \in \Omega$. We say that f is *starlike* on Ω with respect to z_0 if f is biholomorphic on Ω and $f(\Omega)$ is a starlike domain with respect to $f(z_0)$ (thus $(1-t)f(z_0) + tf(z) \in f(\Omega)$ for all $z \in \Omega$ and $t \in [0,1]$). We shall use the term starlike to mean starlike with respect to 0. That is, a starlike mapping f on a domain $\Omega \subset X$ with $0 \in \Omega$ is a biholomorphic mapping from Ω into Y such that $f(0) = 0$ and $f(\Omega)$ is a starlike domain in Y.

If B is the unit ball of X, we let $S^*(B)$ denote the subset of $S(B)$ consisting of normalized starlike mappings from B into X.

We begin with the Euclidean unit ball in $\mathbb{C}^n$. In 1955 Matsuno [Mat] obtained (essentially) the following basic result:

Theorem 6.2.2. *Let $f : B \to \mathbb{C}^n$ be a locally biholomorphic mapping such that $f(0) = 0$. Then f is starlike if and only if*

$$(6.2.1) \qquad \mathrm{Re}\, \langle [Df(z)]^{-1} f(z), z \rangle > 0, \quad z \in B \setminus \{0\}.$$

Proof. We shall prove the necessity here; the proof of sufficiency will be given in a more general context (Theorem 6.2.5).

Assuming that f is starlike, we may define a map $v : B \times [0,1] \to B$ by $v(z,t) = f^{-1}((1-t)f(z))$. Then for fixed t, v is holomorphic in z and $v(0,t) =$

0, so the Schwarz lemma gives $\|v(z,t)\| \le \|z\|$, $0 \le t \le 1$, $z \in B$. For fixed z we obtain

$$0 \ge \lim_{t \to 0^+} \frac{\|v(z,t)\|^2 - \|z\|^2}{t} = \frac{d}{dt}\|v(z,t)\|^2\Big|_{t=0}$$

$$= -2\mathrm{Re}\,\langle [Df(z)]^{-1}f(z), z\rangle,$$

and this shows that $\mathrm{Re}\,\langle [Df(z)]^{-1}f(z), z\rangle \ge 0$.

To see that the inequality is strict, let $g(z) = [Df(z)]^{-1}f(z)$ for $z \in B$. Since g is normalized holomorphic on B and $\mathrm{Re}\,\langle g(z), z\rangle \ge 0$ for $z \in B\backslash\{0\}$, it follows from Lemma 6.1.30 that $g \in \mathcal{M}$, and hence $\mathrm{Re}\,\langle g(z), z\rangle > 0$ on $B\backslash\{0\}$. Therefore (6.2.1) is necessary for starlikeness.

In 1970 Suffridge [Su2] considered the case of the unit polydisc P in $\mathbb{C}^n$ and obtained the following necessary and sufficient condition for starlikeness in this context:

Theorem 6.2.3. *Let $f : P \to \mathbb{C}^n$ be a locally biholomorphic mapping such that $f(0) = 0$. Then f is starlike if and only if*

$$(6.2.2) \qquad\qquad f(z) = Df(z)w(z), \quad z \in P,$$

where $w \in \mathcal{M}$.

We recall that for the case of the maximum norm $\|\cdot\|_\infty$ in $\mathbb{C}^n$, the class $\mathcal{M}$ consists of those normalized holomorphic mappings $w : P \to \mathbb{C}^n$ that satisfy

$$\mathrm{Re}\left[\frac{w_j(z)}{z_j}\right] > 0, \quad \|z\|_\infty = |z_j| > 0, \quad 1 \le j \le n.$$

More generally, in the same paper [Su2] Suffridge obtained the following necessary and sufficient condition for starlikeness on the unit ball $B(p)$ with respect to a p-norm $\|\cdot\|_p$, with $1 \le p < \infty$.

Theorem 6.2.4. *Let $f : B(p) \to \mathbb{C}^n$, $1 \le p < \infty$, be a locally biholomorphic mapping such that $f(0) = 0$. Then f is starlike if and only if*

$$(6.2.3) \qquad\qquad f(z) = Df(z)w(z) \text{ on } B(p),$$

where $w \in \mathcal{M}(p)$.

The extension of these results to the case of an arbitrary norm in $\mathbb{C}^n$ was also carried out by Suffridge [Su3]. Gurganus [Gur] gave a different proof using a differential equations argument. We shall present a version of Suffridge's proof.

Theorem 6.2.5. Let $f : B \to \mathbb{C}^n$ be a locally biholomorphic mapping with $f(0) = 0$. Then f is starlike if and only if there is $w \in \mathcal{M}$ such that

$$(6.2.4) \qquad\qquad f(z) = Df(z)w(z) \text{ on } B.$$

Proof. Necessity. First assume that f is starlike. Let $F(z,t) = (1-t)f(z)$, $z \in B$, $t \in [0,1]$. Since

$$\lim_{t \to 0^+} \frac{F(z,0) - F(z,t)}{t} = f(z),$$

we may apply Lemma 6.1.35 to conclude that there is a mapping $w \in \mathcal{N}_0$ such that

$$f(z) = Df(z)w(z) \text{ on } B.$$

This equality implies that $Dw(0) = I$, and in view of Lemma 6.1.30 we conclude that $w \in \mathcal{M}$, as desired.

Sufficiency. Conversely, assume that f is locally biholomorphic on the unit ball B and there is a map $w \in \mathcal{M}$ such that equality (6.2.4) holds on B. We shall show that f is starlike.

We first observe that $f(z) \neq 0$ if $z \neq 0$. Otherwise, $f(z) = 0$ would imply that $Df(z)w(z) = 0$, and hence $w(z) = 0$, using the fact that f is locally biholomorphic on B. On the other hand, since $z \neq 0$ and $w \in \mathcal{M}$, we have $\mathrm{Re}\,[l_z(w(z))] > 0$ for $l_z \in T(z)$. However, this is impossible if $w(z) = 0$.

First step. For $z \in B \setminus \{0\}$, let V be a neighbourhood of z on which f is biholomorphic, and let f^{-1} denote the corresponding inverse of f on $f(V)$. Also let $v(z,t) = f^{-1}((1-t)f(z))$ for $|t| < \varepsilon$, where ε is a positive number such that $(1-t)f(z) \in f(V)$ for $|t| < \varepsilon$. We shall show that $\|v(z,t)\|$ is a strictly decreasing function of t for $|t|$ sufficiently small. First we observe that

$$v(z,t) = v(z,0) + [Df(z)]^{-1}(-tf(z)) + o(t) = z - tw(z) + o(t),$$

and hence for $t < 0$ with $|t|$ sufficiently small and $l_z \in T(z)$, we obtain

$$\|v(z,t)\| \geq \mathrm{Re}\,[l_z(v(z,t))] = \|z\| - t\,\mathrm{Re}\,[l_z(w(z))] + o(t) > \|z\|.$$

Moreover, applying this argument to $y = v(z, t)$ (in this case we consider $v(y, s) = f^{-1}((1-s)(1-t)f(z)))$, we deduce that $\|v(z, t)\|$ is strictly decreasing for $|t|$ sufficiently small, as claimed.

Second step. Next, for each $\tau \in [0, 1]$ consider the line segment

$$L_\tau = \left\{(1 - t)f(z) : \ 0 \le t \le \tau\right\}.$$

Let $T = \left\{\tau \in [0, 1] : f^{-1} \text{ can be analytically continued along } L_\tau \text{ with the}\right.$ initial value $\left. f^{-1}(f(z)) = z\right\}$.

We shall prove that T is nonempty, open and closed, and hence $T = [0, 1]$. It is clear that T is nonempty and open since f is locally biholomorphic. To show that T is closed, suppose that $\tau_0 \in [0, 1]$ is such that $\tau \in T$ for $\tau \in [0, \tau_0)$. Consider the curve $\gamma \colon [0, \tau_0) \to B$ defined by

$$\gamma(t) = f^{-1}((1 - t)f(z)), \quad 0 \le t < \tau_0.$$

We wish to argue that $\gamma(t)$ converges to a point $z_0 \in B$ as $t \to \tau_0^-$. But this follows from a Cauchy sequence argument using the fact that $\|\gamma(t)\|$ is a decreasing function of t and γ has finite arc length. We note that the length of γ is given by the (improper) integral

$$\int_0^{\tau_0} \|[Df(f^{-1}((1 - t)f(z)))]^{-1}(-f(z))\| dt.$$

This is finite because $\|\gamma(t)\| \le \|z\|$ for $t \in [0, \tau_0)$ and for each $r \in (0, 1)$ there exists a positive number $M = M(r)$ such that

$$\|[Df(z)]^{-1}\| \le M(r), \quad \|z\| \le r.$$

Is is now clear that f^{-1} can be analytically continued along L_{τ_0} and $f^{-1}((1 - \tau_0)f(z)) = z_0$. Hence T is closed, as asserted.

Third step. Finally we show that f is univalent. If we assume that this is not true, then there are distinct points $z, z' \in B$ such that $f(z) = f(z')$. Now f^{-1} can be analytically continued along the segment $L = \{(1 - t)f(z) : 0 \le t \le 1\}$ with either $f^{-1}(f(z)) = z$ or $f^{-1}(f(z)) = z'$ as the initial value. But these analytic continuations must coincide in a neighbourhood of $0 \in \mathbb{C}^n$, because $f(\zeta) = 0$ if and only if $\zeta = 0$. Hence they must coincide everywhere along L and $z = z'$. Since this is a contradiction, the proof is complete.

Suffridge [Su3] observed that his proof remained valid for the unit ball of a complex Banach space if one explicitly added the assumption that for each r, $0 < r < 1$, there exists $M = M(r) > 0$ such that $\|[Df(z)]^{-1}\| \leq M(r)$ when $\|z\| \leq r$. Gurganus [Gur] asserted that this assumption was not needed with his methods; a slight gap in his argument was observed and rectified by Hamada and Kohr [Ham-Koh10]. Thus we have

Theorem 6.2.6. *Let X and Y be two complex Banach spaces and let B denote the unit ball of X. Let $f : B \to Y$ be a locally biholomorphic mapping with $f(0) = 0$. Then f is starlike if and only if*

$$(6.2.5) \qquad f(z) = Df(z)w(z), \quad z \in B,$$

where $w \in \mathcal{M}$.

6.2.2 Starlikeness criteria on more general domains in $\mathbb{C}^n$

It is natural to study starlike mappings of domains which are somewhat more general than the unit ball with respect to some norm. The first person to do so was Kikuchi [Kik] in 1973. His assumptions on the domain are formulated in terms of the Bergman kernel function, which is defined for example in [Ber], [Kran], [Ran].

Theorem 6.2.7. *Let Ω be a bounded pseudoconvex domain in $\mathbb{C}^n$ for which the Bergman kernel function $K_\Omega(z, z)$ assumes its minimum value only at $z = 0$, tends to ∞ at all boundary points, and satisfies $K_\Omega(g(z), g(z)) \leq K_\Omega(z, z)$ for any holomorphic mapping g of Ω into Ω such that $g(0) = 0$. A locally biholomorphic mapping $f : \Omega \to \mathbb{C}^n$ such that $f(0) = 0$ is starlike if and only if*

$$\mathrm{Re}\,\left\langle [Df(z)]^{-1}f(z), \frac{\partial K_\Omega}{\partial \bar{z}}(z, z) \right\rangle > 0, \quad z \in \Omega \setminus \{0\}.$$

However, in practice it may be difficult to check whether a given domain satisfies Kikuchi's assumptions.

Starlike mappings on Reinhardt domains were studied by Gong, Wang, and Yu [Gon-Wa-Yu2]. However, perhaps the most natural class of domains to consider in this context is the class of bounded balanced pseudoconvex domains whose Minkowski function is of class C^1 in $\mathbb{C}^n \setminus \{0\}$. The following result is due independently to Liu and Ren [Liu-Ren2] (necessity) and Gong, Wang, and Yu [Gon-Wa-Yu5] (sufficiency), and to Hamada [Ham4]. Other proofs appear in [Gon4], [Ham-Koh3], and [Ham-Koh-Lic1].

Theorem 6.2.8. *Let Ω be a bounded balanced pseudoconvex domain in $\mathbb{C}^n$ whose Minkowski function h is of class C^1 in $\mathbb{C}^n \setminus \{0\}$. Let $f : \Omega \to \mathbb{C}^n$ be a locally biholomorphic mapping such that $f(0) = 0$. Then f is starlike if and only if*

$$(6.2.6) \qquad \operatorname{Re} \left\langle [Df(z)]^{-1} f(z), \frac{\partial h^2}{\partial \bar{z}}(z) \right\rangle > 0, \quad z \in \Omega \setminus \{0\}.$$

Proof. Assume that f is starlike. Since $(1 - t)f(z) \in f(\Omega)$ for all $z \in \Omega$ and $t \in [0, 1]$, we may define a mapping $v : \Omega \times [0, 1] \to \Omega$ by $v(z, t) = f^{-1}((1-t)f(z))$. Then $v(\cdot, t)$ is holomorphic on Ω, $v(0, t) = 0$ for each $t \in [0, 1]$, and from Lemma 6.1.22 one concludes that $h(v(z, t)) \leq h(z)$ for $z \in \Omega$ and $t \in [0, 1]$. Fixing $z \in \Omega \setminus \{0\}$, we therefore obtain

$$0 \geq \lim_{t \to 0^+} \frac{h^2(v(z, t)) - h^2(z)}{t} = \frac{d}{dt} h^2(v(z, t)) \Big|_{t=0}$$

$$= -2\operatorname{Re} \left\langle [Df(z)]^{-1} f(z), \frac{\partial h^2}{\partial \bar{z}}(z) \right\rangle,$$

making use of the chain rule.

At this point we have shown that

$$\operatorname{Re} \left\langle [Df(z)]^{-1} f(z), \frac{\partial h^2}{\partial \bar{z}}(z) \right\rangle \geq 0, \quad z \in \Omega \setminus \{0\}.$$

Now since $h^2(\zeta w) = |\zeta|^2 h^2(w)$ for $\zeta \in \mathbb{C}$ and $w \in \mathbb{C}^n$, it follows that

$$\left\langle \frac{\partial h^2}{\partial z}(\zeta w), \overline{w} \right\rangle = \bar{\zeta} h^2(w), \quad \zeta \in \mathbb{C} \setminus \{0\}, \ w \in \mathbb{C}^n \setminus \{0\},$$

and setting $\zeta = 1$ in the above, we have

$$\left\langle \frac{\partial h^2}{\partial z}(w), \overline{w} \right\rangle = h^2(w), \quad w \in \mathbb{C}^n \setminus \{0\}.$$

For fixed $w \in \partial\Omega$, let

$$q_w(\zeta) = \begin{cases} \left\langle \dfrac{g(\zeta w)}{\zeta}, \dfrac{\partial h^2}{\partial \overline{z}}(w) \right\rangle, & 0 < |\zeta| < 1 \\ 1, & \zeta = 0, \end{cases}$$

where $g(z) = [Df(z)]^{-1} f(z)$, $z \in \Omega$. Since g is holomorphic on Ω, $g(0) = 0$ and $Dg(0) = I$, it follows that q_w is holomorphic on U and

$$\mathrm{Re}\, q_w(\zeta) = \frac{1}{|\zeta|^2} \mathrm{Re}\, \left\langle g(\zeta w), \frac{\partial h^2}{\partial \overline{z}}(\zeta w) \right\rangle \geq 0, \quad 0 < |\zeta| < 1.$$

Since $q_w(0) = 1$ and $\mathrm{Re}\, q_w(\zeta) \geq 0$ on U, we must have $\mathrm{Re}\, q_w(\zeta) > 0$ on U, by the minimum principle for harmonic functions. Consequently, we have proved that

$$\mathrm{Re}\, \left\langle g(\zeta w), \frac{\partial h^2}{\partial \overline{z}}(\zeta w) \right\rangle > 0, \quad 0 < |\zeta| < 1.$$

The desired conclusion (6.2.6) follows, completing the proof of necessity.

The proof of sufficiency is similar to that in Theorem 6.2.5 and we leave it for the reader. (The argument is sketched again in Theorem 6.2.10.)

Further criteria for starlikeness in several complex variables may be found in the recent monograph of Gong [Gon4].

6.2.3 Sufficient conditions for starlikeness for mappings of class C^1

We next indicate how the idea of starlikeness may be extended to the case of mappings of class C^1 (which are not necessarily holomorphic). In this context we define a starlike map as follows.

Definition 6.2.9. Let Ω be a domain in $\mathbb{C}^n$ such that $0 \in \Omega$ and let $C^1(\Omega)$ denote the set of mappings of class C^1 from Ω into $\mathbb{C}^n$. If $f \in C^1(\Omega)$ we say that f is starlike if f is injective on Ω, $f(0) = 0$ and $f(\Omega)$ is a starlike domain in $\mathbb{C}^n$ (with respect to zero).

For a C^1 mapping f from a domain Ω in $\mathbb{C}^n$ into $\mathbb{C}^n$, let

$$J_r f(z) = \det \frac{\partial(u_1, v_1, \ldots, u_n, v_n)}{\partial(x_1, y_1, \ldots, x_n, y_n)}$$

denote the determinant of the real Jacobian matrix of f at z, where $z_j = x_j + iy_j$ and $f_j = u_j + iv_j$, $j = 1, \ldots, n$. When f is holomorphic on Ω we have $J_r f(z) = |J_f(z)|^2$, $z \in \Omega$.

Let $f \in C^1(\Omega)$ and suppose $J_r f(z) \neq 0$, $z \in \Omega$. Then for each $z \in \Omega$ there exists a neighbourhood $V_z \subset \Omega$ of z such that f is a diffeomorphism of class C^1 from V_z onto $f(V_z)$. Therefore we can define the matrices

$$D_w f^{-1}(w) = \left[\frac{\partial (f|_{V_z})_j^{-1}}{\partial w_k}(w) \right]_{1 \leq j,k \leq n}$$

and

$$D_{\overline{w}} f^{-1}(w) = \left[\frac{\partial (f|_{V_z})_j^{-1}}{\partial \overline{w}_k}(w) \right]_{1 \leq j,k \leq n},$$

where $w = f(z)$.

Recently Kohr [Koh4] obtained a sufficient condition for starlikeness for mappings of class C^1 on domains for which the Bergman kernel function satisfies Kikuchi's conditions. Also Hamada and Kohr [Ham-Koh3] obtained the following result on bounded balanced pseudoconvex domains for which the Minkowski function is C^1 on $\mathbb{C}^n \setminus \{0\}$:

Theorem 6.2.10. *Let Ω be a bounded balanced pseudoconvex domain in $\mathbb{C}^n$ whose Minkowski function h is of class C^1 in $\mathbb{C}^n \setminus \{0\}$. Let $f \in C^1(\Omega)$ be such that $f(0) = 0$. If $J_r f(z) \neq 0$, $z \in \Omega$, and if*

$$(6.2.7) \quad \mathrm{Re} \left\langle [D_w f^{-1}(f(z))]f(z) + [D_{\overline{w}} f^{-1}(f(z))]\overline{f(z)}, \frac{\partial h^2}{\partial \overline{z}}(z) \right\rangle > 0$$

on $\Omega \setminus \{0\}$, then f is starlike.

Proof. We shall give only a sketch of the proof, since it is similar to the proof of sufficiency in Theorem 6.2.5. Since $J_r f(z) \neq 0$ on Ω, f is a local diffeomorphism.

As in the first step in Theorem 6.2.5, we may define $v(z,t) = f^{-1}((1 - t)f(z))$ for $z \in \Omega \setminus \{0\}$ and $|t|$ sufficiently small. Using (6.2.7) and the chain rule, we deduce that $h^2(v(z,t))$ is strictly decreasing for $|t|$ sufficiently small.

The second and third steps may also be carried out as before. We are no longer dealing with *analytic* continuation, but we may still speak of *continuing*

f^{-1} along a line segment since f is a local diffeomorphism. This completes the proof.

Of course if $f \in H(\Omega)$, then Theorem 6.2.10 reduces to the sufficiency condition in Theorem 6.2.8. Further, if Ω is the unit ball $B(p)$ with respect to a p-norm $\| \cdot \|_p$, where $1 < p < \infty$, then $h(z) = \|z\|_p$, $z \in \mathbb{C}^n$, and we obtain the sufficient condition for starlikeness in Theorem 6.2.4. On the other hand, if $n = 1$ and Ω is the unit disc, then Theorem 6.2.10 gives the following result of Mocanu [Moc2]:

Corollary 6.2.11. *Let $f \in C^1(U)$ satisfy the conditions*
(i) $f(0) = 0$ and $f(z) \neq 0$, $z \in U \setminus \{0\}$,
(ii) $J_r f(z) > 0$, $z \in U$,

$$\text{(iii) } \mathrm{Re} \left[\frac{z \dfrac{\partial f}{\partial z}(z) - \overline{z} \dfrac{\partial f}{\partial \overline{z}}(z)}{f(z)} \right] > 0, \ z \in U \setminus \{0\}.$$

Then f is starlike.

Proof. For all $\zeta \in U$ there exists a neighbourhood $V_\zeta \subset U$ of ζ such that $f : V_\zeta \to f(V_\zeta)$ is a diffeomorphism of class C^1. Let $g(w) = f^{-1}(w)$ on $f(V_\zeta)$. After simple computations, one deduces the relations

$$\frac{\partial g}{\partial w} = \frac{1}{J_r f} \cdot \frac{\partial \overline{f}}{\partial \overline{z}} \quad \text{and} \quad \frac{\partial g}{\partial \overline{w}} = -\frac{1}{J_r f} \cdot \frac{\partial f}{\partial \overline{z}}.$$

The result now follows from Theorem 6.2.10, making use of the fact that $h(z) = |z|$ for all $z \in \mathbb{C}$.

6.2.4   Starlikeness of order γ in $\mathbb{C}^n$

In Chapter 2 we considered functions in S which are starlike of a given order, and proved the Marx-Strohhäcker theorem which states that a normalized convex function is starlike of order $1/2$. This result can be generalized to several complex variables [Cu2], [Koh2] (also see [Ham-Koh-Lic2]), as we shall see in Section 6.3.3.

In this section we shall introduce the necessary definitions. Let B be the unit ball of $\mathbb{C}^n$ with respect to an arbitrary norm.

Definition 6.2.12. Let $f: B \to \mathbb{C}^n$ be a locally biholomorphic mapping such that $f(0) = 0$, and let $\gamma \in (0,1)$. We say that f is starlike of order γ if

$$\left| \frac{1}{\|z\|} l_z([Df(z)]^{-1}f(z)) - \frac{1}{2\gamma} \right| < \frac{1}{2\gamma}, \quad z \in B \setminus \{0\}, \ l_z \in T(z).$$

Equivalently we may express this as

$$(6.2.8) \qquad \mathrm{Re} \left\{ \frac{\|z\|}{l_z([Df(z)]^{-1}f(z))} \right\} > \gamma, \quad z \in B \setminus \{0\}, \ l_z \in T(z),$$

or as

$$(6.2.9) \qquad \qquad \mathrm{Re} \left\{ \frac{\alpha}{l_u([Df(\alpha u)]^{-1}f(\alpha u))} \right\} > \gamma$$

for all $\alpha \in U \setminus \{0\}$, $u \in \mathbb{C}^n$, $\|u\| = 1$, and $l_u \in T(u)$.

It is easy to see that in one variable the condition (6.2.8) reduces to $\mathrm{Re} \left\{ \frac{zf'(z)}{f(z)} \right\} > \gamma$, that is the usual condition for starlikeness of order γ in the unit disc U.

Notes. There are now many references for starlike mappings in higher dimensions. For more information see the recent monographs [Gon4], [Koh-Lic2]. Some of the original papers are [Bar-Fit-Gon1], [Gon-Wa-Yu1,2,4,5], [Gra-Koh1], [Gra-Koh-Koh], [Gur], [Ham4], [Ham-Koh1,3], [Kik], [Koh1,4,7], [Liu-Ren2], [Pfa-Su3], [Por5], [Rop-Su2], [Su2,3,4], [Chu2], [Kub-Por].

Problems

6.2.1. Let $1 < p < \infty$ and let $f : B(p) \subset \mathbb{C}^2 \to \mathbb{C}^2$ be given by $f(z) = (z_1 + az_2^2, z_2)$, $z = (z_1, z_2) \in B(p)$. Show that f is starlike if and only if

$$|a| \le \left(\frac{p^2 - 1}{4} \right)^{1/p} \left(\frac{p+1}{p-1} \right).$$

(Suffridge, 1975 [Su4], Roper and Suffridge, 1999 [Rop-Su2].)

6.2.2. Let $1 \le p \le \infty$ and $f : B(p) \subset \mathbb{C}^2 \to \mathbb{C}^2$ be given by $f(z) = (z_1 + az_1z_2, z_2)$, $z = (z_1, z_2) \in B(p)$. Show that f is starlike if and only if $|a| \le 1$.

(Roper and Suffridge, 1999 [Rop-Su2].)

6.2.3. Let

$$f_1(z) = \left(\frac{z_1}{(1-z_1)^2}, \frac{z_2}{(1-z_2)^2}\right) \quad \text{and} \quad f_2(z) = \left(\frac{z_1}{(1+z_1)^2}, \frac{z_2}{(1+z_2)^2}\right)$$

for $z = (z_1, z_2) \in B$. Show that $f = (f_1 + f_2)/2$ is not starlike on the Euclidean unit ball of $\mathbb{C}^2$. (In fact f does not belong to $S(B)$.)

6.2.4. Let X be a complex Hilbert space with the inner product $\langle \cdot, \cdot \rangle$ and the induced norm $\| \cdot \|$. Let f be a normalized starlike function on the unit disc U. Also let $z_0 \in X$ with $\|z_0\| = 1$ and let $F : B \to X$ be given by

$$F(z) = \frac{f(\langle z, z_0 \rangle)z}{\langle z, z_0 \rangle}, \quad z \in B.$$

Show that F is starlike.
(Suffridge, 1975 [Su4].)

Hint. Apply Theorem 6.2.6.

6.2.5. Let $f_1, \ldots, f_n$ be normalized starlike functions on the unit disc U. Let $1 \le p \le \infty$ and $f : B(p) \to \mathbb{C}^n$ be given by $f(z) = (f_1(z_1), \ldots, f_n(z_n))$, $z = (z_1, \ldots, z_n) \in B(p)$. Show that f is starlike.

6.2.6. Let $p_j > 1$, $j = 1, \ldots, n$, and let $B(p_1, \ldots, p_n) \subset \mathbb{C}^n$ be the domain

$$B(p_1, \ldots, p_n) = \left\{ z = (z_1, \ldots, z_n) \in \mathbb{C}^n : \sum_{j=1}^{n} |z_j|^{p_j} < 1 \right\}.$$

Show that if $f : B(p_1, \ldots, p_n) \to \mathbb{C}^n$ is given by

$$f(z) = \left(\frac{z_1}{(1-z_1)^2}, \ldots, \frac{z_n}{(1-z_n)^2}\right),$$

then f is starlike.
(Hamada, 2000 [Ham4].)

6.3   Criteria for convexity

6.3.1   Criteria for convexity on the unit polydisc and the Euclidean unit ball

The present section deals with the study of certain necessary and sufficient conditions for convexity on the Euclidean unit ball and unit polydisc of $\mathbb{C}^n$. In

Section 6.3.2 we shall study the notion of convexity on the unit ball of complex Hilbert and Banach spaces.

We begin with the definition of a convex mapping from a domain in a complex Banach space X into another complex Banach space Y.

Definition 6.3.1. Let X and Y be two complex Banach spaces and $\Omega \subset X$ be a domain. Also let $f : \Omega \to Y$ be a holomorphic mapping. We say that f is *convex* on Ω if f is biholomorphic on Ω and $f(\Omega)$ is a convex domain in Y (thus $(1 - t)f(z) + tf(w) \in f(\Omega)$ for $z, w \in \Omega$ and $t \in [0,1]$).

If B is the unit ball of X, we let $K(B)$ denote the subset of $S(B)$ consisting of normalized convex mappings from B into X.

In 1970 Suffridge [Su2] obtained a remarkable structure theorem for convex maps of the unit polydisc P of $\mathbb{C}^n$. (Thus we consider $\mathbb{C}^n$ with the maximum norm $\| \cdot \|_\infty$.) This result gives a complete characterization of convexity in the polydisc.

Theorem 6.3.2. *Let $f : P \to \mathbb{C}^n$ be a locally biholomorphic mapping such that $f(0) = 0$. Then f is convex if and only if there exist convex functions φ_j, $1 \leq j \leq n$, on the unit disc U such that*

$$(6.3.1) \qquad f(z) = M(\varphi_1(z_1), \ldots, \varphi_n(z_n)), \quad z = (z_1, \ldots, z_n) \in P,$$

where $M \in L(\mathbb{C}^n, \mathbb{C}^n)$ is a nonsingular linear transformation.

Proof. It is obvious that if f satisfies equality (6.3.1) with $\varphi_1, \ldots, \varphi_n$ convex functions on U, then f is convex. Therefore, we have only to show the converse.

Assume that $f = (f_1, \ldots, f_n)$ is a convex mapping of P. We shall show that $\dfrac{\partial^2 f_m}{\partial z_k \partial z_l} \equiv 0$ on P, for $m, k, l = 1, \ldots, n$, $k \neq l$. For $A_j \geq 0$, $1 \leq j \leq n$, let

$$A_t(z) = (z_1 e^{iA_1 t}, z_2 e^{iA_2 t}, \ldots, z_n e^{iA_n t}), \quad -1 \leq t \leq 1.$$

Then $\|A_t(z)\|_\infty = \|z\|_\infty$ and if

$$F(z,t) = \frac{1}{2}[f(A_t(z)) + f(A_{-t}(z))], \quad z \in P, \quad 0 \leq t \leq 1,$$

then $F(\cdot, t) \in H(P)$, $F(0, t) = 0$, $t \geq 0$, and $F(z, 0) = f(z)$, $z \in P$. Moreover, since f is convex, $F(P, t) \subseteq f(P)$. Straightforward computations yield that the limit

$$\lim_{t\to 0+} \frac{F(z,0) - F(z,t)}{t^2} = g(z) = (g_1(z), \ldots, g_n(z)), \quad z \in P,$$

exists, where

(6.3.2)
$$2g_j = \sum_{k=1}^{n} A_k^2 \left[z_k^2 \frac{\partial^2 f_j}{\partial z_k^2} + z_k \frac{\partial f_j}{\partial z_k} \right]$$

$$+ 2 \sum_{k=2}^{n} \sum_{l=1}^{k-1} A_k A_l z_k z_l \frac{\partial^2 f_j}{\partial z_l \partial z_k}.$$

Since this limit is holomorphic on P, it follows from Lemma 6.1.35 (for $\rho = 2$) that there is a mapping $w \in \mathcal{N}_0$ such that

$$g(z) = Df(z)w(z), \quad z \in P.$$

Let $J_f(z) = \det Df(z)$, $z \in P$. Then $w_j = \dfrac{\det D^{(j)}}{J_f}$ where $D^{(j)}$ is obtained from Df by replacing the jth column by g written as a column vector.

Now for fixed k, $1 \le k \le n$, let $A_k = 1$ and $A_l = 0$, $l \ne k$, $1 \le l \le n$. For $z \in P$ such that $\|z\|_\infty = |z_j| > 0$, $j \ne k$ and $z_k = 0$, we have $g(z) = 0$ and hence $w_j(z) = 0$. Since Re $[w_j/z_j] \ge 0$ for $\|z\|_\infty = |z_j| > 0$ and w_j/z_j assumes the value 0, we deduce by the minimum principle for harmonic functions that $w_j \equiv 0$. Consequently, we conclude for $1 \le j \le n$ and $1 \le k \le n$ that

(6.3.3)
$$z_k^2 \frac{\partial^2 f_j}{\partial z_k^2} + z_k \frac{\partial f_j}{\partial z_k} = \frac{\partial f_j}{\partial z_k} h_k$$

(replacing w_k by h_k), where Re $[h_k(z)/z_k] \ge 0$ for $\|z\|_\infty = |z_k| > 0$.

With k as before, fix l, $1 \le l \le n$, $l \ne k$, and let $A_k = 1$, $A_l = \varepsilon > 0$ and $A_m = 0$, $1 \le m \le n$, $m \ne k, l$. In view of (6.3.3) we deduce that there is a (different) mapping $w \in \mathcal{N}_0$ such that

$$w_j = \varepsilon \frac{z_k z_l G_j}{J_f} + O(\varepsilon^2), \quad j \ne k,$$

where G_j is obtained from J_f by replacing the jth column by the column $\partial^2 f_m/\partial z_l \partial z_k$, $1 \le m \le n$. Letting $\varepsilon \to 0$ we see that

$$\text{Re} \left[\frac{z_k z_l G_j}{z_j J_f} \right] \ge 0 \text{ for } \|z\|_\infty = |z_j| > 0,$$

and since Re $\left[\dfrac{z_k z_l G_j}{z_j J_f}\right] = 0$ for $z_k z_l = 0$, we deduce that $G_j \equiv 0$ for $j \neq k$. In fact this relation is true for $j = k$ also, for we may interchange the roles of l and k without affecting the value of G_j.

On the other hand, because the system of equations

$$\sum_{j=1}^{n} \frac{\partial f_m}{\partial z_j}\psi_j = \frac{\partial^2 f_m}{\partial z_k \partial z_l}, \quad 1 \leq m \leq n$$

has the solution

$$\psi_j = \frac{G_j}{J_f} = 0, \; 1 \leq j \leq n,$$

we deduce that

$$\frac{\partial^2 f_m}{\partial z_k \partial z_l} = 0, \quad 1 \leq m \leq n.$$

Solving this equation, we obtain

(6.3.4) $$f_m(z) = \sum_{j=1}^{n} c_{jm}\varphi_{jm}(z_j), \quad 1 \leq m \leq n,$$

where φ_{jm} is a holomorphic function on the unit disc U.

Substituting (6.3.4) into (6.3.3), we obtain

$$c_{km}z_k^2\varphi''_{km} + c_{km}z_k\varphi'_{km} = c_{km}h_k\varphi'_{km}, \quad k, m = 1, \dots, n.$$

We may assume that f is normalized since this may be achieved by multiplying f by a nonsingular matrix. Then we may take $c_{mm} = 1$ and $\varphi'_{mm}(0) = 1$, and we obtain

$$z_m^2\varphi''_{mm} + z_m\varphi'_{mm} = h_m\varphi'_{mm}.$$

Using the fact that $\varphi'_{mm} \not\equiv 0$ and Re $(h_m(z)/z_m) \geq 0$ when $\|z\|_\infty = |z_m| > 0$, it is not hard to deduce from this equation that h_m is a function only of z_m and that φ_{mm} is convex.

It remains to show that $\varphi_{km} = \varphi_{kk}$ for $k \neq m$. If $c_{km} = 0$ we may replace φ_{km} by φ_{kk}. Otherwise we obtain

$$z_k^2\varphi''_{km} + z_k\varphi'_{km} = h_k\varphi'_{km},$$

and since we may take $\varphi_{km}(0) = 0$, φ_{km} differs from φ_{kk} at most by a multiplicative constant, which we may absorb into c_{km}.

We remark that Liu and Ren [Liu-Ren3] obtained an extension of Suffridge's theorem to the case of products of bounded convex circular domains in $\mathbb{C}^n$. We only state their result and we leave the proof for the reader.

Theorem 6.3.3. Let $\Omega_1 \subset \mathbb{C}^{n_1}, \ldots, \Omega_k \subset \mathbb{C}^{n_k}$ be bounded convex circular domains, whose Minkowski functions $h_1, \ldots, h_k$ are real-analytic (except for a lower dimensional set). Let $Z^j = (z_1^j, \ldots, z_{n_j}^j)$ denote the coordinates in $\mathbb{C}^{n_j}$, $j = 1, \ldots, k$. Suppose that $f : \Omega_1 \times \cdots \times \Omega_k \to \mathbb{C}^{n_1 + \cdots + n_k}$ is a locally biholomorphic mapping such that $f(0, \ldots, 0) = 0$. Then f is a convex mapping if and only if there exist convex mappings $\phi_j : \Omega_j \to \mathbb{C}^{n_j}$, $j = 1, \ldots, k$, such that

$$f(Z^1, \ldots, Z^k) = T(\phi_1(Z^1), \ldots, \phi_k(Z^k))$$

where T is a nonsingular linear transformation.

Another case in which the convex maps have a very rigid structure is that of the unit ball with respect to the 1-norm in $\mathbb{C}^n$. Suffridge showed that any convex map of $B(1)$ must be linear [Su3,4].

In 1970 Suffridge [Su2] obtained other necessary and sufficient conditions for convexity on the unit ball of $\mathbb{C}^n$ with respect to a p-norm, where $1 \leq p < \infty$, and in 1973 Kikuchi [Kik] obtained a necessary and sufficient condition for convexity on bounded pseudoconvex domains in $\mathbb{C}^n$ for which the Bergman kernel function becomes infinite everywhere on the boundary. In particular Kikuchi [Kik] obtained the following fundamental criterion for convexity for holomorphic mappings of the Euclidean unit ball. We remark that in 1993 Gong, Wang, and Yu [Gon-Wa-Yu3] obtained an equivalent condition via a different method.

Theorem 6.3.4. Let $f : B \to \mathbb{C}^n$ be a locally biholomorphic mapping. Then f is convex if and only if

(6.3.5) $$1 - \mathrm{Re}\, \langle [Df(z)]^{-1} D^2 f(z)(v, v), z \rangle > 0,$$

for all $z \in B$ and $v \in \mathbb{C}^n$ with $\|v\| = 1$ and $\mathrm{Re}\, \langle z, v \rangle = 0$.

Proof. First assume that f is a convex mapping. We need to make use of the fact that in this case $f(B_r)$ is a convex domain for any $r \in (0, 1)$. This we shall prove more generally for convex maps of the unit ball of a Banach space (Lemma 6.3.7).

We now use an argument of Hamada and Kohr [Ham-Koh10]. Let $z \in B \backslash \{0\}$ and $v \in \mathbb{C}^n \backslash \{0\}$ be such that Re $\langle z, v \rangle = 0$. Also let $r = \|z\|$ and define

$$g(t) = \|f^{-1}(f(z) + tDf(z)v)\|^2$$

$$= \langle f^{-1}(f(z) + tDf(z)v), f^{-1}(f(z) + tDf(z)v) \rangle$$

for t in a neighbourhood of zero. Then $g(0) = \|z\|^2 = r^2$ and $g'(0) = 2\text{Re } \langle z, v \rangle = 0$. If $g''(0) < 0$, then we would have $g(t) < g(0) = r^2$ for $t \neq 0$ near zero, and $f(z) + tDf(z)v \in f(B_r)$ for nonzero t near 0. However, since $f|_{B_r}$ is convex, this would imply that $f(z) \in f(B_r)$ which is false.

Therefore, we must have $g''(0) \geq 0$. This gives

$$g''(0) = 2\|v\|^2 + 2\text{Re } \langle D^2 f^{-1}(f(z))(Df(z)v, Df(z)v), z \rangle$$

$$= 2\|v\|^2 - 2\text{Re } \langle [Df(z)]^{-1}D^2 f(z)(v, v), z \rangle \geq 0,$$

which is equivalent to (6.3.5) except that we have not shown that the inequality is strict. To obtain strictness of the inequality, we argue as follows:

Since Re $\langle z, v \rangle = 0$, one can replace $z \neq 0$ by αZ and v by $\alpha V/|\alpha|$ with $0 < |\alpha| < 1$, $\|Z\| = \|V\| = 1$ and Re $\langle Z, V \rangle = 0$, to obtain

$$1 - \text{Re } \langle [Df(z)]^{-1}D^2 f(z)(v, v), z \rangle$$

$$= 1 - \text{Re } \left\{ \alpha \langle [Df(\alpha Z)]^{-1}D^2 f(\alpha Z)(V, V), Z \rangle \right\}.$$

Since Re $\left\{ \alpha \langle [Df(\alpha Z)]^{-1}D^2 f(\alpha Z)(V, V), Z \rangle \right\}$ is a harmonic function of α, one concludes using the minimum principle for harmonic functions that the inequality (6.3.5) is strictly satisfied for all $z \in B \backslash \{0\}$ and $v \in \mathbb{C}^n$, $\|v\| = 1$ such that Re $\langle z, v \rangle = 0$.

Before proving the converse, we make the following additional remark on the proof of necessity:

Letting $S_r = \partial B_r$ for $0 < r < 1$, we note that $f(S_r)$ is a hypersurface with defining function $\psi(w) = \|f^{-1}(w)\|^2 - r^2$, i.e. $f(S_r) = \{w \in f(B) : \psi(w) = 0\}$. Moreover, if $z \in S_r$ and v varies over unit vectors which are tangent to S_r at

z, i.e. such that Re $\langle z, v \rangle = 0$, then $Df(z)v$ varies over all directions in the (real) tangent space to the hypersurface $f(S_r)$. Hence the condition $g''(0) > 0$, for all such v, says precisely that the (real) Hessian of ψ is positive definite on the real tangent space to $f(S_r)$ at $f(z)$. Equivalently, the second fundamental form of $f(S_r)$ is positive definite. In other words, $f(S_r)$ is a strongly convex hypersurface.

We now prove that the condition (6.3.5) is also sufficient for convexity. To this end, we shall use an argument similar to that in [Gon-Wa-Yu3].

First, we show that if f is biholomorphic on B_r with $0 < r < 1$, then $f(B_r)$ is a convex domain. Without loss of generality, we may assume that $f(0) = 0$.

Since f is biholomorphic on B_r, $f(S_\mu)$ is a real hypersurface for any μ with $0 < \mu < r$. Reversing the argument in the first part of the proof, we conclude from (6.3.5) that for any $z \in S_\mu$ and $v \in \mathbb{C}^n$, $\|v\| = 1$, satisfying Re $\langle z, v \rangle = 0$, we have $g''(0) > 0$. Hence $f(S_\mu)$ is a strongly convex hypersurface for any μ, $0 < \mu < r$. Therefore $f(B_r)$ is a convex domain (see [Kran, Propositions 3.1.6 and 3.1.7]).

Next we show that if f is biholomorphic on B_r then f is injective on $\overline{B}_r$. It is elementary to show that $f(\overline{B}_r)$ is a convex set, in particular a starlike set. If f is not injective on $\overline{B}_r$, then we can find distinct points $z, z' \in \overline{B}_r$ and $w \in f(\overline{B}_r)$ such that $f(z) = f(z') = w$. Since f is univalent on B_r and is locally biholomorphic on B, it follows that $w \neq 0$. The argument in the third step of the proof of Theorem 6.2.5 now shows that $z = z'$, contrary to the assumption.

Finally we consider the set

$$\mathcal{R} = \left\{ r \in (0, 1] : f \text{ is biholomorphic on } B_r \right\}.$$

We shall show that $\mathcal{R} = (0, 1]$. It is clear that $\mathcal{R}$ is nonempty, and it is elementary to see that $\mathcal{R}$ is closed in $(0, 1]$. The proof will be complete if we can show that $\mathcal{R}$ is open in $(0, 1]$.

If $\mathcal{R}$ is not open, then there exists $r \in \mathcal{R}$, a sequence $\{\varepsilon_p\}_{p \in \mathbb{N}}$, $\varepsilon_p > 0$, $\lim_{p \to \infty} \varepsilon_p = 0$, and two sequences $\{z_p\}_{p \in \mathbb{N}}$ and $\{z'_p\}_{p \in \mathbb{N}}$ such that $z_p, z'_p \in B_{r+\varepsilon_p}$, $z_p \neq z'_p$, and $f(z_p) = f(z'_p)$, for all $p = 1, 2, \ldots$. Since $\{z_p\}_{p \in \mathbb{N}}$ and $\{z'_p\}_{p \in \mathbb{N}}$ are bounded sequences, there exist subsequences $\{z_{p_k}\}_{k \in \mathbb{N}}$, $\{z'_{p_k}\}_{k \in \mathbb{N}}$ of $\{z_p\}_{p \in \mathbb{N}}$

and $\{z_p'\}_{p\in\mathbb{N}}$, which converge to z and z' respectively. It is clear that $z, z' \in \overline{B}_r$ and that $f(z) = f(z')$. If $z \neq z'$, we obtain a contradiction to the fact that f is injective on $\overline{B}_r$. However if $z = z'$ we have a contradiction to the assumption that f is locally biholomorphic on B. Hence we must have $\mathcal{R} = (0, 1]$ as claimed, and f is biholomorphic on B.

We remark that in the case $n = 1$ the condition (6.3.5) reduces to the usual condition for convexity on the unit disc U, given in Theorem 2.2.3. Indeed, let $z \in U$ and $v \in \mathbb{C}$ with $|v| = 1$ and Re $[z\overline{v}] = 0$. Then $\overline{z}v + z\overline{v} = 0$ and substituting from this equation in (6.3.5), we obtain

$$0 < 1 - \text{Re}\left\{\overline{z}v^2\frac{f''(z)}{f'(z)}\right\} = 1 + \text{Re}\left\{\frac{zf''(z)}{f'(z)}\right\},$$

i.e.

$$\text{Re}\left\{1 + \frac{zf''(z)}{f'(z)}\right\} > 0, \quad z \in U.$$

6.3.2 Necessary and sufficient conditions for convexity in complex Banach spaces

We next obtain some necessary and sufficient conditions for convexity on the unit ball of a complex Banach space. First we note a condition considered by Suffridge [Su3] and Roper and Suffridge [Rop-Su2], which is necessary but not sufficient for convexity. (On the Euclidean unit ball in $\mathbb{C}^n$ it says that the second fundamental form of $f(\partial B_r)$ at $f(z)$ is positive in the direction of $Df(z)(iz)$.)

Theorem 6.3.5. *Let X and Y be two complex Banach spaces and let B denote the unit ball of X. If $f : B \to Y$ is convex then*

(6.3.6) $D^2f(z)(z, z) + Df(z)z = Df(z)w(z), \quad z \in B,$

where $w \in \mathcal{M}$.

Proof. Without loss of generality, we may assume that $f(0) = 0$. Let

$$F(z, t) = \frac{1}{2}\left[f\left(e^{it}z\right) + f\left(e^{-it}z\right)\right], \quad z \in B, \quad t \in [0, 1].$$

Then $F(\cdot, t)$ is holomorphic on B, $F(z, 0) = f(z)$, $F(0, t) = 0$, and $F(z, t) \in f(B)$ for $z \in B$, $t \in [0, 1]$. Moreover, two applications of L'Hôpital's rule yield

$$g(z) = \lim_{t \to 0^+} \frac{f(z) - \dfrac{1}{2} \left[f\left(e^{it}z\right) + f\left(e^{-it}z\right) \right]}{t^2}$$

$$= \frac{1}{2}[Df(z)(z) + D^2 f(z)(z, z)].$$

Hence $F(z, t)$ satisfies the hypotheses of Lemma 6.1.35 with $\rho = 2$, and (6.3.6) follows with $w \in \mathcal{N}_0$. It is clear from (6.3.6) that $Dw(0) = I$, and Lemma 6.1.30 yields that $w \in \mathcal{M}$. This completes the proof.

An elementary example of Suffridge [Su3] shows that the condition (6.3.6) is not sufficient for convexity in higher dimensions:

Example 6.3.6. Let $\mathbb{C}^2$ be the space of two complex variables with the maximum norm $\| \cdot \|_\infty$, and let $f : P \subset \mathbb{C}^2 \to \mathbb{C}^2$ be given by

$$f(z) = (z_1 + z_2^2/2, z_2), \quad z = (z_1, z_2) \in P.$$

Then f satisfies the relation (6.3.6). However f is not convex on P, because f is not of the form (6.3.1).

A basic result about convex maps is the following lemma due to Suffridge [Su3]. The notations are the same as in Theorem 6.3.5.

Lemma 6.3.7. *If $f : B \to Y$ is a convex mapping then $f(B_r)$ is a convex domain, for all $r \in (0, 1)$.*

Proof. The space $X \times X$ is a Banach space with respect to the norm $\|(z, z')\| = \max\{\|z\|, \|z'\|\}$. The unit ball of this space is $B \times B$. For fixed $t \in [0, 1]$ we consider the map $\varphi : B \times B \to B$ given by

$$\varphi(z, z') = f^{-1}((1 - t)f(z) + tf(z')).$$

Then φ satisfies the hypotheses of the Schwarz lemma (Lemma 6.1.28) and hence $\|\varphi(z, z')\| \leq \max\{\|z\|, \|z'\|\}$. Thus if $z, z' \in B_r$, $r \in (0, 1)$, then $\varphi(z, z') \in B_r$, and $f(B_r)$ is convex.

When X is a complex Hilbert space with inner product $\langle \cdot, \cdot \rangle$ and Y is a complex Banach space, Hamada and Kohr [Ham-Koh10] have recently obtained the following characterization of convexity. In the proof of sufficiency,

we do not know whether the assumption (6.3.7) (which is automatically satisfied in the finite dimensional case) and the assumption that f be biholomorphic
are essential.

Theorem 6.3.8. *(i) Let $f : B \to Y$ be a convex mapping. Then*

$$\|v\|^2 - \text{Re} \ \langle [Df(z)]^{-1} D^2 f(z)(v,v), z \rangle > 0$$

for $z \in B$, $v \in X \setminus \{0\}$ with $\text{Re} \ \langle z, v \rangle = 0$.

*(ii) Let $f : B \to Y$ be a biholomorphic mapping on B. Assume that for
each $r \in (0,1)$, there exists a constant $M = M(r)$ such that*

(6.3.7) $\|[Df(z)]^{-1}(f(z) - f(w))\| \leq M(r), \quad z, w \in B_r.$

If

$$\|v\|^2 - \text{Re} \ \langle [Df(z)]^{-1} D^2 f(z)(v,v), z \rangle > 0$$

for $z \in B$ and $v \in X \setminus \{0\}$ with $\text{Re} \ \langle z, v \rangle = 0$, then f is convex.

Proof. The argument in the first part of the proof of Theorem 6.3.4 proves
the first statement, and indeed was originally given in the context of complex
Hilbert spaces in [Ham-Koh10].

To prove the second statement, let $r \in (0,1)$ be given and let $\Omega_r = f(B_r)$.
It suffices to prove that Ω_r is convex. For this purpose, let

$$S = \Big\{ (P,Q) \in \Omega_r \times \Omega_r : \ [P,Q] \subset \Omega_r \Big\},$$

where $[P,Q]$ denotes the straight line segment joining P and Q. It is obvious
that S is open and nonempty. Since Ω_r is connected, the proof will be complete
if we can show that S is a closed subset of $\Omega_r \times \Omega_r$. Now if we suppose that
S is not closed, then we can find points $P, Q \in \Omega_r$ and $P_\nu, Q_\nu \in \Omega_r$ such that
$P_\nu \to P$, $Q_\nu \to Q$, $[P_\nu, Q_\nu] \subset \Omega_r$ and $[P,Q]$ is not contained in Ω_r. In this
case there exists a point $w = f(z) \in \partial \Omega_r \cap [P,Q]$. Indeed, let

$$t_0 = \sup \Big\{ \tau \in [0,1] : (1-t)P + tQ \in \Omega_r, 0 \leq t \leq \tau \Big\}.$$

Since $P, Q \in \Omega_r$, $t_0 > 0$ and $(1-t)P + tQ \in \Omega_r$ for $0 \leq t < t_0$. Let

$$v(t) = f^{-1}((1-t)P + tQ), \quad 0 \leq t < t_0.$$

Then $\|v(t)\| < r$ for $0 \le t < t_0$. Since

$$\frac{dv(t)}{dt} = [Df(v(t))]^{-1}(Q - P) = \frac{1}{t}[Df(v(t))]^{-1}(f(v(t)) - P),$$

it follows that $\|dv/dt\|$ is integrable on $[t_0/2, t_0)$ by (6.3.7). Let $t_0/2 < s_1 < s_2 < t_0$. Then

$$\|v(s_2) - v(s_1)\| = \left\| \int_{s_1}^{s_2} \frac{dv(t)}{dt} dt \right\| \le \int_{s_1}^{s_2} \left\| \frac{dv(t)}{dt} \right\| dt \le M(r) \log(s_2/s_1).$$

Let $\{\tau_k\}_{k \in \mathbb{N}}$ be an increasing sequence such that $\tau_k > 0$ and $\tau_k \to t_0$. A Cauchy sequence argument shows that there is a point $z \in X$ such that $v(\tau_k) \to z$ as $k \to \infty$. Moreover, since $\|v(\tau_k)\| < r$ we have $\|z\| \le r$, and by the continuity of f, we must have $f(z) = (1 - t_0)P + t_0 Q$. We now see that $\|z\| = r$, for otherwise we obtain a contradiction with the definition of t_0.

Let

$$g(t) = \langle f^{-1}(f(z) + t(Q - P)), f^{-1}(f(z) + t(Q - P)) \rangle$$

for t in a neighbourhood of 0. Then $g(0) = r^2$ and obviously $g(t) \le r^2$ for t in a neighbourhood of 0, since $[P, Q] \subset \overline{\Omega}_r$. Hence $g'(0) = 2\mathrm{Re}\,\langle z, v \rangle = 0$, where $v = [Df(z)]^{-1}(Q - P)$.

Short computations yield that

$$g''(0) = 2\|v\|^2 + 2\mathrm{Re}\,\langle D^2 f^{-1}(f(z))(Df(z)v, Df(z)v), z \rangle$$

$$= 2\|v\|^2 - 2\mathrm{Re}\,\langle [Df(z)]^{-1}D^2 f(z)(v, v), z \rangle > 0,$$

which implies that $g(t) > r^2$ for $t \ne 0$ near 0. However, this is a contradiction with $g(t) \le r^2$. Thus S must be closed, as claimed. This completes the proof. (If we assume only that f is locally biholomorphic, the conclusion is that $f(B)$ is a convex domain.)

In the finite-dimensional situation, we have the following result of Hamada and Kohr [Ham-Koh7]:

Theorem 6.3.9. *Let B be the unit ball of $\mathbb{C}^n$ with respect to a norm which is of class C^2 in $\mathbb{C}^n \setminus \{0\}$. Denote this norm by h. Suppose that $f : B \to \mathbb{C}^n$ is a locally biholomorphic mapping. Then f is convex if and only if*

$$v^* \frac{\partial^2 h^2}{\partial z \partial \overline{z}}(z)v + \mathrm{Re}\left\{ v^t \frac{\partial^2 h^2}{\partial z^2}(z)v \right\}$$

$$\geq \operatorname{Re}\left\{\left(\frac{\partial h^2}{\partial z}(z)\right)^t [Df(z)]^{-1} D^2 f(z)(v,v)\right\},$$

for all $z \in B \setminus \{0\}$ and $v \in \mathbb{C}^n \setminus \{0\}$ with $\operatorname{Re}\left\{v^t \frac{\partial h^2}{\partial z}(z)\right\} = 0.$

We leave the proof for the reader. Note that the quantity on the left-hand side of this inequality is just the second fundamental form of the hypersurface ∂B_r (up to a scalar factor), where $r = \|z\|$.

We now consider a two-point condition for convexity, analogous to (2.2.3), obtained by Suffridge [Su3,4] for the unit ball of a complex Banach space.

Theorem 6.3.10. Let $f : B \to Y$ be a holomorphic mapping.

(i) If f is convex then

$$f(x) - f(y) = Df(x)w(x,y), \quad x, y \in B,$$

where $\operatorname{Re}\{l_x(w(x,y))\} > 0$ for all $x, y \in B$, $\|y\| < \|x\|$ and $l_x \in T(x)$.

(ii) Assume f is locally biholomorphic and

$$f(x) - f(y) = Df(x)w(x,y), \quad x, y \in B,$$

where $\operatorname{Re}\{l_x(w(x,y))\} > 0$ for all $x, y \in B$, $\|y\| < \|x\|$ and $l_x \in T(x)$. Also assume that for each $r \in (0,1)$, there exists $M = M(r) > 0$ such that $\|[Df(x)]^{-1}\| \leq M(r)$ when $\|x\| \leq r$. Then f is convex.

Proof. Necessity. Because $(1-t)f(x) + tf(y) \in f(B)$ for all $x, y \in B$ and $t \in [0,1]$, we can find a mapping $v : B \times B \times [0,1] \to B$ such that $f(v(x,y,t)) = (1-t)f(x) + tf(y)$. As in the proof of Lemma 6.3.7, we have

$$\|v(x,y,t)\| \leq \max\{\|x\|, \|y\|\}, \ x, y \in B, \ t \in [0,1].$$

On the other hand,

$$f(v(x,y,t)) = f(x) + Df(x)(v(x,y,t) - x) + R(v(x,y,t),x),$$

where $R(v(x,y,t),x)/t \to 0$ as $t \to 0^+$. Hence

$$f(x) - f(y) = \lim_{t \to 0^+} \frac{f(x) - f(v(x,y,t))}{t}$$

$$= \lim_{t \to 0^+}\left[Df(x)\left(\frac{x - v(x,y,t)}{t}\right) - \frac{R(v(x,y,t),x)}{t}\right] = Df(x)w(x,y).$$

Since the limit

$$w(x, y) = \lim_{t \to 0^+} \frac{x - v(x, y, t)}{t}$$

exists and is holomorphic, we conclude in view of Lemma 6.1.33 that $\mathrm{Re}\,[l_x(w(x, y))] \geq 0$, for $x, y \in B$, $\|y\| \leq \|x\|$, $x \neq 0$, and $l_x \in T(x)$. Moreover, since $Dw(0, 0)(x, y) = x - y$, we deduce as in the proof of Lemma 6.1.30 that $\mathrm{Re}\,[l_x(w(x, y))] > 0$ for $\|y\| < \|x\|$ and $l_x \in T(x)$. This completes the proof of necessity.

Sufficiency. Since $w(x, 0) \in \mathcal{N}$ it follows from Theorem 6.2.6 that $f(x) - f(0)$ is starlike and in particular biholomorphic. Fix x and $y \in B$ and let

$$v(x, y, t) = f^{-1}((1 - t)f(x) + tf(y)), \quad t \in [0, t_0],$$

where t_0 is chosen such that $(1 - t)f(x) + tf(y) \in f(B)$ when $0 \leq t \leq t_0$. Using similar arguments as in the proof of Theorem 6.2.5, we obtain

$$v(x, y, t) = x - tw(x, y) + o(t),$$

and from this the fact that $\|v(x, y, t)\|$ is a decreasing function of t when $\|v(x, y, t)\| > \|y\|$. Next, for each $\tau \in [0, 1]$ we consider the line segment

$$L_\tau = \Big\{ (1 - t)f(x) + tf(y) : 0 \leq t \leq \tau \Big\}$$

and the set $T = \Big\{ \tau \in [0, 1] : f^{-1} \text{ can be analytically continued along } L_\tau \text{ with the initial value } f^{-1}(f(x)) = x \Big\}$. As in the proof of Theorem 6.2.5 we can show that T is nonempty, open, and closed, and this completes the proof.

We remark that Suffridge [Su3,4] gave infinite-dimensional versions of the characterizations of convex maps of the polydisc and of the unit ball with respect to the 1-norm in $\mathbb{C}^n$:

Theorem 6.3.11. *a) Let $X = \ell^1$ be the space of summable complex sequences and let Y be a complex Banach space. If $f : B \subset \ell^1 \to Y$ is convex then $f(z) - f(0)$ is linear.*

b) Let $X = \ell^\infty$ be the space of bounded complex sequences and let Y be a complex Banach space. If $f : B \subset \ell^\infty \to Y$ is convex then

$$f(z) - f(0) = Df(0)(g_1(z_1), \ldots, g_n(z_n), \ldots),$$

where $g_k(z_k)$ is a normalized convex function on the unit disc $|z_k| < 1$, $k \in \mathbb{N}$.

Theorem 6.3.10 is of course useful in n dimensions as well as in the infinite-dimensional case. An interesting application is the following theorem of Roper and Suffridge [Rop-Su2], which gives a sufficient condition for convexity on the Euclidean unit ball in $\mathbb{C}^n$. Such a result cannot hold for normed linear spaces in general because of Theorem 6.3.11. Since the proof is long and rather complicated, we omit it.

Theorem 6.3.12. *Let $f : B \to \mathbb{C}^n$ be a normalized holomorphic mapping on the Euclidean unit ball of $\mathbb{C}^n$. Assume $\sum_{k=2}^{\infty} \dfrac{k^2}{k!} \|D^k f(0)\| \le 1$. Then f is convex.*

The examples below of convex mappings, due to Suffridge [Su4] (see also [Rop-Su2]), make use of Theorem 6.3.10 or Theorem 6.3.4.

Example 6.3.13. Let B be the Euclidean unit ball in $\mathbb{C}^2$ and let $f : B \subset \mathbb{C}^2 \to \mathbb{C}^2$ be given by $f(z) = (z_1 + az_2^2, z_2)$, $z = (z_1, z_2) \in B$. Then f is convex if and only if $|a| \le 1/2$.

First proof. We first make use of Theorem 6.3.10 to prove the above assertion. For this purpose, we have to show that

$$\text{Re}\,\langle [Df(z)]^{-1}(f(z) - f(u)), z \rangle > 0$$

for $z = (z_1, z_2)$ and $u = (u_1, u_2) \in \mathbb{C}^2$ with $\|u\| < \|z\| < 1$.

Short computations yield that f is locally biholomorphic on B and

$$\text{Re}\,\langle [Df(z)]^{-1}(f(z) - f(u)), z \rangle = \|z\|^2 - \text{Re}\,\langle z, u \rangle - \text{Re}\,\{a\bar{z}_1(z_2 - u_2)^2\}$$

$$\ge \|z\|^2 - \text{Re}\,\langle z, u \rangle - |a| \cdot |z_1| \cdot |z_2 - u_2|^2$$

$$= \|z\|^2(1 - |a| \cdot |z_1|) - \text{Re}\,\langle z, u \rangle (1 - 2|a| \cdot |z_1|) - |a| \cdot |z_1|(\|u\|^2 - |z_1 - u_1|^2)$$

$$\ge \|z\|^2(1 - |a| \cdot |z_1|) - \text{Re}\,\langle z, u \rangle (1 - 2|a| \cdot |z_1|) - |a| \cdot |z_1|(\|z\|^2 - |z_1 - u_1|^2)$$

$$= (\|z\|^2 - \text{Re}\,\langle z, u \rangle)(1 - 2|a| \cdot |z_1|) + |a| \cdot |z_1| \cdot |z_1 - u_1|^2 > 0$$

for $|a| \le 1/2$ and $z, u \in B$, $\|u\| < \|z\|$.

If $|a| > 1/2$ we may find $z_1 \in U$ such that $\bar{z}_1 a > 1/2$. Choosing $z \in B$ with first coordinate z_1, $u_1 = z_1$ and $u_2 = -z_2 \in \mathbb{R}$, we deduce that

$$\text{Re}\,\langle [Df(z)]^{-1}(f(z) - f(u)), z \rangle$$

$$= \|z\|^2 - \operatorname{Re} \langle z, u \rangle - \operatorname{Re} \{a\bar{z}_1(z_2 - u_2)^2\}$$
$$< \|z\|^2 - \operatorname{Re} \{z_1\bar{z}_1 - z_2\bar{z}_2\} - \frac{1}{2}(2z_2)^2 = 0.$$

The result now follows from Theorem 6.3.10.

Second proof. In the second proof we use Theorem 6.3.4. Straightforward computations yield the relation

$$[Df(z)]^{-1}D^2f(z)(v,v) = (2av_2^2, 0), \quad z \in B, \quad v = (v_1, v_2) \in \mathbb{C}^2.$$

Therefore for $|a| \leq 1/2$ we obtain

$$1 - \operatorname{Re} \langle [Df(z)]^{-1}D^2f(z)(v,v), z \rangle = 1 - \operatorname{Re} [2a\bar{z}_1 v_2^2]$$

$$\geq 1 - 2|a||z_1||v_2|^2 \geq 1 - 2|a| \geq 0$$

for all $z \in B$, $v \in \mathbb{C}^2$, $\|v\| = 1$ and $\operatorname{Re} \langle z, v \rangle = 0$. Hence in view of Theorem 6.3.4 we conclude that f is convex.

As in the first proof, if $|a| > 1/2$, there is $z_1 \in U$ such that $\bar{z}_1 a > 1/2$. Moreover, if $z = (z_1, 0)$ and $v = (0, 1)$, then $\operatorname{Re} \langle z, v \rangle = 0$ and

$$1 - \operatorname{Re} \langle [Df(z)]^{-1}D^2f(z)(v,v), z \rangle = 1 - 2a\bar{z}_1 < 0.$$

This completes the proof.

With similar arguments we can verify the following ([Su4], [Rop-Su2]):

Example 6.3.14. Let B be the Euclidean unit ball of $\mathbb{C}^2$ and let $f : B \to \mathbb{C}^2$ be given by $f(z) = (z_1 + az_1z_2, z_2)$, $z = (z_1, z_2) \in B$. Then f is convex if and only if $|a| \leq 1/\sqrt{2}$.

Remark 6.3.15. These examples can easily be used to show that Alexander's theorem ($f \in K$ if and only if $zf'(z) \in S^*$, see Theorem 2.2.6) is not true in higher dimensions.

If f is a normalized convex mapping on the unit ball B of $\mathbb{C}^n$, $n \geq 2$, then $Df(z)z$ need not be starlike on B, except for certain special norms. (For example, it is true for the 1-norm since convex maps are linear in this case, and it is true for the polydisc as a simple consequence of Theorem 6.3.2 and Problem 6.2.5 (see [Lic3] and [Koh-Koh]).) To see this, consider the mapping in Example 6.3.14. Then $Df(z)z = (z_1 + 2az_1z_2, z_2)$, $z = (z_1, z_2) \in B$, and for $a = 1/\sqrt{2}$ this is not a starlike map in view of Problem 6.2.2. In fact, $Df(z)z$ is not even univalent, because $\left(z_1 \left(1 + \sqrt{2}z_2\right), z_2\right) = \left(0, -1/\sqrt{2}\right)$ for all $z = \left(z_1, -1/\sqrt{2}\right) \in B$.

Similarly, if $Df(z)z$ is starlike then f need not be convex on the unit ball of $\mathbb{C}^n$, $n \geq 2$. To see this, it suffices to let $n = 2$ and $\mathbb{C}^2$ be the Euclidean space of two complex variables. Also let $f(z) = (z_1 + (a/2)z_2^2, z_2)$ for $z = (z_1, z_2) \in B$. Then $Df(z)z = (z_1 + az_2^2, z_2)$ and in view of Problem 6.2.1, we deduce that $Df(z)z$ is starlike for $|a| \leq 3\sqrt{3}/2$. However, if $|a| > 1$, f is not convex because of Example 6.3.13. This example was considered in [Rop-Su2].

6.3.3 Quasi-convex mappings on the unit ball of $\mathbb{C}^n$

For various reasons, it is appropriate to consider new classes of univalent mappings in several variables which are not simply analogs of familiar subclasses of S. We have seen that the convex mappings of the unit ball in $\mathbb{C}^n$ have a very rigid structure for certain norms. On the other hand, even for the Euclidean norm it may not be easy to verify whether a given mapping is or is not convex. It is also surprisingly difficult to construct convex mappings of the Euclidean unit ball from convex functions on the unit disc. See problems 6.3.2 and 6.3.3 (i). (We shall return to this question in Chapter 11.) In particular, in contrast to the situation of the polydisc, if $f_1, \ldots, f_n \in K$, $n \geq 2$, then the mapping $F: B \to \mathbb{C}^n$ defined by

$$F(z) = (f_1(z_1), \ldots, f_n(z_n)), \quad z = (z_1, \ldots, z_n) \in B,$$

is in general not convex.

Let $f : U \to \mathbb{C}$ be a normalized holomorphic function. As shown in Theorem 2.2.4, $f \in K$ if and only if

$$\mathrm{Re} \left[\frac{2zf'(z)}{f(z) - f(\zeta)} - \frac{z + \zeta}{z - \zeta} \right] \geq 0, \quad z, \zeta \in U.$$

Trying to generalize this idea in several variables, Roper and Suffridge [Rop-Su2] introduced the following class of mappings on the unit ball of $\mathbb{C}^n$ with respect to an arbitrary norm $\| \cdot \|$.

Let $\mathcal{LS}(B)$ be the class of normalized locally biholomorphic mappings on B. Recall that $\hat{\mathbb{C}}$ denotes the extended complex plane.

Definition 6.3.16. Let $u \in \mathbb{C}^n$ with $\|u\| = 1$, and let $l_u \in T(u)$. For $f \in \mathcal{LS}(B)$ define $G_f : U \times U \to \hat{\mathbb{C}}$ by

$$(6.3.8) \qquad G_f(\alpha, \beta) = \frac{2\alpha}{l_u([Df(\alpha u)]^{-1}(f(\alpha u) - f(\beta u)))} - \frac{\alpha + \beta}{\alpha - \beta}.$$

We define the class of *quasi-convex mappings of type A* by

$$\mathcal{G} = \Big\{ f \in \mathcal{LS}(B) : \operatorname{Re} G_f(\alpha, \beta) > 0 \text{ for all } \alpha, \beta \in U, |\beta| < |\alpha|,$$

$$l_u \in T(u), \text{ and } u \in \mathbb{C}^n, \|u\| = 1 \Big\}.$$

Actually this is a slight modification of [Rop-Su2, Definition 3].

Lemma 6.3.17. *Let* $f \in \mathcal{LS}(B)$. *Also let* $u \in \mathbb{C}^n$, $\|u\| = 1$, $l_u \in T(u)$, *and* $\alpha, \beta \in U$, $\alpha \neq 0$. *Then*

$$(6.3.9) \qquad \lim_{\beta \to \alpha} G_f(\alpha, \beta) = 1 + \alpha l_u([Df(\alpha u)]^{-1} D^2 f(\alpha u)(u, u))$$

$$= \frac{1}{|\alpha|} l_{\alpha u}\Big([Df(\alpha u)]^{-1}(D^2 f(\alpha u)(\alpha u, \alpha u) + Df(\alpha u)(\alpha u))\Big).$$

In particular, $G_f(\alpha, \beta)$ *is continuous in* β *at* $\beta = \alpha$.

Proof. Expanding $f(\beta u)$ about αu, we obtain

$$f(\beta u) = f(\alpha u) + Df(\alpha u)((\beta - \alpha)u)$$

$$+ \frac{1}{2} D^2 f(\alpha u)((\beta - \alpha)u, (\beta - \alpha)u) + O((\beta - \alpha)^3).$$

Hence

$$[Df(\alpha u)]^{-1}(f(\alpha u) - f(\beta u))$$

$$= (\alpha - \beta)\Big(u + \frac{1}{2}(\beta - \alpha)[Df(\alpha u)]^{-1} D^2 f(\alpha u)(u, u) + O((\beta - \alpha)^2)\Big).$$

Taking into account the above equality, we deduce that

$$\frac{2\alpha}{l_u([Df(\alpha u)]^{-1}(f(\alpha u) - f(\beta u)))} - \frac{\alpha + \beta}{\alpha - \beta}$$

$$= \frac{(\alpha - \beta)\Big\{1 + \frac{1}{2}(\beta + \alpha)l_u([Df(\alpha u)]^{-1} D^2 f(\alpha u)(u, u) + O(\beta - \alpha))\Big\}}{(\alpha - \beta)\Big\{1 + \frac{1}{2}(\beta - \alpha)l_u([Df(\alpha u)]^{-1} D^2 f(\alpha u)(u, u) + O(\beta - \alpha))\Big\}}.$$

The first equality in (6.3.9) now follows if we let $\beta \to \alpha$. To obtain the second equality we use the fact that $l_{\alpha u}(\cdot) = \dfrac{|\alpha|}{\alpha} l_u(\cdot) \in T(\alpha u)$ for $u \in \mathbb{C}^n$, $\|u\| = 1$, and $\alpha \in U \setminus \{0\}$.

Roper and Suffridge [Rop-Su2] proved that $K(B) \subseteq \mathcal{G} \subseteq S^*(B)$. In fact we shall show that if $f \in \mathcal{G}$, then f is starlike of order $1/2$. These results are given in the next two theorems.

Theorem 6.3.18. *Let $f : B \to \mathbb{C}^n$ be a normalized convex mapping. Then $f \in \mathcal{G}$.*

Proof. Since f is convex, it follows that

$$\mathrm{Re}\left\{ l_z([Df(z)]^{-1}(f(z) - f(v))) \right\} > 0, \quad \|v\| < \|z\| < 1,\ l_z \in T(z),$$

by Theorem 6.3.10. Let $u \in \mathbb{C}^n$, $\|u\| = 1$, and let $l_u \in T(u)$. By the preceding inequality, we have

$$(6.3.10)\quad \mathrm{Re}\left\{ l_{\alpha u}([Df(\alpha u)]^{-1}(f(\alpha u) - f(\beta u))) \right\} > 0, \quad |\beta| < |\alpha| < 1.$$

Using the one-to-one correspondence between $T(\alpha u)$ and $T(u)$ given by $l_{\alpha u}(\cdot) = \dfrac{|\alpha|}{\alpha} l_u(\cdot)$, we obtain from (6.3.10) that

$$\mathrm{Re}\left\{ \frac{1}{\alpha} l_u([Df(\alpha u)]^{-1}(f(\alpha u) - f(\beta u))) \right\} > 0,$$

or equivalently,

$$\mathrm{Re}\left\{ \frac{2\alpha}{l_u([Df(\alpha u)]^{-1}(f(\alpha u) - f(\beta u)))} \right\} > 0, \quad |\beta| < |\alpha| < 1.$$

Letting $|\beta| = |\alpha| < 1$ with $\beta \neq \alpha$, we obtain

$$\mathrm{Re}\left\{ \frac{2\alpha}{l_u([Df(\alpha u)]^{-1}(f(\alpha u) - f(\beta u)))} \right\} \geq 0.$$

Moreover, if $|\alpha| = |\beta|$, $\alpha \neq \beta$, then it is easily checked that $\mathrm{Re}\left\{ \dfrac{\alpha + \beta}{\alpha - \beta} \right\} = 0$, and therefore we obtain

$$\mathrm{Re}\left\{ \frac{2\alpha}{l_u([Df(\alpha u)]^{-1}(f(\alpha u) - f(\beta u)))} - \frac{\alpha + \beta}{\alpha - \beta} \right\} \geq 0.$$

Consequently we have shown that Re $G_f(\alpha, \beta) \geq 0$ for $|\alpha| = |\beta|$, $\alpha \neq \beta$. Keeping $\alpha \in U \setminus \{0\}$ fixed, we note that $G_f(\alpha, \cdot)$ is analytic on the disc $U_{|\alpha|}$ and, using Lemma 6.3.17, continuous on $\overline{U}_{|\alpha|}$. Moreover, Re $G_f(\alpha, \alpha) > 0$ by an argument similar to the proof that strict inequality occurs in (6.3.5). Therefore, by the minimum principle for harmonic functions, we conclude that Re $G_f(\alpha, \beta) > 0$ when $|\beta| < |\alpha| < 1$, i.e. $f \in \mathcal{G}$, as desired. This completes the proof.

The next result is a generalization of the Marx-Strohhäcker theorem (see Theorem 2.3.2) to several complex variables.

Theorem 6.3.19. *If $f \in \mathcal{G}$ then f is starlike of order $1/2$.*

Proof. Since $f \in \mathcal{G}$, Re $G_f(\alpha, \beta) > 0$ for $u \in \mathbb{C}^n$, $\|u\| = 1$, $l_u \in T(u)$, and $|\beta| < |\alpha| < 1$. For $\beta = 0$ one deduces that

$$\text{Re } G_f(\alpha, 0) = \text{Re} \left\{ \frac{2\alpha}{l_u([Df(\alpha u)]^{-1} f(\alpha u))} - 1 \right\} > 0,$$

and therefore

$$\text{Re} \left\{ \frac{\alpha}{l_u([Df(\alpha u)]^{-1} f(\alpha u))} \right\} > \frac{1}{2}.$$

But this is exactly the condition (6.2.9) for $\gamma = 1/2$, so we are done.

The class $\mathcal{G}$ is defined by a two-point condition. There is a related one-point condition which we have already seen in Theorem 6.3.5. Roper and Suffridge [Rop-Su2] made use of it to introduce another subclass of locally univalent mappings on the unit ball in $\mathbb{C}^n$ which extends at least some of the properties of K to several variables.

Definition 6.3.20. For $f \in \mathcal{LS}(B)$, $z \in B \setminus \{0\}$, and $l_z \in T(z)$, let

$$F_f(z, l_z) = l_z([Df(z)]^{-1}(D^2 f(z)(z, z) + Df(z)z)).$$

The class of *quasi-convex mappings of type B* is defined by

$$\mathcal{F} = \left\{ f \in \mathcal{LS}(B) : \text{Re } F_f(z, l_z) > 0, \text{ for all } z \in B \setminus \{0\} \text{ and } l_z \in T(z) \right\}.$$

Roper and Suffridge [Rop-Su2] proved the following inclusion:

Theorem 6.3.21. *If $f \in \mathcal{G}$ then $f \in \mathcal{F}$.*

Proof. This follows from Lemma 6.3.17.

It is easier to work with the two-point condition defining the class $\mathcal{G}$ than with the definition of $\mathcal{F}$, but we do not know whether the inclusion in Theorem 6.3.21 is proper. Roper and Suffridge proved that in the Euclidean unit ball of $\mathbb{C}^n$ the upper and lower bounds on the growth of mappings in $\mathcal{G}$ are the same as for normalized convex maps. We shall return to this result in Chapter 8, and also we shall investigate other properties of these holomorphic mappings in higher dimensions. We remark that in contrast to the situation of convex mappings, if $f_1(z_1), \ldots, f_n(z_n)$ are functions in K, then $F(z) = (f_1(z_1), \ldots, f_n(z_n))$ is a mapping in $\mathcal{G}$ and $\mathcal{F}$ on the unit ball of $\mathbb{C}^n$ with respect to a p-norm, $1 \leq p \leq \infty$ (see Problem 6.3.8). Hence $K(B) \subsetneqq \mathcal{G}$ and $K(B) \subsetneqq \mathcal{F}$ in $\mathbb{C}^n$, $n \geq 2$.

Notes. In both Sections 6.2 and 6.3 we have studied some well known and also some new criteria for starlikeness and convexity for holomorphic (or non-holomorphic) mappings in the unit ball and certain pseudoconvex domains in $\mathbb{C}^n$. Hamada and Kohr [Ham-Koh7] obtained necessary and sufficient conditions for convexity of locally biholomorphic mappings on bounded balanced convex domains whose Minkowski function is of class C^1 in $\mathbb{C}^n \setminus \{0\}$. For some similar results, the reader may consult [Gon4]. For other results about starlikeness and convexity, the reader may use the following references: [Su3], [Su4], [Rop-Su2], [Gon2], [Gon4], [Ham-Koh7], [Ham-Koh3], [Ham-Koh2], [Gon-Wa-Yu2-3], [Koh2], [Koh7-8], [Gra3-5].

Problems

6.3.1. Let $f(z) = (z_1 + az_2^2, z_2)$, $z = (z_1, z_2) \in \mathbb{C}^2$. Prove the following assertions using Theorem 6.3.10:

a) if $\|z\|_1 = |z_1| + |z_2|$, then f is convex on the unit ball $B(1) \subset \mathbb{C}^2$ with respect to this norm if and only if $a = 0$.

b) if $\|z\|_\infty = \max\{|z_1|, |z_2|\}$ then f is convex on the unit polydisc P of $\mathbb{C}^2$ if and only if $a = 0$.

(Suffridge, 1970 [Su2] and 1975 [Su4].)

6.3.2. Let $B(p)$ be the unit ball of $\mathbb{C}^2$ with respect to a p-norm, where

$1 \leq p < \infty$. Let $f : B(p) \to \mathbb{C}^2$ be given by $f(z) = (z_1/(1 - z_1), z_2)$, $z = (z_1, z_2) \in B(p)$. Show that f cannot be convex.
(Suffridge, 1975 [Su4].)

6.3.3. (i) Let B denote the Euclidean unit ball of $\mathbb{C}^n$, $n \geq 2$, and let

$$f(z) = \left(\frac{z_1}{1 - z_1}, \frac{z_2}{1 - z_2}, \ldots, \frac{z_n}{1 - z_n} \right), \quad z = (z_1, \ldots, z_n) \in B.$$

Show that f is not convex. (However, f is convex on the unit polydisc of $\mathbb{C}^n$.)
(Gong, Wang and Yu, 1993 [Gon-Wa-Yu3], Roper and Suffridge, 1995 [Rop-Su1].)

(ii) Show that if $f : B \to \mathbb{C}^2$ is given by

$$f(z) = \left(\frac{1}{2} \log \frac{1 + z_1}{1 - z_1}, \frac{z_2}{(1 - z_1^2)^{1/2}} \right), \quad z = (z_1, z_2) \in B,$$

where B is the Euclidean unit ball of $\mathbb{C}^2$, then f is convex.
(Roper and Suffridge, 1995 [Rop-Su1].)

6.3.4. Show that $f(z) = z/(1 - z_1)$, $z = (z_1, \ldots, z_n) \in B$, is a convex mapping on the Euclidean unit ball B of $\mathbb{C}^n$.

6.3.5. Let X be a complex Hilbert space with inner product $\langle \cdot, \cdot \rangle$ and let u be a unit vector in X. Show that $f(z) = \dfrac{z}{1 - \langle z, u \rangle}$, $z \in B$, is a convex mapping on B.
(Hamada and Kohr, 2002 [Ham-Koh9].)

6.3.6. Let $B(p)$ be the unit ball of $\mathbb{C}^n$ with respect to a p-norm, where $1 \leq p \leq \infty$. Let $F : B(p) \to \mathbb{C}^n$ be a mapping with one of its coordinate maps, f_k, a function of one complex variable only. Show that it is a necessary condition for $F \in \mathcal{F}$ that $f_k \in K$.
(Roper and Suffridge, 1999 [Rop-Su2].)

6.3.7. Let $B(p)$ denote the unit ball of $\mathbb{C}^2$ with a p-norm, where $1 < p < \infty$. Also let $f : B(p) \to \mathbb{C}^2$ be given by $f(z) = (z_1 + az_2^2, z_2)$, $z = (z_1, z_2) \in B(p)$. Show that $f \in \mathcal{G}$ (or $f \in \mathcal{F}$) if and only if

$$|a| \leq \frac{1}{2} \left(\frac{p^2 - 1}{4} \right)^{1/p} \left(\frac{p + 1}{p - 1} \right).$$

(Roper and Suffridge, 1999 [Rop-Su2].)

6.3.8. Let $f : B \subset \mathbb{C}^n \to \mathbb{C}^n$ be given by $f(z) = (f_1(z_1), \ldots, f_n(z_n))$, $z = (z_1, \ldots, z_n) \in B$, where $f_j \in K$, $j = 1, \ldots, n$. Show that $f \in \mathcal{G}$ in any absolute norm. (That is, any norm for which $|z_j| \leq |w_j|$ for each j implies $\|z\| \leq \|w\|$.)

(Roper and Suffridge, 1999 [Rop-Su2].)

6.3.9. Let $B(p)$ denote the unit ball of $\mathbb{C}^2$ with respect to a p-norm, $1 \leq p \leq \infty$, and $f(z) = (z_1 + a z_1 z_2, z_2)$, $z = (z_1, z_2) \in B(p)$. Show that $f \in \mathcal{G}$ (or $f \in \mathcal{F}$) if and only if $|a| \leq \left[\frac{2}{3}(p+1)\right]^{1/p} \left(\frac{p+1}{3p}\right)$.

(Roper and Suffridge, 1999 [Rop-Su2].)

6.3.10. Let B be the unit ball of a complex Hilbert space X and suppose that $f : B \to X$ is a locally biholomorphic mapping. Show that the condition

$$\mathrm{Re}\, \langle [Df(x)]^{-1}(f(x) - f(y)), x \rangle > 0, \quad \|y\| < \|x\| < 1,$$

implies the condition

$$\|v\|^2 - \mathrm{Re}\, \langle [Df(x)]^{-1} D^2 f(x)(v, v), x \rangle > 0$$

for $x \in B$, $v \in X \setminus \{0\}$, and $\mathrm{Re}\, \langle x, v \rangle = 0$.

6.3.11. Consider the class $\mathcal{F}$ introduced in Definition 6.3.20. Must any $f \in \mathcal{F}$ be biholomorphic on the unit ball of $\mathbb{C}^n$, $n \geq 2$?

6.4 Spirallikeness and Φ-likeness in several complex variables

In this section we are going to study the sets of spirallike and Φ-like mappings in the unit ball of $\mathbb{C}^n$.

In 1975 Gurganus [Gur] gave a beautiful extension of the theory of Φ-like holomorphic functions in the unit disc to locally biholomorphic mappings on the unit ball of $\mathbb{C}^n$ and complex Banach spaces. The case in which Φ is linear leads to the notion of spirallikeness with respect to a linear operator, considered by Gurganus and (in somewhat greater generality) by Suffridge [Su4]. More recently, Hamada and Kohr [Ham-Koh1,4] considered spirallike mappings of type α, which corresponds to the case in which the linear operator is a scalar multiple of the identity. This case is probably the most natural generalization of spirallikeness to several variables; however, interesting phenomena arise when one considers a more general linear operator. Kohr [Koh3] and Hamada and Kohr [Ham-Koh3] also gave some sufficient conditions for diffeomorphism and spirallikeness for mappings which are locally diffeomorphisms of class C^1 on the unit ball, as well as on bounded balanced pseudoconvex domains in $\mathbb{C}^n$.

We consider the Euclidean structure on $\mathbb{C}^n$ and we let B denote the unit ball of $\mathbb{C}^n$. As usual, by $\langle \cdot, \cdot \rangle$ we denote the Euclidean inner product in $\mathbb{C}^n$. We recall that for the Euclidean case the class $\mathcal{N}$ consists of those holomorphic mappings $h : B \to \mathbb{C}^n$ such that $h(0) = 0$ and Re $\langle h(z), z \rangle > 0$, $z \in B \setminus \{0\}$. Gurganus [Gur] gave the definitions below of a Φ-like holomorphic mapping and a Φ-like region in $\mathbb{C}^n$. These definitions generalize Brickman's notion [Bri] of Φ-likeness in the unit disc (see Section 2.4.3).

Definition 6.4.1. Let $f : B \to \mathbb{C}^n$ be a normalized locally biholomorphic mapping. Let $\Phi \in H(f(B))$ be such that $\Phi(0) = 0$. We say that f is Φ-*like* if

$$(6.4.1) \qquad [Df(z)]^{-1}\Phi(f(z)) \in \mathcal{N}.$$

In the case $n = 1$ we do not need the assumption $f'(\zeta) \neq 0$, $\zeta \in U$, but in higher dimensions the open mapping theorem fails (see Problem 6.4.4), and this is the reason why in Definition 6.4.1 we assume that f is locally biholomorphic on B.

Definition 6.4.2. Let Ω be a region in $\mathbb{C}^n$ containing 0 and let $\Phi \in H(\Omega)$ be such that $\Phi(0) = 0$ and Re $\langle D\Phi(0)z, z \rangle > 0$ for all $z \in \mathbb{C}^n \setminus \{0\}$ (i.e. $D\Phi(0) + D\Phi(0)^*$ is a positive definite operator). We say that Ω is Φ-*like* if for each $w_0 \in \Omega$, the initial value problem

$$(6.4.2) \qquad \frac{dw}{dt} = -\Phi(w), \quad w(0) = w_0$$

has a solution $w(t)$ defined for all $t \geq 0$ such that $w(t) \in \Omega$ for all $t \geq 0$ and $w(t) \to 0$ as $t \to \infty$.

Remark 6.4.3. The local existence and uniqueness of solutions of the initial value problem (6.4.2) follows from standard theorems in ordinary differential equations. The existence of solutions defined for all $t \geq 0$ for a given differential equation requires special features of the equation, for example the assumption $h \in \mathcal{N}$ in Lemma 6.4.4. We shall return to this question in Chapter 8.

We now study some properties of Φ-like mappings on the unit ball of $\mathbb{C}^n$. For generalizations of some of these results to the case of complex Banach spaces, the reader may consult [Gur].

We begin with the following lemma of Gurganus [Gur]:

Lemma 6.4.4. *Let $h \in \mathcal{N}$. Then for each $z \in B$, the initial value problem*

$$(6.4.3) \qquad \frac{dv}{dt} = -h(v), \quad v(0) = z,$$

has a unique solution $v(t) = v(z, t)$ defined for all $t \geq 0$. For fixed t, $v_t(z) = v(z, t)$ is a univalent Schwarz mapping which satisfies the inequality

$$(6.4.4) \qquad \|v(z, t)\| \leq \|z\| \exp\left\{ -\frac{1 - \|z\|}{1 + \|z\|} k_h t \right\}, \quad z \in B,$$

where $k_h = \min\limits_{\|z\|=1} \operatorname{Re} \langle Dh(0)z, z \rangle > 0$.

Proof. We only make some remarks on the proof since we do not need the full force of the lemma. The local existence and uniqueness of solutions is clear; the existence of solutions defined for all $t \geq 0$ is established in more generality in Theorem 8.1.3. (The normalization $Dh(0) = I$ assumed there plays no role in the existence and uniqueness theorem.)

Since $\dfrac{\partial}{\partial t}\|v(z, t)\|^2 = -2\operatorname{Re}\langle h(v), v \rangle \leq 0$, $v(z, 0) = z$, and $v(0, t) = 0$, we see that $v_t(z) = v(z, t)$ is a Schwarz mapping for any fixed $t \geq 0$. We do not need the univalence of $v_t(z)$ at this stage, but we do need to know that $v(z, t) \to 0$ as $t \to \infty$ for any fixed $z \in B$. This follows from the estimate (6.4.4); to prove it we make use of a generalization of Lemma 6.1.32 to the class $\mathcal{N}$. We leave this as an exercise for the reader.

Gurganus [Gur] showed that, as in one variable, Φ-likeness is equivalent to univalence. The reader may compare the following results with Theorems 2.4.16, 2.4.18 and Corollary 2.4.17.

Theorem 6.4.5. *Let f be Φ-like. Then f is univalent on B and $f(B)$ is a Φ-like domain.*

Proof. First step. Since f is Φ-like, there exists a mapping $h \in \mathcal{N}$ such that

$$\Phi(f(z)) = Df(z)h(z), \quad z \in B.$$

Moreover, by the normalization of f we obtain $D\Phi(0) = Dh(0)$, and hence Re $\langle D\Phi(0)(z), z \rangle > 0$ for all $z \in \mathbb{C}^n \setminus \{0\}$, by Lemma 6.1.30. Next, fix $z \in B$. In view of Lemma 6.4.4, the initial value problem

$$\frac{dv}{dt} = -h(v), \quad t \geq 0, \quad v(0) = z,$$

has a unique solution $v_z(t) = v(z,t)$. Let $w_z(t) = w(t) = f(v_z(t))$, $t \geq 0$. Then $w_z(0) = f(z)$ and a simple computation yields

$$\frac{dw_z}{dt} = Df(v_z(t))\frac{dv_z}{dt} = -Df(v_z(t))h(v_z(t)) = -\Phi(f(v_z(t))).$$

Hence

(6.4.5) $$\frac{dw_z}{dt} = -\Phi(w_z), \quad w_z(0) = f(z).$$

It is clear that $w_z(t) \in f(B)$, and in view of (6.4.4) it follows that $v_z(t) \to 0$ as $t \to \infty$. Therefore

$$\lim_{t \to \infty} w_z(t) = \lim_{t \to \infty} f(v_z(t)) = f(0) = 0.$$

Consequently, $f(B)$ is a Φ-like domain.

Second step. We now prove that f is univalent on B. To this end, let $a, b \in B$ and assume $f(a) = f(b)$. Then the solutions $w_a(t)$ and $w_b(t)$ of the initial value problem (6.4.5) satisfy $w_a(0) = w_b(0)$. Using the uniqueness of solutions of this equation, we deduce that $w_a(t) = w_b(t)$ for all $t \geq 0$, and consequently $f(v_a(t)) = f(v_b(t))$ for all $t \geq 0$. Now, since f is locally biholomorphic on B, f has a local holomorphic inverse at zero and since $v_a(t) \to 0$, $v_b(t) \to 0$ as $t \to \infty$, there exists $t_0 \geq 0$ such that $v_a(t) = v_b(t)$ for all $t \geq t_0$. However, since $v_a(t)$ and $v_b(t)$ are solutions of the initial value problem

$$\frac{dv}{dt} = -h(v), \quad v(t_0) = v_a(t_0) = v_b(t_0),$$

we conclude by the uniqueness of solutions that $v_a(t) = v_b(t)$ for $t \geq 0$. Hence $a = v_a(0) = v_b(0) = b$ and this completes the proof of Theorem 6.4.5.

Corollary 6.4.6. *Let $f : B \to \mathbb{C}^n$ be a normalized locally biholomorphic mapping. Then f is univalent on B if and only if f is Φ-like for some Φ.*

Proof. In the previous theorem we have seen that if f is Φ-like, then f is univalent. Thus it suffices to prove the converse. For this purpose, let $h \in \mathcal{N}$ be arbitrary and consider the equation

$$\Phi(f(z)) = Df(z)h(z), \quad z \in B.$$

Since f is biholomorphic on B, this equation determines a mapping $\Phi \in H(f(B))$ such that

$$\mathrm{Re}\,\langle [Df(z)]^{-1}\Phi(f(z)), z \rangle > 0, \quad z \in B \setminus \{0\}.$$

Thus f is Φ-like.

The next result of Gurganus [Gur] is the converse of Theorem 6.4.5.

Theorem 6.4.7. *Let $f : B \to \mathbb{C}^n$ be a normalized univalent mapping on B such that $f(B)$ is a Φ-like domain in $\mathbb{C}^n$. Then f is Φ-like.*

Proof. Let $h(z) = [Df(z)]^{-1}\Phi(f(z))$, $z \in B$. Then $h \in H(B)$, $h(0) = [Df(0)]^{-1}\Phi(f(0)) = 0$, $Dh(0) = D\Phi(0)$ and we claim that $h \in \mathcal{N}$. To see this, fix $z \in B$ and let $w_z(t) = w(z,t)$ be the unique solution of the initial value problem

$$(6.4.6) \qquad \frac{dw_z}{dt} = -\Phi(w_z), \quad t \geq 0, \quad w_z(0) = f(z).$$

By hypothesis, we have $w_z(t) \in f(B)$ for $t \geq 0$ and $w_z(t) \to 0$ as $t \to \infty$. Also let $v_z(t) = v(z,t) = f^{-1}(w_z(t))$. Clearly, if $z = 0$ then $w_z(t) = 0$ for $t \geq 0$ by the uniqueness of solutions of (6.4.6). If $z \neq 0$ then $w_z(t) \neq 0$ for all $t \geq 0$, again in view of the uniqueness of solutions. Hence in this case $v_z(t) \neq 0$ for $z \neq 0$ and $t \geq 0$. Further, since $\Phi(z)$ and $f(z) = w_z(0)$ are holomorphic mappings, it follows as in the proof of Theorem 2.4.18 that $w(\cdot, t)$ is holomorphic in z for fixed t. Since f is univalent on B, $v_z(t)$ is also holomorphic in z for fixed t, and moreover $\|v_z(t)\| < 1$, $z \in B$, $t \geq 0$. Consequently, in view of the Schwarz lemma, $\|v_z(t)\| \leq \|z\|$ for all $z \in B$ and $t \geq 0$. We therefore obtain

$$\left. \frac{d\|v_z(t)\|}{dt} \right|_{t=0} = \lim_{t \to 0+} \frac{\|v_z(t)\| - \|v_z(0)\|}{t} = \lim_{t \to 0+} \frac{\|v_z(t)\| - \|z\|}{t} \leq 0.$$

Moreover, since

$$\|v_z(t)\| \frac{d\|v_z(t)\|}{dt} = -\mathrm{Re}\,\langle h(v_z(t)), v_z(t) \rangle, \quad t \in [0, \infty),$$

we must have

$$\text{Re }\langle h(z), z \rangle \geq 0, \quad z \in B.$$

Finally, in view of Lemma 6.1.30 and the relation

$$\text{Re }\langle Dh(0)z, z \rangle = \text{Re }\langle D\Phi(0)z, z \rangle > 0,$$

we deduce that Re $\langle h(z), z \rangle > 0$ on $B \setminus \{0\}$. This completes the proof.

Theorems 6.4.5 and 6.4.7 may be used to obtain an alternate proof of the necessary and sufficient condition for starlikeness given in Theorem 6.2.2, and also to characterize spirallikeness in higher dimensions. The proof below is due to Gurganus [Gur].

Corollary 6.4.8. *Let $f : B \to \mathbb{C}^n$ be a locally biholomorphic mapping such that $f(0) = 0$. Then f is starlike if and only $[Df(z)]^{-1}f(z) \in \mathcal{M}$.*

Proof. We may assume that $Df(0) = I$. Obviously if we take $\Phi = I$ then the solution of (6.4.2) is $w(t) = w_0 e^{-t}$. Hence a domain Ω containing the origin is I-like if and only if Ω is starlike.

Now assume that $f : B \to \mathbb{C}^n$ is a locally biholomorphic mapping such that $f(0) = 0$ and $[Df(z)]^{-1}f(z) \in \mathcal{M}$. Then Theorem 6.4.5 implies that f is starlike.

Conversely, if f is starlike then Theorem 6.4.7 implies that $[Df(z)]^{-1}f(z) \in \mathcal{N}$. Using the fact that $[Df(z)]^{-1}f(z)$ is a normalized mapping, we conclude that $[Df(z)]^{-1}f(z) \in \mathcal{M}$ and we are done.

We next present some ideas about spirallikeness on the unit ball of $\mathbb{C}^n$. Gurganus [Gur] introduced the notion of spirallikeness with respect to a normal linear operator (i.e. an operator $A \in L(\mathbb{C}^n, \mathbb{C}^n)$ such that $AA^* = A^*A$) whose eigenvalues have positive real part. Suffridge [Su4] enlarged the class of operators considered and also extended the definition to complex Banach spaces. We shall give Suffridge's definition, but we consider only the Euclidean unit ball in $\mathbb{C}^n$.

For a linear operator $A \in L(\mathbb{C}^n, \mathbb{C}^n)$ we introduce the notation

$$(6.4.7) \qquad m(A) = \min \left\{ \text{Re }\langle A(z), z \rangle : \|z\| = 1 \right\}.$$

Definition 6.4.9. Let $f : B \to \mathbb{C}^n$ be a normalized univalent mapping on B. Let $A \in L(\mathbb{C}^n, \mathbb{C}^n)$ be such that $m(A) > 0$. We say that f *is spirallike* *relative to A if $e^{-tA}f(B) \subseteq f(B)$ for all $t \geq 0$, where*

$$e^{-tA} = \sum_{k=0}^{\infty} \frac{(-1)^k}{k!} t^k A^k.$$

The following generalization of Theorem 2.4.10 is due to Suffridge [Su4]; the important special case when A is normal was considered by Gurganus [Gur].

Theorem 6.4.10. *Let $A \in L(\mathbb{C}^n, \mathbb{C}^n)$ be such that $m(A) > 0$ and let $f : B \to \mathbb{C}^n$ be a normalized locally biholomorphic mapping. Then f is spirallike relative to A if and only if*

(6.4.8) $[Df(z)]^{-1} Af(z) \in \mathcal{N}.$

Proof. First step. Assume that f is locally biholomorphic and satisfies (6.4.8). Then there exists a mapping $h \in \mathcal{N}$ such that $Dh(0) = A$ and

$$Df(z)h(z) = Af(z), \quad z \in B.$$

In view of Lemma 6.4.4, for each $z \in B$ there exists a unique solution $v_z(t) = v(z, t)$ to the initial value problem

(6.4.9) $\dfrac{dv_z}{dt} = -h(v_z), \quad v_z(0) = z.$

From (6.4.4) we conclude that $\|v_z(t)\| \to 0$ as $t \to \infty$. Hence f is a Φ-like mapping with $\Phi = A$. Therefore f is univalent on B, and using the result of Theorem 6.4.5 we deduce that $f(B)$ is a Φ-like domain in $\mathbb{C}^n$. If we let $u(z, t) = f^{-1}(e^{-tA}f(z))$ for $z \in B$ and $t \geq 0$ sufficiently small such that $e^{-tA}f(z) \in f(B)$, then $u = u(z, t)$ is a solution of the initial value problem (6.4.9). In view of the uniqueness of solutions of this equation, we conclude that $u(z, t) = v(z, t)$ for $z \in B$ and $t \geq 0$. Since $f(B)$ is a Φ-like domain with $\Phi = A$, it follows that $f(u(z, t)) \in f(B)$, $z \in B$, $t \geq 0$. Hence

$$e^{-tA}f(z) \in f(B), \quad z \in B, \quad t \geq 0,$$

which means that $f(B)$ is a spirallike domain. Consequently f is spirallike, as desired.

Second step. Conversely, assume that f is spirallike relative to A. Let $F(z,t) = e^{-tA}f(z)$, $z \in B$ and $t \geq 0$. Then $F(\cdot, t) \in H(B)$, $F(B, t) \subset f(B)$, $F(z, 0) = f(z)$, $z \in B$, and $F(0, t) = 0$ for all $t \geq 0$. Since

$$\lim_{t \to 0^+} \frac{F(z, 0) - F(z, t)}{t} = Af(z), \quad z \in B,$$

we conclude from Lemma 6.1.35 that there exists a mapping $h \in \mathcal{N}_0$ such that

$$Af(z) = Df(z)h(z), \quad z \in B.$$

Since $Dh(0) = A$ and $\text{Re}\,\langle A(z), z \rangle > 0$, $z \in \mathbb{C}^n \setminus \{0\}$, we deduce in view of Lemma 6.1.30 that $h \in \mathcal{N}$. Thus $[Df(z)]^{-1}Af(z) \in \mathcal{N}$ and this completes the proof.

Remark 6.4.11. As already noted, Gurganus [Gur] originally defined spirallikeness in several variables as spirallikeness with respect to a normal matrix A whose eigenvalues $\lambda_1, \ldots, \lambda_n$ have positive real part, or equivalently such that $m(A) > 0$. If A is a normal matrix, there exists a unitary matrix V such that VAV^{-1} is diagonal with diagonal entries $\lambda_1, \ldots, \lambda_n$. In this case a domain Ω in $\mathbb{C}^n$ containing 0 is spirallike with respect to A if and only if whenever $w = (w_1, \ldots, w_n) \in V(\Omega)$, the spiral $(w_1 e^{-\lambda_1 t}, \ldots, w_n e^{-\lambda_n t})$, $t \geq 0$, is contained in $V(\Omega)$. In particular if $A = e^{-i\alpha}I$ for some $\alpha \in (-\pi/2, \pi/2)$, we obtain a class of mappings studied by Hamada and Kohr [Ham-Koh4] and called *spirallike of type α*. In this case the condition (6.4.8) reduces to

$$(6.4.10) \qquad e^{-i\alpha}[Df(z)]^{-1}f(z) \in \mathcal{N}.$$

This is the most straightforward generalization of the notion of spirallikeness to several variables, and it is this class whose behaviour is closest to that of spirallike functions of one variable. In particular we shall see in Chapter 8 that spirallike mappings of type α can be characterized in terms of Loewner chains.

The following example of Hamada and Kohr [Ham-Koh4] shows that the class of spirallike maps with respect to an arbitrary linear operator need not be a normal family, even when the operator is diagonal.

Example 6.4.12. In the Euclidean space $\mathbb{C}^2$, let $f(z) = (z_1 + az_2^2, z_2)$ and let $A(z) = (2z_1, z_2)$, $z = (z_1, z_2) \in \mathbb{C}^2$. Then for all $a \in \mathbb{C}$, $[Df(z)]^{-1}Af(z) = A(z)$. Since $A \in \mathcal{N}$, it follows that f is spirallike with respect to A for all $a \in \mathbb{C}$.

Open Problem 6.4.13. *Find conditions on the linear operator A such that the class of spirallike maps relative to A be a normal family on the unit ball of $\mathbb{C}^n$, $n \geq 2$.*

Notes. Recently Hamada, Kohr, and Liczberski [Ham-Koh-Lic1] extended the above notions and results concerning Φ-likeness and spirallikeness in the unit ball of $\mathbb{C}^n$ to bounded balanced pseudoconvex domains whose Minkowski function is of class C^1 in $\mathbb{C}^n \setminus \{0\}$. Also in [Ham-Koh3] the authors obtained some conditions for diffeomorphism and spirallikeness for a local diffeomorphism of class C^1 on the above domains. Kohr [Koh3] has recently studied the notion of spirallikeness on bounded domains Ω in $\mathbb{C}^n$ whose Bergman kernel function becomes infinite everywhere on the boundary. For further material concerning spirallike and Φ-like holomorphic mappings in higher dimensions, the reader may consult the instructive papers [Gur], [Ham3] and [Su4]. Also see the monograph [Koh-Lic2].

Problems

6.4.1. Let $f_j(z_j)$, $j = 1, \ldots, n$, be a spirallike function of type α on the unit disc U, where $\alpha \in (-\pi/2, \pi/2)$. Let $\lambda_j \geq 0$, $\sum_{j=1}^{n} \lambda_j = 1$. Show that

$$F(z) = z \prod_{j=1}^{n} \left(\frac{f_j(z_j)}{z_j} \right)^{\lambda_j}, \quad z = (z_1, \ldots, z_n) \in B,$$

is spirallike of type α on the Euclidean unit ball of $\mathbb{C}^n$.
(Cf. Pfaltzgraff and Suffridge, 1999 [Pfa-Su3]; Hamada and Kohr, 2001 [Ham-Koh6].)

6.4.2. Let $f : B \subset \mathbb{C}^2 \to \mathbb{C}^2$ be given by

$$f(z) = \frac{z}{(1 + e^{i\alpha} z_1 z_2)^{\frac{1+e^{-i\alpha}}{2}}}, \quad z = (z_1, z_2) \in B,$$

where $\alpha \in (-\pi, \pi)$. Show that f is spirallike relative to $e^{-i\alpha/2}I$ on the Euclidean unit ball of $\mathbb{C}^2$ (i.e. f is spirallike of type $\alpha/2$).
(Suffridge, 1975 [Su4].)

6.4.3. Let $f_j(z_j)$, $j = 1, \ldots, n$, be a spirallike function of type $\alpha \in (-\pi/2, \pi/2)$ on the unit disc U. Show that $f(z) = (f_1(z_1), \ldots, f_n(z_n))$, $z = (z_1, \ldots, z_n) \in B$, is spirallike of type α on the Euclidean unit ball B of $\mathbb{C}^n$.
(Suffridge, 1975 [Su4].)

6.4.4. Let B be the Euclidean unit ball in $\mathbb{C}^2$ and let $f(z) = (z_1, z_1 z_2)$, $z = (z_1, z_2) \in B$. Show that the image of B is not open.

Chapter 7

Growth, covering and distortion results for starlike and convex mappings in $\mathbb{C}^n$ and complex Banach spaces

The aim of this chapter is to give growth, covering and distortion results for subclasses of normalized biholomorphic mappings on the unit ball in $\mathbb{C}^n$, and to some extent on the unit ball of a Banach or Hilbert space or on certain bounded pseudoconvex domains in $\mathbb{C}^n$. Also we shall consider bounds for the coefficients of starlike and convex mappings on the unit ball of $\mathbb{C}^n$. Such bounds are expressed in terms of the kth Fréchet derivative of the mapping, using either the norm or the linear functionals introduced in Section 6.1.6.

We have seen in Section 6.1.7 that the full class $S(B)$ of normalized biholomorphic mappings on the unit ball in $\mathbb{C}^n$ is not a normal family. Moreover, in higher dimensions the growth and covering theorems fail for the full class $S(B)$. Similarly, there is no distortion theorem involving the differential $Df(z)$ or the Jacobian determinant $J_f(z)$ of a mapping $f \in S(B)$. Finally, there are no bounds for the Taylor series coefficients of order two or higher for mappings in $S(B)$.

Because of such examples, we are obliged to restrict to proper subclasses of $S(B)$ in studying growth, covering, and distortion theorems or coefficient estimates. In this chapter we shall focus primarily on the normalized starlike mappings and the normalized convex mappings.

7.1 Growth, covering and distortion results for starlike mappings in several complex variables and complex Banach spaces

7.1.1 Growth and covering results for starlike mappings on the unit ball and some pseudoconvex domains in $\mathbb{C}^n$. Extensions to complex Banach spaces

We begin this section with the growth theorem for normalized starlike mappings on the unit ball of $\mathbb{C}^n$ with the usual Euclidean structure. This result was obtained by Barnard, FitzGerald and Gong [Bar-Fit-Gon1] using the analytical characterization of starlikeness, and by Kubicka and Poreda [Kub-Por] (and later Chuaqui [Chu2]) using the method of Loewner chains. We shall give a proof using the analytical characterization of starlikeness; the idea is due to Liu and Ren [Liu-Ren2] (cf. [Gon4]). In the next chapter we shall give some generalizations to the case of normalized biholomorphic mappings which have parametric representation on the unit ball of $\mathbb{C}^n$.

Theorem 7.1.1. Let $f : B \to \mathbb{C}^n$ be a normalized starlike mapping on the Euclidean unit ball B. Then

$$(7.1.1) \qquad \frac{\|z\|}{(1 + \|z\|)^2} \leq \|f(z)\| \leq \frac{\|z\|}{(1 - \|z\|)^2} \quad on \quad B.$$

These estimates are sharp. Consequently, $f(B) \supseteq B_{1/4}$.

Proof. Fixing $z_0 \in B \setminus \{0\}$, we consider the curve γ in B parametrized by

$$z(t) = f^{-1}(tf(z_0)), \quad t \in [0, 1].$$

Thus γ is the inverse image under f of the straight line segment joining 0 to $f(z_0)$. We have

$$\frac{dz(t)}{dt} = Df^{-1}(tf(z_0))f(z_0) = \frac{1}{t}[Df(z(t))]^{-1}f(z(t)),$$

and hence

(7.1.2)
$$\frac{d}{dt}\|z(t)\|^2 = 2\mathrm{Re}\left\langle \frac{dz(t)}{dt}, z(t)\right\rangle$$

$$= \frac{2}{t}\mathrm{Re}\left\langle [Df(z(t))]^{-1}f(z(t)), z(t)\right\rangle, \quad t \in (0,1].$$

Since f is starlike, $(Df)^{-1} \circ f \in \mathcal{M}$, and hence

(7.1.3) $\|z\|^2\dfrac{1-\|z\|}{1+\|z\|} \leq \mathrm{Re}\left\langle [Df(z)]^{-1}f(z), z\right\rangle \leq \|z\|^2\dfrac{1+\|z\|}{1-\|z\|}, \; z \in B,$

by (6.1.27). Combining (7.1.2) with (7.1.3) and using the fact that

$$\frac{d}{dt}\|z(t)\|^2 = 2\|z(t)\|\frac{d}{dt}\|z(t)\|,$$

we obtain

$$\frac{1}{t}\|z(t)\|\frac{1-\|z(t)\|}{1+\|z(t)\|} \leq \frac{d}{dt}\|z(t)\| \leq \frac{1}{t}\|z(t)\|\frac{1+\|z(t)\|}{1-\|z(t)\|},$$

or

$$\frac{1-\|z(t)\|}{\|z(t)\|(1+\|z(t)\|)}d\|z(t)\| \leq \frac{dt}{t} \leq \frac{1+\|z(t)\|}{\|z(t)\|(1-\|z(t)\|)}d\|z(t)\|.$$

Integrating both sides of this inequality from ε to 1, where $0 < \varepsilon < 1$, we deduce that

$$\log\frac{\|z_0\|}{(1+\|z_0\|)^2} - \log\frac{\|z(\varepsilon)\|}{(1+\|z(\varepsilon)\|)^2} \leq \log\frac{1}{\varepsilon}$$

$$\leq \log\frac{\|z_0\|}{(1-\|z_0\|)^2} - \log\frac{\|z(\varepsilon)\|}{(1-\|z(\varepsilon)\|)^2},$$

or equivalently,

$$\frac{\|z_0\|}{(1+\|z_0\|)^2} \leq \frac{\|z(\varepsilon)\|}{\varepsilon(1+\|z(\varepsilon)\|)^2}$$

and

$$\frac{\|z(\varepsilon)\|}{\varepsilon(1-\|z(\varepsilon)\|)^2} \leq \frac{\|z_0\|}{(1-\|z_0\|)^2}.$$

Since $\displaystyle\lim_{\varepsilon\to 0^+}\frac{z(\varepsilon)}{\varepsilon} = f(z_0)$ by the normalization of f, we obtain (7.1.1) by letting $\varepsilon \to 0^+$ in the last two inequalities.

In order to show that these estimates are sharp, let

$$f(z) = \left(\frac{z_1}{(1-z_1)^2}, \ldots, \frac{z_n}{(1-z_n)^2}\right), \quad z = (z_1, \ldots, z_n) \in B.$$

From Problem 6.2.5, we know that f is starlike on B. Moreover, for $r \in [0,1)$ and $z = (r, 0, \ldots, 0)$, we have $\|f(z)\| = \dfrac{r}{(1-r)^2} = \dfrac{\|z\|}{(1-\|z\|)^2}$. Similarly, for $r \in [0,1)$ and $z = (-r, 0, \ldots, 0)$, we have $\|f(z)\| = \dfrac{r}{(1+r)^2} = \dfrac{\|z\|}{(1+\|z\|)^2}$. This completes the proof.

The same example shows that the covering result given in Theorem 7.1.1 is sharp. Thus the *Koebe constant* for normalized starlike mappings of the Euclidean unit ball in $\mathbb{C}^n$ is $1/4$, as in one variable.

When f is k-fold symmetric (i.e. $\exp(-2\pi i/k)f(e^{2\pi i/k}z) = f(z)$ for $z \in B$, where k is a positive integer), we obtain the following refinement of the above result, due to Barnard, FitzGerald and Gong [Bar-Fit-Gon1]. We leave the proof of Corollary 7.1.2 for the reader.

Corollary 7.1.2. *If $f : B \to \mathbb{C}^n$ is a normalized starlike mapping which is k-fold symmetric, then*

$$(7.1.4) \qquad \frac{\|z\|}{(1+\|z\|^k)^{2/k}} \leq \|f(z)\| \leq \frac{\|z\|}{(1-\|z\|^k)^{2/k}}.$$

These estimates are sharp. Consequently, $f(B)$ contains a ball centered at zero and of radius $2^{-2/k}$.

This result leads to another proof [Bar-Fit-Gon1] of the well known inequivalence between the ball and the polydisc in $\mathbb{C}^n$, $n \geq 2$ (Theorem 6.1.21).

Corollary 7.1.3. *The image of the unit ball B under a normalized starlike mapping f is a balanced domain in $\mathbb{C}^n$ if and only if $f(B) = B$. Consequently, the Euclidean unit ball and the unit polydisc of $\mathbb{C}^n$, $n \geq 2$, are not biholomorphically equivalent.*

Proof. First assume that $f(B)$ is a balanced domain. Then this domain is k-fold symmetric for all $k \in \mathbb{N}$. Letting $k \to \infty$ in (7.1.4), we deduce that $\|f(z)\| = \|z\|$ on B. Together with the normalization of f, this implies that $f(z) \equiv z$. (Without using Cartan's theorem on fixed points, this can be seen as follows: Let $u \in \mathbb{C}^n$, $\|u\| = 1$, and let π_u denote the orthogonal projection of $\mathbb{C}^n$ onto the subspace $\mathbb{C}u$. Let $\psi_u : \mathbb{C} \to \mathbb{C}^n$ be given by $\psi_u(\zeta) = \zeta u$. Then the Schwarz lemma in one variable implies that $\psi_u^{-1} \circ \pi_u \circ f(\zeta u) = \zeta$, $\zeta \in U$. Hence $\pi_u \circ f(\zeta u) = \zeta u$, which implies that $f(\zeta u) = \zeta u$, for otherwise we could not have $\|f(\zeta u)\| = \|\zeta u\|$.)

The converse is obvious.

Next suppose that $F : B \to P$ is a biholomorphic mapping such that $F(B) = P$. By composing with an automorphism of P we may assume that $F(0) = 0$. There exist unitary matrices W and V such that $W[DF(0)]V$ is diagonal. After a linear change of coordinates in the target space we may assume that $W[DF(0)]V = I$. Now the mapping $G = W \circ F \circ V$ is a normalized biholomorphic mapping of B onto a balanced domain, and hence $G(z) \equiv z$. On the other hand, the image of B under G is $W(P)$. Hence $W(P) = B$, which is impossible.

A similar growth theorem holds for starlike mappings on the unit polydisc P in $\mathbb{C}^n$. We leave the proof for the reader.

The growth theorem for starlike mappings has been studied in other situations, for example on the unit ball for a p-norm, $1 < p < \infty$ ([Gon-Wa-Yu1], [Pfa3]), and classical bounded symmetric domains ([Gon-Wa-Yu4], [LiuT2]; see also [Gon4]). Further results for starlike maps including generalizations of the covering theorem can be found in [Gon4]. A case we shall consider in more detail is the case of bounded balanced pseudoconvex domains whose Minkowski function h is of class C^1 in $\mathbb{C}^n \setminus \{0\}$. (It is easily seen that h is continuous at 0 and nonzero except at 0 for any bounded balanced domain.) This case was studied by Liu and Ren [Liu-Ren2] using the analytical characterization of starlikeness (see also [Gon4, Theorem 3.4.1]), and by Hamada [Ham4] using the method of Loewner chains.

Theorem 7.1.4. *Let Ω be a bounded balanced pseudoconvex domain whose Minkowski function h is of class C^1 on $\mathbb{C}^n \setminus \{0\}$. If $f : \Omega \to \mathbb{C}^n$ is a normalized starlike mapping, then*

$$(7.1.5) \qquad \frac{h(z)}{(1 + h(z))^2} \leq h(f(z)) \leq \frac{h(z)}{(1 - h(z))^2} \quad on \quad \Omega.$$

Consequently, $f(\Omega) \supseteq 1/4\Omega$.

Proof. Let $z \in \Omega \setminus \{0\}$ and $u = z/h(z) \in \partial\Omega$. Also let

$$g(\zeta) = \begin{cases} \dfrac{1}{\zeta} \left\langle [Df(\zeta u)]^{-1} f(\zeta u), \dfrac{\partial h^2}{\partial \overline{z}}(u) \right\rangle, & 0 < |\zeta| < 1 \\ 1, & \zeta = 0. \end{cases}$$

Using the fact that $\left\langle u, \dfrac{\partial h^2}{\partial \overline{z}}(u) \right\rangle = h^2(u) = 1$ (see the proof of Theorem 6.2.8) we see that g is holomorphic on U. We also note that the Minkowski function satisfies

$$\frac{\partial h^2}{\partial \bar{z}_j}(\zeta u) = \zeta \frac{\partial h^2}{\partial \bar{z}_j}(u), \quad \zeta \neq 0, \ j = 1, \ldots, n.$$

This allows us to rewrite g in the form

$$g(\zeta) = \frac{1}{|\zeta|^2} \left\langle [Df(\zeta u)]^{-1} f(\zeta u), \frac{\partial h^2}{\partial \bar{z}}(\zeta u) \right\rangle, \quad \zeta \in U \setminus \{0\},$$

and using Theorem 6.2.8 we see that Re $g(\zeta) > 0$, $\zeta \in U$. Hence $g \in \mathcal{P}$, and by Theorem 2.1.3 we have

$$|\zeta|^2 \frac{1 - |\zeta|}{1 + |\zeta|} \leq \text{Re} \left\langle [Df(\zeta u)]^{-1} f(\zeta u), \frac{\partial h^2}{\partial \bar{z}}(\zeta u) \right\rangle \leq |\zeta|^2 \frac{1 + |\zeta|}{1 - |\zeta|}.$$

Letting $\zeta = h(z)$ in the above, we obtain

$$h^2(z) \frac{1 - h(z)}{1 + h(z)} \leq \text{Re} \left\langle [Df(z)]^{-1} f(z), \frac{\partial h^2}{\partial \bar{z}}(z) \right\rangle \leq h^2(z) \frac{1 + h(z)}{1 - h(z)}.$$

Now let $z(t) = f^{-1}(tf(z))$, $0 \leq t \leq 1$, be the inverse image of the straight line segment between 0 and $f(z)$. Then $z(t)$ is a curve of class C^1 contained in the domain Ω such that $z(0) = 0$ and $z(1) = z$. A short computation yields that

$$\frac{dh^2(z(t))}{dt} = 2\text{Re} \left\langle \frac{dz(t)}{dt}, \frac{\partial h^2}{\partial \bar{z}}(z(t)) \right\rangle$$

$$= \frac{2}{t}\text{Re} \left\langle [Df(z(t))]^{-1} f(z(t)), \frac{\partial h^2}{\partial \bar{z}}(z(t)) \right\rangle, \ t \in (0, 1],$$

and thus we obtain the relation

$$\frac{2}{t} h^2(z(t)) \frac{1 - h(z(t))}{1 + h(z(t))} \leq \frac{dh^2(z(t))}{dt} \leq \frac{2}{t} h^2(z(t)) \frac{1 + h(z(t))}{1 - h(z(t))}.$$

Integrating both sides of the above inequality from ε to 1, with $0 < \varepsilon < 1$, and using the fact that $dh^2(z(t)) = 2h(z(t))dh(z(t))$, we deduce that

$$\int_\varepsilon^1 \frac{1 - h(z(t))}{h(z(t))(1 + h(z(t)))} dh(z(t)) \leq \int_\varepsilon^1 \frac{dt}{t} \leq \int_\varepsilon^1 \frac{1 + h(z(t))}{h(z(t))(1 - h(z(t)))} dh(z(t)).$$

After short computations, we obtain the relations

$$\log \frac{h(z)}{(1 + h(z))^2} - \log \frac{h(z(\varepsilon))}{(1 + h(z(\varepsilon)))^2} \leq \log \frac{1}{\varepsilon}$$

$$\leq \log \frac{h(z)}{(1 - h(z))^2} - \log \frac{h(z(\varepsilon))}{(1 - h(z(\varepsilon)))^2},$$

or equivalently,

$$\frac{h(z)}{(1 + h(z))^2} \leq \frac{h(z(\varepsilon))}{(1 + h(z(\varepsilon)))^2 \varepsilon}$$

and

$$\frac{h(z)}{(1 - h(z))^2} \geq \frac{h(z(\varepsilon))}{(1 - h(z(\varepsilon)))^2 \varepsilon}.$$

Since $\displaystyle\lim_{\varepsilon \to 0^+} \frac{h(z(\varepsilon))}{\varepsilon} = h(f(z))$ by the normalization of f, we obtain (7.1.5) by letting $\varepsilon \to 0^+$ in the two previous relations. This completes the proof.

Finally we remark that the 1/4-growth result given in Theorem 7.1.1 can be extended to the case of normalized starlike mappings on the unit ball of complex Banach spaces (see [Don-Zha], cf. [Gon4]).

Theorem 7.1.5. *Let X be a complex Banach space and let B be the unit ball of X. Let $f : B \to X$ be a normalized starlike mapping. Then*

$$(7.1.6) \qquad \frac{\|z\|}{(1 + \|z\|)^2} \leq \|f(z)\| \leq \frac{\|z\|}{(1 - \|z\|)^2}, \quad z \in B.$$

Remark 7.1.6. When X is a complex Hilbert space the growth result (7.1.6) is sharp. To see this, let $f \in S^*$ and let $u \in X$ be a unit vector. Also let $\langle \cdot, \cdot \rangle$ be the inner product of X and consider the holomorphic map $F_u : B \to X$ given by

$$(7.1.7) \qquad F_u(z) = f(\langle z, u \rangle)u + \sqrt{f'(\langle z, u \rangle)}(z - \langle z, u \rangle u), \quad z \in B.$$

Graham and Kohr [Gra-Koh1] proved that F_u is a starlike mapping of B. (This extension operator was introduced by Roper and Suffridge [Rop-Su1] and will be the subject of Chapter 11.)

If we begin with the function $f(\zeta) = \dfrac{\zeta}{(1 - \zeta)^2}$, $\zeta \in U$, a straightforward computation in (7.1.7) yields that

$$F_u(z) = \frac{\langle z, u \rangle u}{(1 - \langle z, u \rangle)^2} + \sqrt{\frac{1 + \langle z, u \rangle}{(1 - \langle z, u \rangle)^3}}(z - \langle z, u \rangle u), \ z \in B.$$

If $z = ru$, where $r \in [0, 1)$, then $\|z\| = r$ and $\|F_u(z)\| = \dfrac{r}{(1-r)^2} = \dfrac{\|z\|}{(1-\|z\|)^2}$.
On the other hand, if $z = -ru$, $r \in [0, 1)$, then $\|z\| = r$ and $\|F_u(z)\| = \dfrac{r}{(1+r)^2} = \dfrac{\|z\|}{(1+\|z\|)^2}$. This completes the proof.

7.1.2   Bounds for coefficients of normalized starlike mappings in $\mathbb{C}^n$

We next consider some bounds for the power series coefficients of normalized starlike mappings on the unit ball of $\mathbb{C}^n$. In this section we work with an arbitrary norm $\|\cdot\|$ in $\mathbb{C}^n$. As usual, B will denote the unit ball with respect to this norm. The coefficient bounds are expressed in terms of bounds for the kth Fréchet derivative of the mapping at 0.

We begin by obtaining some estimates for the kth Fréchet derivative of mappings in the class $\mathcal{M}$ (cf. [Koh6], [Gra-Ham-Koh]). These results are of independent interest; in particular the norm estimates lead to another proof of the compactness of the class $\mathcal{M}$ [Gra-Ham-Koh] (cf. Theorem 6.1.39).

Theorem 7.1.7. *If $p \in \mathcal{M}$ then*

$$(7.1.8) \quad \left| l_w\left(\frac{1}{k!} D^k p(0)(w^k)\right) \right| \leq 2, \quad k \geq 2, \quad \|w\| = 1, \quad l_w \in T(w).$$

These estimates are sharp when B is the unit ball of $\mathbb{C}^n$ with respect to a p-norm, $1 \leq p \leq \infty$.

Moreover,

$$(7.1.9) \quad \left\| \frac{1}{m!} D^m p(0)(w^m) \right\| \leq 2k_m, \quad m \geq 2, \quad \|w\| = 1,$$

where $k_m = m^{m/(m-1)}$. Consequently, $\mathcal{M}$ is a compact set.

Proof. Fix $w \in \mathbb{C}^n$ with $\|w\| = 1$ and let $l_w \in T(w)$. Also let

$$q(\zeta) = \begin{cases} \frac{1}{\zeta} l_w(p(\zeta w)), & \zeta \in U \setminus \{0\} \\ 1, & \zeta = 0. \end{cases}$$

Since p is normalized and holomorphic on B, it is clear that q is a holomorphic function on U. Moreover, since $p \in \mathcal{M}$ it follows that $\mathrm{Re}\, q(\zeta) > 0$ on U. Indeed, since $l_{\alpha w}(\cdot) = \dfrac{|\alpha|}{\alpha} l_w(\cdot) \in T(\alpha w)$ for each $\alpha \in \mathbb{C} \setminus \{0\}$, it follows that for $\zeta \in U \setminus \{0\}$ we have

$$q(\zeta) = \frac{1}{\zeta} l_w(p(\zeta w)) = \frac{1}{|\zeta|} l_{\zeta w}(p(\zeta w)),$$

and hence Re $q(\zeta) > 0$ on U. Therefore $q \in \mathcal{P}$, and using Theorem 2.1.5 we conclude that

$$\text{(7.1.10)} \qquad \left| \frac{1}{k!} q^{(k)}(0) \right| \le 2, \quad k \ge 1.$$

On the other hand, expanding $p(\zeta w)/\zeta$ in a power series in ζ we deduce that

$$q(\zeta) = 1 + \sum_{k=2}^{\infty} l_w \left(\frac{1}{k!} D^k p(0)(w^k) \right) \zeta^{k-1}.$$

Hence by the uniqueness of Taylor series expansions, we conclude that

$$\text{(7.1.11)} \qquad q^{(k-1)}(0) = l_w \left(\frac{1}{k} D^k p(0)(w^k) \right), \quad k \ge 2.$$

Using (7.1.10) and (7.1.11), we obtain (7.1.8), as desired.

We next prove that these estimates are sharp on the unit ball of $\mathbb{C}^n$ with respect to a p-norm, $1 \le p \le \infty$. To this end, let $\lambda \in \mathbb{C}$, $|\lambda| = 1$, and let $p : B \to \mathbb{C}^n$ be given by

$$p(z) = \left(z_1 \frac{1 + \lambda z_1}{1 - \lambda z_1}, \ldots, z_n \frac{1 + \lambda z_n}{1 - \lambda z_n} \right),$$

for $z = (z_1, \ldots, z_n) \in B$. Then $p \in H(B)$, $p(0) = 0$, $Dp(0) = I$, and it is easy to deduce that Re $l_z(p(z)) > 0$ for $z \in B \setminus \{0\}$ and $l_z \in T(z)$ (see Section 6.1.6). Hence $p \in \mathcal{M}$. On the other hand, a short computation yields that

$$\frac{1}{k!} D^k p(0)(w^k) = 2\lambda^{k-1}(w_1^k, \ldots, w_n^k),$$

for $w = (w_1, \ldots, w_n) \in \mathbb{C}^n$ and $k \ge 2$. Therefore, for $w = (1, 0, \ldots, 0)$ and $l_w \in T(w)$, we obtain

$$\left| l_w \left(\frac{1}{k!} D^k p(0)(w^k) \right) \right| = 2.$$

We next prove (7.1.9). For this purpose, let $P_m(z) = \frac{1}{m!} D^m p(0)(z^m)$. Then $P_m(z)$ is a homogeneous polynomial of degree m and recalling Lemma 6.1.37, we have

$$\text{(7.1.12)} \qquad \|P_m\| \le k_m |V(P_m)|,$$

where $k_m = m^{m/(m-1)}$ for $m \geq 2$.

We recall that
$$\|P_m\| = \sup\left\{\|P_m(z)\| : z \in \overline{B}\right\},$$

and since P_m is continuous on $\overline{B}$, the numerical radius of P_m is just
$$|V(P_m)| = \sup\left\{|l_z(P_m(z))| : l_z \in T(z), \|z\| = 1\right\}.$$

Taking into account (7.1.8) and (7.1.12), we easily deduce that
$$|V(P_m)| \leq 2, \quad m \geq 2,$$

and consequently,
$$\|P_m\| \leq 2k_m, \quad m \geq 2.$$

We finally prove the compactness of $\mathcal{M}$. For this purpose it suffices to show the local uniform boundedness of $\mathcal{M}$, since the fact that $\mathcal{M}$ is closed has been proved in Theorem 6.1.39. Indeed, from (7.1.9) we deduce that
$$\|p(z)\| \leq r + \sum_{m=2}^{\infty} \|P_m(z)\| \leq r\left(1 + 2\sum_{m=2}^{\infty} k_m r^{m-1}\right)$$
$$\leq r\left(1 + 4\sum_{m=2}^{\infty} m r^{m-1}\right) = M(r) \leq \frac{4r}{(1-r)^2}, \quad \|z\| \leq r < 1,$$

making use of the fact that $k_m \leq 2m$ for $m \geq 2$. This estimate yields the fact that $\mathcal{M}$ is locally uniformly bounded on B, and hence it is a normal family, as claimed.

As an application of Theorem 7.1.7 we shall obtain some estimates for the second order Fréchet derivative of a normalized starlike mapping (cf. [Koh6], [Gra-Ham-Koh]). It seems to be more difficult to estimate the higher order Fréchet derivatives by this method.

Theorem 7.1.8. *If $f \in S^*(B)$ then*

(7.1.13) $\left|l_w\left(\frac{1}{2}D^2 f(0)(w^2)\right)\right| \leq 2, \quad \|w\| = 1, \quad l_w \in T(w).$

This estimate is sharp when B is the unit ball of $\mathbb{C}^n$ with respect to a p-norm, $1 \leq p \leq \infty$.

Proof. Let $p(z) = [Df(z)]^{-1}f(z)$, $z \in B$. Then $p \in \mathcal{M}$ since f is starlike, and from (7.1.8) we have

$$(7.1.14) \qquad \left| l_w(D^2 p(0)(w^2)) \right| \le 2, \quad \|w\| = 1, \quad l_w \in T(w).$$

Fix $w \in \mathbb{C}^n \setminus \{0\}$. Since $f(z) = Df(z)p(z)$, $z \in B$, a simple computation yields that

$$(7.1.15) \qquad D^2 f(0)(w^2) = -D^2 p(0)(w^2).$$

To this end, it suffices to expand the mappings $f(z)$ and $Df(z)p(z)$ in power series and then to compare the second order terms. Using (7.1.14) and (7.1.15), one deduces (7.1.13), as desired. In order to see that (7.1.13) is sharp on the unit ball of $\mathbb{C}^n$ with respect to a p-norm, $1 \le p \le \infty$, it suffices to consider

$$f(z) = \left(\frac{z_1}{(1 - z_1)^2}, \ldots, \frac{z_n}{(1 - z_n)^2} \right), \quad z = (z_1, \ldots, z_n) \in B.$$

Then f is a normalized starlike mapping on B (see Problem 6.2.5), and for $w = e_1 = (1, 0, \ldots, 0)$ and $l_w \in T(w)$ we easily deduce that

$$\left| l_w \left(\frac{1}{2} D^2 f(0)(w^2) \right) \right| = 2.$$

This completes the proof.

By a similar argument, one can also estimate the norm of the second order Fréchet derivative of a normalized starlike mapping on the unit ball [Gra-Ham-Koh].

Theorem 7.1.9. *If $f \in S^*(B)$ then*

$$\left\| \frac{1}{2} D^2 f(0)(w^2) \right\| \le 8, \quad \|w\| = 1.$$

Proof. Again letting $p(z) = [Df(z)]^{-1}f(z)$, we have

$$\left\| \frac{1}{2} D^2 p(0)(w^2) \right\| \le 2k_2 = 8, \quad \|w\| = 1,$$

from (7.1.9), and together with (7.1.15) this gives the desired bound.

For higher order Fréchet derivatives of starlike mappings on B, we have the following result due to Poreda [Por5].

Theorem 7.1.10. *Let* $f : B \to \mathbb{C}^n$ *be a normalized starlike mapping. Then*

(7.1.16)
$$\left\| \frac{1}{k!} D^k f(0)(w^k) \right\| \leq \left[\frac{e(k+1)}{2} \right]^2,$$

for all $w \in \mathbb{C}^n$, $\|w\| = 1$, *and* $k \in \mathbb{N}$, $k \geq 2$.

Proof. Fix $w \in \mathbb{C}^n$ with $\|w\| = 1$ and $k \geq 2$. Using the Cauchy integral formula for vector-valued holomorphic functions

$$\frac{1}{k!} D^k f(0)(w^k) = \frac{1}{2\pi i} \int\limits_{|\zeta| = r} \frac{f(\zeta w)}{\zeta^{k+1}} d\zeta, \quad 0 < r < 1,$$

we deduce that

$$\left\| \frac{1}{k!} D^k f(0)(w^k) \right\| \leq \frac{1}{2\pi} \int_0^{2\pi} \frac{\|f(re^{i\theta} w)\|}{r^k} d\theta \leq \frac{1}{r^{k-1}(1-r)^2},$$

where the last step uses the growth theorem for starlike maps (Theorem 7.1.5).

Now, if we let

$$g(r) = \frac{1}{r^{k-1}(1-r)^2}, \quad 0 < r < 1,$$

then an elementary computation yields that

$$\min \left\{ g(r) : 0 < r < 1 \right\} = g\left(\frac{k-1}{k+1} \right)$$

$$= \left[\left(1 + \frac{2}{k-1} \right)^{\frac{k-1}{2}} \right]^2 \left(\frac{k+1}{2} \right)^2 \leq \left[\frac{e(k+1)}{2} \right]^2,$$

and hence we obtain the desired estimate (7.1.16). $\blacksquare$

The following could be regarded as a generalization of the Bieberbach conjecture to several complex variables. More precisely, it would generalize Theorem 2.2.16.

Conjecture 7.1.11. Let $f : B \to \mathbb{C}^n$ be a normalized starlike mapping on the unit ball of $\mathbb{C}^n$, $n \geq 2$. Then

$$\left| l_w \left(\frac{1}{k!} D^k f(0)(w^k) \right) \right| \leq k, \, k \geq 2, \, \|w\| = 1, \, l_w \in T(w).$$

Remark 7.1.12. The corresponding estimate for the norm of the kth order Fréchet derivative of a starlike mapping on the unit ball of $\mathbb{C}^n$, $n \geq 2$, i.e.

(7.1.17)
$$\left\| \frac{1}{k!} D^k f(0)(w^k) \right\| \leq k, \quad \|w\| = 1, \quad k \geq 2,$$

does not hold in general.

To see this, let $n = 2$ and let $\mathbb{C}^2$ be the space of two complex variables with the Euclidean norm. Also let

$$f(z) = (z_1 + az_2^2, z_2), \quad z = (z_1, z_2) \in B.$$

If $|a| = 3\sqrt{3}/2$ then $f \in S^*(B)$ by Problem 6.2.1. However for $w = e_2 = (0, 1)$, we have

$$\left\| \frac{1}{2} D^2 f(0)(w^2) \right\| = |a| > 2.$$

It is still possible that the norm estimates (7.1.17) are true in the case of the polydisc. Gong [Gon5, Theorem 5.3.1] gave a proof that (7.1.17) holds for $k = 2, 3$, and formulated the following:

Conjecture 7.1.13. If $n \geq 2$ and $f \colon P \to \mathbb{C}^n$ is a normalized starlike mapping of the unit polydisc then

$$\left\| \frac{1}{k!} D^k f(0)(w^k) \right\|_\infty \leq k, \quad k \geq 2, \quad \|w\|_\infty = 1.$$

It follows easily from Theorem 6.3.2 that (7.1.17) is true for all $k \in \mathbb{N}$, $k \geq 2$, for maps of the form $f(z) = Dg(z)z$, $z \in P$, where $g \colon P \to \mathbb{C}^n$ is a normalized convex map.

Problems

7.1.1. Show that if $h : B(p) \to \mathbb{C}^n$ is given by

$$h(z) = \left(z_1 \frac{1 + \lambda z_1}{1 - \lambda z_1}, \ldots, z_n \frac{1 + \lambda z_n}{1 - \lambda z_n} \right), \quad z = (z_1, \ldots, z_n) \in B(p),$$

where $1 \leq p \leq \infty$ and $|\lambda| = 1$, then $h \in \mathcal{M}$.

7.1.2. Prove the estimate (7.1.17) for $k = 2, 3$ in the case of normalized starlike mappings of the unit polydisc.

7.1.3. Prove Corollary 7.1.2.

7.1.4. Prove Theorem 7.1.5.

7.1.3   A distortion result for a subclass of starlike mappings in $\mathbb{C}^n$

We finish this section with an estimate for the Jacobian determinant of a particular subclass of the normalized starlike mappings on the Euclidean unit ball in $\mathbb{C}^n$, $n \geq 2$. This result is due to Pfaltzgraff and Suffridge [Pfa-Su3], and they conjectured that it is valid for the full class $S^*(B)$. (In this subsection, B denotes the Euclidean unit ball in $\mathbb{C}^n$.) First we discuss a procedure from [Pfa-Su3] for generating mappings in $S^*(B)$ from mappings in S^*. We have already seen one method of doing this in Problem 6.2.5.

Theorem 7.1.14. *Suppose that $g \in H(B, \mathbb{C})$ and $g(0) = 1$. Let $F(z) = zg(z)$, $z \in B$. Then $F \in S^*(B)$ if and only if*

$$(7.1.18) \qquad \text{Re}\left\{1 + \frac{Dg(z)(z)}{g(z)}\right\} > 0, \quad z \in B.$$

Proof. For the purposes of this proof, vectors in $\mathbb{C}^n$ will be written as column vectors, and the transpose of such a vector is a row vector. Note that if $F \in S^*(B)$ or if (7.1.18) holds then $g(z) \neq 0$, $z \in B$. It is easy to see that

$$DF(z) = g(z)(I + zL(z)^t),$$

where $L(z) = \dfrac{Dg(z)}{g(z)}$, $z \in B$. Hence using the formula

$$(I + ab^t)^{-1} = I - \frac{ab^t}{1 + b^t a}, \quad a, b \in \mathbb{C}^n, \quad b^t a \neq -1,$$

we obtain

$$[DF(z)]^{-1} = \frac{1}{g(z)}\left[I - \frac{zL(z)^t}{1 + L(z)^t z}\right], \quad 1 + L(z)^t z \neq 0.$$

From this it follows that

$$w(z) = [DF(z)]^{-1}F(z) = z\left[1 - \frac{L(z)^t z}{1 + L(z)^t z}\right] = z\left[\frac{1}{1 + L(z)^t z}\right],$$

and the condition $\text{Re}\langle w(z), z\rangle > 0$ is equivalent to $\text{Re}\{1 + L(z)^t z\} > 0$, which is (7.1.18). (Moreover, if this condition is satisfied then $DF(z)$ is clearly invertible.)

Theorem 7.1.15. *For each $j = 1, \ldots, n$, let f_j be a normalized starlike function on the unit disc U. If $\lambda_j \geq 0$ and $\sum_{j=1}^{n} \lambda_j = 1$, then*

$$(7.1.19) \qquad F(z) = z \prod_{j=1}^{n} \left(\frac{f_j(z_j)}{z_j} \right)^{\lambda_j}, \qquad z = (z_1, \ldots, z_n) \in B,$$

is a starlike mapping on B.

Proof. Let $\varphi_j(\zeta) = f_j(\zeta)/\zeta$, $\zeta \in U$, $j = 1, \ldots, n$, and let $g(z) = \prod_{j=1}^{n} (\varphi_j(z_j))^{\lambda_j}$ and $L(z) = Dg(z)/g(z)$, $z \in B$. By logarithmic differentiation, we have

$$L(z)^t = \left(\lambda_1 \frac{\varphi_1'(z_1)}{\varphi_1(z_1)}, \ldots, \lambda_n \frac{\varphi_n'(z_n)}{\varphi_n(z_n)} \right),$$

and hence the condition (7.1.18) becomes

$$\mathrm{Re} \left\{ 1 + \sum_{j=1}^{n} \lambda_j \frac{z_j \varphi_j'(z_j)}{\varphi_j(z_j)} \right\} > 0, \qquad z = (z_1, \ldots, z_n) \in B.$$

Since $\sum_{j=1}^{n} \lambda_j = 1$, this is equivalent to

$$\mathrm{Re} \sum_{j=1}^{n} \lambda_j \frac{z_j f_j'(z_j)}{f_j(z_j)} > 0, \qquad z = (z_1, \ldots, z_n) \in B.$$

This condition is clearly satisfied since $f_1, \ldots, f_n \in S^*$ and $\lambda_j \geq 0, j = 1, \ldots, n$, and the proof is complete.

Let $S_\pi^*(B)$ denote the set of starlike mappings F defined by (7.1.19). When $n = 1$, $S_\pi^*(U) = S^*$, but in higher dimensions $S_\pi^*(B) \subsetneqq S^*(B)$. To see this, let $n = 2$ and let $f(z) = (z_1 + az_2^2, z_2)$ for $z = (z_1, z_2) \in B$, where $0 < |a| \leq 3\sqrt{3}/2$. Taking into account Problem 6.2.1, one concludes that $f \in S^*(B)$, however $f \notin S_\pi^*(B)$.

Pfaltzgraff and Suffridge [Pfa-Su3] obtained the following bounds for the Jacobian determinant of mappings in $S_\pi^*(B)$:

Theorem 7.1.16. *If $F \in S_\pi^*(B)$ then*

$$(7.1.20) \qquad \frac{1 - \|z\|}{(1 + \|z\|)^{2n+1}} \leq |J_F(z)| \leq \frac{1 + \|z\|}{(1 - \|z\|)^{2n+1}}, \qquad z \in B.$$

These estimates are sharp.

Proof. Let $F \in S_\pi^*(B)$. Then there exist $\lambda_j \geq 0$, $\sum_{j=1}^{n} \lambda_j = 1$, and normal-ized starlike functions $f_j(z_j)$, $j = 1, \ldots, n$, such that $F(z) = zg(z)$, where

$$g(z) = \prod_{j=1}^{n} (\varphi_j(z_j))^{\lambda_j}, \quad z = (z_1, \ldots, z_n) \in B,$$

and $\varphi_j(z_j) = f_j(z_j)/z_j$, $j = 1, \ldots, n$.

As shown in Theorems 7.1.14 and 7.1.15, we have

$$DF(z) = g(z)\{I + zL(z)^t\}, \quad z \in B,$$

where

$$L(z)^t = \left(\lambda_1 \frac{\varphi_1'(z_1)}{\varphi_1(z_1)}, \ldots, \lambda_n \frac{\varphi_n'(z_n)}{\varphi_n(z_n)} \right).$$

Note that the $n \times n$ matrix $zL(z)^t$ has proportional columns, and hence its rank is 1. Therefore,

$$J_F(z) = \det DF(z) = (g(z))^n \left\{ 1 + \sum_{j=1}^{n} \lambda_j \frac{z_j \varphi_j'(z_j)}{\varphi_j(z_j)} \right\},$$

or equivalently

$$(7.1.21) \qquad J_F(z) = \prod_{j=1}^{n} \left(\frac{f_j(z_j)}{z_j} \right)^{n\lambda_j} \sum_{j=1}^{n} \lambda_j \frac{z_j f_j'(z_j)}{f_j(z_j)}.$$

On the other hand, because f_j is starlike on U, $j = 1, \ldots, n$, Theorem 2.2.7 gives

$$(7.1.22) \qquad \frac{1}{(1 + \|z\|)^2} \leq \frac{1}{(1 + |z_j|)^2} \leq \left| \frac{f_j(z_j)}{z_j} \right|$$

$$\leq \frac{1}{(1 - |z_j|)^2} \leq \frac{1}{(1 - \|z\|)^2}.$$

Also since $\dfrac{z_j f_j'(z_j)}{f_j(z_j)} \in P$, $j = 1, \ldots, n$, Theorem 2.1.3 implies that

$$(7.1.23) \qquad \frac{1 - \|z\|}{1 + \|z\|} \leq \left| \sum_{j=1}^{n} \lambda_j \frac{z_j f_j'(z_j)}{f_j(z_j)} \right| \leq \frac{1 + \|z\|}{1 - \|z\|}.$$

Finally, from (7.1.21), (7.1.22) and (7.1.23), we obtain the bounds (7.1.20), as desired. Equality holds in (7.1.20) if we take $F(z) = \dfrac{z}{(1-z_1)^2}$ for $z = (z_1, \ldots, z_n) \in B$. To see this, let $\lambda_1 = 1$ and $f_1(\zeta) = \dfrac{\zeta}{(1-\zeta)^2}$, $\zeta \in U$, which yield in (7.1.21) that

$$J_F(z) = \frac{1}{(1-z_1)^{2n}} \left(\frac{1+z_1}{1-z_1} \right).$$

Therefore, for $r \in [0,1)$ and $z = (r, 0, \ldots, 0)$, we obtain equality in the upper bound in (7.1.20), and for $r \in [0,1)$ and $z = (-r, 0, \ldots, 0)$, we obtain equality in the lower bound in (7.1.20). This completes the proof.

The problem of finding the sharp bounds for $|J_F(z)|$ for the full class $S^*(B)$ is still an open problem in several complex variables. However, Pfaltzgraff and Suffridge [Pfa-Su3] proposed the following:

Conjecture 7.1.17. The distortion bounds (7.1.20) hold for all normalized starlike mappings of B, when B is the Euclidean unit ball of $\mathbb{C}^n$, $n \geq 2$.

7.2 Growth, covering and distortion results for convex mappings in several complex variables and complex Banach spaces

7.2.1 Growth and covering results for convex mappings

In this section we shall prove a number of growth and covering results for normalized convex mappings on the unit ball of $\mathbb{C}^n$ and complex Banach spaces.

We first discuss the case of the unit polydisc P of $\mathbb{C}^n$ (i.e. the unit ball with respect to the maximum norm $\| \cdot \|_\infty$).

Using Suffridge's criterion for convexity of biholomorphic maps of the polydisc (Theorem 6.3.2), it is not hard to show the following (see [Lic1] and [Gon4]):

Theorem 7.2.1. If $f : P \to \mathbb{C}^n$ is a normalized convex mapping, then

$$(7.2.1) \qquad \frac{\|z\|_\infty}{1 + \|z\|_\infty} \leq \|f(z)\|_\infty \leq \frac{\|z\|_\infty}{1 - \|z\|_\infty} \quad on \quad P.$$

These estimates are sharp. Consequently, $f(P) \supseteq P_{1/2}$.

Proof. Since f is normalized convex, it follows from Theorem 6.3.2 that f has the form

$$f(z) = (f_1(z_1), \ldots, f_n(z_n)), \quad z = (z_1, \ldots, z_n) \in P,$$

where $f_1, \ldots, f_n$ are normalized convex functions on U. Therefore from Theorem 2.2.8 we have

$$\frac{|z_j|}{1 + |z_j|} \le |f_j(z_j)| \le \frac{|z_j|}{1 - |z_j|}, \quad z_j \in U, \quad j = 1, \ldots, n,$$

and thus

$$\max_{1 \le j \le n} \frac{|z_j|}{1 + |z_j|} = \frac{\|z\|_\infty}{1 + \|z\|_\infty} \le \max_{1 \le j \le n} |f_j(z_j)| = \|f(z)\|_\infty$$

$$\le \max_{1 \le j \le n} \frac{|z_j|}{1 - |z_j|} = \frac{\|z\|_\infty}{1 - \|z\|_\infty},$$

as desired.

Sharpness follows by considering the mapping $f : P \to \mathbb{C}^n$ given by

$$f(z) = \left(\frac{z_1}{1 - z_1}, \ldots, \frac{z_n}{1 - z_n} \right), \quad z = (z_1, \ldots, z_n) \in P,$$

which is a normalized convex mapping of P. For $z = (r, 0, \ldots, 0)$, $0 \le r < 1$, we have $\|z\|_\infty = r$ and $\|f(z)\|_\infty = r/(1 - r)$, and for $z = (-r, 0, \ldots, 0)$, $0 \le r < 1$, we have $\|z\|_\infty = r$ and $\|f(z)\|_\infty = r/(1 + r)$. This completes the proof.

When B is the Euclidean unit ball of $\mathbb{C}^n$, we have the following result obtained with different methods by Suffridge [Su5], FitzGerald and Thomas [Fit-Th], and Liu [LiuT2]. (See also [Lic2] for the upper bound in (7.2.2).) In the next chapter we shall give another proof of this result, using the method of Loewner chains. Kohr [Koh8] gave a two-point version of this theorem.

Theorem 7.2.2. *Let $f : B \to \mathbb{C}^n$ be a normalized convex mapping. Then*

$$(7.2.2) \qquad \frac{\|z\|}{1 + \|z\|} \le \|f(z)\| \le \frac{\|z\|}{1 - \|z\|} \quad on \quad B.$$

These estimates are sharp. Consequently, $f(B) \supseteq B_{1/2}$.

The above results show that the Koebe constant for normalized convex mappings of the unit ball and the unit polydisc of $\mathbb{C}^n$, $n \geq 2$, is equal to $1/2$, as in the case of one complex variable.

We shall give the proof of this result in the context of complex Banach spaces (Theorem 7.2.3).

In Section 6.3 we have introduced the class of quasi-convex mappings of type A, as another natural extension to higher dimensions of the normalized convex functions on the unit disc U. Roper and Suffridge [Rop-Su2] showed that the $1/2$-growth result, given in Theorem 7.2.2, remains true for this larger class, and Graham, Hamada and Kohr [Gra-Ham-Koh] have recently shown that the result remains true on the unit ball of $\mathbb{C}^n$ with respect to an arbitrary norm. In fact it is true for normalized starlike mappings of order $1/2$, as we shall see in the proof of the theorem below.

We now prove the extension of Theorem 7.2.2 to the case of complex Banach spaces. This result was stated without proof by Gong [Gon4, Theorem 4.3.3]. Complete proofs of the growth result for normalized convex mappings in complex Banach spaces were given by Hamada and Kohr [Ham-Koh9] and Chen [CheHB]. The covering result was considered by Hamada and Kohr [Ham-Koh10].

Theorem 7.2.3. *Let X be a complex Banach space with norm $\|\cdot\|$ and let B be the unit ball of X. If $f : B \to X$ is a normalized convex mapping, then*

$$\frac{\|z\|}{1 + \|z\|} \leq \|f(z)\| \leq \frac{\|z\|}{1 - \|z\|}, \quad z \in B.$$

Consequently, $f(B) \supseteq B_{1/2}$.

Proof.

Step 1. We begin by observing that the Marx-Strohhäcker theorem (a convex mapping is starlike of order $1/2$) remains valid on the unit ball of a complex Banach space. To see this, we first remark that the definitions of starlikeness of order γ (Definition 6.2.12), and of quasi-convex mappings of type A (Definition 6.3.16), extend easily to complex Banach spaces. The proofs of Lemma 6.3.17, Theorem 6.3.18, and Theorem 6.3.19, in which we showed that a quasi-convex mapping of type A is starlike of order $1/2$, are likewise valid on the unit ball of a complex Banach space.

The condition of starlikeness of order $1/2$ can be expressed using (6.2.8) as

$$\text{Re}\left\{\frac{\|z\|}{l_z([Df(z)]^{-1}f(z))}\right\} > 1/2, \quad z \in B \setminus \{0\}, \; l_z \in T(z).$$

Equivalently, we may write this as

$$\text{Re}\left\{\frac{\alpha}{l_u([Df(\alpha u)]^{-1}f(\alpha u))}\right\} > \frac{1}{2},$$

for all $\alpha \in U \setminus \{0\}$, $u \in X$, $\|u\| = 1$ and $l_u \in T(u)$.

Now fix $u \in X$, $\|u\| = 1$, and let $l_u \in T(u)$. Also let

$$g(\alpha) = \frac{1}{\alpha} l_u([Df(\alpha u)]^{-1}f(\alpha u)), \quad \alpha \in U.$$

Then g is holomorphic on U and $g(0) = 1$.

The above inequality is equivalent to

$$|g(\alpha) - 1| < 1, \quad \alpha \in U,$$

and since $g(0) = 1$ we deduce from the Schwarz lemma that

(7.2.3) $$|g(\alpha) - 1| \leq |\alpha|, \quad \alpha \in U.$$

The relation (7.2.3) implies that

$$1 - |\alpha| \leq \text{Re } g(\alpha) \leq 1 + |\alpha| \quad \text{on} \quad U,$$

and hence

(7.2.4) $$(1 - \|z\|)\|z\| \leq \text{Re } \{l_z([Df(z)]^{-1}f(z))\} \leq (1 + \|z\|)\|z\|,$$

for all $z \in B \setminus \{0\}$ and $l_z \in T(z)$.

Now fix $z_0 \in B \setminus \{0\}$ and let γ be the curve in B parametrized by

$$z(t) = f^{-1}(tf(z_0)), \quad t \in [0, 1].$$

We easily compute that

$$\frac{dz(t)}{dt} = Df^{-1}(tf(z_0))f(z_0) = \frac{1}{t}[Df(z(t))]^{-1}f(z(t)).$$

Since $\lim\limits_{t\to 0^+}\dfrac{dz(t)}{dt} = f(z_0)$, $\dfrac{dz}{dt}$ is continuous on $[0,1]$, and hence there exists a constant $C > 0$ such that $\left\|\dfrac{dz(t)}{dt}\right\| \le C$ for $t \in [0,1]$. Therefore, for $0 \le t_1 < t_2 \le 1$, we have

$$\left| \|z(t_2)\| - \|z(t_1)\| \right| \le \|z(t_2) - z(t_1)\| = \left\| \int_{t_1}^{t_2} \frac{dz(t)}{dt}\, dt \right\|$$

$$\le \int_{t_1}^{t_2} \left\| \frac{dz(t)}{dt} \right\|\, dt \le C(t_2 - t_1).$$

It follows that $\|z(t)\|$ is absolutely continuous on $[0,1]$, and hence $\dfrac{d}{dt}\|z(t)\|$ exists a.e. and is integrable on $[0,1]$.

When $\dfrac{d}{dt}\|z(t)\|$ exists, it is given by

(7.2.5) $$\frac{d}{dt}\|z(t)\| = \mathrm{Re}\,\left\{ l_{z(t)}\left(\frac{dz(t)}{dt}\right) \right\}$$

$$= \frac{1}{t}\mathrm{Re}\,\left\{ l_{z(t)}([Df(z(t))]^{-1}f(z(t))) \right\}$$

by [Kat2, Lemma 1.3]. Indeed, let $\tilde{t}$ be such that $\dfrac{d}{dt}\|z(t)\|\Big|_{t=\tilde{t}}$ exists. Then

$$\mathrm{Re}\,\{l_{z(\tilde{t})}(z(t))\} \le \|z(t)\| \cdot \|l_{z(\tilde{t})}\| = \|z(t)\|$$

and

$$\mathrm{Re}\,\{l_{z(\tilde{t})}(z(\tilde{t}))\} = \|z(\tilde{t})\|,$$

and therefore

$$\mathrm{Re}\,\{l_{z(\tilde{t})}(z(t) - z(\tilde{t}))\} \le \|z(t)\| - \|z(\tilde{t})\|.$$

Dividing both sides of this inequality by $t - \tilde{t}$ and letting $t \to \tilde{t}$ from above and from below respectively, we obtain (7.2.5), as claimed.

We now use the relations (7.2.4) and (7.2.5) to obtain

(7.2.6) $$\frac{1}{t}(1 - \|z(t)\|)\|z(t)\| \le \frac{d}{dt}\|z(t)\| \le \frac{1}{t}(1 + \|z(t)\|)\|z(t)\|.$$

In view of (7.2.6) we easily deduce that

$$\frac{d\|z(t)\|}{(1 + \|z(t)\|)\|z(t)\|} \le \frac{dt}{t} \le \frac{d\|z(t)\|}{(1 - \|z(t)\|)\|z(t)\|},$$

and integrating both sides of this inequality between ε and 1, where $0 < \varepsilon < 1$, we obtain

(7.2.7)
$$\log \frac{\|z_0\|}{1 + \|z_0\|} - \log \frac{\|z(\varepsilon)\|}{1 + \|z(\varepsilon)\|} \leq \log \frac{1}{\varepsilon}$$

$$\leq \log \frac{\|z_0\|}{1 - \|z_0\|} - \log \frac{\|z(\varepsilon)\|}{1 - \|z(\varepsilon)\|}.$$

Now it is obvious that (7.2.7) is equivalent to the relations

$$\frac{\|z_0\|}{1 + \|z_0\|} \leq \frac{\|z(\varepsilon)\|}{\varepsilon(1 + \|z(\varepsilon)\|)}$$

and

$$\frac{\|z(\varepsilon)\|}{\varepsilon(1 - \|z(\varepsilon)\|)} \leq \frac{\|z_0\|}{1 - \|z_0\|}.$$

Using the fact that $\lim\limits_{\varepsilon \to 0^+} \dfrac{z(\varepsilon)}{\varepsilon} = f(z_0)$, we obtain the growth result by letting $\varepsilon \to 0^+$ in the last two inequalities.

Step 2. We next prove the covering result, using similar arguments as in [Ham-Koh10].

It is clear that there exists $\rho > 0$ such that f maps the ball B_ρ biholomorphically onto a neighbourhood V of 0. Given $w \in V \setminus \{0\}$, there exists a unique $z \in B_\rho$ such that $f(z) = w$. Let r, $\rho < r < 1$, be fixed and let

$$t_0 = \sup\{t > 0 : tf(z) \in f(B_r)\}.$$

Using the growth theorem we see that $1 < t_0 < \infty$. Also if we let $v(t) = f^{-1}(tf(z))$, then the convexity of $f(B_r)$ implies that $\|v(t)\| < r$ for $0 \leq t < t_0$, and we have

$$\frac{dv(t)}{dt} = \frac{1}{t}[Df(v(t))]^{-1}f(v(t)).$$

Since $f \in K(B)$ it follows that $[Df(v(t))]^{-1}f(v(t)) \in M$, and in view of Theorem 6.1.39 we deduce that there exists $L = L(r) > 0$ such that

$$\left\| \frac{dv(t)}{dt} \right\| \leq \frac{2}{t_0}L(r), \quad \frac{t_0}{2} \leq t < t_0.$$

Therefore $\|dv/dt\|$ is integrable on $[t_0/2, t_0)$. Next, let s_1, s_2 be such that $t_0/2 < s_1 < s_2 < t_0$. Then

$$\|v(s_1) - v(s_2)\| = \left\| \int_{s_1}^{s_2} \frac{dv(t)}{dt} dt \right\|$$

$$\le \int_{s_1}^{s_2} \left\| \frac{dv(t)}{dt} \right\| dt \le \frac{2}{t_0} L(r)(s_2 - s_1) < \infty.$$

Moreover, if $\{t_k\}_{k \in \mathbb{N}}$ is an increasing sequence of positive numbers such that $t_k \to t_0$ as $k \to \infty$, then we obtain

$$\|v(t_k) - v(t_m)\| \to 0 \quad \text{as} \quad k, m \to \infty.$$

Therefore $\{v(t_k)\}_{k \in \mathbb{N}}$ is a Cauchy sequence and there exists a point $z_0' \in X$ such that $v(t_k) \to z_0'$ as $k \to \infty$. Since $\|v(t_k)\| < r$ and f is biholomorphic on B_r we must have $\|z_0'\| = r$, and the continuity of f at z_0' gives $f(z_0') = t_0 f(z)$. Using the first step of the proof, we have $\|t_0 f(z)\| \ge r/(1 + r)$, and thus $f(B_r)$ contains the ball $B_{r/(1+r)}$. Letting $r \nearrow 1$, we conclude that $f(B)$ contains the ball $B_{1/2}$, as desired. This completes the proof.

Remark 7.2.4. When X is a complex Hilbert space with inner product $\langle \cdot, \cdot \rangle$, the above result is sharp.

To see this, we remark that Roper and Suffridge [Rop-Su1] proved that if $f : U \to \mathbb{C}$ is a convex function and $u \in X$ is a unit vector, then the mapping $F_u : B \to X$ given by (7.1.7), i.e.

$$F_u(z) = f(\langle z, u \rangle)u + \sqrt{f'(\langle z, u \rangle)}(z - \langle z, u \rangle u), \quad z \in B,$$

is a convex mapping of B. We shall discuss this result in Theorem 11.1.5.

If we take the function $f(\zeta) = \zeta/(1 - \zeta)$, $\zeta \in U$, a straightforward computation using the above mapping yields that

$$F_u(z) = \frac{z}{1 - \langle z, u \rangle}, \quad z \in B.$$

If $z = ru$ where $r \in [0, 1)$, then $\|F_u(z)\| = r/(1 - r) = \|z\|/(1 - \|z\|)$, while if $z = -ru$, $r \in [0, 1)$, then $\|F_u(z)\| = r/(1 + r) = \|z\|/(1 + \|z\|)$.

On the other hand, since $-u/2 \notin F_u(B)$, the constant $1/2$ cannot be replaced by a larger constant in the covering result. This completes the proof.

Remark 7.2.5. Various results obtained in the finite dimensional case are special cases of Theorem 7.2.3. Gong and Liu [Gon-Liu1] obtained the growth theorem for convex mappings of the unit ball $B(p)$ with respect to a p-norm, $1 \le p \le \infty$. Also Hamada [Ham2] obtained the upper bound in the growth result of Theorem 7.2.3 when $X = \mathbb{C}^n$ with respect to an arbitrary norm.

Let Ω be a bounded balanced convex domain in $\mathbb{C}^n$. Then the Minkowski function of Ω is a norm on $\mathbb{C}^n$ and Ω is the unit ball of $\mathbb{C}^n$ with respect to this norm. Therefore, Theorem 7.2.3 holds for Ω. Liu and Ren [Liu-Ren4] obtained this result, and the corresponding covering theorem, in the case when the Minkowski function is C^1 except for a lower dimensional manifold in $\overline{\Omega}$. Their theorem is also presented in [Gon4]. In particular, this result applies to the complex ellipsoid $B(p_1, \ldots, p_n)$ where $p_1, \ldots, p_n \geq 1$.

7.2.2 Covering theorem and the translation theorem in the case of nonunivalent convex mappings in several complex variables

We now come back to the case of convex mappings on the Euclidean unit ball B of $\mathbb{C}^n$. We have seen that the Koebe constant for normalized convex mappings of the ball B is $1/2$. This result may be deduced as a direct consequence of the lower estimate (7.2.2). There is also a more general covering result [Gra4] which does not require univalence. To state it, let $\widehat{\Omega}$ denote the convex hull of a set $\Omega \subset \mathbb{C}^n$.

Theorem 7.2.6. Let $f : B \to \mathbb{C}^n$ be a normalized holomorphic mapping. Then $\widehat{f(B)}$ contains the ball $B_{1/2}$.

Proof. Let $u \in \partial B$ and let π_u denote the orthogonal projection of $\mathbb{C}^n$ onto the 1-dimensional subspace of $\mathbb{C}^n$ generated by u. Let $\psi_u(\zeta) = \zeta u$ and let

$$(7.2.8) \qquad g(\zeta) = \psi_u^{-1} \circ \pi_u \circ f(\zeta u), \quad \zeta \in U.$$

Then g is holomorphic on U, $g(0) = 0$ and $g'(0) = 1$, since f is normalized. Therefore, using the result of Theorem 2.2.10, we deduce that $\widehat{g(U)}$ contains the disc $U_{1/2}$. Moreover, by (7.2.8), $\psi_u(\widehat{g(U)}) = (\psi_u(g(U)))\widehat{\ } \subset \pi_u(\widehat{f(B)})$. Theorem 7.2.6 now follows from the following elementary result [Gra4], by choosing $\Omega = \widehat{f(B)}$.

Lemma 7.2.7. Let Ω be a convex set in $\mathbb{C}^n$ which contains a neighbourhood of 0. Assume Ω has the following property: whenever $u \in \partial B$, $\pi_u(\Omega)$ contains the disc centered at zero and of radius $1/2$ in the subspace $\mathbb{C}u$. Then Ω contains the ball $B_{1/2}$.

Proof. Let $w_0 \in \partial\Omega$ be a point at minimum distance from 0. Let H be a supporting hyperplane through w_0. We may assume that $w_0 = (r, 0, \ldots, 0)$, where $r > 0$, because otherwise we can use a unitary transformation to reduce to this situation. The equation of H is therefore Re $w_1 = r$, since H must be perpendicular to w_0. Since Ω is a convex set we have Re $w_1 < r$, for all $w \in \Omega$.

The orthogonal projection of $\mathbb{C}^n$ onto the subspace generated by w_0 is $(w_1, \ldots, w_n) \mapsto (w_1, 0, \ldots, 0)$. Hence all points in the image of Ω under this projection satisfy Re $w_1 \leq r$. In view of the hypothesis we now conclude that $r \geq 1/2$. This completes the proof.

There is a refinement of this result in the presence of k-fold symmetry, or of the following weaker notion of symmetry [Gra-Var2]:

Definition 7.2.8. Let $f : B \to \mathbb{C}^n$ be a normalized holomorphic mapping such that $f(B)$ is a convex domain. We say that f has *critical-slice symmetry of order k* if there is a point $w \in \partial f(B)$ at minimum distance from 0 such that, on setting $a = w/\|w\|$, the function $\phi(\zeta) = \langle f(\zeta a), a \rangle$ has symmetry of order k.

The following result, obtained by Graham and Varolin [Gra-Var2], is a generalization to higher dimensions of Theorem 2.2.15.

Theorem 7.2.9. *Let $f : B \to \mathbb{C}^n$ have critical-slice symmetry of order k. Then $f(B) \supseteq B_{r_k}$, where*

$$(7.2.9) \qquad r_k = \int_0^1 \frac{dt}{(1 + t^k)^{2/k}}.$$

This result is sharp.

Proof. Let $w \in \partial f(B)$ be a point which satisfies all of the conditions of Definition 7.2.8, and let $a = w/\|w\|$. If $\phi(\zeta) = \langle f(\zeta a), a \rangle$, then by hypothesis ϕ is k-fold symmetric. Let π_a denote the orthogonal projection of $\mathbb{C}^n$ onto the one-dimensional subspace $\mathbb{C}a$ through 0 containing a. By Theorem 2.2.15, $\widehat{\phi(U)} \supseteq U_{r_k}$ and since $\widehat{\phi(U)} \subseteq \pi_a(f(B))$, we deduce that $\pi_a(f(B)) \supseteq U_{r_k}$. Let H be the (unique) supporting hyperplane at w. Then $a \perp H$, and hence $\pi_a(H)$ is a line. Since $\pi_a(w) = w$, we must have $\|w\| \geq r_k$.

The sharpness of the radius r_k follows from the extension theorem of Roper and Suffridge which will be discussed in Theorem 11.1.2: if $f : U \to \mathbb{C}$ is a convex function such that $f(0) = f'(0) - 1 = 0$, then there exists a normalized convex mapping $F : B \to \mathbb{C}^n$ such that $F(z_1, 0, \ldots, 0) = (f(z_1), 0, \ldots, 0)$. (Also if f is k-fold symmetric then so is F.) This completes the proof.

From Theorem 7.2.9 we conclude that the Koebe constant for normalized convex mappings of B with k-fold symmetry is equal to r_k given by (7.2.9). In particular, for odd convex mappings this constant is $\pi/4$.

We next discuss a theorem of Graham and Varolin [Gra-Var1] about the translations of the image of a convex map of B. It is an analog of a theorem of MacGregor [Mac2,4], which in one variable holds in much greater generality than for convex functions. MacGregor [Mac2] (see also [Mac4]) proved the following:

Theorem 7.2.10. *Suppose that $f \in H(U)$ and that the Taylor series expansion of f has the form $f(z) = a_0 + z^k + O(|z|^{k+1})$. Let $\Omega = f(U)$. Then any translation of Ω through a distance less than $\pi/2$ has nonempty intersection with Ω.*

In particular the theorem holds for all functions in S. A conformal map onto an infinite strip of width $\pi/2$ is extremal.

In dimension greater than one the theorem cannot hold for all mappings in $S(B)$, because of the existence of disjoint Fatou-Bieberbach domains which are translates of each other [Ros-Rud]. However, it does hold for mappings in $K(B)$. To prove this, we need to recall some properties of convex sets. Let M be a convex set in $\mathbb{R}^m$ (not necessarily compact) and let u be a unit vector in $\mathbb{R}^m$. There are two natural notions of the width of M in the direction of u.

Definition 7.2.11. (i) $w_M(u) = \sup |c_1 - c_2|$, where the hyperplanes $x \cdot u = c_i$, $i = 1, 2$, have nonempty intersections with M.

(ii) $d_M(u) = \sup_{x \in \mathbb{R}^m} l(\sigma_{x,u} \cap M)$, where $\sigma_{x,u} = \{tx + u : t \in \mathbb{R}\}$ and l is the Lebesgue measure in $\mathbb{R}^m$.

We also introduce

Definition 7.2.12. $W_M = \inf_{\|u\|=1} w_M(u)$, $D_M = \inf_{\|u\|=1} d_M(u)$.

If W_M or D_M is finite, the corresponding infimum is actually a minimum.

When M is a compact convex set we have $D_M = W_M$, as shown in [Val, Theorem 12.18]. An elementary argument [Gra-Var1] shows that this remains valid for arbitrary convex sets.

Lemma 7.2.13. $D_M = W_M$.

Proof. It suffices to prove this equality for an arbitrary closed convex set M. For $0 < r < \infty$, let $\overline{B}_r^m = \{x \in \mathbb{R}^m : \|x\| \leq r\}$ and let $M_r = M \cap \overline{B}_r^m$. Clearly $D_{M_r} \nearrow D_M$ and $W_{M_r} \nearrow W_M$ as $r \to \infty$. Since $D_{M_r} = W_{M_r}$ for all r, we deduce that $D_M = W_M$.

We are now able to prove the analog of MacGregor's result [Mac2] as obtained by Graham and Varolin [Gra-Var1].

Theorem 7.2.14. *Let* $f : B \to \mathbb{C}^n$ *be a holomorphic mapping such that* $Df(0) = I$ *and* $\Omega = f(B)$ *is a convex domain. Then any translation of* Ω *through a distance less than* $\pi/2$ *meets* Ω.

Proof. We may assume $f(0) = 0$. Suppose $b \in \mathbb{C}^n$ is such that $\Omega \cap (\Omega + b) = \emptyset$. Then $W_\Omega = D_\Omega \leq \|b\|$. Let $u \in \partial B$ be such that $w_\Omega(u) = W_\Omega$. Also let $\mathbb{C}u$ denote the one-dimensional complex subspace of $\mathbb{C}^n$ generated by u, and let π_u be the orthogonal projection of $\mathbb{C}^n$ onto $\mathbb{C}u$. Let $U_{\mathbb{C}u}$ denote the unit disc in $\mathbb{C}u$ and let $g = \pi_u \circ f|_{U_{\mathbb{C}u}}$. Then g may be regarded as a function of one variable from the unit disc U to $\mathbb{C}$. Moreover, $g(0) = 0$ and $g'(0) = 1$, because $Df(0) = I$. Also $g(U)$ is contained in a strip of width $W_\Omega = D_\Omega$. Hence, by MacGregor's result in one variable (see Theorem 7.2.10), we deduce that $W_\Omega = D_\Omega \geq \pi/2$, and thus we conclude that $\|b\| \geq \pi/2$.

Open Problem 7.2.15. *Does the translation theorem (with the constant* $\pi/2$*) hold for other classes of maps in* $\mathbb{C}^n$*,* $n \geq 2$*, in particular starlike maps?*

This problem was formulated by Graham and Varolin [Gra-Var1].

7.2.3 Bounds for coefficients of convex mappings in $\mathbb{C}^n$ and complex Hilbert spaces

We now turn to the study of coefficient bounds and distortion theorems for convex mappings in higher dimensions. The known results are somewhat more complete than for starlike mappings, but there still remain a number of unsolved problems.

We begin with the unit ball $B(p)$ in $\mathbb{C}^n$ with respect to a p-norm, $1 \leq p \leq \infty$, and we give an estimate for the kth Fréchet derivative of a normalized convex mapping at the origin which was proved by Gong and Liu [Gon-Liu1]. In the Euclidean case the result was also obtained by Liczberski [Lic2], rediscovered by Kohr [Koh6], and obtained in a slightly stronger

form by Pfaltzgraff and Suffridge [Pfa-Su4]. We mention that FitzGerald and
Thomas [Fit-Th] gave coefficient bounds for normalized convex mappings of
the Euclidean unit ball of $\mathbb{C}^n$ that involve combinations of partial derivatives
different from those of the homogeneous polynomials $A_k(z)$ which occur in
Theorem 7.2.16. Also Liczberski [Lic1] obtained sharp bounds for the coefficients of normalized convex mappings of the unit polydisc in $\mathbb{C}^n$. Our argument
is similar to [Pfa-Su4, Theorem 5.1].

Theorem 7.2.16. *Let* $f(z) = z + \sum_{k=2}^{\infty} \dfrac{1}{k!} D^k f(0)(z^k)$ *be a normalized convex
mapping on the unit ball* $B(p)$ *with* $1 \leq p \leq \infty$. *Then*

$$(7.2.10) \qquad \left\| \frac{1}{k!} D^k f(0)(w^k) \right\|_p \leq 1, \quad k \in \mathbb{N}, \quad k \geq 2, \quad \|w\|_p = 1.$$

Proof. Let $A_k(z) = \dfrac{1}{k!} D^k f(0)(z^k)$ for $k \geq 2$. Then $A_k(z)$ is a homogeneous
polynomial in z of degree k. Now fix k and let $g_k : B(p) \to \mathbb{C}^n$ be given by

$$g_k(z) = \frac{1}{k} \sum_{j=1}^{k} f\left(e^{\frac{2\pi i j}{k}} z\right), \quad z \in B(p).$$

Since f is a convex mapping on $B(p)$ it follows that $g_k(B(p)) \subseteq f(B(p))$ and
hence if $v(z) = f^{-1}(g_k(z))$ for $z \in B(p)$, then v is a holomorphic mapping from
$B(p)$ into $B(p)$ such that $v(0) = 0$. We conclude from the Schwarz lemma (see
Lemma 6.1.28) that

$$\|v(z)\|_p \leq \|z\|_p, \quad z \in B(p).$$

Further, using the Cauchy estimates for Schwarz mappings on $B(p)$, we deduce
that

$$(7.2.11) \qquad \left\| \frac{1}{k!} D^k v(0)(z^k) \right\|_p \leq 1, \quad \|z\|_p \leq 1.$$

Moreover, since $e^{2\pi i/k}$ is a k-th root of unity, we see that

$$g_k(z) = \sum_{j=1}^{\infty} A_{jk}(z), \quad z \in B(p).$$

Next, let $z \in \mathbb{C}^n$, $\|z\|_p = 1$, and consider

$$\frac{g_k(\beta z)}{\beta^k} = \frac{f(v(\beta z))}{\beta^k}, \quad |\beta| < 1.$$

This function is holomorphic on the unit disc, since $\beta = 0$ is a removable singularity. Letting $\beta \to 0$ in the above, we deduce from (7.2.11) that

$$\left\| A_k(z) \right\|_p = \left\| \frac{1}{k!} D^k v(0)(z^k) \right\|_p \le 1.$$

This completes the proof.

In the Euclidean case or in a Hilbert space, one may strengthen the conclusion of Theorem 7.2.16 using a result of Hörmander on the norm of polynomials defined on Hilbert spaces. We shall briefly consider polynomials and multilinear mappings in complex vector spaces in order to explain Hörmander's result. The basic source is [Hör1]. For more details about this subject, the reader may consult [Fra-Ve], [Hil-Phi] and [Muj].

Let V be a vector space over $\mathbb{C}$.

Definition 7.2.17. A function $A : V \to \mathbb{C}$ is called an (abstract) homogeneous polynomial of degree k if for any $z, w \in V$, $A(sz + tw)$ is a homogeneous polynomial of degree k in $s, t \in \mathbb{C}$ in the algebraic sense.

Obviously, if the dimension of V is finite, the abstract and the algebraic definitions of a polynomial coincide. We have [Hör1]:

Lemma 7.2.18. *To each homogeneous polynomial $A(z)$ of degree k in V there exists a unique function $\widetilde{A}(z^{(1)}, \ldots, z^{(k)})$ with values in $\mathbb{C}$, which is defined whenever $z^{(i)} \in V$, $i = 1, \ldots, k$, and has the following properties:*

(i) $\widetilde{A}(z^{(1)}, \ldots, z^{(k)})$ is linear in $z^{(i)}$ if the other arguments are fixed, $i = 1, \ldots, k$;

(ii) $\widetilde{A}(z^{(1)}, \ldots, z^{(k)})$ is symmetric in its arguments;

(iii) $\widetilde{A}(z, \ldots, z) = A(z)$.

The function $\widetilde{A}(z^{(1)}, \ldots, z^{(k)})$, defined in Lemma 7.2.18, is called the kth-*polar form* of $A(z)$.

We remark that Definition 7.2.17 and Lemma 7.2.18 can be easily extended to the case when the polynomial is defined in V with values in another vector space over $\mathbb{C}$. For further information, see [Hör1]. See also [Hil-Phi, Chapter 26].

We can now give the basic result of Hörmander [Hör1] about the norm of polynomials defined in Hilbert spaces.

Theorem 7.2.19. *Let X be a complex Hilbert space with inner product*

$\langle \cdot, \cdot \rangle$ *and the induced norm* $\|z\|_1 = \sqrt{\langle z, z \rangle}$. *Also let* Y *be a complex Banach space with respect to a norm* $\| \cdot \|_2$. *If* $A: X \to Y$ *is a homogeneous polynomial of degree* k, *then*

$$\sup_{z \in X \setminus \{0\}} \frac{\|A(z)\|_2}{\|z\|_1^k} = \sup_{\substack{z^{(i)} \in X \setminus \{0\} \\ i=1,\ldots,k}} \frac{\|\tilde{A}(z^{(1)}, \ldots, z^{(k)})\|_2}{\|z^{(1)}\|_1 \cdots \|z^{(k)}\|_1}.$$

Using Hörmander's result, Pfaltzgraff and Suffridge [Pfa-Su4] obtained the coefficient bounds in the next theorem for the case of the Euclidean unit ball in $\mathbf{C}^n$. Hamada and Kohr [Ham-Koh9] gave the generalization to the Hilbert space case.

Theorem 7.2.20. *Let* X *be a complex Hilbert space with the inner product* $\langle \cdot, \cdot \rangle$, *and let* B *denote the unit ball of* X *with respect to the induced norm* $\| \cdot \|$. *Let* $f: B \to X$ *be a normalized convex mapping. Then*

(7.2.12) $$\left\| \frac{1}{k!} D^k f(0) \right\| \le 1, \quad k \in \mathbf{N}, \quad k \ge 2.$$

These estimates are sharp. Consequently,

$$\left\| \frac{1}{k!} D^k f(z)(z^k) \right\| \le \frac{\|z\|^k}{(1 - \|z\|)^{k+1}}, \quad z \in B, \quad k \in \mathbf{N}.$$

Proof. For $z \in B$ and $k \ge 2$ we define $A_k(z)$, $g_k(z)$, and $v(z)$ analogously to the proof of Theorem 7.2.16. Then $A_k(z)$ is a homogeneous polynomial in z of degree k. Also v is a Schwarz mapping and

(7.2.13) $$\left\| \frac{1}{k!} D^k v(0)(w^k) \right\| \le 1, \quad w \in X, \ \|w\| = 1, \ k \ge 2.$$

Since

$$g_k(z) = \frac{1}{k!} D^k f(0)(z^k) + O(\|z\|^{2k})$$

and

$$f^{-1}(z) = z + O(\|z\|^2)$$

in a neighbourhood of zero, we conclude that

$$D^k f(0)(w^k) = D^k v(0)(w^k).$$

From (7.2.13) and the above equality we deduce that

$$(7.2.14) \qquad \left\| \frac{1}{k!} D^k f(0)(w^k) \right\| \leq 1, \quad w \in X, \quad \|w\| = 1.$$

Now, if we apply Theorem 7.2.19 to the multilinear mapping $\tilde{A}_k = \frac{1}{k!} D^k f(0)$, we obtain

$$\left\| \frac{1}{k!} D^k f(0) \right\| = \sup_{\substack{\|z^{(j)}\|=1 \\ 1 \leq j \leq k}} \left\| \frac{1}{k!} D^k f(0)(z^{(1)}, \ldots, z^{(k)}) \right\| = \sup_{\|z\|=1} \left\| \frac{1}{k!} D^k f(0)(z^k) \right\|.$$

Consequently, from (7.2.14) and the above equality we conclude (7.2.12), as desired. Next we show that the bounds (7.2.12) are sharp. For this purpose, let u be a unit vector in X and define $F_u : B \to X$ by

$$F_u(z) = \frac{z}{1 - \langle z, u \rangle}, \quad z \in B.$$

Then F_u is a convex mapping, as will be shown in Theorem 11.1.5 (see also Remark 7.2.4). In addition, it is easy to see that

$$\frac{1}{k!} D^k F_u(0)(w^k) = \langle w, u \rangle^{k-1} w,$$

for all $w \in X$, $\|w\| = 1$ and $k \in \mathbb{N}$, $k \geq 2$. Hence for $w = u$ in the above equality, we deduce that

$$\left\| \frac{1}{k!} D^k F_u(0) \right\| = \|u\|^{2k-1} = 1, \quad k \geq 2.$$

On the other hand, since f is normalized, it has the power series expansion

$$f(z) = z + \sum_{k=2}^{\infty} A_k(z), \quad z \in B.$$

Also since f is bounded on each ball B_r, $0 < r < 1$, the above series converges uniformly on any such ball (see [Muj, Theorem 7.13]), and we deduce that

$$Df(z)(z) = z + \sum_{k=2}^{\infty} k A_k(z), \quad z \in B.$$

(The term-by-term differentiation of power series in infinite dimensional spaces is discussed in [Din2, p.150-151].)

Further, it is easy to prove by induction that

$$D^m f(z)(z^m) = \sum_{k=m}^{\infty} k(k-1)\cdots(k-m+1)A_k(z),$$

for all $z \in B$ and $m \geq 1$. Consequently, from (7.2.12) we obtain

$$\left\| \frac{1}{m!} D^m f(z)(z^m) \right\|$$

$$\leq \frac{1}{m!} \sum_{k=m}^{\infty} k(k-1)\cdots(k-m+1)\|z\|^k = \frac{\|z\|^m}{(1-\|z\|)^{m+1}},$$

as asserted. This completes the proof.

Some interesting consequences of Theorem 7.2.20 including estimates for individual coefficients have been obtained by Kohr [Koh6] in the case of two complex variables. Also see [Fit-Th] and [Gon4].

7.2.4   Distortion results for convex mappings in $\mathbb{C}^n$ and complex Hilbert spaces

In this section we derive upper and lower bounds for $\|Df(z)\|$ for normalized convex mappings on the unit ball of a complex Hilbert space X. The upper bound is a simple consequence of Theorem 7.2.20 (see [Ham-Koh9]). However, the lower bound is more difficult and its proof uses an ingenious argument of Gong and Liu, who obtained the corresponding result in the finite dimensional case for any norm which satisfies a certain smoothness condition ([Gon-Liu2]; see also [Gon4]). On the Euclidean unit ball in $\mathbb{C}^n$, Pfaltzgraff and Suffridge [Pfa-Su4] obtained the upper estimate for $\|Df(z)\|$ and a lower estimate for $\|Df(z)(z)\|$ as a consequence of a distortion theorem for linear-invariant families of given norm order. The class of normalized convex mappings of B is a linear-invariant family of norm order 1. When $X = \mathbb{C}^n$, $n \geq 2$, the upper bound in (7.2.17) is attained by the Cayley transform, but the lower bound in (7.2.17) is not sharp (see [Lic-St3]). We shall come back to these results in Chapter 10. The generalization of the lower estimate for $\|Df(z)\|$ to complex Hilbert spaces was obtained by Hamada and Kohr [Ham-Koh10].

For the proof of this theorem we need to introduce the infinitesimal Carathéodory and Kobayashi-Royden metrics.

Definition 7.2.21. Let Ω be a domain in a complex Hilbert space X. For $z \in \Omega$ and $v \in X$, the infinitesimal Carathéodory metric is defined by

$$(7.2.15) \qquad E_\Omega(z, v) = \sup \Big\{ |Dg(z)v| : g \in H(\Omega, U),\ g(z) = 0 \Big\}.$$

The infinitesimal Kobayashi-Royden metric is defined by

$$(7.2.16) \qquad F_\Omega(z, v) = \inf \Big\{ |\alpha| :\ \exists\, g \in H(U, \Omega) \text{ such that}$$

$$g(0) = z,\ \alpha g'(0) = v \Big\}.$$

(Strictly speaking, these are pseudometrics, i.e. they may not be positive definite.)

A basic property of these metrics is the contractive property under holomorphic mappings. That is, if Ω, Ω' are domains in X and $h \colon \Omega \to \Omega'$ is a holomorphic mapping, then

$$E_{\Omega'}(h(z), Dh(z)v) \le E_\Omega(z, v),$$

and similarly for the infinitesimal Kobayashi-Royden metric. It follows that these metrics are preserved by biholomorphic mappings. Using this fact it is not hard to compute these metrics on the unit ball B of X and to show that they coincide on B (cf. [Fra-Ve]). (They also coincide on any convex domain in X by a basic result of Lempert (see [Lem1,2]), extended to infinite dimensions by Dineen, Timoney, and Vigué [Din-Tim-Vig].) For further details about invariant metrics, see [Fra-Ve], [Gra1], [Jar-Pf], [Kob], [Kran], [Lem1], [Pol-Sh].

Lemma 7.2.22. $E_B(z, v)^2 = F_B(z, v)^2 = \dfrac{\|v\|^2}{1 - \|z\|^2} + \dfrac{|\langle z, v \rangle|^2}{(1 - \|z\|^2)^2}.$

Proof. For any $b \in B$, let

$$T_b(z) = \frac{\langle z, b \rangle}{1 + s_b} b + s_b z, \quad z \in X,$$

where $s_b = \sqrt{1 - \|b\|^2}$. We have seen in Chapter 6 that up to multiplication by unitary transformations of X, the biholomorphic automorphisms of B are of the form

$$\phi(z) = \phi_b(z), \quad z \in B,$$

where
$$\phi_b(z) = T_b\left(\frac{z-b}{1-\langle z, b\rangle}\right), \quad z \in B.$$

Since $\phi_b(b) = 0$ and the Carathéodory differential metric is a biholomorphic invariant, we deduce that

$$E_B(b, v)^2 = E_B(0, D\phi_b(b)v)^2.$$

On the other hand, since $E_B(0, v) = \|v\|$ and

$$D\phi_b(b)v = \frac{v}{s_b} + \frac{\langle v, b\rangle b}{s_b^2(1 + s_b)},$$

we obtain the desired result for E_B. A similar argument holds for F_B.

Theorem 7.2.23. Let $f : B \to X$ be a normalized convex mapping. Then

(7.2.17) $$\frac{1}{(1 + \|z\|)^2} \le \|Df(z)\| \le \frac{1}{(1 - \|z\|)^2} \quad on \quad B.$$

Proof. First we prove the upper estimate. Since f is normalized, it has the power series expansion

$$f(z) = z + \sum_{k=2}^{\infty} \frac{1}{k!} D^k f(0)(z^k), \quad z \in B.$$

Note that the above series converges uniformly on any ball B_r, $0 < r < 1$, and

$$Df(z) = I + \sum_{k=2}^{\infty} \frac{k}{k!} D^k f(0)(z^{k-1}, \cdot), \quad z \in B.$$

Since $\frac{1}{k!} D^k f(0)(z^{k-1}, \cdot)$ is a continuous linear operator, as the restriction of $\frac{1}{k!} D^k f(0)$ to $\{(\underbrace{z, \dots, z}_{k-1 \, times})\} \times X$, we conclude in view of (7.2.12) that

$$\left\|\frac{1}{k!} D^k f(0)(z^{k-1}, \cdot)\right\| \le \|z\|^{k-1} \quad on \quad B.$$

Therefore we obtain

$$\|Df(z)\| \le 1 + \sum_{k=2}^{\infty} k\|z\|^{k-1} = \frac{1}{(1 - \|z\|)^2}, \quad z \in B.$$

We now give the proof of the lower estimate in (7.2.17), using a similar method as in [Gon-Liu2] (also see [Gon4]).

Fix $z \in B_r, 0 < r < 1$. Let $t_0 = \sup\{t > 0 : tf(z) \in f(B_r)\}$. From Theorem 7.2.3 we deduce that $1 < t_0 < \infty$. Moreover, using similar arguments as in the proof of Theorem 6.3.8, it is not difficult to see that there exists a point $z' \in \partial B_r$ such that $f(z') = t_0 f(z)$. Let $\mu = 1/t_0$ and $z(t) = f^{-1}(tf(z'))$ for $0 \le t \le 1$. Also let

$$g(t) = \begin{cases} \dfrac{\|z(t)\|}{t(1 + \|z(t)\|)}, & 0 < t \le 1 \\ \|f(z')\|, & t = 0. \end{cases}$$

Since $\|\cdot\|^2$ is of class C^∞ and $z(t)$ is a C^1 mapping from $[0,1]$ to B, it follows that g is of class C^1 on $(0,1]$. It is also clear that g is continuous on $[0,1]$. We prove that g is decreasing on $[0,1]$ (cf. [Liu-Ren4]). A simple computation yields that

$$g'(t) = \frac{t\dfrac{d}{dt}\|z(t)\| - \|z(t)\|(1 + \|z(t)\|)}{t^2(1 + \|z(t)\|)^2}.$$

On the other hand, from (7.2.4) we have

(7.2.18)
$$(1 - \|z\|)\|z\|^2 \le \mathrm{Re}\,\langle [Df(z)]^{-1}f(z), z \rangle$$

$$\le (1 + \|z\|)\|z\|^2, \quad z \in B.$$

Also
(7.2.19)
$$\frac{d\|z(t)\|}{dt} = \frac{1}{\|z(t)\|}\mathrm{Re}\,\left\langle \frac{dz(t)}{dt}, z(t) \right\rangle$$

$$= \frac{1}{t\|z(t)\|}\mathrm{Re}\,\langle [Df(z(t))]^{-1}f(z(t)), z(t) \rangle.$$

From (7.2.18) and (7.2.19) we deduce that $g'(t) \le 0$ on $(0,1)$, and hence g is decreasing on $[0,1]$. This implies that

$$\frac{\|z'\|}{1 + \|z'\|} \le \frac{\|z(t)\|}{t(1 + \|z(t)\|)}, \quad 0 < t < 1,$$

and hence

$$\frac{r}{1+r} = \frac{\|z'\|}{1 + \|z'\|} \le \frac{\|z(\mu)\|}{\mu(1 + \|z(\mu)\|)} = \frac{\|z\|}{\mu(1 + \|z\|)},$$

which may be rewritten as

(7.2.20)
$$1 - \mu \geq \frac{r - \|z\|}{r(1 + \|z\|)}.$$

Now, let $G(y) = f^{-1}((1 - \mu)f(y) + \mu f(z'))$, for $y \in B$. Since f is a normalized convex mapping, G is biholomorphic on B, $G(B) \subset B$ and $G(0) = z$. Also, since the Carathéodory differential metric is invariant under biholomorphic mappings, we have

$$E_{G(B)}(z, DG(0)v) = E_B(0, v), \quad v \in X.$$

Further, since $G(B) \subset B$, the contractive property of the Carathéodory differential metric gives

$$E_{G(B)}(z, DG(0)v) \geq E_B(z, DG(0)v).$$

Thus

(7.2.21)
$$E_B(z, DG(0)v) \leq E_B(0, v).$$

Using the definition of G, we have

$$DG(0) = (1 - \mu)[Df(z)]^{-1},$$

and therefore from (7.2.21) we have

$$(1 - \mu)E_B(z, [Df(z)]^{-1}v) \leq E_B(0, v).$$

Setting $v = Df(z)\zeta$ in this inequality, we obtain

$$(1 - \mu)E_B(z, \zeta) \leq E_B(0, Df(z)\zeta).$$

From (7.2.20) and this relation we conclude that

$$\frac{r - \|z\|}{r(1 + \|z\|)} E_B(z, \zeta) \leq E_B(0, Df(z)\zeta),$$

and letting $r \to 1$ we obtain

(7.2.22)
$$\frac{1 - \|z\|}{1 + \|z\|} E_B(z, \zeta) \leq \|Df(z)\zeta\|.$$

From Lemma 7.2.22 it can be seen that for fixed $z \neq 0$ and for $\|v\| = 1$, $E_B(z, v)$ is maximized when $v = e^{i\theta} z / \|z\|$, $\theta \in \mathbb{R}$. Using this observation and the inequality (7.2.22), we finally deduce the relation

$$\|Df(z)\| \geq \frac{1}{(1 + \|z\|)^2}.$$

The estimates (7.2.23) below are immediate consequences of Theorems 7.2.23 and 7.2.26. In the finite-dimensional case these estimates were obtained by Gong and Liu [Gon-Liu2] (see also [Gon4]), on the unit ball with respect to any norm which is C^1 except for a lower dimensional manifold. Pfaltzgraff and Suffridge [Pfa-Su4] also obtained these bounds on the Euclidean unit ball in $\mathbb{C}^n$, and formulated (the finite-dimensional version of) Conjecture 7.2.25. (See also [Lic3] in the case of the maximum norm in $\mathbb{C}^n$.) We present these results in the context of complex Hilbert spaces.

Corollary 7.2.24. *Let $f : B \to X$ be a normalized convex mapping. Then*

$$(7.2.23) \qquad \frac{\|z\|}{(1 + \|z\|)^2} \leq \|Df(z)z\| \leq \frac{\|z\|}{(1 - \|z\|)^2}, \qquad z \in B.$$

Proof. The upper bound follows directly from Theorem 7.2.23. The lower bound follows from Theorem 7.2.26 for $\zeta = z/\|z\|$, $z \in B \setminus \{0\}$.

Conjecture 7.2.25. *If $f: B \to X$ is a normalized convex mapping of the unit ball B in a complex Hilbert space X then*

$$\|Df(z)v\| \geq \frac{\|v\|}{(1 + \|z\|)^2}, \qquad z \in B, \, v \in X.$$

One may also formulate a distortion theorem which takes account of the behaviour of $Df(z)$ in different directions at z by using the infinitesimal Carathéodory or Kobayashi-Royden metric. In the finite-dimensional case this result was obtained by Gong and Liu [Gon-Liu2] for convex mappings on bounded convex circular domains in $\mathbb{C}^n$. The generalization to Hilbert spaces is again due to Hamada and Kohr [Ham-Koh10]. We omit the proof and we leave it for the reader.

Theorem 7.2.26. *Let $f : B \to X$ be a normalized convex mapping. Then*

$$\frac{1 - \|z\|}{1 + \|z\|} E_B(z, \zeta) \leq \|Df(z)\zeta\| \leq \frac{1 + \|z\|}{1 - \|z\|} E_B(z, \zeta),$$

for all $z \in B$ and $\zeta \in X$.

For the case of the Euclidean unit ball in $\mathbb{C}^n$, there is an earlier reformulation of this theorem due to Gong, Wang, and Yu [Gon-Wa-Yu3] (see also [Gon4]). Again we leave the proof for the reader.

Theorem 7.2.27. *Let $f : B \to \mathbb{C}^n$ be a normalized convex mapping. Then the following estimates hold for all $z = (z_1, \ldots, z_n) \in B$:*

$$\left(\frac{1 - \|z\|}{1 + \|z\|}\right)^2 G(z) \leq [Df(z)][Df(z)]^* \leq \left(\frac{1 + \|z\|}{1 - \|z\|}\right)^2 G(z).$$

Here G is the matrix of the Bergman metric of the unit ball B (up to a constant factor), i.e.

$$G(z) = (g_{jk})_{1 \leq j,k \leq n} = \left(\frac{(1 - \|z\|^2)\delta_{j,k} + \overline{z}_j z_k}{(1 - \|z\|^2)^2}\right)_{1 \leq j,k \leq n},$$

and $\delta_{j,k}$ has the value 1 when $j = k$ and is 0 when $j \neq k$.

To see that this result is indeed a reformulation of Theorem 7.2.26 in the finite-dimensional case, it suffices to observe that the square of the infinitesimal Carathéodory or Kobayashi-Royden metric gives the metric with matrix $G(z)$.

The extent to which this result is sharp is discussed in [Gon-Wa-Yu3] and [Gon4]. Gong and Liu also gave estimates for the eigenvalues of $Df(z)$ in the finite-dimensional case ([Gon-Liu2], [Gon4]).

We conclude this section with a two-point distortion theorem that provides a necessary and sufficient condition for univalence on the Euclidean unit ball of $\mathbb{C}^n$, using properties of the Carathéodory distance (see [Koh8]). For other results of this type, the reader may consult [Gon4, Section 7.5].

If Ω is a domain in $\mathbb{C}^n$, then the Carathéodory distance is defined by

$$C_{\Omega}(z, w) = \sup_{g \in H(\Omega, U)} d_h(g(z), g(w)), \quad z, w \in \Omega,$$

where d_h is the hyperbolic distance on U. For further information about the Carathéodory distance, the reader may consult [Jar-Pf], [Kob], [Kran], [Pol-Sh].

Theorem 7.2.28. *Let $f : B \to \mathbb{C}^n$ be a convex mapping. Also let*

$$\mathcal{D}_1 f(z) = (1 - \|z\|^2)\|[Df(z)]^{-1}\|^{-1}, \quad z \in B.$$

Then

(7.2.24)
$$\|f(a) - f(b)\|$$

$$\geq \frac{1}{2}\Big[1 - \exp(-2C_B(a,b))\Big] \max\big\{\mathcal{D}_1 f(a), \mathcal{D}_1 f(b)\big\},$$

for all a, $b \in B$. Conversely, let $f : B \to \mathbb{C}^n$ be a locally biholomorphic mapping which satisfies (7.2.24). Then f is univalent on B.

Proof. Fix $b \in B$ and define $F : B \to \mathbb{C}^n$ by

$$F(z) = [D\phi_{-b}(0)]^{-1}[Df(b)]^{-1}(f(\phi_{-b}(z)) - f(b)), \quad z \in B,$$

where ϕ_b is given by (6.1.9), i.e.

$$\phi_b(z) = T_b\left(\frac{z - b}{1 - \langle z, b \rangle}\right), \quad z \in B.$$

Since F is normalized convex, Theorem 7.2.2 gives

$$\frac{\|z\|}{1 + \|z\|} \leq \|F(z)\| \leq \frac{\|z\|}{1 - \|z\|}, \quad z \in B,$$

and since

$$C_B(z, 0) = \frac{1}{2}\log\left[\frac{1 + \|z\|}{1 - \|z\|}\right], \quad z \in B,$$

we obtain

$$\frac{1}{2}\Big[1 - \exp(-2C_B(z,0))\Big] \leq \|F(z)\| \leq \frac{1}{2}\Big[\exp(2C_B(z,0)) - 1\Big], \quad z \in B.$$

For $z = (\phi_{-b})^{-1}(a) = \phi_b(a)$, we obtain

$$\frac{1}{2}\Big[1 - \exp\Big(-2C_B((\phi_{-b})^{-1}(a), (\phi_{-b})^{-1}(b))\Big)\Big]$$

$$\leq \|[D\phi_{-b}(0)]^{-1}[Df(b)]^{-1}(f(a) - f(b))\|$$

$$\leq \frac{1}{2}\Big[\exp\Big(2C_B((\phi_{-b})^{-1}(a), (\phi_{-b})^{-1}(b))\Big) - 1\Big].$$

On the other hand, since $[D\phi_{-b}(0)]^{-1} = D\phi_b(b)$ and the Carathéodory distance is invariant under biholomorphic mappings, we obtain from the above that

$$\frac{1}{2}(1 - \exp(-2C_B(a,b))) \leq \|[D\phi_b(b)][Df(b)]^{-1}(f(a) - f(b))\|$$

$$\leq \frac{1}{2}(\exp(2C_B(a,b)) - 1).$$

Moreover, it is not difficult to check that $\|[D\phi_{-b}(0)]^{-1}\| = 1/(1 - \|b\|^2)$ (see (10.2.7)), and since

$$\|[D\phi_{-b}(0)]^{-1}[Df(b)]^{-1}(f(a) - f(b))\|$$

$$\leq \frac{1}{1 - \|b\|^2}\|[Df(b)]^{-1}(f(a) - f(b))\|,$$

the relation (7.2.24) easily follows, as desired.

Conversely, assume that f is locally biholomorphic on B and satisfies (7.2.24). Let a, $b \in B$ such that $f(a) = f(b)$. In view of (7.2.24) we deduce that $\exp(2C_B(a,b)) = 1$ or $\mathcal{D}_1 f(a) = \mathcal{D}_1 f(b) = 0$. However, the latter condition cannot hold since f is locally biholomorphic. Thus we must have $a = b$. This completes the proof.

Problems

7.2.1. Prove Theorem 7.2.26.

7.2.2. Prove Theorem 7.2.27.

Chapter 8

Loewner chains in several complex variables

8.1 Loewner chains and the Loewner differential equation in several complex variables

8.1.1 The Loewner differential equation in $\mathbb{C}^n$

In this chapter we shall consider the generalization of the Loewner differential equation to n dimensions. This subject was first studied by Pfaltzgraff [Pfa1]; later contributions, refining Pfaltzgraff's results in $\mathbb{C}^n$ and permitting generalizations to the unit ball of a Banach space, were made by Poreda [Por3]. We also include recent improvements in the existence theorems for the Loewner equation which depend on the fact that the class $\mathcal{M}$ is compact [Gra-Ham-Koh]. We give various applications, including univalence criteria and characterizations of subclasses of $S(B)$, such as spirallike and close-to-starlike maps. Finally, we shall obtain some quasiconformal extension results to $\mathbb{C}^n$ of univalent mappings which can be embedded in Loewner chains.

We shall see that in several variables the class $S^0(B)$ of mappings which admit a parametric representation remains well-behaved, and satisfies growth theorems and coefficient estimates. For the case of the polydisc these mappings were studied by Poreda [Por1-2]; on the Euclidean ball they were studied

295

extensively by Kohr [Koh9]; and they have recently been studied on the unit
ball with respect to an arbitrary norm by Graham, Hamada, and Kohr [Gra-
Ham-Koh]. Somewhat surprisingly, we shall see that in higher dimensions there
exist univalent mappings which can be embedded in Loewner chains, but which
do not admit a parametric representation.

We shall work on the Euclidean unit ball B of $\mathbb{C}^n$, and we begin this
section by giving some basic facts about subordination and Loewner chains in
this context. As usual, by $\langle \cdot, \cdot \rangle$ we denote the inner product on $\mathbb{C}^n$ and by $\| \cdot \|$
the induced norm. Also let $H(B)$ be the set of holomorphic mappings from B
into $\mathbb{C}^n$.

Definition 8.1.1. Let $f, g \in H(B)$. We say that f is *subordinate* to g ($f \prec$
g) if there exists a Schwarz mapping v on B such that $f(z) = g(v(z))$, $z \in B$.

As in the case of one variable, if $f \prec g$ then $f(0) = g(0)$ and $f(B) \subseteq g(B)$.
Also if g is univalent on B, then $f \prec g$ if and only if $f(0) = g(0)$ and $f(B) \subseteq$
$g(B)$. Furthermore, if $f \prec g$ then $f(B_r) \subseteq g(B_r)$ for each r, $0 < r < 1$, and
$\|Df(0)\| \leq \|Dg(0)\|$.

In this chapter, if g is a mapping which depends holomorphically on $z \in B$
and also depends on other (real) variables, we shall write $Dg(z, \cdot)$ for the
differential of g in the z variable.

Definition 8.1.2. A mapping $f : B \times [0, \infty) \to \mathbb{C}^n$ is called a *subordination
chain* if it satisfies the following conditions:

(i) $f(\cdot, t) \in H(B)$ and $Df(0, t) = \varphi(t)I$, $t \geq 0$, where $\varphi : [0, \infty) \to \mathbb{C}$ is
a continuous function on $[0, \infty)$ such that $\varphi(t) \neq 0$, $t \geq 0$, $|\varphi(\cdot)|$ is strictly
increasing on $[0, \infty)$, and $|\varphi(t)| \to \infty$ as $t \to \infty$;

(ii) $f(\cdot, s) \prec f(\cdot, t)$, whenever $0 \leq s \leq t < \infty$, i.e. there exists a Schwarz
mapping $v_{s,t}(\cdot) = v(\cdot, s, t)$, called the *transition mapping* associated to $f(z, t)$,
such that $f(z, s) = f(v(z, s, t), t)$, $0 \leq s \leq t < \infty$, $z \in B$.

A subordination chain is called a *Loewner chain* (or a *univalent subordination chain*) if in addition $f(\cdot, t)$ is univalent on B for all $t \geq 0$.

As in one variable we may *normalize* a Loewner chain so that $f(0, t) = 0$
and $Df(0, t) = e^t I$, $t \geq 0$, and we shall assume henceforth that all Loewner
chains are normalized in this way.

Note that the requirement that $Df(0, t) = \varphi(t)I$ in Definition 8.1.2 is somewhat restrictive, but without it the chain cannot necessarily be normalized.

Also as in one variable, we shall see that a normalized Loewner chain satisfies a strong regularity condition in t: it is locally Lipschitz in t locally uniformly with respect to z.

This normalization forces the normalization $Dv(0, s, t) = e^{s-t}I$, $0 \leq s \leq t < \infty$, on the Schwarz mapping $v(z, s, t)$. Moreover, using the assumption (ii) of Definition 8.1.2, it is not difficult to see that if $f(z, t)$ is a Loewner chain, then the mapping $v(z, s, t)$, given by the equality $f(z, s) = f(v(z, s, t), t)$, satisfies the *semigroup property*

$$v(z, s, u) = v(v(z, s, t), t, u), \quad z \in B, \quad 0 \leq s \leq t \leq u < \infty.$$

This property combined with the fact that $\|v(z, s, t)\| \leq \|z\|$ implies that

$$\|v(z, s, u)\| \leq \|v(z, s, t)\|, \quad z \in B, \quad 0 \leq s \leq t \leq u < \infty,$$

and thus $\|v(z, s, \cdot)\|$ is a decreasing function on $[s, \infty)$.

Moreover, the univalence of $f(\cdot, t)$ implies that $v(\cdot, s, t)$ is uniquely determined and is also univalent on B.

Recall the class $\mathcal{M}$ given by

$$\mathcal{M} = \Big\{h \in H(B): \ h(0) = 0, \ Dh(0) = I, \ \mathrm{Re}\,\langle h(z), z \rangle > 0, \ z \in B \setminus \{0\}\Big\},$$

which is related in higher dimensions to the Carathéodory class of functions with positive real part in the unit disc. This class will play a fundamental role throughout this chapter. As we have seen in Theorem 6.1.39, the class $\mathcal{M}$ is compact, and for each $r \in (0, 1)$ there is a number $M = M(r) \leq \dfrac{4r}{(1-r)^2}$ such that $\|h(z)\| \leq M(r)$ for $\|z\| \leq r$ and $h \in \mathcal{M}$, by the proof of Theorem 7.1.7.

The basic existence theorem for the Loewner differential equation on B is due to Pfaltzgraff [Pfa1, Theorem 2.1]. His assumption concerning the boundedness of $h(z, t)$ can be omitted since it follows automatically from the compactness of $\mathcal{M}$. (Pfaltzgraff assumed that for each $r \in (0, 1)$ and $T > 0$ there exists a number $K = K(r, T) > 0$ such that $\|h(z, t)\| \leq K(r, T)$ for $\|z\| \leq r$ and $0 \leq t \leq T$.)

Later Poreda [Por3] studied the initial value problem (8.1.1) on the unit ball of a Banach space under the assumption that the mapping h is bounded on $B_r \times [0, \infty)$ for each $r \in (0, 1)$. This condition is at first sight more restrictive than Pfaltzgraff's, but again it is automatically satisfied in view of Theorem 6.1.39 or Theorem 7.1.7. The reader may compare the result below, due to Pfaltzgraff [Pfa1], with Theorem 3.1.10.

Theorem 8.1.3. Let $h : B \times [0, \infty) \to \mathbb{C}^n$ satisfy the following conditions:
(i) $h(\cdot, t) \in \mathcal{M}$ for each $t \geq 0$;
(ii) for each $z \in B$, $h(z, t)$ is a measurable function of $t \in [0, \infty)$.
Then for each $s \geq 0$ and $z \in B$, the initial value problem

$$(8.1.1) \qquad \frac{\partial v}{\partial t} = -h(v, t), \quad a.e. \quad t \geq s, \quad v(s) = z,$$

has a unique locally absolutely continuous solution $v(t) = v(z, s, t) = e^{s-t}z + \ldots$. Furthermore, for fixed s and t, $0 \leq s \leq t < \infty$, $v_{s,t}(z) = v(z, s, t)$ is a univalent Schwarz mapping, and for fixed $s \geq 0$ and $z \in B$ it is a Lipschitz function of $t \geq s$ locally uniformly with respect to z.

Proof. As in the proof of Theorem 3.1.10 we shall apply the classical method of successive approximations to construct the solution. Fix $s \geq 0$ and $r \in (0, 1)$. We shall show that if $z \in \overline{B}_r$, then (8.1.1) has a unique solution on any interval $[s, T]$ with $T > s$.

Estimating with Cauchy's integral formula in several variables and using the fact that for each $\rho \in (0, 1)$ there is $K = K(\rho) > 0$ such that $\|h(z, t)\| \leq K(\rho)$, $z \in \overline{B}_\rho$, $t \geq s$, we deduce the existence of a constant $M = M(r) > 0$ such that

$$(8.1.2) \qquad \|h(w, t) - h(z, t)\| \leq M(r)\|z - w\|$$

for $z, w \in \overline{B}_r$ and $t \geq s$.

Let $R = (1 + r)/2$, and let

$$c = \min\left\{\frac{1 - r}{2K(R)}, T - s\right\}.$$

Fix $z \in \overline{B}_r$ and consider the following successive approximations

$$v_0(t) = v_0(z,t) = z,$$

(8.1.3)

$$v_m(t) = v_m(z,t) = z - \int_s^t h(v_{m-1}(z,\tau),\tau)d\tau, \quad m = 1,2,\ldots,$$

on $s \leq t \leq s+c$. If we assume that $v_{m-1}(z,t) \in \overline{B}_R$ for $s \leq t \leq s+c$, then we obtain

$$\|v_m(z,t) - z\| \leq K(R)(t-s) \leq K(R)c \leq (1-r)/2,$$

and therefore $\|v_m(z,t)\| \leq R < 1$ $(m = 0,1,\ldots)$. Since $v_0(z,t) = z \in \overline{B}_r \subseteq \overline{B}_R$, we see by induction that the successive approximations are well defined. Using (8.1.2) and the locally uniform boundedness of $\mathcal{M}$, we obtain by induction that

$$\|v_m(z,t) - v_{m-1}(z,t)\| \leq K(R)(M(R))^{m-1}(t-s)^m/m!,$$

when $s \leq t \leq s+c$ and $\|z\| \leq r$. Hence the mapping

(8.1.4)

$$v(t) = v(z,t) = \lim_{m \to \infty} v_m(z,t)$$

is well defined and continuous on $s \leq t \leq s+c$. Also since the convergence is uniform on $\overline{B}_r$ for fixed t, $v(z,t)$ is holomorphic as a function of z. Clearly $v(z,t) \in \overline{B}_R$ when $z \in \overline{B}_r$ and $s \leq t \leq s+c$. Moreover, by using (8.1.2), (8.1.3), and (8.1.4) we deduce that $v(z,t)$ satisfies the integral equation

(8.1.5)

$$v(z,t) = z - \int_s^t h(v(z,\tau),\tau)d\tau, \quad v(z,s) = z,$$

on $s \leq t \leq s+c$. This equation, the hypothesis (ii) and the boundedness of $\|h(z,t)\|$ on $\overline{B}_R \times [s,\infty)$ show that $v(z,t)$ is a Lipschitz continuous (hence absolutely continuous) function of $t \in [s, s+c]$ uniformly with respect to $z \in \overline{B}_r$, and that

(8.1.6) $\quad \dfrac{\partial v}{\partial t}(z,t) = -h(v(z,t),t), \quad a.e. \quad t \in [s, s+c], \quad v(z,s) = z.$

We shall prove that $\|v(z,t)\| \leq \|z\|$ for $s \leq t \leq s+c$. Clearly, if we establish this relation on $[s, s+c]$ then the solution of the initial value problem (8.1.6)

can be continued to $s+c \le t \le s+2c$ with the initial condition $v(t) = v(z, s+c)$ when $t = s + c$, by applying the method of successive approximations with the same constants r, R, c, $K(R)$, $M(R)$. By iterating this procedure we may extend the solution to the full interval $[s, T]$ in a finite number of steps. Since T is arbitrary, the above arguments show the existence of a solution to the initial value problem (8.1.1) on $s \le t < \infty$. This solution will be a Lipschitz continuous function of t locally uniformly with respect to $z \in B$, and holomorphic in $z \in B$.

Thus our objective is to prove that

$$(8.1.7) \qquad\qquad \|v(z,t)\| \le \|z\| \quad \text{on} \quad [s, s+c].$$

Since $v(z, \cdot)$ is a Lipschitz continuous function of $t \in [s, s+c]$, the same is true of $\|v(z, \cdot)\|$. Therefore $\partial\|v(z,t)\|/\partial t$ exists a.e., is an integrable function and

$$(8.1.8) \qquad \|v(z,t)\| = \|z\| + \int_s^t \frac{\partial\|v(z,\tau)\|}{\partial\tau} d\tau, \quad s \le t \le s+c.$$

On the other hand, because

$$\|v(z,t)\| \frac{\partial\|v(z,t)\|}{\partial t} = \operatorname{Re}\left\langle \frac{\partial v}{\partial t}(z,t), v(z,t) \right\rangle, \quad a.e. \quad t \in [s, s+c],$$

we may substitute from (8.1.6) and use the fact that $h(\cdot, t) \in \mathcal{M}$ for $t \ge 0$, to deduce that

$$\|v(z,t)\| \frac{\partial\|v(z,t)\|}{\partial t} = -\operatorname{Re}\left\langle h(v(z,t),t), v(z,t) \right\rangle \le 0.$$

Therefore, $\partial\|v(z,t)\|/\partial t \le 0$ a.e. on $[s, s+c]$, and from (8.1.8) we deduce that $\|v(z,t)\| \le \|z\|$ on $[s, s+c]$.

Let s and t be fixed, $0 \le s < t$. We have already noted that the solution $v(z, s, t)$ of (8.1.6) is holomorphic in z, and hence $v(z, s, t)$ is a Schwarz mapping.

We next prove that the solution $v(z, s, t)$ of (8.1.1) is unique. Assume $u(z, s, t)$ is another locally absolutely continuous solution of (8.1.1) such that $u(z, s, s) = z$. Fix $T > s$. Then for $t \in [s, T]$, we have

$$v(t) = v(z, s, t) = v(z, s, s) - \int_s^t h(v(z, s, \tau), \tau) d\tau$$

and

$$u(t) = u(z, s, t) = u(z, s, s) - \int_s^t h(u(z, s, \tau), \tau)d\tau.$$

Also, for $\|z\| \le r < 1$ and $s \le t \le T$, we have $\|v(z, s, t)\| \le r$ and $\|u(z, s, t)\| \le r$. Hence from (8.1.2) we deduce that

$$\|v(z, s, t) - u(z, s, t)\| \le \int_s^t \|h(v(z, s, \tau), \tau) - h(u(z, s, \tau), \tau)\|d\tau$$

$$\le M(r) \int_s^t \|v(z, s, \tau) - u(z, s, \tau)\|d\tau.$$

Let $L = \sup\limits_{s \le t \le T} \|u(t) - v(t)\|$. Using the previous estimate, one may prove inductively that

$$\|v(t) - u(t)\| \le (M(r))^m L \frac{(t-s)^m}{m!}, \quad t \in [s, T],$$

for all $m \in \mathbb{N}$ and $\|z\| \le r$. This implies that $u(t)$ and $v(t)$ coincide on $[s, T]$ for each $T > s$. Hence the solution of (8.1.1) is unique.

We next show that $v(\cdot, s, t)$ is univalent on B whenever $t \ge s$. Since $v(z, s, s) = z$ is univalent, it suffices to prove this assertion for $t > s$. For this purpose, let $t_0 > s$ be fixed and let z', z'' be such that $\|z'\| \le r$, $\|z''\| \le r < 1$, and $v(z', s, t_0) = v(z'', s, t_0)$. Also let $w(t) = v_1(t) - v_2(t)$ for $t \ge s$, where $v_1(t) = v(z', s, t)$ and $v_2(t) = v(z'', s, t)$. Then $w(t_0) = 0$. Moreover, from (8.1.2) we obtain

$$\left\| \frac{\partial w(t)}{\partial t} \right\| = \left\| h(v_1(t), t) - h(v_2(t), t) \right\| \le M(r)\|w(t)\|, \quad a.e. \quad t \in [s, t_0].$$

Let

$$A = \sup \left\{ \|w(t)\| : s \le t \le t_0 \right\}.$$

Then for $t \in [s, t_0]$ we obtain the estimate

$$\|w(t)\| = \left\| \int_t^{t_0} \frac{\partial w(\tau)}{\partial \tau} d\tau \right\| \le \int_t^{t_0} M(r)A d\tau = M(r)A(t_0 - t),$$

and using an inductive argument, we deduce that

$$\|w(t)\| \le \frac{(M(r))^m}{m!} A(t_0 - t)^m, \quad t \in [s, t_0], \quad m \in \mathbb{N}.$$

Consequently, we must have $w(t) = 0$ for $t \in [s, t_0]$. But the condition $w(s) = 0$ implies that $z' = z''$.

It remains to show that $Dv(0, s, t) = e^{s-t}I$. For this purpose, let $s \geq 0$ be fixed and let $V(t) = Dv(0, s, t)$ for $t \geq s$. From (8.1.5) and the fact that $\mathcal{M}$ is locally uniformly bounded, we deduce that $v(z, s, t)$ satisfies the Lipschitz condition

$$(8.1.9) \qquad \|v(z, s, t_1) - v(z, s, t_2)\| = \left\| \int_{t_1}^{t_2} h(v(z, s, \tau), \tau)d\tau \right\|$$

$$\leq K(r)(t_2 - t_1), \quad s \leq t_1 \leq t_2 < \infty, \quad \|z\| \leq r.$$

Now Cauchy's integral formula for vector-valued holomorphic functions of a single variable gives

$$Dv(0, s, t)(w) = \frac{1}{2\pi i} \int_{|\zeta|=\rho} \frac{v(\zeta w, s, t)}{\zeta^2} d\zeta, \quad \rho < 1,$$

for all $w \in \mathbb{C}^n$, $\|w\| = 1$. Together with (8.1.9) this gives (taking the supremum over w)

$$\|V(t_1) - V(t_2)\| \leq C|t_1 - t_2|, \quad t_1, t_2 \in [s, \infty),$$

for some constant $C = C(r)$. Hence V is a Lipschitz continuous function of $t \in [s, \infty)$, so that $\partial V(t)/\partial t$ exists a.e. $t \geq s$. Also since

$$\frac{\partial}{\partial t} Dv(0, s, t) = -Dh(v(0, s, t), t)Dv(0, s, t), \quad a.e. \quad t \geq s,$$

we obtain, using $Dh(0, t) = I$ and $v(0, s, t) = 0$, that

$$(8.1.10) \qquad \frac{\partial V}{\partial t}(t) = -V(t), \quad a.e. \quad t \geq s, \quad V(s) = I.$$

Solving the initial value problem (8.1.10), we obtain the unique solution $V(t) = e^{s-t}I$. Hence $Dv(0, s, t) = e^{s-t}I$, as desired. This completes the proof.

The following growth result is due to Pfaltzgraff [Pfa1]:

Lemma 8.1.4. *Suppose that $h(z, t)$ satisfies the hypotheses of Theorem 8.1.3, and let $v(z, s, t)$ be the solution of the initial value problem (8.1.1). Then*

$$(8.1.11) \qquad \frac{e^t \|v(z, s, t)\|}{(1 - \|v(z, s, t)\|)^2} \leq \frac{e^s \|z\|}{(1 - \|z\|)^2},$$

and

(8.1.12)
$$\frac{e^s\|z\|}{(1+\|z\|)^2} \le \frac{e^t\|v(z,s,t)\|}{(1+\|v(z,s,t)\|)^2},$$

for all $z \in B$ and $0 \le s \le t < \infty$.

Proof. Fix $z \in B \setminus \{0\}$ and $s \ge 0$, and let $v(t) = v(z,s,t)$ for $t \ge s$. For a.e. $t \ge s$ we have

(8.1.13)
$$\frac{d\|v(t)\|}{dt} = -\frac{1}{\|v(t)\|}\mathrm{Re}\,\langle h(v(t),t), v(t)\rangle$$

$$\le -\frac{\|v(t)\|(1-\|v(t)\|)}{1+\|v(t)\|} < 0,$$

making use of (6.1.27) and the fact that

$$\frac{d}{dt}\|v(t)\|^2 = 2\mathrm{Re}\,\left\langle \frac{dv(t)}{dt}, v(t)\right\rangle, \quad a.e. \quad t \ge s.$$

Hence $\|v(t)\|$ is strictly decreasing from $\|z\| = \|v(s)\|$ to 0 as t increases from s to ∞. Also we know from the proof of Theorem 8.1.3 that $\|v(t)\|$ is Lipschitz (and hence absolutely continuous) in t for $t \ge s$. Thus we may integrate in (8.1.13) and change variables to deduce that

$$-\int_{\|z\|}^{\|v\|} \left(\frac{1+x}{1-x}\right)\frac{dx}{x} = -\int_s^t \left(\frac{1+\|v(\tau)\|}{1-\|v(\tau)\|}\right)\frac{1}{\|v(\tau)\|}\frac{d\|v(\tau)\|}{d\tau}d\tau$$

$$\ge \int_s^t d\tau = t - s.$$

From this we obtain

$$\log\left\{\frac{\|z\|}{(1-\|z\|)^2} \cdot \frac{(1-\|v(z,s,t)\|)^2}{\|v(z,s,t)\|}\right\} \ge t - s,$$

which is equivalent to (8.1.11). Similar reasoning gives the estimate (8.1.12). This completes the proof.

Next we show that if $v(t) = v(z,s,t)$ is the unique solution of the initial value problem (8.1.1), then $\lim_{t\to\infty} e^t v(z,s,t)$ exists and gives rise to a Loewner chain. Our proof expands on ideas of Poreda [Por3, Theorems 2 and 3]. Poreda obtained this result on the unit ball of a Banach space, with slightly stronger regularity assumptions in t. (The use of normal families in the first step of the argument below is not essential, and the univalence of $f(\cdot, t)$ can be established by another argument in infinite dimensions [Por3, Theorem 4].)

Theorem 8.1.5. *Let $h : B \times [0, \infty) \to \mathbb{C}^n$ satisfy the conditions (i)-(ii) of Theorem 8.1.3. Let $v(t) = v(z, s, t)$ be the solution of the initial value problem (8.1.1). Then for each $s \geq 0$, the limit*

$$(8.1.14) \qquad \lim_{t \to \infty} e^t v(z, s, t) = f(z, s), \quad s \geq 0,$$

exists locally uniformly on B. Moreover, $f(\cdot, s)$ is univalent on B and $f(z, s) = f(v(z, s, t), t)$ for all $z \in B$ and $0 \leq s \leq t < \infty$. Hence $f(z, t)$ is a Loewner chain, and this Loewner chain has the property that $\{e^{-t} f(z, t)\}_{t \geq 0}$ is a normal family on B. In addition, $f(z, \cdot)$ is a locally Lipschitz function on $[0, \infty)$ locally uniformly with respect to $z \in B$. Finally, for a.e. $t \geq 0$,

$$\frac{\partial f}{\partial t}(z, t) = Df(z, t)h(z, t), \quad \forall z \in B.$$

Proof. Fix $s \geq 0$ and let $\varphi(z, t) = e^t v(z, s, t)$ for $t \geq s$. Then from (8.1.11) we have

$$\|\varphi(z, t)\| \leq e^s \frac{\|z\|}{(1 - \|z\|)^2}, \quad z \in B,$$

and hence the mappings $\{\varphi(z, t)\}_{t \geq s}$ form a normal family. Thus there exists a sequence $\{t_m\}$, $t_m \to \infty$ such that the limit

$$\lim_{m \to \infty} \varphi(z, t_m) = f(z, s)$$

exists locally uniformly on B. Moreover, since $\varphi(z, t_m)$ is a univalent mapping and $D\varphi(0, t_m) = e^s I$ for each t_m, the Jacobian of $f(z, s)$ cannot be identically zero. Consequently, $f(\cdot, s)$ is a univalent mapping on B such that $f(0, s) = 0$ and $Df(0, s) = e^s I$. However, we shall establish the existence of the limit (8.1.14) independently of this sequence.

Since $v(t) = v(z, s, t)$ is the solution of (8.1.1), we have

$$\frac{\partial \varphi}{\partial t}(z, t) = \varphi(z, t) - e^t h(e^{-t}\varphi(z, t), t), \quad a.e. \quad t \geq s,$$

for all $z \in B$. Let $g(z, t) = h(z, t) - z$ for $z \in B$ and $t \geq 0$. Then $g(\cdot, t) \in H(B)$, $g(0, t) = 0$, and $Dg(0, t) = 0$ for $t \geq 0$. Moreover, we have

$$(8.1.15) \qquad \frac{\partial \varphi}{\partial t}(z, t) = -e^t g(e^{-t}\varphi(z, t), t), \quad a.e. \quad t \geq s,$$

for all $z \in B$. Since for each $r \in (0,1)$ there exists $M = M(r) > 0$ such that $\|h(z,t)\| \leq M(r)$, $\|z\| \leq r$, $t \geq 0$, we deduce that

$$(8.1.16) \qquad \|g(z,t)\| \leq r + M(r) = K(r), \quad \|z\| \leq r, \quad t \geq 0.$$

Let $r_0 \in (0,1)$ be fixed. Since $g(0,t) = 0$ and $Dg(0,t) = 0$, it follows from Lemma 6.1.28 and the relation (8.1.16) that

$$(8.1.17) \qquad \|g(z,t)\| \leq \frac{K(r_0)}{r_0^2}\|z\|^2, \quad \|z\| \leq r_0, \quad t \geq 0.$$

Next, using Lemma 8.1.4, we obtain

$$(8.1.18) \qquad \|v(z,s,t)\| \leq e^{s-t}\frac{r_0}{(1-r_0)^2}, \quad \|z\| \leq r_0, \quad t \geq s.$$

Hence, from (8.1.17) and (8.1.18) we deduce that

$$(8.1.19) \qquad \|g(e^{-t}\varphi(z,t),t)\| \leq K(r_0)e^{-2t}\frac{e^{2s}}{(1-r_0)^4}$$

for $\|z\| \leq r_0$ and $t \geq s$, where we have used the fact that $\|e^{-t}\varphi(z,t)\| \leq \|z\| \leq r_0$. Next, let $t_0 = \dfrac{e^s}{(1-r_0)^2} > s$. From (8.1.19) we therefore obtain

$$\|g(e^{-t}\varphi(z,t),t)\| \leq K(r_0)e^{-2t}t^2, \quad \|z\| \leq r_0, \quad t \geq t_0.$$

Further, using this inequality and the relation (8.1.15), one concludes by integration that

$$\|\varphi(z,t_2) - \varphi(z,t_1)\| \leq K(r_0)\int_{t_1}^{t_2} t^2 e^{-t}dt$$

for $t_2 \geq t_1 \geq t_0$ and $\|z\| \leq r_0$. The convergence of the integral $\int_0^\infty t^2 e^{-t}dt$ implies that for any $\varepsilon > 0$ there exists $T_0 > 0$ such that for $t_1, t_2 > T_0$,

$$\|\varphi(z,t_2) - \varphi(z,t_1)\| < \varepsilon, \quad \|z\| \leq r_0.$$

Hence the limit (8.1.14) exists locally uniformly on B, and by Weierstrass' theorem gives a holomorphic mapping on B.

We now prove that the mapping $f : B \times [0,\infty) \to \mathbb{C}^n$, constructed in this way, is a Loewner chain. As in one variable, the semigroup property

$$v(z,s,t) = v(v(z,s,\tau),\tau,t), \quad 0 \leq s \leq \tau \leq t < \infty,$$

follows from the uniqueness of solutions to initial value problems of the form
(8.1.1). Using it, we deduce that

$$f(z,s) = \lim_{t\to\infty} e^t v(z,s,t)$$

$$= \lim_{t\to\infty} e^t v(v(z,s,\tau),\tau,t) = f(v(z,s,\tau),\tau),$$

for $z \in B$ and $0 \le s \le \tau \le \infty$. We also have $f(0,t) = 0$ and $Df(0,t) = e^t I$
for $t \ge 0$, and hence by Hurwitz's theorem $f(\cdot,t)$ is univalent for each $t \ge 0$.
Hence $f(z,t)$ satisfies all of the conditions needed to be a Loewner chain.

Moreover, $f(z,t)$ is locally Lipschitz in t locally uniformly in z. Indeed, in
view of Lemma 8.1.4 and the definition of $f(z,t)$, we have

$$\|f(z,t)\| \le e^t \frac{\|z\|}{(1-\|z\|)^2},$$

and using the Cauchy integral formula (6.1.18) and the mean value theorem
(applied to the real and imaginary parts of the components of f), it is easy to
deduce that for all $r \in (0,1)$ and $T > 0$ there is a constant $L = L(r,T)$ such
that

$$\|f(z,t) - f(w,t)\| \le L(r,T)\|z - w\|,$$

for all $z, w \in \overline{B}_r$ and $0 \le t \le T$. We may therefore conclude that

$$\|f(z_0,s) - f(z_0,t)\| = \|f(v(z_0,s,t),t) - f(z_0,t)\|$$

$$\le L(\|z_0\|,T)\|v(z_0,s,t) - z_0\| \le L(\|z_0\|,T)M(\|z_0\|)(t-s),$$

for $0 \le s \le t \le T$ and $z_0 \in B$, where we have used the fact that

$$\|v(z_0,s,t) - z_0\| = \left\|\int_s^t h(v(z_0,s,\tau),\tau)d\tau\right\| \le M(\|z_0\|)(t-s).$$

In particular, $f(z,t)$ is locally absolutely continuous in t locally uniformly
with respect to z.

In addition, (8.1.14) and (8.1.11) imply that $\{e^{-t}f(z,t)\}_{t\ge 0}$ is a normal
family on B.

It remains to prove that for almost all $t \ge 0$,

(8.1.20) $$\frac{\partial f}{\partial t}(z,t) = Df(z,t)h(z,t), \quad z \in B.$$

By using the subordination property and applying the mean value theorem to the real and imaginary parts of the components of f, we obtain

(8.1.21)
$$\frac{1}{\eta}\Big[f(z, t+\eta) - f(z, t)\Big]$$

$$= \frac{1}{\eta}\Big[f(z, t+\eta) - f(v(z, t, t+\eta), t+\eta)\Big]$$

$$= A(z, t, \eta)\Big(\frac{1}{\eta}\big[z - v(z, t, t+\eta)\big]\Big), \quad z \in B, \ t \geq 0, \ \eta > 0,$$

where $A(z, t, \eta)$ is a real-linear operator which tends to the invertible complex-linear operator $Df(z, t)$ as $\eta \to 0^+$. In view of this we deduce that the difference quotient in the first member of (8.1.21) has a limit as $\eta \to 0^+$ if and only if the same is true of the difference quotient in the last member of (8.1.21). But as just remarked, $f(z, t)$ is locally Lipschitz in t locally uniformly in z. Using this fact and Vitali's theorem (Theorem 6.1.16) we deduce that, except for a set of measure 0 in t, the two sided-limit of the left-hand side of (8.1.21) exists for all $z \in B$ as $\eta \to 0$. Hence we obtain for a.e. $t \geq 0$,

$$\frac{\partial f}{\partial t}(z, t) = \frac{\partial f}{\partial t^+}(z, t) = Df(z, t)\widetilde{h}(z, t),$$

where $\widetilde{h}(\cdot, t) \in \mathcal{M}$ in view of Lemmas 6.1.30 and 6.1.33 and the normalization of the transition mappings. To complete the proof, we apply Corollary 8.1.10 to conclude that $\widetilde{h}(\cdot, t) = h(\cdot, t)$ for a.e. $t \geq 0$.

Next we show that solutions of the Loewner differential equation in n dimensions which satisfy a growth condition in t give Loewner chains. This is one of the basic results in the theory of univalence in several complex variables. It is originally due to Pfaltzgraff [Pfa1, Theorem 2.3], and is a generalization to higher dimensions of Theorem 3.1.11. We have incorporated improvements in the assumptions on h resulting from the compactness of the class $\mathcal{M}$ [Gra-Ham-Koh], and we have added some details/variations in other parts of the argument. We also combine Pfaltzgraff's result with [Por3, Theorem 6] to conclude that the mapping $f(z, t)$ which solves the differential equation (8.1.22) coincides with the mapping defined by (8.1.14). This situation corresponds to the one variable case given in Theorem 3.1.12. Poreda [Por3] established the analogous result in the context of complex Banach spaces.

Theorem 8.1.6. *Suppose that $r \in (0,1]$ and that $f_t(z) = f(z,t) = e^t z + \ldots$ is a mapping from $B_r \times [0,\infty)$ into $\mathbb{C}^n$ such that $f_t \in H(B_r)$ and $f(z,t)$ is a locally absolutely continuous function of $t \in [0,\infty)$ locally uniformly with respect to $z \in B_r$. Suppose $h : B \times [0,\infty) \to \mathbb{C}^n$ satisfies the conditions (i) and (ii) of Theorem 8.1.3, and that for all $z \in B_r$,*

$$(8.1.22) \qquad \frac{\partial f}{\partial t}(z,t) = Df(z,t)h(z,t), \quad a.e. \quad t \geq 0.$$

Further, suppose there exists a mapping $F \in H(B_r)$ and a sequence $\{t_m\}_{m \in \mathbb{N}}$, $t_m > 0$, increasing to ∞, such that

$$(8.1.23) \qquad \lim_{m \to \infty} e^{-t_m} f(z,t_m) = F(z),$$

locally uniformly on B_r. Then $f(z,t)$ extends to a Loewner chain $g : B \times [0,\infty) \to \mathbb{C}^n$ and

$$\lim_{t \to \infty} e^t v(z,s,t) = g(z,s)$$

locally uniformly on B for each $s \geq 0$, where $v(t) = v(z,s,t)$ is the solution of the initial value problem

$$(8.1.24) \qquad \frac{\partial v}{\partial t} = -h(v,t), \quad a.e. \quad t \geq s, \quad v(s) = z,$$

for all $z \in B$.

Proof. First we show that $f(z,t)$ is a locally Lipschitz continuous function of $t \in [0,\infty)$ locally uniformly with respect to $z \in B_r$. Indeed, since $f(z,t)$ is a locally absolutely continuous function of $t \in [0,\infty)$ locally uniformly with respect to $z \in B_r$, it is obvious that $f(z,t)$ is continuous on $B_r \times [0,\infty)$. Therefore, for each $\rho \in (0,r)$ and $T > 0$ there exists $K = K(\rho,T) > 0$ such that

$$\|f(z,t)\| \leq K(\rho,T), \quad \|z\| \leq \rho, \quad 0 \leq t \leq T.$$

Moreover, using the Cauchy integral formula, we deduce that there exists $L = L(\rho,T) > 0$ such that

$$(8.1.25) \qquad \|Df(z,t)\| \leq L(\rho,T), \quad \|z\| \leq \rho, \quad 0 \leq t \leq T.$$

Indeed, since

$$Df(z,t)(u) = \frac{1}{2\pi i} \int\limits_{|\zeta| = \delta} \frac{f(z + \zeta u, t)}{\zeta^2} d\zeta,$$

for $\|u\| = 1$ and $\delta \in (0, r - \rho)$, we deduce for $\delta = (r - \rho)/2$ that

$$\|Df(z,t)(u)\| \leq K\left(\frac{r+\rho}{2}, T\right)\frac{2}{r-\rho} = L(\rho, T),$$

for all z, u with $\|z\| \leq \rho < r$, $\|u\| = 1$, and $t \in [0, T]$, as claimed.

Hence if $\|z\| \leq \rho$, $\|w\| \leq \rho$ and $t \in [0, T]$, then by (8.1.25) we obtain

$$\|f(z,t) - f(w,t)\| = \left\|\int_0^1 Df((1-\tau)z + \tau w, t)(w - z)d\tau\right\|$$

$$\leq \int_0^1 \|Df((1-\tau)z + \tau w, t)\|d\tau \cdot \|z - w\| \leq L(\rho, T)\|z - w\|.$$

On the other hand, taking into account the relations (8.1.22) and (8.1.25), and the local uniform boundedness of $\mathcal{M}$, we easily deduce that for each $\rho \in (0, r)$ and $T > 0$, there exists $N = N(\rho, T) > 0$ such that

$$\left\|\frac{\partial f}{\partial t}(z, t)\right\| \leq N(\rho, T), \quad \|z\| \leq \rho, \quad a.e. \quad t \in [0, T].$$

Together with the absolute continuity hypothesis, this implies that for $0 \leq t_1 \leq t_2 \leq T$ and $\|z\| \leq \rho < r$,

$$\|f(z, t_1) - f(z, t_2)\| = \left\|\int_{t_1}^{t_2} \frac{\partial f}{\partial t}(z, t)dt\right\| \leq N(\rho, T)(t_2 - t_1).$$

Next, fix $s \geq 0$. We prove that $f(v(z, s, t), t) = f(z, s)$ for all $z \in B_r$ and $t \geq s$, where $v = v(z, s, t)$ ($t \geq s$, $z \in B$) is the univalent Schwarz mapping that gives the solution of the initial value problem (8.1.24). For this purpose, let $\tilde{f}(z, s, t) = f(v(z, s, t), t)$, for $t \geq s$ and $z \in B_r$. In the proof of Theorem 8.1.3 we have seen that $v(z, s, t)$ is a Lipschitz continuous function of t locally uniformly with respect to z. Since $f(z, t)$ is also locally Lipschitz continuous in $t \geq 0$ locally uniformly with respect to $z \in B_r$, it is not difficult to verify that the same is true of $\tilde{f}(z, s, t)$. Indeed, for all $\rho \in (0, r)$ and $T > s$, we obtain the relations

$$\|\tilde{f}(z, s, t_1) - \tilde{f}(z, s, t_2)\| = \|f(v(z, s, t_1), t_1) - f(v(z, s, t_2), t_2)\|$$

$$\leq \|f(v(z, s, t_1), t_1) - f(v(z, s, t_1), t_2)\| + \|f(v(z, s, t_1), t_2) - f(v(z, s, t_2), t_2)\|$$

$$\leq N(\rho, T)(t_2 - t_1) + L(\rho, T)\|v(z, s, t_1) - v(z, s, t_2)\|$$

$$\leq N(\rho, T)(t_2 - t_1) + M(\rho, T)(t_2 - t_1), \ s \leq t_1 < t_2 \leq T, \ \|z\| \leq \rho,$$

where for the last inequality we have used the Lipschitz continuity of $v(z, s, \cdot)$ on $[s, \infty)$ locally uniformly with respect to z.

Hence $\dfrac{\partial f}{\partial t}(z, s, t)$ exists a.e. $t \geq s$ and for all $z \in B_r$. For $\eta > 0$ we have

$$\frac{1}{\eta}\Big[f(z, s, t + \eta) - f(z, s, t)\Big]$$

$$= \frac{1}{\eta}\Big[f(v(z, s, t + \eta), t + \eta) - f(v(z, s, t), t + \eta)\Big]$$

$$+ \frac{1}{\eta}\Big[f(v(z, s, t), t + \eta) - f(v(z, s, t), t)\Big].$$

The difference quotient on the left of the preceding equation has the limit $\dfrac{\partial f}{\partial t}(z, s, t)$ a.e. $t \geq s$ as $\eta \to 0^+$, and the second difference quotient on the right has the limit $\dfrac{\partial f}{\partial t}(v(z, s, t), t)$ a.e. $t \geq s$ as $\eta \to 0^+$. Applying the mean value theorem to the real and imaginary parts of the components of f, we may replace the first difference quotient on the right by

$$A(z, s, t, \eta)\Big(\frac{1}{\eta}\big[v(z, s, t + \eta) - v(z, s, t)\big]\Big),$$

where $A(z, s, t, \eta)$ is a real-linear operator which tends to the complex-linear operator $Df(v(z, s, t), t)$ as $\eta \to 0^+$. The difference quotient involving v has the limit $\dfrac{\partial v}{\partial t}(z, s, t)$ a.e. $t \geq s$ as $\eta \to 0^+$. Hence for almost all $t \geq s$, we obtain

$$\frac{\partial f}{\partial t}(z, s, t) = Df(v(z, s, t), t)\frac{\partial v}{\partial t}(z, s, t) + \frac{\partial f}{\partial t}(v(z, s, t), t)$$

$$= Df(v(z, s, t), t)\Big(\frac{\partial v}{\partial t}(z, s, t) + h(v(z, s, t), t)\Big) = 0,$$

making use of relations (8.1.22) and (8.1.24). It follows that $f(z, s, t)$ is a constant function with respect to t, and thus

$$f(v(z, s, t), t) = f(v(z, s, s), s) = f(z, s), \ z \in B_r, \ t \geq s,$$

as claimed.

Next, let $\psi(z,t) = e^t v(z,s,t) = e^s z + \ldots$, $t \geq s$, $z \in B$. The limit

$$\lim_{t \to \infty} e^t v(z,s,t) = \lim_{t \to \infty} \psi(z,t) = g(z,s)$$

exists locally uniformly on B and the mapping $g : B \times [0,\infty) \to \mathbb{C}^n$ is a Loewner chain, by Theorem 8.1.5. Moreover, we can show that $g(z,s)$ gives the desired holomorphic extension of $f(z,s)$. Indeed, it follows from (8.1.23) and the Schwarz lemma (Lemma 6.1.28) that for each $\rho \in (0,r)$, there exists $K_0 = K_0(\rho) > 0$ such that

$$\|e^{-t_m} f(z,t_m)\| \leq K_0(\rho), \ \|z\| \leq \rho, \ m \in \mathbb{N},$$

and thus

$$\|e^{-t_m} f(z,t_m) - z\| \leq (K_0(\rho) + \rho)\frac{\|z\|^2}{\rho^2}, \ \|z\| \leq \rho, \ m \in \mathbb{N}.$$

Hence, by Lemma 8.1.4 and the equality $f(z,s) = f(v(z,s,t),t)$, $z \in B_r$, $0 \leq s \leq t < \infty$, we deduce that

$$\|f(z,s) - e^{t_m} v(z,s,t_m)\| \leq \frac{K_0(\rho) + \rho}{(1-\rho)^4} e^{2s - t_m}, \ m \in \mathbb{N},$$

and therefore

$$\lim_{m \to \infty} e^{t_m} v(z,s,t_m) = f(z,s)$$

uniformly on compact subsets of B_r. Combining the above observations, we deduce that

$$f(z,s) = g(z,s), \quad z \in B_r, \quad s \geq 0.$$

This completes the proof.

Remark 8.1.7. (i) The assumption (8.1.23) in Theorem 8.1.6 can be replaced, when $r \in (0,1)$, by the assumption that $\|e^{-t} f(z,t)\| \leq C$ on B_r for $t \geq 0$. A priori the exceptional null set in (8.1.22) may depend on z, but the local Lipschitz property of $f(z,t)$ and Theorem 8.1.9 show that there is no loss of generality in assuming that it is independent of z (see [Gra-Ham-Koh]).

(ii) In higher dimensions, univalent solutions of the generalized Loewner equation (8.1.22) need not be unique. For if $f(z,t)$ is a Loewner chain which satisfies the differential equation (8.1.22) for a.e. $t \geq 0$ and all $z \in B$, and if

$\Phi : \mathbb{C}^n \to \mathbb{C}^n$ is a normalized entire biholomorphic mapping, not the identity, then $g(z,t) = \Phi(f(z,t))$ is another Loewner chain satisfying (8.1.22) for almost all $t \geq 0$ and all $z \in B$ ([Gra-Ham-Koh]; compare with [Bec3-4], [Pom3]).

(iii) However, if $h(z,t)$ is a mapping which satisfies the conditions (i) and (ii) of Theorem 8.1.3, then there is a unique Loewner chain $f(z,t)$ which satisfies the generalized Loewner differential equation (8.1.22) for almost all $t \geq 0$ and all $z \in B$, and which has the property that $\{e^{-t}f(z,t)\}_{t\geq 0}$ is a normal family on B. (Compare with Theorem 3.1.13.)

Proof. The existence of such a Loewner chain follows from Theorem 8.1.5. To show uniqueness it suffices to use Corollaries 8.1.10 and 8.1.11.

8.1.2 Transition mappings associated to Loewner chains on the unit ball of $\mathbb{C}^n$

In this section we study properties of transition mappings associated to Loewner chains and deduce some consequences. We recall that all Loewner chains are assumed to be normalized. We start with a Lipschitz regularity result (see [Cu-Koh1,2], [Gra-Koh-Koh2]); we remark that part (ii) is proved under weaker assumptions than in the source papers.

Theorem 8.1.8. *Let $f(z,t)$ be a Loewner chain and let $v(z,s,t)$ be the transition mapping associated to $f(z,t)$. Then the following statements hold:*

(i) For each $r \in (0,1)$ there exists $M = M(r) \leq 4r/(1-r)^2$ such that

$$\|v(z,s,t) - v(z,s,u)\| \leq M(r)(1 - e^{t-u}), \ \|z\| \leq r, \ 0 \leq s \leq t \leq u < \infty.$$

Thus for each $s \geq 0$, $v(z,s,t)$ is a Lipschitz continuous function of $t \geq s$ locally uniformly with respect to $z \in B$.

(ii) For each $r \in (0,1)$ and $T > 0$ there exists $N = N(r,T) > 0$ such that

$$\|f(z,s) - f(z,t)\| \leq N(r,T)(1 - e^{s-t}), \ \|z\| \leq r, \ 0 \leq s \leq t \leq T.$$

Thus $f(z,t)$ is a locally Lipschitz continuous function of $t \in [0,\infty)$ locally uniformly with respect to $z \in B$.

Proof. First we show (i). For this purpose, we show that for each $r \in (0,1)$ there exists $M = M(r) \leq 4r/(1-r)^2$ (independent of v) such that

(8.1.26) $\|z - v(z,s,t)\| \leq M(r)(1 - e^{s-t}), \ \|z\| \leq r, \ 0 \leq s \leq t < \infty.$

In fact we can take $M(r) = 4r/(1-r)^2 - 3r$, $0 < r < 1$. (This is an increasing function of $r \in (0,1)$.) To see this, fix s, t with $0 \leq s < t < \infty$, and let

$$g_{s,t}(z) = \frac{z - v(z,s,t)}{1 - e^{s-t}}, \quad z \in B.$$

Then it is obvious that $g_{s,t} \in \mathcal{M}$. Indeed, $g_{s,t} \in H(B)$, $g_{s,t}(0) = 0$, $Dg_{s,t}(0) = I$ and

$$\mathrm{Re}\, \langle g_{s,t}(z), z \rangle = \frac{1}{1 - e^{s-t}} \Big[\|z\|^2 - \mathrm{Re}\, \langle v(z,s,t), z \rangle \Big]$$

$$\geq \frac{1}{1 - e^{s-t}} \Big[\|z\|^2 - \|v(z,s,t)\| \|z\| \Big] \geq 0, \quad z \in B.$$

From Lemma 6.1.30, one deduces that $g_{s,t} \in \mathcal{M}$. Also using the proof of Theorem 7.1.7 we deduce that for each $r \in (0,1)$ there exists $M = M(r) > 0$ given by $M(r) = 4r/(1-r)^2 - 3r$, such that $\|g_{s,t}(z)\| \leq M(r)$, $\|z\| \leq r$. Hence the relation (8.1.26) follows.

Next, using the semigroup property of the transition mapping $v(z,s,t)$, we obtain for $0 \leq s \leq t \leq u < \infty$ and $\|z\| \leq r$ that

$$\|v(z,s,t) - v(z,s,u)\| = \|v(z,s,t) - v(v(z,s,t),t,u)\|$$

$$\leq M(r)(1 - e^{t-u}) \leq M(r)(u-t),$$

where we have used the fact that $\|v(z,s,t)\| \leq \|z\| \leq r$. Therefore $v(z,s,t)$ is Lipschitz continuous in t for $t \geq s$, locally uniformly with respect to $z \in B$.

We now prove the second statement. To this end, we remark that a similar argument to the above yields that for each $r \in (0,1)$ there exists $R = R(r) > 0$ such that

$$\|v(z,s,t) - v(z,\tau,t)\| \leq R(r)(1 - e^{s-\tau}), \quad \|z\| \leq r, \ 0 \leq s \leq \tau \leq t < \infty.$$

(See also Problem 8.1.4.)

Next, fix $T > 0$ and $r \in (0,1)$. Since $f(\cdot, T)$ is holomorphic on B, for each $\rho \in (0,1)$ there exists $L = L(\rho, T) > 0$ such that

$$\|f(z,T)\| \leq L(\rho, T), \quad \|z\| \leq \rho.$$

Applying the Cauchy integral formula, it is not difficult to deduce that there exists $K = K(r,T) > 0$ such that

$$\|Df(z,T)\| \leq K(r,T), \quad \|z\| \leq r.$$

Now, let $0 \le s \le t \le T$ and $z \in \overline{B}_r$. Since $f(z, s) = f(v(z, s, T), T)$ and $f(z, t) = f(v(z, t, T), T)$, we obtain

$$\|f(z, s) - f(z, t)\| = \|f(v(z, s, T), T) - f(v(z, t, T), T)\|$$

$$\le \int_0^1 \|Df((1 - \lambda)v(z, s, T) + \lambda v(z, t, T), T)\| d\lambda \cdot \|v(z, s, T) - v(z, t, T)\|.$$

Combining the above arguments, we conclude that

$$\|f(z, s) - f(z, t)\| \le K(r, T)R(r)(1 - e^{s-t}) = N(r, T)(1 - e^{s-t}),$$

as desired. This completes the proof.

We are now able to prove that any Loewner chain on the Euclidean unit ball in $\mathbb{C}^n$ satisfies the generalized Loewner differential equation (8.1.27) ([Gra-Ham-Koh]; compare with [Cu-Koh2] and Theorem 3.1.12).

Theorem 8.1.9. *Let $f(z, t)$ be a Loewner chain. Then there is a mapping $h = h(z, t)$ such that $h(\cdot, t) \in \mathcal{M}$ for each $t \ge 0$, $h(z, t)$ is measurable in t for each $z \in B$, and for a.e. $t \ge 0$,*

$$(8.1.27) \qquad \frac{\partial f}{\partial t}(z, t) = Df(z, t)h(z, t), \quad \forall z \in B.$$

(That is, there is a null set $E \subset [0, \infty)$ such that for all $t \in [0, \infty) \setminus E$, $\frac{\partial f}{\partial t}(\cdot, t)$ exists and is holomorphic on B. Also, for $t \in [0, \infty) \setminus E$ and $z \in B$, (8.1.27) holds. We interpret the left-hand side of (8.1.27) as a right-hand derivative when $t = 0$.)

Moreover, if there exists a sequence $\{t_m\}_{m \in \mathbb{N}}$ such that $t_m > 0$, $t_m \to \infty$, and

$$(8.1.28) \qquad \lim_{m \to \infty} e^{-t_m} f(z, t_m) = F(z)$$

locally uniformly on B, then $f(z, s) = \lim_{t \to \infty} e^t w(z, s, t)$ locally uniformly on B for each $s \ge 0$, where $w(t) = w(z, s, t)$ is the solution of the initial value problem

$$\frac{\partial w}{\partial t} = -h(w, t), \quad a.e. \quad t \ge s, \quad w(s) = z,$$

for all $z \in B$.

Proof. Let $v = v(z, s, t)$ be the transition mapping defined by the chain $f(z, t)$, i.e. $f(z, s) = f(v(z, s, t), t)$, $z \in B$, $0 \le s \le t < \infty$. Taking into account the normalization of $f(z,t)$, we deduce that $Dv(0,s,t) = e^{s-t}I$ for $t \geq s \geq 0$. By using the subordination property and applying the mean value theorem to the real and imaginary parts of the components of f, we obtain

$$(8.1.29) \qquad \frac{1}{r}\Big[f(z,t+r) - f(z,t)\Big]$$

$$= \frac{1}{r}\Big[f(z,t+r) - f(v(z,t,t+r),t+r)\Big]$$

$$= A(z,t,r)\Big(\frac{1}{r}\big[z - v(z,t,t+r)\big]\Big), \quad z \in B, \quad t \geq 0, \quad r > 0,$$

where $A(z,t,r)$ is a real-linear operator which tends to the invertible complex linear operator $Df(z,t)$ as $r \to 0^+$. In view of this we deduce that the difference quotient in the first member of (8.1.29) has a limit as $r \to 0^+$ if and only if the same is true of the difference quotient in the last member of (8.1.29). Since $f(z,t)$ is locally Lipschitz in t locally uniformly with respect to $z \in B$ by Theorem 8.1.8, the difference quotients on the left-hand side of (8.1.29) are a family of locally uniformly bounded holomorphic functions of $z \in B$. Let Q be a countable set of uniqueness for the holomorphic functions on B. For each $z \in Q$ the limit as $r \to 0$ of the difference quotients on the left side of (8.1.29) exists except when $t \in E_z$, where E_z is a subset of $[0,\infty)$ of measure 0. The set $E = \bigcup\{E_z : z \in Q\}$ also has measure 0, and Vitali's theorem (Theorem 6.1.16) implies that for $t \notin E$, the difference quotient on the left-hand side of (8.1.29) has a limit as $r \to 0$ which is holomorphic in z. Therefore, $\frac{\partial f}{\partial t}(\cdot,t)$ is holomorphic on B for each $t \in [0,\infty) \setminus E$. Moreover, since $v(z,s,t)$ is a Schwarz mapping and $Dv(0,s,t) = e^{s-t}I$, the difference quotient on the right has a limit $h(z,t)$ in $\mathcal{M}$ as $r \to 0^+$ for $t \notin E$, by Lemmas 6.1.30 and 6.1.33. For $t \in E$ and $z \in B$, we let $h(z,t) = z$. Then it is clear that $h(\cdot,t) \in \mathcal{M}$ for $t \geq 0$.

The mapping $h(z,t)$ is measurable in $t \in [0,\infty)$ for each $z \in B$, since $\frac{\partial f}{\partial t}(z,t)$ and $[Df(z,t)]^{-1}$ are measurable in t. To deduce the measurability of $[Df(z,t)]^{-1}$, it suffices to use the Cauchy integral formula and the local Lipschitz continuity of $f(z,t)$ in t, and to prove that $Df(z,t)$ is also locally Lipschitz continuous in t locally uniformly with respect to $z \in B$.

Finally it suffices to note that if (8.1.28) holds locally uniformly on B, then the final statement follows from Theorem 8.1.6. This completes the proof.

Next we give some consequences of the above result. (Compare with Theorem 3.1.12 and [Gra-Koh-Koh2]. Another proof of (8.1.30) is given in [Cu-Koh2].)

Corollary 8.1.10. *Let $f(z,t)$ be a Loewner chain and let $v(t) = v(z,s,t)$ be the transition mapping associated to $f(z,t)$. Also let $h(z,t)$ be the mapping given by Theorem 8.1.9. Then for each $s \geq 0$ and $z \in B$, $v = v(z,s,t)$ satisfies the initial value problem*

$$(8.1.30) \qquad \frac{\partial v}{\partial t} = -h(v,t), \quad a.e. \quad t \geq s, \quad v(z,s,s) = z.$$

Moreover, on any interval $(0,t]$, $v = v(z,s,t)$ satisfies the initial value problem

$$(8.1.31) \quad \frac{\partial v}{\partial s}(z,s,t) = Dv(z,s,t)h(z,s), \quad a.e. \quad s \in (0,t], \ v(z,t,t) = z.$$

Proof. Fix $s \geq 0$ and let $w(t) = w(z,s,t)$ be the unique locally absolutely continuous solution of the initial value problem

$$(8.1.32) \qquad \frac{\partial w}{\partial t} = -h(w,t), \quad a.e. \quad t \geq s, \quad w(s) = z,$$

for all $z \in B$. Also let $\tau > s$ and let $g(z,s,t) = f(w(z,s,t),t)$ for $t \in [s,\tau]$ and $z \in B$. Since both mappings $f(z,t)$ and $w(z,s,t)$ are locally Lipschitz continuous in t locally uniformly with respect to $z \in B$, it follows by a similar argument as in the proof of Theorem 8.1.6 that the same is true for $g(z,s,t)$. Then this mapping is locally absolutely continuous in t, so is differentiable with respect to t for almost all $t \in [s,\tau]$. Using (8.1.32) and (8.1.27), we obtain for almost all $t \in [s,\tau]$ that

$$\frac{\partial g}{\partial t}(z,s,t) = \frac{\partial f}{\partial t}(w(z,s,t),t) + Df(w(z,s,t),t)\frac{\partial w}{\partial t}(z,s,t)$$

$$= Df(w(z,s,t),t)h(w(z,s,t),t) - Df(w(z,s,t),t)h(w(z,s,t),t) = 0.$$

Hence $g(z,s,t) = g(z,s,s) = f(z,s)$ for $t \in [s,\tau]$, that is $f(w(z,s,t),t) = f(z,s)$. Since τ is arbitrary, we deduce that this equality holds for all s,t with $0 \leq s \leq t < \infty$, and $z \in B$.

On the other hand, since $v = v(z, s, t)$ is the transition mapping associated to $f(z, t)$, we have

$$f(v(z, s, t), t) = f(z, s), \quad t \geq s, \quad z \in B.$$

Combining the above arguments, we obtain

$$f(v(z, s, t), t) = f(w(z, s, t), t), \quad z \in B, \quad 0 \leq s \leq t < \infty.$$

Since $f(\cdot, t)$ is univalent on B, we must have $w(z, s, t) = v(z, s, t)$ for all $z \in B$ and $t \geq s$.

In order to prove (8.1.31), we use the equality $f(z, s) = f(v(z, s, t), t)$. Fix $t > 0$. If we differentiate both sides of this equality with respect to $s \in (0, t]$ (this is possible since $v(z, s, t)$ is Lipschitz continuous in s for $s \in [0, t]$ locally uniformly with respect to $z \in B$ (see Problem 8.1.4)), we obtain

$$\frac{\partial f}{\partial s}(z, s) = Df(v(z, s, t), t)\frac{\partial v}{\partial s}(z, s, t), \quad a.e. \quad s \in (0, t].$$

On the other hand, in view of (8.1.27) and the above equality, we deduce that

$$Df(z, s)h(z, s) = Df(v(z, s, t), t)\frac{\partial v}{\partial s}(z, s, t), \quad a.e. \quad s \in (0, t],$$

and since $Df(z, s) = Df(v(z, s, t), t)Dv(z, s, t)$, we obtain

$$Df(v(z, s, t), t)Dv(z, s, t)h(z, s) = Df(v(z, s, t), t)\frac{\partial v}{\partial s}(z, s, t).$$

Since $f(\cdot, t)$ is univalent on B, $Df(v(z, s, t), t)$ is a nonsingular matrix, and therefore the above equality implies (8.1.31), as desired. This completes the proof.

Another consequence of Theorem 8.1.9 is given below. (Compare with Theorem 3.1.12 and [Gra-Koh-Koh2]. Another proof of Corollary 8.1.11 is given in [Cu-Koh1].)

Corollary 8.1.11. *Let* $f(z, t)$ *be a Loewner chain such that the family* $\{e^{-t}f(z, t)\}_{t \geq 0}$ *is a normal family on* B. *Then for each* $s \geq 0$, *the limit*

$$f(z, s) = \lim_{t \to \infty} e^t v(z, s, t)$$

exists locally uniformly on B, where $v = v(z, s, t)$ is the transition mapping associated to $f(z, t)$.

Proof. It suffices to apply Theorem 8.1.9 and Corollary 8.1.10.

We next give the growth result for Loewner chains which satisfy the conditions of Corollary 8.1.11 (see [Gra-Ham-Koh], [Cu-Koh1], [Gra-Koh-Koh2]).

Corollary 8.1.12. *Let $f(z, t)$ be a Loewner chain such that $\{e^{-t}f(z, t)\}_{t \geq 0}$ is a normal family on B. Then*

$$(8.1.33) \quad \frac{\|z\|}{(1 + \|z\|)^2} \leq \|e^{-t}f(z, t)\| \leq \frac{\|z\|}{(1 - \|z\|)^2}, \ z \in B, \ 0 \leq t < \infty.$$

In particular, if $f(z) = f(z, 0)$ then

$$\frac{\|z\|}{(1 + \|z\|)^2} \leq \|f(z)\| \leq \frac{\|z\|}{(1 - \|z\|)^2}, \quad z \in B.$$

Proof. It suffices to apply Corollaries 8.1.10, 8.1.11 and Lemma 8.1.4, to obtain (8.1.33).

Remark 8.1.13. In several complex variables there exist Loewner chains $f(z, t)$ such that $f(z, 0)$ does not satisfy the 1/4-growth result above (see Example 8.3.12).

We conclude this section with a compactness result for the set of Loewner chains (see [Gra-Koh-Koh2] and compare with Theorem 3.1.9).

Theorem 8.1.14. *Every sequence of Loewner chains $\{f_k(z, t)\}_{k \in \mathbb{N}}$, such that $\{e^{-t}f_k(z, t)\}_{t \geq 0}$ is a normal family on B for each $k \in \mathbb{N}$, contains a subsequence that converges locally uniformly on B for each fixed $t \geq 0$ to a Loewner chain $f(z, t)$, such that $\{e^{-t}f(z, t)\}_{t \geq 0}$ is a normal family on B.*

Proof. Using the upper bound in (8.1.33) and the argument in part (ii) of Theorem 8.1.8, we deduce that for each $r \in (0, 1)$ and $T > 0$ there exists $N = N(r, T) > 0$ such that

$$(8.1.34) \ \|f_k(z, t) - f_k(z, s)\| \leq N(r, T)(t - s), \ \|z\| \leq r, \ 0 \leq s \leq t \leq T,$$

for all $k \in \mathbb{N}$.

The upper bound in (8.1.33) also implies that for $0 < r < 1$ and $T > 0$ there exists a constant $L = L(r, T) > 0$ such that for each $k \in \mathbb{N}$,

$$(8.1.35) \qquad \qquad \|f_k(z, t) - f_k(w, t)\| \leq L(r, T)\|z - w\|$$

for $\|z\| \le r$, $\|w\| \le r$, $0 \le t \le T$ and $k \in \mathbb{N}$.

The estimates (8.1.34) and (8.1.35) together imply that the mappings $f_k(z,t)$, $k \in \mathbb{N}$, are equicontinuous on $\{(z,t) : \|z\| \le 1 - 1/m, 0 \le t \le m\}$, $m = 2,3,\ldots$. The Arzela-Ascoli theorem (see e.g. [Roy, p.167-169]) implies that for m fixed, there is a subsequence $\{f_{k_p}(z,t)\}_{p \in \mathbb{N}}$ which converges pointwise on $\overline{B}_{1-1/m} \times [0,m]$ to a limit $\tilde{f}_m(z,t)$, and furthermore the convergence is uniform on $\overline{B}_{1-1/m} \times [0,m]$. A diagonal sequence argument then shows that there exists a subsequence, which we denote again by $\{f_{k_p}(z,t)\}_{p \in \mathbb{N}}$, which converges pointwise on $B \times [0,\infty)$ to a limit $f(z,t)$, and furthermore the convergence is uniform on each compact subset of $B \times [0,\infty)$. In particular $f_{k_p}(z,t) \to f(z,t)$ locally uniformly on B for each $t \ge 0$, and $f(z,t)$ is holomorphic in z. Since $f_{k_p}(0,t) = 0$ and $Df_{k_p}(0,t) = e^t I$, $p \in \mathbb{N}$, it follows that the same is true for $f(z,t)$, and hence this limit must be univalent on B. Furthermore, since $f_{k_p}(z,s) \prec f_{k_p}(z,t)$, $0 \le s \le t < \infty$, $z \in B$, it follows that there exist univalent Schwarz mappings $v_{k_p} = v_{k_p}(z,s,t)$ such that $Dv_{k_p}(0,s,t) = e^{s-t}I$ and

$$(8.1.36) \qquad f_{k_p}(z,s) = f_{k_p}(v_{k_p}(z,s,t),t), \quad p = 1,2,\ldots.$$

Since $\|v_{k_p}(z,s,t)\| \le \|z\|$, $p = 1,2,\ldots$, we deduce that $\{v_{k_p}(z,s,t)\}_{p \in \mathbb{N}}$ is a normal family. Therefore there exists a subsequence $\{k'_p\}_{p \in \mathbb{N}}$ of $\{k_p\}_{p \in \mathbb{N}}$ such that $v_{k'_p}(z,s,t) \to v(z,s,t)$ locally uniformly on B. This limit is univalent on B and satisfies the following conditions:

$$v(0,s,t) = 0, \quad Dv(0,s,t) = e^{s-t}I, \quad \|v(z,s,t)\| \le \|z\|.$$

Taking the limit in (8.1.36) through the subsequence $\{k'_p\}_{p \in \mathbb{N}}$, one deduces that

$$f(z,s) = f(v(z,s,t),t), \quad z \in B, \quad 0 \le s \le t < \infty.$$

We have therefore proved that $f(z,t)$ is a Loewner chain.

Moreover, in view of (8.1.33), we have

$$\|e^{-t}f_{k'_p}(z,t)\| \le \frac{\|z\|}{(1 - \|z\|)^2}, \quad z \in B, \quad 0 \le t < \infty,$$

for $p = 1,2,\ldots$, and thus

$$\|e^{-t}f(z,t)\| \le \frac{\|z\|}{(1 - \|z\|)^2}, \quad z \in B, \quad 0 \le t < \infty.$$

Hence $\{e^{-t}f(z,t)\}_{t\geq0}$ is a normal family, as desired. This completes the proof.

Notes. As noted by Pfaltzgraff, Theorem 8.1.6 can be generalized to the unit ball with respect to an arbitrary norm in $\mathbb{C}^n$, if one introduces the appropriate generalization of the class $\mathcal{M}$ (see Section 6.1.6). However, the proof does not generalize to the unit ball of an infinite-dimensional Banach space because of the reliance on Montel's theorem. Poreda [Por3] later showed that one could get around this difficulty by introducing slightly stronger assumptions. Also Poreda studied univalent mappings on the polydisc which admit parametric representations, and which therefore can be embedded in Loewner chains [Por1,2]. The study of the Loewner differential equation has recently been extended to bounded balanced pseudoconvex domains whose Minkowski function is of class C^1 in $\mathbb{C}^n \setminus \{0\}$ by Hamada [Ham5] and Hamada and Kohr [Ham-Koh1,4]. Loewner chains on the unit ball are also studied in [Cu3].

At an earlier stage, Pfaltzgraff [Pfa2] used subordination chains to study the quasiconformal extension of holomorphic maps in several complex variables. More recently, Chuaqui [Chu2] studied a subclass of the normalized starlike mappings of B, called strongly starlike mappings and showed that quasiconformal extensions to $\mathbb{C}^n$ exist for such mappings. Hamada [Ham4] extended the above results to strongly starlike mappings on bounded balanced pseudoconvex domains Ω whose Minkowski function is of class C^1 on $\mathbb{C}^n\setminus\{0\}$, and Hamada and Kohr [Ham-Koh6] obtained some similar results in the case of strongly spirallike mappings of type α on Ω. We shall discuss these results in Section 8.5.

Problems

8.1.1. In Theorem 8.1.3, prove that for each $t > 0$, there is a null set $E_t \subset (0,t]$ such that for all $s \in (0,t] \setminus E_t$, $\frac{\partial v}{\partial s}(z,s,t)$ exists for all $z \in B$ and is holomorphic on B. Also prove that $v(z,s,t)$ satisfies the initial value problem

$$\frac{\partial v}{\partial s}(z,s,t) = Dv(z,s,t)h(z,s), \quad a.e. \quad s \in (0,t], \quad v(z,t,t) = z.$$

Hint. Use similar arguments as in the proofs of Theorem 8.1.9 and Corollary 8.1.10.

8.1.2. Let $f = f(z,t) : B \times [0,\infty) \to \mathbb{C}^n$ be such that $f(0,t) = 0$ and $Df(0,t) = a_1(t)I, t \geq 0$, where $a_1(t) \neq 0, t \geq 0$. Assume that $f(z,t)$ is a locally absolutely continuous function of $t \geq 0$ locally uniformly with respect to $z \in B$, and $f(\cdot,t)$ is holomorphic on B for each $t \geq 0$. Let $h = h(z,t) : B \times [0,\infty) \to \mathbb{C}^n$ be such that

(i) $h(\cdot,t) \in H(B)$, $h(0,t) = 0$ and Re $\langle h(z,t), z\rangle > 0$ for $z \in B \setminus \{0\}$ and $t \geq 0$;

(ii) $h(z,\cdot)$ is a measurable function on $[0,\infty)$ for each $z \in B$.

Also assume that

$$\frac{\partial f}{\partial t}(z,t) = Df(z,t)h(z,t), \quad a.e. \quad t \geq 0, \quad \forall z \in B.$$

Further, if $a_1(\cdot) \in C^1([0,\infty))$, $|a_1(\cdot)|$ is strictly increasing on $[0,\infty)$, $|a_1(t)| \to \infty$ as $t \to \infty$, and if there is a sequence $\{t_m\}_{m\in\mathbb{N}}, t_m > 0, t_m \to \infty$, such that

$$\lim_{m\to\infty} \frac{f(z,t_m)}{a_1(t_m)} = F(z)$$

locally uniformly on B, then prove that $f(z,t)$ is a (non-normalized) Loewner chain.

(Curt, 1994 [Cu1]; see also Chen and Ren, 1994 [Che-Ren].)

8.1.3. Let $f(z,t)$ be a (normalized) Loewner chain and Φ be an entire normalized biholomorphic mapping. Also let $g(z,t) = \Phi(f(z,t)), z \in B, t \geq 0$. Show that $g(z,t)$ is a Loewner chain.

8.1.4. Let $f(z,t)$ be a Loewner chain and $v = v(z,s,t)$ be the transition mapping associated to $f(z,t)$. Prove that for each $r \in (0,1)$, there exists $R = R(r) > 0$ such that

$$\|v(z,s,t) - v(z,\tau,t)\| \leq R(r)(1 - e^{s-\tau})$$

for $\|z\| \leq r$ and $0 \leq s \leq \tau \leq t < \infty$.

(Curt and Kohr, 2001 [Cu-Koh1].)

Hint. Let $r \in (0,1)$. Since

$$\|v(z,\tau,t) - v(w,\tau,t)\| \leq \int_0^1 \|Dv((1-\sigma)z + \sigma w, \tau,t)\|d\sigma \cdot \|z - w\|$$

for $\|z\| \leq r$, $\|w\| \leq r$, $0 \leq \tau \leq t < \infty$, we obtain in view of the Cauchy integral formula,

$$\|v(z,\tau,t) - v(w,\tau,t)\| \leq N(r)\|z - w\|,$$

for some $N = N(r) > 0$. Replacing w by $v(z,s,\tau)$ in the above inequality, using the semigroup property of transition mappings and the relation (8.1.26), the conclusion follows.

8.2   Close-to-starlike and spirallike mappings of type alpha on the unit ball of $\mathbb{C}^n$

8.2.1   An alternative characterization of spirallikeness of type alpha in terms of Loewner chains

In this section we shall give a characterization of normalized spirallike mappings of type α on the Euclidean unit ball of $\mathbb{C}^n$ in terms of Loewner chains. The results of this section can be generalized to the case of an arbitrary norm in $\mathbb{C}^n$.

Let $\alpha \in \mathbb{R}$ be such that $|\alpha| < \pi/2$. Recall that a normalized locally biholomorphic mapping $f : B \to \mathbb{C}^n$ is called spirallike of type α if

$$(8.2.1) \qquad \mathrm{Re}\,[e^{-i\alpha}\langle [Df(z)]^{-1}f(z), z\rangle] > 0, \quad z \in B \setminus \{0\}.$$

Let
$$(8.2.2) \qquad f(z,t) = e^{(1-ia)t}f(e^{iat}z), \quad z \in B, \quad t \geq 0,$$

where $a = \tan \alpha$.

The following theorem of Hamada and Kohr [Ham-Koh4] is the n-dimensional version of Corollary 3.2.9, and shows that spirallike mappings of type α can be embedded in Loewner chains.

Theorem 8.2.1. Let $f : B \to \mathbb{C}^n$ be a normalized locally biholomorphic mapping and let $\alpha \in \mathbb{R}$, $|\alpha| < \pi/2$. Then f is a spirallike mapping of type α iff $f(z,t)$ given by (8.2.2) is a Loewner chain.

Proof. First, assume that f is spirallike of type α. Then it is easy to check that $f_t(z) = f(z,t) \in H(B)$, $f_t(0) = 0$, and $Df_t(0) = e^t I$, $t \geq 0$. We note that $f(z,t)$ is a C^∞ mapping on $B \times [0, \infty)$.

A short computation yields the equality

(8.2.3) $\qquad \dfrac{\partial f}{\partial t}(z,t) = Df(z,t)h(z,t), \quad z \in B, \quad t \geq 0,$

where

(8.2.4) $\qquad h(z,t) = iaz + (1-ia)e^{-iat}[Df(e^{iat}z)]^{-1}f(e^{iat}z).$

It is clear that $h(z,t)$ is a measurable function of $t \in [0,\infty)$ for each $z \in B$, $h(0,t) = 0$ and $Dh(0,t) = I$. Moreover,

(8.2.5) $\;$ Re $\langle h(z,t), z \rangle = $ Re $\left[\dfrac{e^{-i\alpha}}{\cos \alpha} \langle [Df(e^{iat}z)]^{-1}f(e^{iat}z), e^{iat}z \rangle \right] > 0,$

for all $z \in B \setminus \{0\}$ and $t \geq 0$, by (8.2.1). Hence $h(\cdot, t) \in \mathcal{M}$.

Let $t_m = m$ if $a = 0$ and $t_m = 2\pi m/a$ if $a \neq 0$. Then $e^{-t_m}f(z,t_m) = f(z)$ holds for any $m \in \mathbb{Z}$. Therefore, using the result of Theorem 8.1.6, we conclude that $f(z,t)$ is a Loewner chain.

Conversely, assume that $f(z,t)$ is a Loewner chain. Since $f(z,t)$ is of class C^∞ on $B \times [0,\infty)$, the mapping $h = h(z,t) \in \mathcal{M}$ given by Theorem 8.1.9 is also of class C^∞ on $B \times [0,\infty)$, and (8.2.3) holds for all $z \in B$ and $t \geq 0$. It is obvious that $h(z,t)$ is given by (8.2.4). Using the fact that

$$\text{Re } \langle h(z,t), z \rangle > 0, \quad z \in B \setminus \{0\}, \quad t \geq 0,$$

we obtain from (8.2.5) that

$$\text{Re } \left[e^{-i\alpha} \langle [Df(z)]^{-1}f(z), z \rangle \right] > 0, \quad z \in B \setminus \{0\}.$$

Thus f is spirallike of type α, as desired.

Theorem 8.2.1 yields the following geometric interpretation of the notion of spirallikeness of type α, as in the case of one complex variable (see [Ham-Koh4] and also Section 2.4.2).

Corollary 8.2.2. *Let $f : B \to \mathbb{C}^n$ be a normalized locally biholomorphic mapping and let $\alpha \in \mathbb{R}$ with $|\alpha| < \pi/2$. Then f is spirallike of type α if and only if f is biholomorphic on B and the spiral $\exp(-e^{-i\alpha}\tau)f(z)$ $(\tau \geq 0)$ is contained in $f(B)$ for any $z \in B$.*

Proof. From Theorem 8.2.1, we know that f is spirallike of type α if and only if

$$f(z,t) = e^{(1-ia)t} f(e^{iat}z)$$

is a Loewner chain. In this case f is biholomorphic on B and

$$f(z) \prec f(z,t) \prec e^{(1-ia)t} f(z), \quad z \in B, \quad t \geq 0,$$

which implies that

$$(8.2.6) \qquad \exp\left(-e^{-i\alpha} \frac{t}{\cos\alpha}\right) f(z) \in f(B), \quad z \in B, \quad t \geq 0.$$

Conversely, if $f \in S(B)$ and each spiral $\exp\left(-e^{-i\alpha}\tau\right) f(z)$, $\tau \geq 0$, is contained in $f(B)$, then (8.2.6) holds and it is easy to see that (8.2.2) defines a Loewner chain. This completes the proof.

The case $\alpha = 0$ in Theorem 8.2.1 yields the familiar characterization of starlikeness in terms of Loewner chains, due to Pfaltzgraff and Suffridge [Pfa-Su1].

Corollary 8.2.3. *Let* $f : B \to \mathbb{C}^n$ *be a normalized locally biholomorphic mapping. Then* f *is starlike if and only if* $f(z,t) = e^t f(z)$ *is a Loewner chain.*

8.2.2 Close-to-starlike mappings on the unit ball of $\mathbb{C}^n$

In Section 2.4 we studied certain properties of close-to-convex functions in the unit disc, and we saw that these functions appear as a natural generalization of starlike functions. The extension to higher dimensions of this class of univalent functions was considered by Pfaltzgraff and Suffridge [Pfa-Su1] and by Suffridge [Su4]. In fact, because of the failure of Alexander's theorem in several variables, there is more than one possible generalization of close-to-convexity. We shall restrict our attention to the Euclidean case in discussing these notions.

Suffridge [Su4] defined close-to-convexity in higher dimensions as follows:

Definition 8.2.4. *Let* $f : B \to \mathbb{C}^n$ *be a holomorphic mapping. We say that* f *is* close-to-convex *if there exists a convex mapping* $g \in H(B)$ *such that*

$$(8.2.7) \qquad \mathrm{Re}\, \langle Df(z)[Dg(z)]^{-1}u, u \rangle > 0, \ z \in B \setminus \{0\}, \ u \in \mathbb{C}^n \setminus \{0\}.$$

Using a version of the Noshiro-Warschawski-Wolff theorem (see Lemma 2.4.1) in higher dimensions, Suffridge [Su4] showed that a close-to-convex mapping of B is univalent. However, examples show that in dimension $n \geq 2$, the close-to-convex mappings do not satisfy a growth theorem and do not form a normal family.

The notion of close-to-starlikeness, introduced by Pfaltzgraff and Suffridge [Pfa-Su1], is more closely connected with Loewner chains. Again we shall just give the definition in the Euclidean case, although Pfaltzgraff and Suffridge worked with an arbitrary norm in $\mathbb{C}^n$.

Definition 8.2.5. Let $f : B \to \mathbb{C}^n$ be a normalized holomorphic mapping. We say that f is *close-to-starlike* if there exist $h \in \mathcal{M}$ and a starlike mapping g on B such that

$$(8.2.8) \qquad Df(z)h(z) = g(z), \quad z \in B.$$

Note that since f and h are normalized, it follows that g must also be normalized. It is also clear that any normalized starlike mapping on B is close-to-starlike (with respect to itself).

When $n = 1$ the above definition is equivalent to the usual definition of close-to-convexity (see Definition 2.4.3), because of Alexander's theorem. Indeed, since $h \in \mathcal{M}$, we may write $h(z) = zp(z)$, $z \in U$, where $p \in \mathcal{P}$. The relation (8.2.8) is therefore equivalent to

$$\frac{zf'(z)}{g(z)} = \frac{1}{p(z)}, \quad z \in U,$$

and we obtain Re $\left[\dfrac{zf'(z)}{g(z)}\right] > 0$ on U. Moreover, if $\varphi(z) = \displaystyle\int_0^z \frac{g(t)}{t}dt$, $z \in U$, then $\varphi \in K$ in view of Alexander's theorem, and thus we obtain the relation

$$\mathrm{Re}\left[\frac{f'(z)}{\varphi'(z)}\right] > 0, \quad z \in U,$$

which gives the usual definition of close-to-convexity.

Pfaltzgraff and Suffridge [Pfa-Su1] studied generalizations of Theorem 3.2.10 to higher dimensions, and showed that the condition of close-to-starlikeness can be used to give characterizations in terms of Loewner chains. In one direction, they proved the result below.

Theorem 8.2.6. *Let $f, g : B \to \mathbb{C}^n$ be normalized holomorphic mappings such that g is starlike. If*

$$F(z, t) = f(z) + (e^t - 1)g(z), \quad z \in B, \quad t \geq 0,$$

is a Loewner chain, then f is close-to-starlike relative to g.

Proof. Since $F(z, t)$ is a Loewner chain, $f(z) = f(z, 0)$ is univalent on B. In view of Theorem 8.1.9 we deduce that there exists a mapping $h = h(z, t)$ such that $h(\cdot, t) \in \mathcal{M}$ for each $t \geq 0$, $h(z, \cdot)$ is measurable on $[0, \infty)$ for all $z \in B$, and

$$\frac{\partial F}{\partial t}(z, t) = DF(z, t)h(z, t), \quad z \in B, \quad t \geq 0.$$

It is easily seen this equation holds for all $t \geq 0$ since the specific chain $F(z, t)$ is a C^∞ mapping with respect to $(z, t) \in B \times [0, \infty)$. Setting $t = 0$ in the above, we obtain $g(z) = Df(z)h(z)$, $z \in B$, where $h(z) = h(z, 0) \in \mathcal{M}$. Thus f is close-to-starlike relative to g, as desired. This completes the proof.

Pfaltzgraff and Suffridge [Pfa-Su1] also obtained the converse of this result. Since the proof is rather long, we leave it for the reader. However, we mention that it is interesting. Here we present only a special case.

Theorem 8.2.7. *Let $\phi, \psi \in H(B, \mathbb{C})$ be such that $\phi(0) = \psi(0) = 1$ and $\phi(z) \neq 0$, $z \in B \setminus \{0\}$. Also let $f, g \in H(B)$ be given by $f(z) = z\phi(z)$, $g(z) = z\psi(z)$, $z \in B$, and assume that f is locally biholomorphic on B and close-to-starlike relative to $g \in S^*(B)$. Then*

$$F(z, t) = f(z) + (e^t - 1)g(z), \quad z \in B, \, t \geq 0,$$

is a Loewner chain, and hence for any $r \in (0, 1)$ the complement of $f(B_r)$ is a union of nonintersecting rays.

Proof. In view of Theorem 7.1.14 we know that the assumption that g be starlike is equivalent to

$$\mathrm{Re}\left\{ \frac{\psi(z) + D\psi(z)z}{\psi(z)} \right\} > 0, \quad z \in B.$$

A computation similar to the proof of Theorem 7.1.14 shows that the assumption that f be close-to-starlike relative to g is equivalent to

$$\mathrm{Re}\left\{ \frac{\phi(z) + D\phi(z)z}{\psi(z)} \right\} > 0, \quad z \in B.$$

Taking into account the above inequalities, we obtain

$$\text{Re}\left\{e^{-t}\frac{\phi(z)+D\phi(z)z}{\psi(z)}+(1-e^{-t})\frac{\psi(z)+D\psi(z)z}{\psi(z)}\right\}>0,$$

for all $z \in B$ and $t \geq 0$. Now let $h : B \times [0,\infty) \to \mathbb{C}^n$ be given by

$$h(z,t)=z\left\{e^{-t}\frac{\phi(z)+D\phi(z)z}{\psi(z)}+(1-e^{-t})\frac{\psi(z)+D\psi(z)z}{\psi(z)}\right\}^{-1}.$$

Then $h(\cdot,t) \in \mathcal{M}$, $t \geq 0$, $h(z,\cdot)$ is measurable on $[0,\infty)$, $z \in B$, and

$$\frac{\partial F}{\partial t}(z,t)=DF(z,t)h(z,t),\quad z\in B,\quad t\geq 0.$$

On the other hand, it is clear that $\lim_{t\to\infty} e^{-t}F(z,t)=g(z)$ locally uniformly on B. Hence in view of Theorem 8.1.6 we conclude that $F(z,t)$ is a Loewner chain.

As in the case of one variable (see Corollary 3.2.11), the rays

$$L(z,r)=\{f(z)+tg(z):t\geq 0\},\quad \|z\|=r<1,$$

are disjoint rays which fill up the complement of $f(B_r)$. This completes the proof.

Poreda [Por4] obtained an integral representation for close-to-starlike mappings in higher dimensions, and in addition showed that if $h \in \mathcal{M}$ and $g \in S^*(B)$ are given, there is a unique normalized holomorphic mapping f which satisfies (8.2.8). We have

Theorem 8.2.8. *Let $h \in \mathcal{M}$ and let $g \in S^*(B)$. Then there exists a unique normalized holomorphic solution f of (8.2.8), and it is given by*

$$(8.2.9) \qquad f(z)=\int_0^\infty g(v(z,t))dt,\quad z\in B,$$

where $v = v(z,t)$ is the solution of the differential equation

$$\frac{\partial v}{\partial t}(z,t)=-h(v(z,t)),\ (z,t)\in B\times[0,\infty),\ v(z,0)=z.$$

Proof. First we show that any normalized holomorphic solution of (8.2.8) must have the form (8.2.9).

For fixed $z \in B$, we have

$$-\frac{d}{dt}f(v(z,t)) = g(v(z,t)), \quad t \geq 0,$$

and integrating with respect to t gives

(8.2.10) $-f(v(z,\tau)) + f(z) = \int_0^\tau g(v(z,t))dt, \ \tau \geq 0.$

By Lemma 8.1.4 and the growth theorem for starlike mappings on B (see Theorem 7.1.1), we have

(8.2.11) $\|g(v(z,t))\| \leq \frac{e^{-t}\|z\|}{(1 - \|z\|)^2}, \quad t \geq 0.$

The improper integral $\int_0^\infty g(v(z,t))dt$ therefore converges absolutely, and (8.2.10) yields

$$f(z) = \int_0^\infty g(v(z,t))dt,$$

as desired. Here we have used the fact that $\lim_{t\to\infty} v(z,t) = 0$ by Lemma 8.1.4.

Conversely, let us show that the mapping f defined by (8.2.9) satisfies (8.2.8). The estimate (8.2.11) implies that the integral in (8.2.9) exists and that f is holomorphic on B. Using the semigroup property

$$v(v(z,t),\tau) = v(z,t+\tau), \ z \in B, \ t,\tau \geq 0,$$

which follows from the uniqueness of solutions of the initial value problem $\partial w/\partial\tau = -h(w(z,\tau))$, $w(z,0) = v(z,t)$, we obtain

$$f(v(z,\tau)) = \int_\tau^\infty g(v(z,t))dt, \ z \in B, \ \tau \geq 0.$$

Differentiating with respect to τ gives

$$Df(v(z,\tau))\frac{\partial v}{\partial\tau}(z,\tau) = -g(v(z,\tau)), \quad z \in B, \ \tau \geq 0.$$

Setting $\tau = 0$ gives the desired result and completes the proof.

Poreda [Por4] showed that it is possible to deduce the growth theorem for close-to-starlike mappings on B from the integral representation (8.2.9) (compare with Corollary 8.3.9).

Corollary 8.2.9. *If $f : B \to \mathbb{C}^n$ is a close-to-starlike mapping then*

$$\|f(z)\| \leq \frac{\|z\|}{(1 - \|z\|)^2}, \quad z \in B.$$

We finish this section with some examples of close-to-starlike mappings on the Euclidean unit ball B of $\mathbb{C}^n$ (see [Pfa-Su1], [Su4]).

Example 8.2.10. Let f_j, g_j be normalized holomorphic functions on U such that $g_j \in S^*$, $j = 1, \ldots, n$. Also let $h_j(z) = \int_0^{z_j} \frac{g_j(t)}{t} dt$, $j = 1, \ldots, n$. Assume f_j is close-to-convex with respect to h_j for $j = 1, \ldots, n$. Then

$$\text{Re} \left[\frac{g_j(\zeta)}{\zeta f_j'(\zeta)} \right] = \text{Re} \left[\frac{h_j'(\zeta)}{f_j'(\zeta)} \right] > 0, \quad \zeta \in U.$$

Also let $f(z) = (f_1(z_1), \ldots, f_n(z_n))$, $z = (z_1, \ldots, z_n) \in B$. Then f is a normalized biholomorphic mapping on B. In view of Problem 6.2.5, we deduce that $g(z) = (g_1(z_1), \ldots, g_n(z_n))$ is a starlike mapping on B. Moreover, a short computation yields that

$$\text{Re} \, \langle [Df(z)]^{-1} g(z), z \rangle = \text{Re} \sum_{j=1}^{n} \frac{g_j(z_j)}{z_j f_j'(z_j)} |z_j|^2 > 0$$

for $z = (z_1, \ldots z_n) \in B \setminus \{0\}$. Consequently, f is close-to-starlike relative to g.

Example 8.2.11. Let f, g be normalized holomorphic functions on U such that $g \in S^*$. Assume that

$$\text{Re} \left[\frac{\zeta f'(\zeta)}{g(\zeta)} \right] > 0, \quad |\zeta| < 1.$$

Hence f is close-to-convex. Also let $u \in \mathbb{C}^n$ with $\|u\| = 1$ and define

$$F(z) = \frac{f(\langle z, u \rangle)}{\langle z, u \rangle} z \quad \text{and} \quad G(z) = \frac{g(\langle z, u \rangle)}{\langle z, u \rangle} z, \, z \in B.$$

Then F is normalized holomorphic on B and it is not difficult to check that G is normalized starlike on B (see Problem 6.2.4). Moreover, letting

$$h(z) = \frac{g(\langle z, u \rangle)}{\langle z, u \rangle f'(\langle z, u \rangle)} z, \quad z \in B,$$

we easily deduce that $h \in \mathcal{M}$ and $DF(z)h(z) = G(z)$, $z \in B$. Therefore F is close-to-starlike relative to G.

Problems

8.2.1. Give an example of a close-to-starlike map on the Euclidean unit ball B of $\mathbb{C}^n$ which is not spirallike on B.

8.2.2. Let B be the Euclidean unit ball of $\mathbb{C}^2$ and let $f : B \to \mathbb{C}^2$ be given by $f(z) = (z_1 + az_2^2, z_2)$, $z = (z_1, z_2) \in B$. Find the values of $a \in \mathbb{C}$ for which f is close-to-convex.

8.3 Univalent mappings which admit a parametric representation

8.3.1 Examples of mappings which admit parametric representation on the unit ball of $\mathbb{C}^n$

Univalent mappings of the unit ball in $\mathbb{C}^n$ which arise as in Theorem 8.1.5 (specifically, the mappings $f(z) = f(z, 0)$, $z \in B$) are said to admit a parametric representation. The purpose of this section is to derive properties of such mappings on the Euclidean unit ball. Growth, covering, and distortion theorems will be given, as well as coefficient estimates and conjectured coefficient estimates. Also we shall show that in higher dimensions, there exist normalized univalent mappings on the unit ball which can be embedded in Loewner chains, but which do not have parametric representation. In fact, the class of mappings which admit a parametric representation is compact, whereas the class of all mappings which can be embedded as the first element of a Loewner chain is not a normal family.

In one variable, as we have seen in Chapter 3, any function $f \in S$ has parametric representation. Thus we have another difference between the one-variable theory and the several-variables theory of univalent mappings.

Poreda [Por1,2] studied univalent mappings of the polydisc which admit parametric representation. Also Kubicka and Poreda [Kub-Por] considered parametric representations of starlike mappings of the Euclidean unit ball of $\mathbb{C}^n$. Most of the theorems presented in this section are due to Kohr [Koh9] in the Euclidean case; they were generalized to the case of an arbitrary norm in

$\mathbb{C}^n$ by Graham, Hamada, and Kohr [Gra-Ham-Koh]. Some of them may also be found, along with further results, in [Koh-Lic2]. We also mention that some of the theorems have recently been extended by Hamada and Kohr [Ham-Koh5] to the case of bounded balanced pseudoconvex domains whose Minkowski function is of class C^1 on $\mathbb{C}^n \setminus \{0\}$. However, we shall restrict our discussion to the case of the Euclidean unit ball.

Definition 8.3.1. Let $g \in H(U)$ be a univalent function such that $g(0) = 1$, $g(\overline{\zeta}) = \overline{g(\zeta)}$ for $\zeta \in U$ (i.e. g has real coefficients), Re $g(\zeta) > 0$ on U, and assume g satisfies the following conditions for $r \in (0,1)$:

$$(8.3.1) \quad \begin{cases} \min_{|\zeta|=r} \text{Re } g(\zeta) = \min\{g(r), g(-r)\} \\ \\ \max_{|\zeta|=r} \text{Re } g(\zeta) = \max\{g(r), g(-r)\}. \end{cases}$$

We define $\mathcal{M}_g$ to be the class of mappings given by

$$\mathcal{M}_g = \Big\{ h : B \to \mathbb{C}^n : h \in H(B),\ h(0) = 0,\ Dh(0) = I, \\ \Big\langle h(z), \frac{z}{\|z\|^2} \Big\rangle \in g(U),\ z \in B \Big\}.$$

Note that $\Big\langle h(z), \dfrac{z}{\|z\|^2} \Big\rangle$ is understood to have the value 1 (its limiting value) when $z = 0$.

Clearly, if $g(\zeta) = (1+\zeta)/(1-\zeta)$, $\zeta \in U$, then $\mathcal{M}_g$ becomes the class $\mathcal{M}$, which plays the role of the Carathéodory class in several complex variables.

Taking into account Theorem 8.1.5, we introduce the following class of univalent mappings of B (see [Koh9]; cf. [Por1]):

Definition 8.3.2. Let $g : U \to \mathbb{C}$ be a univalent function satisfying the assumptions of Definition 8.3.1. Also let $f \in H(B)$. We say that $f \in S_g^0(B)$ if there exists a mapping $h : B \times [0, \infty) \to \mathbb{C}^n$ which satisfies the conditions

(i) for each $t \geq 0$, $h(\cdot, t) \in \mathcal{M}_g$;

(ii) for each $z \in B$, $h(z,t)$ is a measurable function of $t \in [0, \infty)$;

(iii) $\lim_{t \to \infty} e^t v(z,t) = f(z)$ locally uniformly on B, where $v = v(z,t)$ is the solution of the initial value problem

$$\frac{\partial v}{\partial t} = -h(v,t), \quad a.e. \quad t \geq 0, \quad v(z,0) = z,$$

for all $z \in B$.

The class $S_g^0(B)$ is called *the class of mappings which have g-parametric representation on B*. If $g(\zeta) = (1 + \zeta)/(1 - \zeta)$, the class $S_g^0(B)$ is denoted by $S^0(B)$, and is called *the class of mappings which have parametric representation on B* (cf. [Por1]).

Remark 8.3.3. Theorems 8.1.5 and 8.1.9 imply that $f \in S^0(B)$ if and only if there is a Loewner chain $f(z,t)$ such that $\{e^{-t}f(z,t)\}_{t\geq 0}$ is a normal family on B and f can be embedded as the first element of this chain, i.e. $f(z) = f(z,0)$, $z \in B$ (compare with [Por1]). It is clear that another class of interest is the subclass $S^1(B)$ of $S(B)$ consisting of maps f which can be embedded as the initial element of a Loewner chain $f(z,t)$ (without the requirement that $\{e^{-t}f(z,t)\}_{t\geq 0}$ be a normal family), i.e.

$$S^1(B) = \Big\{ f \in S(B) : \exists\, f(z,t)\, \text{Loewner chain},\, f(z) = f(z,0),\, z \in B \Big\}.$$

We have

$$S_g^0(B) \subseteq S^0(B) \subseteq S^1(B) \subseteq S(B).$$

When $n = 1$ we have $S^0(U) = S$, by Theorems 3.1.8 and 3.1.12. In this chapter we shall see that in higher dimensions, $S^0(B) \subsetneq S(B)$ and $S^0(B) \subsetneq S^1(B)$.

Theorem 8.1.6 can be used to generate a large number of examples of mappings in $S_g^0(B)$ if the mapping $h = h(z,t)$ satisfies conditions (i) and (ii) of Definition 8.3.2. In connection with this result, we say that a mapping $f : B \times [0, \infty) \to \mathbb{C}^n$ is a *g-Loewner chain* if $f(z,t)$ satisfies the assumptions of Theorem 8.1.6 for $r = 1$, and the mapping $h = h(z,t)$, for which (8.1.22) holds a.e. $t \geq 0$ and for all $z \in B$, satisfies conditions (i) and (ii) of Definition 8.3.2 ([Koh9], see also [Gra-Ham-Koh]). Combining Theorems 8.1.6 and 8.1.9 and Corollary 8.1.10, we deduce that $f(z,t)$ is a g-Loewner chain if and only if $f(z,t)$ is a Loewner chain such that $\{e^{-t}f(z,t)\}_{t\geq 0}$ is a normal family on B, and the mapping $h = h(z,t)$ which occurs in the Loewner differential equation

$$\frac{\partial f}{\partial t}(z,t) = Df(z,t)h(z,t), \quad a.e. \quad t \geq 0,$$

satisfies $h(\cdot, t) \in \mathcal{M}_g$ for a.e. $t \geq 0$.

Obviously, if $f(z,t)$ is a g-Loewner chain and $f(z) = f(z,0)$, $z \in B$, then $f \in S_g^0(B)$ by Theorem 8.1.6, and conversely, if $f \in S_g^0(B)$ there exists a g-

Loewner chain $f(z, t)$ such that $f(z) = f(z, 0)$, $z \in B$, in view of Theorem 8.1.5 and the above arguments.

Example 8.3.4. Let $f \in H(B)$ be a normalized starlike mapping. Then $f(z, t) = e^t f(z)$ is a Loewner chain with transition mapping

$$v(z, s, t) = f^{-1}(e^{s-t} f(z)), \quad z \in B, \quad t \geq s \geq 0.$$

It is easy to see that

$$\frac{\partial v}{\partial t} = -h(v, t), \quad t \geq s, \quad z \in B,$$

where

$$(8.3.2) \qquad\qquad h(z, t) = [Df(z)]^{-1} f(z).$$

Moreover,

$$f^{-1}(e^{-t} f(z)) = e^{-t} f(z) + O(e^{-2t}) \quad \text{as} \quad t \to \infty$$

locally uniformly in z, and hence $\lim_{t \to \infty} e^t v(z, 0, t) = f(z)$ locally uniformly on B. Since f is starlike, $h(\cdot, t) \in \mathcal{M}$ and thus $f \in S^0(B)$ (see also [Kub-Por]).

This result can also be deduced from Theorem 8.1.6, using the fact that the Loewner chains for starlike mappings are g-Loewner chains with $g(\zeta) = (1+\zeta)/(1-\zeta)$. We note that for the case of a starlike map, the mapping $h(z, t)$ in Loewner's equation

$$\frac{\partial f}{\partial t}(z, t) = Df(z, t)h(z, t) \quad t \geq 0, \ z \in B,$$

is given by (8.3.2).

Remark 8.3.5. We may use Theorems 8.1.6, 8.1.9 and 8.2.1 to conclude that the class of spirallike maps of type α, $|\alpha| < \pi/2$, is also a subclass of $S^0(B)$. The Loewner chains which characterize the above mappings are g-Loewner chains with $g(\zeta) = (1+\zeta)/(1-\zeta)$, $\zeta \in U$.

A case of particular interest in our study is the set $S_g^0(B)$ with $g(\zeta) = 1+\zeta$, $\zeta \in U$. In this case $g(U)$ is the open disc centered at 1 and of radius 1. Therefore

$$\mathcal{M}_g = \Big\{ h \in H(B) : h(0) = 0, \ Dh(0) = I,$$

$$\Big| \frac{1}{\|z\|^2} \langle h(z), z \rangle - 1 \Big| < 1, \ z \in B \setminus \{0\} \Big\}.$$

Let $u \in \mathbb{C}^n$ with $\|u\| = 1$. For a normalized locally biholomorphic mapping f on B, let

$$G_f(\alpha, \beta) = \frac{2\alpha}{\langle [Df(\alpha u)]^{-1}(f(\alpha u) - f(\beta u)), u \rangle} - \frac{\alpha + \beta}{\alpha - \beta}, \quad \alpha, \beta \in U.$$

In Section 6.3, we have defined the set $\mathcal{G}$ of quasi-convex mappings of type A, as the set of normalized locally biholomorphic mappings $f : B \to \mathbb{C}^n$ such that $\text{Re } G_f(\alpha, \beta) > 0$ for all $\alpha, \beta \in \mathbb{C}$, $|\beta| < |\alpha| < 1$, and any $u \in \mathbb{C}^n$, $\|u\| = 1$. We shall refer in this section to the set $\mathcal{G}$ as the set of quasi-convex mappings.

We also proved that

$$K(B) \subset \mathcal{G} \subset S^*(B).$$

In fact we proved in Theorem 6.3.19 that if $f \in \mathcal{G}$ then f is starlike of order $1/2$, that is f satisfies the relation

$$\text{Re} \left\{ \frac{\|z\|^2}{\langle [Df(z)]^{-1}f(z), z \rangle} \right\} > \frac{1}{2}, \quad z \in B \setminus \{0\},$$

which is equivalent to

$$(8.3.3) \qquad \left| \frac{1}{\|z\|^2} \langle [Df(z)]^{-1}f(z), z \rangle - 1 \right| < 1, \quad z \in B \setminus \{0\}.$$

If we set $h(z) = [Df(z)]^{-1}f(z)$ and use the above inequality, we see that $h \in \mathcal{M}_g$. Now, let $f(z, t) = e^t f(z)$. Since $f \in \mathcal{G}$, it follows that $f \in S^*(B)$, and thus $f(z, t)$ is a Loewner chain. Using similar reasoning as in Example 8.3.4, we deduce that $f(z, t)$ is a g-Loewner chain and $f \in S_g^0(B)$. Therefore we have proved that

$$K(B) \subset \mathcal{G} \subset S_g^0(B) \quad \text{with} \quad g(\zeta) = 1 + \zeta, \zeta \in U.$$

These inclusion relations between $K(B)$, $\mathcal{G}$ and $S_g^0(B)$ with $g(\zeta) = 1 + \zeta$, were one of the motivations for studying the set $S_g^0(B)$. Thus some important subclasses of $S(B)$ are in fact subclasses of $S_g^0(B)$ for certain choices of g.

8.3.2 Growth results and coefficient bounds for mappings in $S_g^0(B)$

One of the main results of this section is the following growth theorem for the class $S_g^0(B)$ ([Koh9], [Gra-Ham-Koh]):

Theorem 8.3.6. *Let* $g : U \to \mathbb{C}$ *satisfy the assumptions of Definition 8.3.1 and let* $f \in S_g^0(B)$. *Then*

$$(8.3.4) \qquad \|z\| \exp \int_0^{\|z\|} \left[\frac{1}{\max\{g(x), g(-x)\}} - 1 \right] \frac{dx}{x} \leq \|f(z)\|$$

$$\leq \|z\| \exp \int_0^{\|z\|} \left[\frac{1}{\min\{g(x), g(-x)\}} - 1 \right] \frac{dx}{x}, \quad z \in B.$$

Proof. First we note that it is not difficult to prove the convergence of the above integrals, using the fact that $g(0) = 1$ and Re $g(\zeta) > 0$ on U. To establish the bounds (8.3.4) we shall need some technical lemmas, as follows.

Lemma 8.3.7. *Let* $g \in H(U)$ *satisfy the assumptions of Definition 8.3.1 and let* $h \in \mathcal{M}_g$. *Then*

$$(8.3.5) \qquad \|z\|^2 \min\{g(\|z\|), g(-\|z\|)\} \leq \text{Re } \langle h(z), z \rangle$$

$$\leq \|z\|^2 \max\{g(\|z\|), g(-\|z\|)\}, \quad z \in B.$$

Proof. Let $z \in B \setminus \{0\}$, and let $p : U \to \mathbb{C}$ be given by

$$p(\zeta) = \begin{cases} \dfrac{1}{\zeta} \left\langle h\left(\zeta \dfrac{z}{\|z\|}\right), \dfrac{z}{\|z\|} \right\rangle, & \zeta \in U \setminus \{0\} \\[2mm] 1, & \zeta = 0. \end{cases}$$

Then $p \in H(U)$, $p(0) = g(0) = 1$, and since $h \in \mathcal{M}_g$ we deduce that $p(U) \subseteq g(U)$. Therefore $p \prec g$, and from the subordination principle it follows that $p(U_r) \subseteq g(U_r)$, $r \in (0, 1)$. On the other hand, combining the maximum and minimum principles for harmonic functions with (8.3.1), we deduce that

$$\min\{g(|\zeta|), g(-|\zeta|)\} \leq \text{Re } p(\zeta) \leq \max\{g(|\zeta|), g(-|\zeta|)\}, \quad \zeta \in U.$$

Setting $\zeta = \|z\|$ in the above relation, we obtain (8.3.5), as desired.

Lemma 8.3.8. *Suppose that* g *satisfies the assumptions of Definition 8.3.1 and that* h *satisfies conditions (i)-(ii) of Definition 8.3.2. For* $z \in B$ *and* $s \geq 0$, *let* $v = v(z, s, t)$ *be the solution of the initial value problem*

$$\frac{\partial v}{\partial t} = -h(v, t) \quad a.e. \quad t \geq s, \quad v(z, s, s) = z.$$

Then

(8.3.6)
$$e^s \|z\| \exp \int_{\|v(z,s,t)\|}^{\|z\|} \left[\frac{1}{\max\{g(x), g(-x)\}} - 1 \right] \frac{dx}{x}$$

$$\leq e^t \|v(z,s,t)\| \leq e^s \|z\| \exp \int_{\|v(z,s,t)\|}^{\|z\|} \left[\frac{1}{\min\{g(x), g(-x)\}} - 1 \right] \frac{dx}{x}$$

for $z \in B$ and $t \geq s \geq 0$.

Proof. Fix $s \geq 0$ and $z \in B \setminus \{0\}$, and let $v(t) = v(z, s, t)$. Then

$$\frac{d\|v\|}{dt} = \frac{1}{\|v\|} \mathrm{Re} \left\langle \frac{dv}{dt}, v \right\rangle, \quad a.e. \quad t \geq s,$$

and using the differential equation satisfied by v, one deduces that

(8.3.7)
$$\frac{d\|v\|}{dt} = -\mathrm{Re} \left\langle h(v, t), \frac{v(t)}{\|v(t)\|} \right\rangle, \quad a.e. \quad t \geq s.$$

From (8.3.5) and the fact that $\mathrm{Re}\, g(\zeta) > 0$ on U, one obtains $\dfrac{d\|v\|}{dt} < 0$ a.e. $t \geq s$, which means that $\|v(t)\|$ is strictly decreasing from $\|z\| = \|v(s)\|$ to 0 as t increases from s to ∞. Integrating both sides of (8.3.7) with respect to t, making a change of variable and using (8.3.5), we obtain

$$-\int_{\|z\|}^{\|v\|} \frac{dx}{x \min\{g(x), g(-x)\}}$$

$$= -\int_s^t \frac{1}{\min\{g(\|v(\tau)\|), g(-\|v(\tau)\|)\} \|v(\tau)\|} \frac{d\|v(\tau)\|}{d\tau} d\tau \geq \int_s^t d\tau = t - s$$

and

$$-\int_{\|z\|}^{\|v\|} \frac{dx}{x \max\{g(x), g(-x)\}}$$

$$= -\int_s^t \frac{1}{\max\{g(\|v(\tau)\|), g(-\|v(\tau)\|)\} \|v(\tau)\|} \frac{d\|v(\tau)\|}{d\tau} d\tau \leq \int_s^t d\tau = t - s.$$

Adding $\log \|v\| - \log \|z\|$ to both sides of the last two equations and exponentiating gives (8.3.6).

We now return to the proof of Theorem 8.3.6. Since $f \in S_g^0(B)$ we have

(8.3.8)
$$\lim_{t \to \infty} e^t v(z, t) = f(z)$$

locally uniformly on B, where $v = v(z,t)$ is the solution of the initial value problem

$$\frac{\partial v}{\partial t} = -h(v,t), \quad a.e. \quad t \geq 0, \quad v(z,0) = z,$$

for all $z \in B$. Setting $s = 0$ in (8.3.6), we conclude that

$$\|z\| \exp \int_{\|v(z,t)\|}^{\|z\|} \left[\frac{1}{\max\{g(x), g(-x)\}} - 1 \right] \frac{dx}{x} \leq e^t \|v(z,t)\|$$

$$\leq \|z\| \exp \int_{\|v(z,t)\|}^{\|z\|} \left[\frac{1}{\min\{g(x), g(-x)\}} - 1 \right] \frac{dx}{x}$$

for $z \in B$ and $t \geq 0$. Since

$$\lim_{t \to \infty} e^t \|v(z,t)\| = \|f(z)\| < \infty,$$

we must have

$$\lim_{t \to \infty} \|v(z,t)\| = \lim_{t \to \infty} e^{-t} \|e^t v(z,t)\| = 0.$$

If we now let $t \to \infty$ in the above inequalities and use (8.3.8), we obtain the bounds (8.3.4), as claimed. This completes the proof of Theorem 8.3.6.

Of particular interest in Theorem 8.3.6 is the case $g(\zeta) = (1+\zeta)/(1-\zeta)$ for $\zeta \in U$. This gives the following growth theorem for mappings in $S^0(B)$: (Cf. [Por1], [Koh9], [Gra-Ham-Koh]. Compare with Corollary 8.1.12.)

Corollary 8.3.9. *If* $f \in S^0(B)$ *then*

$$\frac{\|z\|}{(1+\|z\|)^2} \leq \|f(z)\| \leq \frac{\|z\|}{(1-\|z\|)^2}, \quad z \in B.$$

These estimates are sharp. Consequently, $f(B) \supseteq B_{1/4}$.

As we have already seen, all normalized starlike mappings of B satisfy this growth result. Similarly, spirallike mappings of type α, $\alpha \in (-\pi/2, \pi/2)$, belong to $S^0(B)$ and hence satisfy the same growth result. However, in general this is not true for spirallike mappings with respect to linear operators. To see this, we give an example of a family of spirallike mappings with respect to a diagonal matrix whose growth cannot be estimated from above [Ham-Koh4].

Example 8.3.10. Let $n = 2$ and $f : B \subset \mathbb{C}^2 \to \mathbb{C}^2$ be given by $f(z) = (z_1 + az_2^2, z_2)$, $z = (z_1, z_2) \in B$. Define $A \in L(\mathbb{C}^2, \mathbb{C}^2)$ by $A(z) = (2z_1, z_2)$ for $z = (z_1, z_2) \in \mathbb{C}^2$. Then it is easy to deduce that

$$[Df(z)]^{-1} Af(z) = (2z_1, z_2),$$

and hence f is a spirallike mapping relative to A for any $a \in \mathbb{C}$. However, if $z_0 = (0, 1/2)$ it is easy to see that $\|f(z_0)\| \to \infty$ as $|a| \to \infty$.

Another consequence of Corollary 8.3.9 is that $S^0(B)$ is a normal family. This implies that $S(B)$ is a larger class than $S^0(B)$, except in the one-dimensional case, since only in dimension one is $S(B)$ a normal family. Actually we can prove more, namely the compactness of $S^0(B)$ [Gra-Koh-Koh2].

Corollary 8.3.11. $S^0(B)$ *is a compact set in the topology of* $H(B)$.

Proof. It suffices to show that $S^0(B)$ is closed. For this purpose, let $\{f_k\}_{k \in \mathbb{N}} \subset S^0(B)$ be such that $f_k \to f$ locally uniformly on B as $k \to \infty$. Then for each $k \in \mathbb{N}$, there is a Loewner chain $f_k(z, t)$ such that $\{e^{-t} f_k(z, t)\}_{t \geq 0}$ is a normal family and $f_k(z) = f_k(z, 0)$, $z \in B$. From Theorem 8.1.14 we deduce that there is a subsequence $\{f_{k_p}(z, t)\}_{p \in \mathbb{N}}$ such that $f_{k_p}(z, t) \to f(z, t)$ locally uniformly on B for each $t \geq 0$, $f(z, t)$ is a Loewner chain and $\{e^{-t} f(z, t)\}_{t \geq 0}$ is a normal family. It is obvious that $f(z) = f(z, 0)$, $z \in B$, and in view of Theorem 8.1.9, one concludes that $f \in S^0(B)$. This completes the proof.

The following example shows that $S^0(B)$ is also a proper subset of $S^1(B)$ in higher dimensions.

Example 8.3.12. (i) As noted in Remark 8.1.7 (ii), if $f(z, t)$ is a (normalized) Loewner chain and $\Phi : \mathbb{C}^n \to \mathbb{C}^n$ is an entire normalized biholomorphic mapping, not the identity, then $\Phi \circ f(z, t)$ is also a (normalized) Loewner chain.

We first remark that for any such Φ there exists a point $z_0 \in \mathbb{C}^n$ such that $\|\Phi(z_0)\| > \|z_0\|$. For otherwise, Φ maps B to itself and Cartan's theorem on fixed points (Theorem 6.1.19) implies that Φ must be the identity. After conjugation with a unitary transformation, we may assume that there exists $\rho > 0$ such that $\|\Phi(\rho, 0, \ldots, 0)\| > \rho$.

Now consider the Loewner chain

$$f(z, t) = \left(\frac{e^t z_1}{(1 - z_1)^2}, \ldots, \frac{e^t z_n}{(1 - z_n)^2} \right), \quad z = (z_1, \ldots, z_n) \in B,$$

whose initial element $f(z) = f(z, 0)$ satisfies

$$\|f(r, 0, \ldots, 0)\| = \frac{r}{(1 - r)^2}, \quad 0 \leq r < 1.$$

Choose r such that $\dfrac{r}{(1 - r)^2} = \rho$, where ρ is as above. It is evident that $\Phi \circ f$ is the first element of the Loewner chain $\Phi \circ f(z, t)$, and hence $\Phi \circ f \in S^1(B)$, but

$$\|\Phi \circ f(r, 0, \dots, 0)\| > \frac{r}{(1-r)^2}.$$

In view of Corollary 8.3.9 we conclude that $\Phi \circ f \notin S^0(B)$.

(ii) For example, let $n = 2$ and let $\Phi(z) = (z_1, z_2 + z_1^2)$, $z = (z_1, z_2) \in \mathbb{C}^2$. Then Φ is an entire normalized biholomorphic mapping of $\mathbb{C}^2$. Also if

$$f(z,t) = \left(\frac{e^t z_1}{(1-z_1)^2}, \frac{e^t z_2}{(1-z_2)^2} \right), \quad z \in B, \quad t \geq 0,$$

then

$$\Phi \circ f(z,t) = \left(\frac{e^t z_1}{(1-z_1)^2}, \frac{e^t z_2}{(1-z_2)^2} + \frac{e^{2t} z_1^2}{(1-z_1)^4} \right), \quad z \in B, \quad t \geq 0,$$

is a Loewner chain. Let $f(z) = f(z, 0)$, $z \in B$. Then for each $r \in (0, 1)$,

$$\|\Phi \circ f(r, 0)\| = \frac{r}{(1-r)^2} \sqrt{1 + \frac{r^2}{(1-r)^4}} > \frac{r}{(1-r)^2},$$

and hence $\Phi \circ f \notin S^0(B)$. We remark that $\{e^{-t} \Phi \circ f(z, t)\}_{t \geq 0}$ is not a normal family.

We have seen that all normalized convex and quasi-convex mappings of B belong to $S_g^0(B)$, with $g(\zeta) = 1 + \zeta$, $\zeta \in U$. The growth and covering theorems for convex and quasi-convex mappings (see Theorem 7.2.2 and [Rop-Su2, Theorem 4.1]) may therefore be deduced from Theorem 8.3.6:

Corollary 8.3.13. *Let* $g(\zeta) = 1 + \zeta$ *and* $f \in S_g^0(B)$. *Then*

$$\frac{\|z\|}{1 + \|z\|} \leq \|f(z)\| \leq \frac{\|z\|}{1 - \|z\|}, \quad z \in B.$$

These estimates are sharp. Consequently, $f(B) \supseteq B_{1/2}$.

We now pass to coefficient estimates for the class $S_g^0(B)$. There are some generalizations of one-variable results, but also some unexpected phenomena. Poreda [Por1] obtained the second order coefficient bounds for mappings which have parametric representation on the unit polydisc of $\mathbb{C}^n$. The estimate (8.3.9) was obtained by Kohr [Koh9, Theorem 2.4]. (The reader may compare with [Gra-Ham-Koh] for the case of an arbitrary norm in $\mathbb{C}^n$.) The estimate (8.3.10) was obtained by Graham, Hamada and Kohr [Gra-Ham-Koh].

Theorem 8.3.14. *Let* $g : U \to \mathbb{C}$ *satisfy the assumptions of Definition 8.3.1 and* $f \in S_g^0(B)$. *Then*

$$(8.3.9) \qquad \left| \frac{1}{2} \langle D^2 f(0)(w,w), w \rangle \right| \le |g'(0)|, \quad w \in \mathbb{C}^n, \quad \|w\| = 1.$$

Consequently,

$$(8.3.10) \qquad \left\| \frac{1}{2} D^2 f(0) \right\| \le 4|g'(0)|.$$

Proof. First we prove (8.3.9), using similar arguments as in the proof of [Por1, Theorem 3].

Since $f \in S_g^0(B)$ there is a g-Loewner chain $f(z,t)$ such that $f(z) = f(z,0)$, $z \in B$. Also there is a mapping $h = h(z,t) \in \mathcal{M}_g$ for each $t \ge 0$, measurable in t for each $z \in B$, such that for almost all $t \ge 0$,

$$(8.3.11) \qquad \frac{\partial f}{\partial t}(z,t) = Df(z,t)h(z,t), \quad \forall z \in B.$$

Moreover, if $s \ge 0$ and $v = v(z,s,t)$ is the unique solution of the initial value problem

$$\frac{\partial v}{\partial t} = -h(v,t), \quad a.e. \quad t \ge s, \quad v(z,s,s) = z,$$

then

$$f(z,s) = \lim_{t \to \infty} e^t v(z,s,t)$$

locally uniformly on B, by Theorems 8.1.6 and 8.1.9.

Fix $z \in B \setminus \{0\}$ and $t_0 \ge 0$. Let

$$p_{t_0}(\zeta) = \begin{cases} \dfrac{1}{\zeta} \left\langle h\left(\zeta \dfrac{z}{\|z\|}, t_0 \right), \dfrac{z}{\|z\|} \right\rangle, & \zeta \in U \setminus \{0\} \\[2ex] 1, & \zeta = 0. \end{cases}$$

Then $p_{t_0} \in H(U)$, $p_{t_0}(U) \subseteq g(U)$, and $p_{t_0}(0) = g(0)$. Hence $p_{t_0} \prec g$, which implies $|p'_{t_0}(0)| \le |g'(0)|$ by the subordination principle.

A Taylor series expansion in ζ gives

$$h\left(\zeta \frac{z}{\|z\|}, t_0 \right) = \frac{z}{\|z\|} \zeta + \frac{1}{2} \left\langle D^2 h(0, t_0)(z, z), \frac{z}{\|z\|^3} \right\rangle \zeta^2 + \cdots, \quad \zeta \in U,$$

and identifying the coefficients yields

$$p'_{t_0}(0) = \frac{1}{2}\left\langle D^2 h(0, t_0)(z, z), \frac{z}{\|z\|^3}\right\rangle.$$

We therefore obtain the estimate

(8.3.12) $$\left|\frac{1}{2}\left\langle D^2 h(0, t_0)(z, z), \frac{z}{\|z\|^3}\right\rangle\right| \leq |g'(0)|.$$

Fix $T > 0$ and integrate both sides of the equality (8.3.11), to obtain

$$f(z, T) - f(z, 0) = \int_0^T Df(z, t)h(z, t)dt, \quad z \in B.$$

For $z \in B \setminus \{0\}$ define $G_z, H_z : U \to \mathbb{C}^n$ respectively by

$$G_z(\zeta) = f(\zeta z, T) - f(\zeta z, 0)$$

and

$$H_z(\zeta) = \int_0^T Df(\zeta z, t)h(\zeta z, t)dt.$$

Then $G_z = H_z$ on U and G_z is clearly holomorphic on U.

Differentiating twice the mapping H_z and noting that $Df(0, t) = e^t I$ and $h(0, t) = 0$, we deduce that

$$\frac{d^2 H_z}{d\zeta^2}(0) = \int_0^T [2D^2 f(0, t)(z, z) + e^t D^2 h(0, t)(z, z)]dt,$$

or equivalently

$$D^2 f(0, T)(z, z) - D^2 f(0, 0)(z, z) = \int_0^T [2D^2 f(0, t)(z, z) + e^t D^2 h(0, t)(z, z)]dt.$$

Next let

$$q_z(t) = e^{-2t} D^2 f(0, t)(z, z) - D^2 f(0, 0)(z, z) - \int_0^t e^{-\tau} D^2 h(0, \tau)(z, z)d\tau,$$

for fixed $z \in B$ and $t \geq 0$. Then it is obvious that $q_z(0) = 0$, and we also obtain for almost all $t \geq 0$ that

$$\frac{d}{dt} q_z(t)$$

$$= e^{-2t}\Big[-2D^2f(0,t)(z,z) + \frac{d}{dt}D^2f(0,t)(z,z) - e^t D^2h(0,t)(z,z)\Big] = 0,$$

making use of the fact that

$$D^2f(0,t)(z,z) - D^2f(0,0)(z,z) = \int_0^t [2D^2f(0,\tau)(z,z) + e^\tau D^2h(0,\tau)(z,z)]d\tau.$$

Consequently $q_z(t) = q_z(0)$, $t \geq 0$, and thus we have the equality

$$e^{-2T}D^2f(0,T)(z,z) - D^2f(0,0)(z,z) = \int_0^T e^{-t}D^2h(0,t)(z,z)dt.$$

From this we conclude that

(8.3.13) $\qquad e^{-2T}\langle D^2f(0,T)(z,z),z\rangle - \langle D^2f(0,0)(z,z),z\rangle$

$$= \int_0^T e^{-t}\langle D^2h(0,t)(z,z),z\rangle dt.$$

Next, we note that Lemma 8.3.8 gives

(8.3.14) $\quad \|f(z,T)\| \leq e^T\|z\| \exp \int_0^{\|z\|} \Big[\frac{1}{\min\{g(x),g(-x)\}} - 1\Big]\frac{dx}{x}.$

Using Cauchy's formula

$$\frac{1}{2}D^2f(0,T)(u,u) = \frac{1}{2\pi i}\int_{|\zeta|=r} \frac{f(\zeta u,T)}{\zeta^3}d\zeta, \quad 0 < r < 1,$$

for $u \in \mathbb{C}^n$, $\|u\| = 1$, and taking into account (8.3.14), we easily obtain that

$$\lim_{T\to\infty} e^{-2T}D^2f(0,T)(z,z) = 0.$$

Next, letting $T \to \infty$ in (8.3.13) and using the above equality and (8.3.12), we conclude that

$$\Big|\frac{1}{2}\Big\langle D^2f(0,0)(z,z), \frac{z}{\|z\|^3}\Big\rangle\Big| \leq |g'(0)|.$$

It is now clear that this inequality is equivalent to (8.3.9), if we take into account the fact that $f(z) = f(z,0)$ for $z \in B$.

Finally we prove (8.3.10). For this purpose, let $P_2(z) = \frac{1}{2}D^2 f(0)(z, z)$. Then $P_2(z)$ is a continuous homogeneous polynomial of degree 2, and hence from Lemma 6.1.37 we deduce that

$$\|P_2\| \leq 4|V(P_2)|.$$

The estimate (8.3.9) yields that

$$|V(P_2)| \leq |g'(0)|,$$

and the result now follows in view of Theorem 7.2.19.

It is interesting to consider the case $g(\zeta) = (1 + \zeta)/(1 - \zeta)$ in Theorem 8.3.14. (Compare with [Por1, Theorem 3] in the case of the polydisc.) As we have seen in Theorems 7.1.8 and 7.1.9, normalized starlike mappings satisfy the following coefficient bounds:

Corollary 8.3.15. If $f \in S^0(B)$ then

$$\left| \frac{1}{2} \langle D^2 f(0)(w, w), w \rangle \right| \leq 2, \quad \|w\| = 1.$$

This estimate is sharp. Consequently,

$$\left\| \frac{1}{2} D^2 f(0) \right\| \leq 8.$$

It would be interesting to determine whether the following conjecture is true for the class $S^0(B)$. This is a version of the Bieberbach conjecture in several complex variables.

Conjecture 8.3.16. If $f \in S^0(B)$ then

$$\left| \frac{1}{k!} \langle D^k f(0)(w^k), w \rangle \right| \leq k,$$

for all $w \in \mathbb{C}^n$, $\|w\| = 1$, and $k \in \mathbb{N}$, $k \geq 2$.

A second possible version of the Bieberbach conjecture for $S^0(B)$, namely

(8.3.15) $$\left\| \frac{1}{k!} D^k f(0)(w^k) \right\| \leq k, \quad \|w\| = 1, \quad k \in \mathbb{N},$$

is false in dimension 2 or greater, in fact it fails for $k = 2$ as Example 8.3.18 shows. However, in the case of the unit polydisc P of $\mathbb{C}^n$, the above inequality is true for $k = 2$ (and open for $k \geq 3$) and was proved by Poreda [Por1, Theorem 3]. Gong [Gon5, Theorem 5.3.1] has recently proved that if f is normalized starlike on the unit polydisc of $\mathbb{C}^n$, then the bound (8.3.15) holds for $k = 2, 3$, and is open for $k \geq 4$ (see also Section 7.1.2). We therefore formulate

Open Problem 8.3.17. Find the sharp upper bounds for

$$\left\| \frac{1}{k!} D^k f(0)(w^k) \right\|, \quad k \geq 2, \quad \|w\| = 1,$$

when $f \in S^0(B)$.

We now give the example which disproves (8.3.15):

Example 8.3.18. Let $n = 2$, $a \in \mathbb{C}$, $|a| = 3\sqrt{3}/2$ and $f : B \subset \mathbb{C}^2 \to \mathbb{C}^2$ be given by

$$f(z) = (z_1 + az_2^2, z_2), \quad z = (z_1, z_2) \in B.$$

Then $f \in S^0(B)$ because f is normalized starlike on B (see Problem 6.2.1). However, a simple computation yields that

$$D^2 f(0)(w, \cdot) = (a_{ij})_{1 \leq i,j \leq 2},$$

where

$$a_{ij} = \begin{cases} 2aw_2, & i = 1, \ j = 2 \\ 0, & \text{otherwise.} \end{cases}$$

Therefore $D^2 f(0)(w, w) = (2aw_2^2, 0)$ for $w = (w_1, w_2) \in \mathbb{C}^2$, and for $w = (0, 1)$ we conclude that

$$\left\| \frac{1}{2} D^2 f(0)(w, w) \right\| = |a| = \frac{3\sqrt{3}}{2} > 2.$$

Another case of special interest in Theorem 8.3.14 is the case $g(\zeta) = 1 + \zeta$. This yields a coefficient estimate which is satisfied by the normalized convex and quasi-convex mappings (cf. Theorem 7.2.16 with $p = 2$):

Corollary 8.3.19. If $f \in S_g^0(B)$ with $g(\zeta) = 1 + \zeta$ then

$$\left| \frac{1}{2} \langle D^2 f(0)(w, w), w \rangle \right| \leq 1, \quad \|w\| = 1.$$

This estimate is sharp. Moreover,

$$\left\| \frac{1}{2} D^2 f(0) \right\| \leq 4,$$

and for $k \in \mathbb{N}$, $k \geq 3$, the estimates

(8.3.16)
$$\left\| \frac{1}{k!} D^k f(0) \right\| < ek$$

hold.

Proof. It suffices to prove the bounds (8.3.16). To this end, fix $k \in \mathbb{N}$, $k \geq 3$, and $w \in \mathbb{C}^n$, $\|w\| = 1$. Using the Cauchy formula

$$\frac{1}{k!} D^k f(0)(w^k) = \frac{1}{2\pi i} \int_{|\zeta|=r} \frac{f(\zeta w)}{\zeta^{k+1}} d\zeta, \quad 0 < r < 1,$$

and taking into account Corollary 8.3.13, we easily obtain

$$\left\| \frac{1}{k!} D^k f(0)(w^k) \right\| \leq \frac{1}{2\pi r^k} \int_0^{2\pi} \|f(re^{i\theta}w)\| d\theta \leq \frac{1}{r^{k-1}(1-r)}.$$

Setting $r = 1 - 1/k$ in this inequality gives

$$\left\| \frac{1}{k!} D^k f(0)(w^k) \right\| \leq k \left(1 + \frac{1}{k-1} \right)^{k-1} < ek.$$

Finally, in view of Theorem 7.2.19 the bounds (8.3.16) follow. This completes the proof.

Remark 8.3.20. Note that if $f : B \to \mathbb{C}^n$ is a normalized convex mapping, then in view of (7.2.12) we have the sharp bounds

(8.3.17)
$$\left\| \frac{1}{k!} D^k f(0) \right\| \leq 1, \quad k \in \mathbb{N}.$$

However, if $f \in S_g^0(B) \setminus K(B)$, with $g(\zeta) = 1 + \zeta$, then (8.3.17) need not be satisfied in dimension $n \geq 2$. To see this, consider the case $n = 2$ and let $f(z) = (z_1 + a z_2^2, z_2)$ for $z = (z_1, z_2) \in B$ and $|a| = 3\sqrt{3}/4$. Then $f \in G$ by Problem 6.3.7. However, since $|a| > 1/2$, it follows from Example 6.3.13 that f is not convex on B. Moreover, if $w = (0, 1)$, then

$$\left\| \frac{1}{2} D^2 f(0)(w, w) \right\| = \frac{3\sqrt{3}}{4} > 1.$$

Hence

$$\left\| \frac{1}{2} D^2 f(0) \right\| > 1.$$

We now give an example of a polynomial map in $S(B)$ which does not satisfy the conclusion of Corollary 8.3.15. This gives another way of seeing that $S^0(B)$ is a proper subclass of $S(B)$ in several variables. We also have seen that $S^0(B)$ is a proper subclass of $S^1(B)$ in higher dimensions.

Example 8.3.21. Let $n \geq 2$ and $f : B \to \mathbb{C}^n$ be given by

$$f(z) = (z_1 + az_2^2, z_2, \ldots, z_n), \; z = (z_1, \ldots, z_n) \in B,$$

where $a \in \mathbb{C}$ with $|a| > 4\sqrt{2}$.

Clearly $f \in S(B)$, and a short computation yields the relation

$$\left| \frac{1}{2} \langle D^2 f(0)(w, w), w \rangle \right| = |a||w_1 w_2^2|, \quad w = (w_1, \ldots, w_n) \in \mathbb{C}^n.$$

Let $w = (r, r, 0, \ldots, 0)$, where $r = 1/\sqrt{2}$. Since $|a| > 4\sqrt{2}$, we obtain

$$\left| \frac{1}{2} \langle D^2 f(0)(w, w), w \rangle \right| = |a|r^3 = \frac{|a|}{2\sqrt{2}} > 2.$$

Another case of special interest in the study of the set $S_g^0(B)$ is the case $g(\zeta) = (1 + c\zeta)/(1 - c\zeta)$, $\zeta \in U$, where $0 < c < 1$. Obviously, g is a univalent function on U that satisfies the assumptions of Definition 8.3.1. The image of the unit disc under g is the disc centered at $(1 + c^2)/(1 - c^2)$ and of radius $2c/(1 - c^2)$. Let $S_c^0(B)$ denote the set $S_g^0(B)$ when $g(\zeta) = (1 + c\zeta)/(1 - c\zeta)$, $c \in (0, 1)$. In the case of one complex variable such g-Loewner chains, with $g(\zeta) = (1 + c\zeta)/(1 - c\zeta)$, were intensively studied by Becker [Bec1] and called c-chains. The interest of a c-chain $f(z, t)$ arises from the fact that its first element $f(z) = f(z, 0)$ can be extended to a quasiconformal homeomorphism of the whole complex plane (see [Bec1,2]).

We next consider some examples of mappings in $S_c^0(B)$, $c \in (0, 1)$. Chuaqui [Chu2] introduced a subset of $S^*(B)$ which he called the set of strongly starlike mappings. He proved that these mappings can be extended quasiconformally to $\mathbb{C}^n$.

Definition 8.3.22. Let $z \in \mathbb{C}^n$, $\|z\| = 1$ and $f \in S^*(B)$. We say that f is *strongly starlike* if the values of

$$q(\zeta) = \frac{1}{\zeta} \langle [Df(\zeta z)]^{-1} f(\zeta z), z \rangle, \quad |\zeta| < 1,$$

lie in a fixed compact subset of the right half-plane, independent of z.

Next, let $c \in (0,1)$ and f be a normalized locally biholomorphic mapping on B such that

$$\left| \frac{1}{\|z\|^2} \langle [Df(z)]^{-1} f(z), z \rangle - \frac{1+c^2}{1-c^2} \right| < \frac{2c}{1-c^2}, \quad z \in B \setminus \{0\}.$$

Clearly if f satisfies the above assumption, then f is strongly starlike. Moreover, since $f(z,t) = e^t f(z)$ is a g-Loewner chain, with $g(\zeta) = (1+c\zeta)/(1-c\zeta)$, one deduces that $f \in S_c^0(B)$. On the other hand, if f is strongly starlike, then it is obvious that f satisfies the above assumption for some absolute constant $c \in (0,1)$, and thus $f \in S_c^0(B)$.

Similarly, Hamada and Kohr [Ham-Koh6] defined the notion of a strongly spirallike mapping of type $\alpha \in (-\pi/2, \pi/2)$ (see (8.5.39)), and obtained a quasiconformal extension result that characterizes these maps. We shall discuss this result in the next section.

From Theorems 8.3.6 and 8.3.14 we obtain the following growth result and coefficient estimates for mappings in $S_c^0(B)$:

Theorem 8.3.23. *Let $f \in S_c^0(B)$ with $c \in (0,1)$. Then*

$$\frac{\|z\|}{(1+c\|z\|)^2} \leq \|f(z)\| \leq \frac{\|z\|}{(1-c\|z\|)^2}, \quad z \in B.$$

These estimates are sharp. Moreover,

$$\left| \frac{1}{2} \langle D^2 f(0)(w,w), w \rangle \right| \leq 2c, \quad \|w\| = 1,$$

and

$$\left\| \frac{1}{2} D^2 f(0) \right\| \leq 8c.$$

Remark 8.3.24. In this section we have studied the class $S_g^0(B)$ for certain choices of the univalent function g, and we have seen that in higher dimensions $S^0(B) \subsetneq S(B)$ and also $S^0(B) \subsetneq S^1(B)$. However, the class $S^0(B)$ contains some important classes of mappings, including all normalized starlike and spirallike mappings of type α, $|\alpha| < \pi/2$. The normalized convex and quasi-convex mappings belong to $S_g^0(B)$ with $g(\zeta) = 1 + \zeta$.

Problems

8.3.1. Show that if $f : B \to \mathbb{C}^n$ is a normalized locally biholomorphic mapping on the Euclidean unit ball B of $\mathbb{C}^n$ such that

$$(1 + \|z\|)\|[Df(z)]^{-1}D^2f(z)(z, \cdot)\| < 2, \quad z \in B,$$

then $f \in S_g^0(B)$ with $g(\zeta) = 1 + \zeta$, $\zeta \in U$.
(Cf. Graham, Hamada and Kohr, 2002 [Gra-Ham-Koh]. See also [Ham-Koh-Lic2].)

Hint. Show the inequality

$$\left| \frac{1}{\|z\|^2} \langle [Df(z)]^{-1}f(z), z \rangle - 1 \right| < 1, \quad z \in B \setminus \{0\}.$$

8.3.2. Let $f \in S$ and $w \in \mathbb{C}^n$ with $\|w\| = 1$. Also let $F(z) = z\dfrac{f(\langle z, w \rangle)}{\langle z, w \rangle}$, $z \in B$. Show that $F \in S^0(B)$. In particular, if $f \in S^*$ then $F \in S^*(B)$.

8.3.3. Let $c \in (0, 1)$ and $f_j \in H(U)$ be such that $f_j(0) = f_j'(0) - 1 = 0$ and
$\dfrac{f_j(\zeta)}{\zeta f_j'(\zeta)} \prec \dfrac{1 + c\zeta}{1 - c\zeta}$, $\zeta \in U$, $j = 1, 2, \ldots, n$. Also let $f(z) = (f_1(z_1), \ldots, f_n(z_n))$, $z = (z_1, \ldots, z_n) \in B$, where B is the Euclidean unit ball of $\mathbb{C}^n$. Show that $f \in S_c^0(B)$.

Hint. Show that

$$\left| \frac{1}{\|z\|^2} \langle [Df(z)]^{-1}f(z), z \rangle - \frac{1 + c^2}{1 - c^2} \right| < \frac{2c}{1 - c^2}, \quad z \in B \setminus \{0\}.$$

8.4 Applications of the method of Loewner chains to univalence criteria on the unit ball of $\mathbb{C}^n$

There are some further applications of the method of Loewner chains to univalence criteria in several complex variables. The main result of this section is an extension of Becker's criterion (Theorem 3.3.1) to higher dimensions. This extension was obtained by Pfaltzgraff in 1974 in his original paper on Loewner chains in several variables [Pfa1].

Theorem 8.4.1. *Let $f : B \to \mathbb{C}^n$ be a normalized locally biholomorphic mapping which satisfies*

$$(8.4.1) \qquad (1 - \|z\|^2)\|[Df(z)]^{-1}D^2f(z)(z, \cdot)\| \le 1, \quad z \in B.$$

Then f is univalent on B.

Proof. We shall prove that the condition (8.4.1) enables us to embed $f(z)$ as the initial element of a Loewner chain. Consider the mapping

$$f(z, t) = f(ze^{-t}) + (e^t - e^{-t})Df(ze^{-t})(z), \quad z \in B, \quad t \ge 0,$$

which obviously satisfies $f(z, 0) = f(z)$. We shall prove that $f(z, t)$ satisfies the hypotheses of Theorem 8.1.6.

Clearly $f(\cdot, t) \in H(B)$, $f(0, t) = 0$ and $Df(0, t) = e^t I$ for $t \ge 0$. We note that $f(z, t)$ is a C^∞ mapping on $B \times [0, \infty)$. It is elementary to verify that

$$\lim_{t \to \infty} e^{-t} f(z, t) = z$$

locally uniformly on B, and hence the condition (8.1.23) holds for $F(z) \equiv z$.

It is also elementary to verify that

$$Df(z, t) = e^t Df(ze^{-t}) + e^t(1 - e^{-2t})D^2f(ze^{-t})(ze^{-t}, \cdot)$$

$$= e^t Df(ze^{-t})[I - E(z, t)],$$

where, for fixed $(z, t) \in B \times [0, \infty)$, $E(z, t)$ is the linear operator

$$(8.4.2) \qquad E(z, t) = -(1 - e^{-2t})[Df(ze^{-t})]^{-1}D^2f(ze^{-t})(ze^{-t}, \cdot).$$

Since $1 - e^{-2t} < 1 - \|ze^{-t}\|^2$ for $z \in B$ and $t \ge 0$, the assumption (8.4.1) immediately implies that $\|E(z, t)\| < 1$. Therefore $I - E(z, t)$ is an invertible operator.

On the other hand, using (8.4.2) we obtain

$$\frac{\partial f}{\partial t}(z, t) = e^t Df(ze^{-t})(z) - (e^t - e^{-t})D^2f(ze^{-t})(ze^{-t}, z)$$

$$= e^t Df(ze^{-t})[I + E(z, t)](z) = Df(z, t)[I - E(z, t)]^{-1}[I + E(z, t)](z).$$

We conclude that $f(z,t)$ satisfies the differential equation

$$\frac{\partial f}{\partial t}(z,t) = Df(z,t)h(z,t), \quad z \in B, \quad t \geq 0,$$

where $h(z,t) = [I - E(z,t)]^{-1}[I + E(z,t)](z)$.

It remains to verify that $h(z,t)$ satisfies the assumptions of Theorem 8.1.3. Clearly $h(z,t)$ is holomorphic in z and measurable (in fact smooth) in t, and satisfies $h(0,t) = 0$, $Dh(0,t) = I$. Moreover, from the inequality

$$\|h(z,t) - z\| = \|E(z,t)(h(z,t) + z)\|$$

$$\leq \|E(z,t)\| \cdot \|h(z,t) + z\| < \|h(z,t) + z\|,$$

we obtain that Re $\langle h(z,t), z \rangle > 0$, for $z \in B \setminus \{0\}$ and $t \geq 0$.

Consequently, the conditions (i) and (ii) of Theorem 8.1.3 are satisfied.

From Theorem 8.1.6 we therefore deduce that $f(z,t)$ is a Loewner chain, and hence $f(z) = f(z,0)$ is univalent on B. This completes the proof.

In the literature there are various generalizations of the preceding result of Pfaltzgraff. One of these is an extension to higher dimensions of Ahlfors' and Becker's result (Theorem 3.3.2), due to Chen and Ren [Che-Ren] and Curt [Cu1]. The proof proceeds along similar lines to the above, and we leave it as an exercise for the reader.

Theorem 8.4.2. Let $f : B \to \mathbb{C}^n$ be a normalized locally biholomorphic mapping. Let $c \in \mathbb{C}$ with $|c| \leq 1$ and $c \neq -1$. Suppose that

$$\|(1 - \|z\|^2)[Df(z)]^{-1}D^2f(z)(z, \cdot) + c\|z\|^2 I\| \leq 1, \quad z \in B.$$

Then f is univalent on B.

We finish this section with an n-dimensional version of Theorem 3.3.9, which generalizes both Theorems 8.4.1 and 8.4.2. This result was obtained in [Cu-Pas]. We only sketch the proof, since it suffices to use similar arguments as in the proofs of Theorems 3.3.9 and 8.4.1.

Theorem 8.4.3. Let $F : B \times \mathbb{C}^n \to \mathbb{C}^n$ satisfy the following assumptions:

(i) The function $L(z,t) = F(e^{-t}z, e^t z)$ is holomorphic on B for all $t \in [0, \infty)$ and is locally absolutely continuous in $t \in [0, \infty)$ locally uniformly with respect to $z \in B$.

(ii) $D_vF(u,v)$ is invertible for all $(u,v) \in B \times \mathbb{C}^n$, and there exists a function $a : [0,\infty) \to \mathbb{C}$ such that $a(t) \neq 0$, $t \geq 0$, $a(\cdot) \in C^1([0,\infty))$, $|a(\cdot)|$ is increasing on $[0,\infty)$, $\lim_{t\to\infty} |a(t)| = \infty$ and

$$e^{-t}D_uF(0,0) + e^tD_vF(0,0) = a(t)I,$$

where $D_uF(u,v)$ (respectively $D_vF(u,v)$) is the $n \times n$ matrix for which the (i,j) entry is $\dfrac{\partial F_i}{\partial u_j}(u,v)$ (respectively $\dfrac{\partial F_i}{\partial v_j}(u,v)$).

(iii) The family of functions $\left\{ \dfrac{F(e^{-t}z, e^tz)}{a(t)} \right\}_{t\geq 0}$ forms a normal family on B.

Let

$$G(u,v) = \frac{\langle u,v\rangle}{\|v\|^2}[D_vF(u,v)]^{-1}[D_uF(u,v)].$$

If

$$\|G(z,z)\| < 1, \quad z \in B,$$

and

$$\left\| G\left(z, \frac{z}{\|z\|^2}\right) \right\| \leq 1, \quad z \in B \setminus \{0\},$$

then $F(z,z)$ is univalent on B.

Proof. Since $L(z,t) = F(e^{-t}z, e^tz)$ it follows that $L(z,t) = a(t)z + \ldots$ for $z \in B$ and $t \geq 0$, and it is obvious that

$$DL(z,t) = e^t D_vF(e^{-t}z, e^tz)(I - E(z,t)),$$

where for each $(z,t) \in B \times [0,\infty)$,

$$E(z,t) = -e^{-2t}[D_vF(e^{-t}z, e^tz)]^{-1}D_uF(e^{-t}z, e^tz).$$

Using the maximum modulus theorem for holomorphic mappings (see Theorem 6.1.27), it is easy to show that $\|E(z,t)\| < 1$ for $z \in B$ and $t \geq 0$, and hence $I - E(z,t)$ is an invertible operator. Moreover, if

$$h(z,t) = [I - E(z,t)]^{-1}[I + E(z,t)](z), \quad z \in B, \quad t \geq 0,$$

then $h(z,t)$ is holomorphic in z and differentiable in t, and a short computation gives

$$\frac{\partial L}{\partial t}(z,t) = DL(z,t)h(z,t), \quad z \in B, \quad t \geq 0.$$

Since

$$\|h(z,t) - z\| = \|E(z,t)(h(z,t) + z)\|$$

$$\leq \|E(z,t)\| \cdot \|h(z,t) + z\| < \|h(z,t) + z\|,$$

we obtain Re $\langle h(z,t), z \rangle > 0$ for $z \in B \setminus \{0\}$ and $t \geq 0$.

Hence from Problem 8.1.2 we deduce that $L(z,t)$ is a Loewner chain, and thus $L(z,0) = F(z,z)$ is univalent on B.

Remark 8.4.4. Let $f : B \to \mathbb{C}^n$ be a normalized locally biholomorphic mapping. Also let $F_k : B \times \mathbb{C}^n \to \mathbb{C}^n$, $k = 1, 2$, be given by

$$F_1(u,v) = f(u) + Df(u)(v - u),$$

$$F_2(u,v) = f(u) + (1+c)^{-1}Df(u)(v - u), \quad |c| \leq 1, \quad c \neq -1.$$

Setting successively $F = F_k$, $k = 1, 2$, in Theorem 8.4.3 yields the results of Theorems 8.4.1 and 8.4.2.

Problems

8.4.1. Prove Theorem 8.4.2.

Hint. Prove that $f(z,t) = f(e^{-t}z) + (1+c)^{-1}(e^t - e^{-t})Df(ze^{-t})(z)$ is a (non-normalized) Loewner chain.

8.4.2. Complete the details in the proof of Theorem 8.4.3.

8.4.3. Let $f : B \to \mathbb{C}^n$ be a normalized holomorphic mapping such that

$$\|Df(z) - I\| < 1, \; z \in B.$$

Prove that f is univalent on B.

(Brodskii, 1983 [Brod].)

Hint. Prove that $f(z,t) = f(e^{-t}z) + (e^t - e^{-t})z$ is a Loewner chain.

8.4.4. Let $f, g \in H(B)$ be such that f and g are normalized and g is locally biholomorphic on B. Assume the following conditions hold for all $z \in B$:

$$\left\| [Dg(z)]^{-1}Df(z) - I \right\| < 1,$$

$$\left\| \|z\|^2 \left[[Dg(z)]^{-1}Df(z) - I \right] + (1 - \|z\|^2)[Dg(z)]^{-1}D^2g(z)(z, \cdot) \right\| \leq 1.$$

Prove that f is univalent on B.

(Cf. Curt and Pascu, 1995 [Cu-Pas].)

   Hint. Use Theorem 8.4.3 for $F(u,v) = f(u) + Dg(u)(v-u)$, $u \in B$, $v \in \mathbb{C}^n$.

   8.4.5. Let $f : B \to \mathbb{C}^n$ be a normalized holomorphic mapping and let $G(z)$ be a nonsingular $n \times n$ matrix which is holomorphic with respect to $z \in B$. Assume $G(0) = I$ and the following conditions hold for all $z \in B$:

$$\left\| [G(z)]^{-1}Df(z) - I \right\| < 1,$$

$$\left\| \|z\|^2 \left[[G(z)]^{-1}Df(z) - I \right] + (1 - \|z\|^2)[G(z)]^{-1}DG(z)(z, \cdot) \right\| \leq 1.$$

Prove that f is univalent on B.

(Ren and Ma, 1995 [Ren-Ma].)

   Hint. Prove that $f(z,t) = f(ze^{-t}) + (e^t - e^{-t})G(ze^{-t})z$ is a Loewner chain.

8.5   Loewner chains and quasiconformal extensions of holomorphic mappings in several complex variables

8.5.1   Construction of quasiconformal extensions by means of Loewner chains

   In this section we obtain sufficient conditions for a holomorphic mapping on the unit ball of $\mathbb{C}^n$, which can be embedded as the first element of a Loewner chain, to have a quasiconformal extension to a mapping of $\mathbb{R}^{2n}$ onto $\mathbb{R}^{2n}$. For this purpose, we briefly recall some definitions and results concerning quasiregular and quasiconformal mappings. For further details, the reader may consult [Boj-Iw], [Car], [Väi], [Vuo], [Ric], [Res], [Mart-Ric-Väi1,2]. In this section we shall consider quasiconformal mappings in higher dimensions.

   Definition 8.5.1. Let G be a domain in $\mathbb{R}^m$, $m \geq 2$, and $W^1_{m,loc}(G)$ be the Sobolev space of maps in $L^m_{loc}(G)$, whose first order weak partial derivatives exist and belong to $L^m_{loc}(G)$. Also let $f : G \to \mathbb{R}^m$ be a mapping. We say that f is *quasiregular* (qr) on G if $f \in W^1_{m,loc}(G)$ and there is a constant $K \geq 1$ such that

(8.5.1) $\|D(f;x)\|^m \leq K \det D(f;x)$, a.e. $x \in G$.

Here $\|D(f;x)\| = \sup\{\|D(f;x)(v)\| : \|v\| = 1\}$ and $D(f;x)$ is the (real) Jacobian matrix of f.

The smallest constant $K \geq 1$ for which (8.5.1) holds is called the *dilatation* (*outer dilatation*) of f in G and is denoted by $K_o(f)$.

If f is quasiregular on G, then the smallest $K \geq 1$ for which the inequality

$$\det D(f;x) \leq Kl(D(f;x))^m$$

holds a.e. in G is called the *inner dilatation* of f and is denoted by $K_i(f)$. Here $l(D(f;x)) = \min_{\|v\|=1} \|D(f;x)(v)\|$.

The *maximal dilatation* of f is the number $K(f) = \max\{K_i(f), K_o(f)\}$. If $K \in [1,\infty)$ is such that $K(f) \leq K$, we say that f is K-*quasiregular*.

We note that if Ω is a domain in $\mathbb{C}^n$, then a holomorphic mapping $f : \Omega \to \mathbb{C}^n$ is quasiregular as a mapping from $\mathbb{R}^{2n}$ to $\mathbb{R}^{2n}$ if and only if there is a constant $K \geq 1$ such that

$$\|Df(z)\|^n \leq K|J_f(z)|, \quad z \in \Omega.$$

This follows from (6.1.8).

The geometric interpretation of quasiregularity is that a nonconstant quasiregular mapping maps infinitesimal spheres to infinitesimal ellipsoids whose major axes are less than or equal to a fixed constant times their minor axes.

It is known that if $f : G \to \mathbb{R}^m$ is a nonconstant quasiregular mapping, then $\det D(f;x) > 0$ a.e. on G ([Mart-Ric-Väi1], [Boj-Iw]). Also if f is a nonconstant quasiregular holomorphic mapping, then f is locally biholomorphic (see [Mard-Ric]). In this section we study quasiregular holomorphic mappings.

A homeomorphism $f : G \to \mathbb{R}^m$ which is quasiregular (K-quasiregular) on G is called *quasiconformal* (qc) (K-quasiconformal (K-qc)).

It is well known that there are many equivalent ways of defining quasiconformal mappings (see [Väi], [Vuo], [Res], [Ric], [Car], [Mart-Ric-Väi1,2]). We recall that if $E = \{x \in \mathbb{R}^m : a_i \leq x_i \leq b_i, i = 1,\ldots,m\}$ is a closed interval, then a map $f : E \to \mathbb{R}^m$ is said to be *absolutely continuous on lines* (ACL) if f is continuous and is absolutely continuous on almost every line segment in E which is parallel to the coordinate axes. Moreover, if V is an open set in

$\mathbb{R}^m$, then a map $f : V \to \mathbb{R}^m$ is ACL if $f|_E$ is ACL for each closed interval E of V. For further details, see [Väi].

The necessary and sufficient condition for K-quasiconformality in the following theorem is known as the analytical definition of K-quasiconformality (see e.g. [Väi]).

Theorem 8.5.2. *Let G_1 and G_2 be domains in $\mathbb{R}^m$. A homeomorphism f of G_1 onto G_2 is a K-quasiconformal mapping if and only if the following conditions hold:*

(i) *f is ACL;*

(ii) *f is differentiable a.e. on G_1;*

(iii) *$\|D(f;x)\|^m/K \leq |\det D(f;x)| \leq Kl(D(f;x))^m$, a.e. on G_1.*

We next prove a beautiful quasiconformal extension result due to Pfaltz-graff [Pfa2], which states that if f is holomorphic and quasiregular on B and satisfies a slightly stronger condition on $[Df(z)]^{-1}D^2f(z)(z,\cdot)$ than (8.4.1), then f extends to a quasiconformal homeomorphism of $\mathbb{C}^n$ onto $\mathbb{C}^n$. For related results in one variable, the reader may consult [Bec1-4]. The quasiconformal extension can be given explicitly in terms of an appropriate Loewner chain. Also we shall discuss a quasiconformal extension result of Hamada and Kohr [Ham-Koh12] for quasiregular holomorphic mappings which can be embedded in Loewner chains $f(z,t)$ such that $\{e^{-t}f(z,t)\}_{t\geq 0}$ is a normal family on B. Finally we shall apply this result to the study of quasiconformal extensions of strongly starlike and strongly spirallike mappings of type alpha on the unit ball of $\mathbb{C}^n$.

Before giving the main result of this section, we need the following lemmas. The first result is an n-dimensional version of a well-known theorem of Hardy and Littlewood (see [Gol4, p.411-413]), and will be useful in obtaining a continuous extension to $\overline{B}$ of a locally biholomorphic quasiconformal map on B. A proof may be found in [Pfa2].

Lemma 8.5.3. *Let $c \in [0,1)$, $M > 0$, and $h : B \to \mathbb{C}$ be a holomorphic function such that*

$$(8.5.2) \qquad \left|\frac{\partial h}{\partial z_j}(z)\right| \leq \frac{M}{(1 - \|z\|)^c}, \quad j = 1,\ldots,n, \; z \in B.$$

Then h has a continuous extension to $\overline{B}$, again denoted by h, and there is *a constant $K > 0$ such that*

(8.5.3) $$|h(z) - h(z')| \leq K\|z - z'\|^{1-c}, \quad z, z' \in \overline{B}.$$

An immediate consequence is the following [Pfa2]:

Lemma 8.5.4. *Let $f : B \to \mathbb{C}^n$ be a holomorphic mapping. Also let $M > 0$ and $c \in [0, 1)$ be such that*

(8.5.4) $$\|Df(z)\| \leq \frac{M}{(1 - \|z\|)^c}, \quad z \in B.$$

Then f has a continuous extension to $\overline{B}$, again denoted by f, and there is a constant $K > 0$ such that

$$\|f(z) - f(z')\| \leq K\|z - z'\|^{1-c}, \quad z, z' \in \overline{B}.$$

We next consider a refinement of Theorem 8.4.1. The growth result (8.5.9) below is contained in [Pfa1, Theorem 2.4] and the other statements can be obtained directly from Theorem 8.1.5 and Corollary 8.1.11. (See also the proof of [Pfa1, Theorem 2.4].) We have

Lemma 8.5.5. *Let $f : B \to \mathbb{C}^n$ be a normalized locally biholomorphic mapping and let $c \in [0, 1)$. Assume*

(8.5.5) $$(1 - \|z\|^2)\|[Df(z)]^{-1}D^2f(z)(z, \cdot)\| \leq c, \quad z \in B.$$

Then f is univalent on B (in fact, f has parametric representation on B) and can be embedded as the first element of the Loewner chain

(8.5.6) $$f(z, t) = f(ze^{-t}) + (e^t - e^{-t})Df(ze^{-t})(z), \quad z \in B, \; t \geq 0.$$

Further, let $v = v(z, s, t)$ be the transition mapping associated to $f(z, t)$. Then

(8.5.7) $$\frac{e^t\|v(z, s, t)\|}{(1 - c\|v(z, s, t)\|)^2} \leq \frac{e^s\|z\|}{(1 - c\|z\|)^2}$$

and

(8.5.8) $$\frac{e^t\|v(z, s, t)\|}{(1 + c\|v(z, s, t)\|)^2} \geq \frac{e^s\|z\|}{(1 + c\|z\|)^2},$$

for all $z \in B$ and $t \geq s \geq 0$. Consequently, $\overline{v(B, s, t)} \subset B$ for $t > s \geq 0$ and

(8.5.9) $$\frac{e^s\|z\|}{(1 + c\|z\|)^2} \leq \|f(z, s)\| \leq \frac{e^s\|z\|}{(1 - c\|z\|)^2}, \quad z \in B, \; s \geq 0.$$

Proof. The fact that $f(z,t)$ is a Loewner chain has been proved in Theorem 8.4.1 since $c < 1$. On the other hand, since $\lim\limits_{t\to\infty} e^{-t}f(z,t) = z$ locally uniformly on B, we deduce from Theorem 8.1.5 and Corollary 8.1.11 (see also Remark 8.3.3) that f has parametric representation on B.

Next, let $E(z,t)$ be the linear operator given by (8.4.2), i.e.

$$E(z,t) = -(1 - e^{-2t})[Df(ze^{-t})]^{-1}D^2 f(ze^{-t})(ze^{-t}, \cdot), \quad z \in B, \ t \geq 0.$$

In view of (8.5.5), we deduce that $\|E(z,t)\| \leq c$, $z \in B$, $t \geq 0$, and using Schwarz's lemma for mappings, we obtain that $\|E(z,t)\| \leq c\|z\|$ for $z \in B$ and $t \geq 0$. Further, let $h : B \times [0, \infty) \to \mathbb{C}^n$ be defined by

$$(8.5.10) \quad h(z,t) = [I - E(z,t)]^{-1}[I + E(z,t)](z), \quad z \in B, \ t \geq 0.$$

Since $\|E(z,t)\| \leq c\|z\|$, it follows that $h(\cdot, t) \in H(B)$, $t \geq 0$, and it is easy to see that

$$\|h(z,t) - z\| = \|E(z,t)(h(z,t) + z)\| \leq \|E(z,t)\| \cdot \|h(z,t) + z\|$$

$$\leq c\|z\| \cdot \|h(z,t) + z\|, \quad z \in B, \ t \geq 0.$$

Consequently, we obtain the relations

$$(8.5.11) \qquad \|z\|\frac{1 - c\|z\|}{1 + c\|z\|} \leq \|h(z,t)\| \leq \|z\|\frac{1 + c\|z\|}{1 - c\|z\|}$$

and

$$(8.5.12) \qquad \|z\|^2\frac{1 - c\|z\|}{1 + c\|z\|} \leq \mathrm{Re}\,\langle h(z,t), z\rangle \leq \|z\|^2\frac{1 + c\|z\|}{1 - c\|z\|},$$

for all $z \in B$ and $t \geq 0$.

Indeed, the relation (8.5.11) is a simple application of the inequality

$$\big|\|h(z,t)\| - \|z\|\big| \leq c\|z\|[\|h(z,t)\| + \|z\|], \quad z \in B, \ t \geq 0.$$

The right inequality in (8.5.12) is obvious and in order to deduce the left hand inequality, we use

$$\|h(z,t) - z\|^2 \leq c^2\|z\|^2 \cdot \|h(z,t) + z\|^2, \quad z \in B, \ t \geq 0,$$

to obtain

$$2\mathrm{Re}\,\langle h(z,t),z\rangle(1+c^2\|z\|^2) \geq (1-c^2\|z\|^2)(\|h(z,t)\|^2+\|z\|^2)$$

$$\geq 2\frac{1-c^2\|z\|^2}{(1+c\|z\|)^2}(1+c^2\|z\|^2)\|z\|^2.$$

The claimed relation now follows.

Further, since $v(z,s,t)$ is the transition mapping associated to $f(z,t)$, it follows from (8.1.30) that $v(z,s,t)$ is the solution of the initial value problem

$$\frac{\partial v}{\partial t} = -h(v,t), \quad t\geq s, \quad v(z,s,s) = z,$$

for all $z\in B$ and $s\geq 0$. Using similar reasoning as in the proof of Lemma 8.1.4 and taking into account the relations (8.5.11) and (8.5.12), we obtain (8.5.7) and (8.5.8). Finally letting $t\to\infty$ in (8.5.7) and (8.5.8), and using the fact that $f(z,s) = \lim\limits_{t\to\infty} e^t v(z,s,t)$ locally uniformly on B, we conclude (8.5.9).

It remains to prove that

$$\overline{v(B,s,t)} \subset B, \quad t>s\geq 0.$$

For this purpose, we use the equality

$$\|v(z,s,t)\|^{-1}\frac{\partial}{\partial t}\|v(z,s,t)\| = -\mathrm{Re}\,\langle h(v(z,s,t),t),v(z,s,t)\rangle, t>s,$$

and in view of (8.5.12) we obtain for $s<t$ that

$$\frac{\partial}{\partial t}\|v(z,s,t)\| \leq -\|v(z,s,t)\|\frac{1-c\|v(z,s,t)\|}{1+c\|v(z,s,t)\|} \leq -\frac{1-c}{1+c}.$$

Integrating this inequality with respect to t and using the fact that $v(z,s,s) = z$, we obtain

$$\|v(z,s,t)\| \leq \|z\|\exp\left\{-\frac{1-c}{1+c}(t-s)\right\}, 0\leq s<t, \|z\|<1.$$

Thus $\overline{v(B,s,t)} \subset B$, as desired. This completes the proof.

The following result of Pfaltzgraff [Pfa2] shows that an estimate on $[Df(z)]^{-1}D^2f(z,\cdot)$ combined with a quasiregularity assumption allows one to extend a holomorphic map continuously from B to $\overline{B}$.

Lemma 8.5.6. *Let $f : B \to \mathbb{C}^n$ be a holomorphic mapping which is qua-siregular on B, and suppose there is a constant $c < 1$ such that*

$$(8.5.13) \qquad (1 - \|z\|^2)\|[Df(z)]^{-1}D^2f(z)(z,\cdot)\| \le 2c, \quad z \in B.$$

Then there is a constant $M > 0$ such that

$$(8.5.14) \qquad \|Df(z)\| \le \frac{M}{(1 - \|z\|)^c}, \quad z \in B,$$

and f has a Hölder continuous extension to $\overline{B}$ (again denoted by f) which satisfies the relation

$$(8.5.15) \qquad \|f(z) - f(z')\| \le K\|z - z'\|^{1-c}, \quad z, z' \in \overline{B},$$

for some constant $K > 0$.

Proof. Taking into account Lemma 8.5.4, it suffices to prove the relation (8.5.14). For this purpose, fix $u \in \overline{B}$ and let $\psi(\zeta) = \det Df(u\zeta) = J_f(u\zeta)$, $|\zeta| < 1$. Then ψ is a nonvanishing holomorphic function on the unit disc U. Considering the trace formula for differentiating the determinant of an $n \times n$ matrix-valued holomorphic function (see [Golb] and also Section 10.2.1), one deduces that

$$(8.5.16) \qquad \zeta\psi'(\zeta) = \psi(\zeta)\mathrm{trace}\{[Df(u\zeta)]^{-1}D^2f(u\zeta)(u\zeta,\cdot)\}, \ |\zeta| < 1.$$

Further, if $A = (a_{ij})_{1 \le i,j \le n} \in L(\mathbb{C}^n, \mathbb{C}^n)$ and $0 \le \alpha_1 \le \alpha_2 \le \ldots \le \alpha_n$ are the eigenvalues of A^*A, then $\|A\|^2 = \alpha_n$. Moreover, since

$$\mathrm{trace}\{A^*A\} = \sum_{j,k=1}^{n} |a_{jk}|^2 = \alpha_1 + \ldots + \alpha_n \le n\alpha_n = n\|A\|^2,$$

we obtain

$$(8.5.17) \qquad |\mathrm{trace}\{A\}| = \left|\sum_{j=1}^{n} a_{jj}\right| \le \left[n\sum_{j=1}^{n} |a_{jj}|^2\right]^{1/2} \le n\|A\|.$$

Hence, from (8.5.13), (8.5.16) and (8.5.17) we obtain

$$\left|\zeta\frac{d}{d\zeta}\log\psi(\zeta)\right| \le \frac{2nc}{1 - \|u\zeta\|^2} \le \frac{2nc}{1 - |\zeta|^2}, \quad |\zeta| < 1.$$

Elementary computations using the above inequality yield that

$$|\psi(\zeta)| = |J_f(u\zeta)| = O\left(\frac{1}{(1 - |\zeta|)^{nc}}\right), \quad |\zeta| < 1.$$

Next, setting $u = z/\|z\|$ and $\zeta = \|z\|$, $z \in B \setminus \{0\}$, in the above relation, we deduce that there is a constant $M_1 > 0$ such that

(8.5.18) $$|J_f(z)| \leq \frac{M_1}{(1 - \|z\|)^{nc}}, \quad z \in B.$$

Finally, using the quasiregularity of f and the relation (8.5.18), we conclude that there is a constant $M_2 > 0$ such that

$$\|Df(z)\|^n \leq M_2|J_f(z)| \leq \frac{M_1 M_2}{(1 - \|z\|)^{nc}}, \quad z \in B.$$

Letting $M = [M_1 M_2]^{1/n}$ in the above relation, we obtain (8.5.14), as desired. This completes the proof.

We are now able to prove the main result of this section, due to Pfaltzgraff [Pfa2]. (For the case of one variable see [Bec1].)

Theorem 8.5.7. *Let $f : B \to \mathbb{C}^n$ be a normalized holomorphic mapping which is quasiregular on B, and let $c \in [0, 1)$. Assume f satisfies the relation (8.5.5). Then f can be extended to a quasiconformal homeomorphism of $\mathbb{R}^{2n}$ onto $\mathbb{R}^{2n}$. The extension F is given by*

(8.5.19) $$F(z) = \begin{cases} f(z, 0), & \|z\| \leq 1 \\ f\left(\frac{z}{\|z\|}, \log \|z\|\right), & \|z\| > 1, \end{cases}$$

where $f(z, t)$ is the Loewner chain given by (8.5.6).

Proof. We divide the proof into the following steps:

Step 1. First we show that if $f(z, t)$ is the Loewner chain given by (8.5.6), then $f_t(z) = f(z, t)$ can be extended to a continuous and injective mapping on $\overline{B}$ (again denoted by $f_t(z)$) for each $t \geq 0$, such that

(8.5.20) $$f(\overline{B}, s) \subseteq f(B, t), \quad 0 \leq s < t.$$

Indeed, it is obvious that $f(\cdot, t)$ is holomorphic on $\overline{B}$ for $t > 0$. Moreover, in view of the relation (8.5.5) and Lemma 8.5.6, we deduce that $f(z) = f(z, 0)$

has a continuous extension to $\overline{B}$ (again denoted by $f(z)$) which satisfies a
Hölder condition

$$\|f(z) - f(z')\| \leq A\|z - z'\|^{1-c}, \quad z, z' \in \overline{B},$$

for some absolute constant $A > 0$.

Moreover, if $v(z, s, t)$ is the transition mapping associated to $f(z, t)$, then
using the inclusion $\overline{v(B, s, t)} \subset B$ for $t > s \geq 0$, which follows from Lemma
8.5.5, the equality $f(z, s) = f(v(z, s, t), t)$ and the continuity of $f(\cdot, t)$ on $\overline{B}$,
$t > 0$, we deduce that

$$\overline{f(v(B, s, t), t)} = f(\overline{v(B, s, t)}, t) \subseteq f(B, t), \quad t > s \geq 0,$$

which yields (8.5.20). Further, the relation (8.5.20) and the continuity of $f_s(\cdot)$
on $\overline{B}$ enable us to extend $v(z, s, t)$ continuously in the first variable to $\overline{B}$ by
the formula

$$(8.5.21) \qquad v(z, s, t) = f_t^{-1}(f_s(z)), \ z \in \overline{B}, t > s \geq 0.$$

For $z \in B$ and $0 \leq s < t$, we obtain in view of (8.5.11) that

$$\|z - v(z, s, t)\| = \left\| \int_s^t h(v(z, s, \tau), \tau) d\tau \right\|$$

$$\leq \int_s^t \|v(\tau)\| \frac{1 + c\|v(\tau)\|}{1 - c\|v(\tau)\|} d\tau \leq \frac{1 + c}{1 - c}(t - s).$$

Using the continuity of $v(\cdot, s, t)$ on $\overline{B}$, we obtain from the above inequality
that

$$(8.5.22) \qquad \|z - v(z, s, t)\| \leq \frac{1 + c}{1 - c}(t - s), \quad \|z\| \leq 1, t > s \geq 0.$$

Now we show the injectivity of $f_s(\cdot)$ on $\overline{B}$. For this purpose, let $z, z' \in \overline{B}$
be such that $f_s(z) = f_s(z')$. Then $f_t(v(z, s, t)) = f_t(v(z', s, t))$ for $t > s \geq 0$,
and since $v(\overline{B}, s, t) \subset B$, $t > s$, and $f_t(\cdot)$ is univalent on B, we deduce that
$v(z, s, t) = v(z', s, t)$. Letting $t \searrow s$, we conclude that $z = z'$, as claimed.

Moreover, using again the fact that $f_s(\overline{B}) \subset f_t(B)$ for $t > s \geq 0$ and the
injectivity of $f(\cdot, t)$ on $\overline{B}$, it is easy to see that the mapping F given by (8.5.19)
is univalent on $\mathbb{C}^n$.

Step 2. We next prove the continuity of F in $\mathbb{C}^n$. For this purpose, we prove the following Hölder conditions

$$(8.5.23) \quad e^{-t}\|f(z,t) - f(z',t)\| \leq A_1\|z - z'\|^{1-c}, \quad z, z' \in \overline{B}, \; t \geq 0,$$

$$(8.5.24) \quad \|f(z,t) - f(z,s)\| \leq A_2 e^t (t-s)^{1-c}, \quad z \in \overline{B}, \; 0 \leq s < t,$$

where A_1 and A_2 are constants which are independent of s and t.

In order to show (8.5.23), we use the equality (see the proof of Theorem 8.4.1)

$$e^{-t}Df(z,t) = Df(ze^{-t})[I - E(z,t)], \quad z \in B, \; t \geq 0,$$

the fact that $\|E(z,t)\| \leq c\|z\|$, $z \in B$, $t \geq 0$, and Lemma 8.5.6, to deduce that there exists a constant $M_1 > 0$ (independent of t) such that

$$(8.5.25) \qquad \|e^{-t}Df(z,t)\| = \|Df(ze^{-t})[I - E(z,t)]\|$$

$$\leq \|Df(ze^{-t})\|(1+c) \leq \frac{M_1(1+c)}{(1-\|z\|)^c}, \quad z \in B, \; t \geq 0.$$

In view of Lemma 8.5.6, we conclude that $e^{-t}f(\cdot,t)$ has a continuous extension to $\overline{B}$ which satisfies (8.5.23) for some absolute constant $A_1 > 0$.

Furthermore, using the relations (8.5.21), (8.5.22) and (8.5.23), we obtain for $z \in \overline{B}$ and $0 \leq s < t$ that

$$\|f(z,t) - f(z,s)\| = \|f(z,t) - f(v(z,s,t),t)\|$$

$$\leq A_1 e^t \|z - v(z,s,t)\|^{1-c} \leq A_1 e^t \left(\frac{1+c}{1-c}\right)^{1-c}(t-s)^{1-c}.$$

Setting $A_2 = A_1 \left(\dfrac{1+c}{1-c}\right)^{1-c}$ gives (8.5.24).

Note that the relations (8.5.23) and (8.5.24) obviously imply the continuity of $f(\cdot,t)$ on $\overline{B}$, and therefore in view of the definition of F, we deduce that F is continuous on $\mathbb{C}^n$.

Step 3. We next prove that F is a homeomorphism of $\mathbb{R}^{2n}$ onto $\mathbb{R}^{2n}$. Taking into account the left-hand side of (8.5.9), we deduce that

$$\|F(z)\| = \left\|f\left(\frac{z}{\|z\|}, \log\|z\|\right)\right\| \to \infty \text{ as } \|z\| \to \infty.$$

Since F is continuous on $\mathbb{R}^{2n}$, injective on $\mathbb{R}^{2n}$ and $\|F(z)\| \to \infty$ as $\|z\| \to \infty$, we obtain that F is also a surjective map on $\mathbb{R}^{2n}$. Let $S^{2n} = \mathbb{R}^{2n} \cup \{\infty\}$ be a one point compactification of $\mathbb{R}^{2n}$. We extend F to S^{2n} by $F(\infty) = \infty$. Then F is a continuous bijective map from S^{2n} onto S^{2n} and since S^{2n} is compact, we deduce that F is a homeomorphism from S^{2n} onto itself. Therefore F is also a homeomorphism of $\mathbb{R}^{2n}$ onto itself, as claimed.

Step 4. We next prove that F is quasiconformal in $\mathbb{R}^{2n}$. For this purpose, we use a standard dilation argument (cf. [Bec1, p.33-34], [Pfa2, p.21-24].)

For $r > 1$, let $f_r(z,t) = rf(z/r,t)$, $h_r(z,t) = rh(z/r,t)$ ($\|z\| < r$, $t \geq 0$) and

$$F_r(z) = \begin{cases} f_r(z,0), & \|z\| \leq 1 \\ f_r\left(\dfrac{z}{\|z\|}, \log\|z\|\right), & \|z\| \geq 1. \end{cases}$$

It is obvious that $f_r(z,t)$ satisfies the differential equation

$$(8.5.26) \qquad \frac{\partial f_r}{\partial t}(z,t) = Df_r(z,t)h_r(z,t), \quad t \geq 0, \; \|z\| < r.$$

In particular the above equation is satisfied for all $t \geq 0$ and $\|z\| \leq 1$. On the other hand, in view of (8.5.23) and the right-hand side inequality in (8.5.9), we obtain for $r > 1$, $0 \leq t \leq T$, $T > 0$, and $z \in \overline{B}$ that

$$\|f_r(z,t) - f(z,t)\| = \|rf(z/r,t) - f(z,t)\|$$

$$\leq \|rf(z/r,t) - f(z/r,t)\| + \|f(z/r,t) - f(z,t)\|$$

$$\leq (r-1)\|f(z/r,t)\| + A_1 e^t \|(z/r) - z\|^{1-c}$$

$$\leq (r-1)e^t \frac{\|z/r\|}{(1-c\|z/r\|)^2} + A_1 e^t \|z/r\|^{1-c}(r-1)^{1-c}$$

$$\leq \frac{r-1}{r}e^T \frac{1}{(1-(c/r))^2} + A_1 e^T \left(\frac{r-1}{r}\right)^{1-c}.$$

Consequently, $\lim\limits_{r \searrow 1} f_r(z,t) = f(z,t)$ uniformly on $\overline{B} \times [0,T]$, and therefore $F_r \to F$ uniformly on compact subsets of $\mathbb{R}^{2n}$ as r decreases to 1.

We next prove that F_r is absolutely continuous on lines, differentiable a.e. and has outer dilatation bounded a.e. by a bound which is independent of r. Since $F_r \to F$ uniformly on compact subsets of $\mathbb{R}^{2n}$ as $r \searrow 1$ and $F|_B = f$

is nonconstant, we will conclude in view of [Väi, Theorem 21.7 and Corollary 37.4] that F is a quasiconformal homeomorphism of $\mathbb{R}^{2n}$ onto $\mathbb{R}^{2n}$.

Step 4.1. We show that F_r is ACL. For this purpose, we prove that F_r satisfies a Lipschitz condition on $\mathbb{C}^n$.

Since

$$\|e^{-t}Df_r(z,t)\| = \|e^{-t}Df(z/r,t)\| = \|Df(e^{-t}z/r)[I - E(e^{-t}z/r,t)]\|$$

$$\leq \|Df(e^{-t}z/r)\|(1+c) \leq \frac{M_1}{(1-\|e^{-t}z/r\|)^c}(1+c) \leq \frac{M_1}{(1-1/r)^c}(1+c),$$

for $z \in \overline{B}$, $t \geq 0$, by (8.5.25), we obtain

$$(8.5.27) \quad e^{-t}\|f_r(z,t) - f_r(w,t)\| \leq \frac{M_1(1+c)}{(1-1/r)^c}\|z-w\| = M_2(r)\|z-w\|,$$

for $z, w \in \overline{B}$ and $t \geq 0$, where $M_2(r)$ is independent of t.

Moreover, using the fact that $f(z,s) = f(v(z,s,t),t)$, $z \in B$, $t \geq s \geq 0$, and the relations (8.5.22) and (8.5.27), we obtain

$$\|f_r(z,t) - f_r(z,s)\| = r\|f(z/r,t) - f(z/r,s)\|$$

$$\leq e^t r M_2(r)\left\|\frac{z}{r} - v\left(\frac{z}{r},s,t\right)\right\|$$

$$\leq e^t r M_2(r)\frac{1+c}{1-c}(t-s) = e^t M_3(r)(t-s), \quad z \in \overline{B}, 0 \leq s < t.$$

Therefore, we have proved that

$$(8.5.28) \quad \|f_r(z,t) - f_r(z,s)\| \leq e^t M_3(r)(t-s), \quad z \in \overline{B}, 0 \leq s < t.$$

Combining the relations (8.5.27) and (8.5.28), and using the definition of F_r, it is not difficult to prove that F_r satisfies a local Lipschitz condition in $\mathbb{C}^n$. Hence F_r is ACL in $\mathbb{R}^{2n}$ and in view of a theorem of Rademacher and Stepanov (see e.g. [Väi, Theorem 29.1]), we deduce that F_r is (real) differentiable a.e. in $\mathbb{R}^{2n}$.

Step 4.2. It remains to show that F_r has outer dilatation bounded a.e. by a bound independent of r. For simplicity, we shall omit the subscript r and set $G(z) = F_r(z)$ for fixed $r > 1$. Since $F_r(z) = f_r(z,0)$ for $z \in B$, and f is quasiregular on B, it suffices to assume $\|z\| \geq 1$.

Let $z = (x, y) = (x_1, y_1, \ldots, x_n, y_n)$, $\|z\| \geq 1$, be a point where the mapping $G = (V, W)$ given by

(8.5.29) $\qquad G : (x_1, y_1, \ldots, x_n, y_n) \mapsto (V_1, W_1, \ldots, V_n, W_n)$,

$$V_k = V_k(x_1, y_1, \ldots, x_n, y_n) = \operatorname{Re} G_k(x, y),$$

$$W_k = W_k(x_1, y_1, \ldots, x_n, y_n) = \operatorname{Im} G_k(x, y),$$

$k = 1, \ldots, n$, is differentiable. Also let $\zeta = z/(r\|z\|) = (\zeta_1, \ldots, \zeta_n)$, $\zeta_k = \xi_k + i\eta_k$, $u_k(\zeta, t) = \operatorname{Re} f_k(\zeta, t)$, $v_k(\zeta, t) = \operatorname{Im} f_k(\zeta, t)$, $V_k(x, y) = ru_k(\xi, \eta, t)$, $W_k(x, y) = rv_k(\xi, \eta, t)$ with $t = \log \|z\|$. After some straightforward computations, based on the chain rule, we obtain

$$D(V, W; x, y) = r \left[D(u, v; \xi, \eta) D(\xi, \eta; x, y) + \frac{\partial}{\partial t} \begin{pmatrix} u(\zeta, t) \\ v(\zeta, t) \end{pmatrix} \operatorname{grad} \, \log \|z\| \right]$$

$$= rD(u, v; \xi, \eta) \left[D(\xi, \eta; x, y) + \begin{pmatrix} \operatorname{Re} h(\zeta, t) \\ \operatorname{Im} h(\zeta, t) \end{pmatrix} \operatorname{grad} \, \log \|z\| \right].$$

Here we have used the relation (8.5.26) and the Cauchy-Riemann equations. (These relations hold since $\|\zeta\| < 1$.) Other elementary computations using the above relation give

(8.5.30) $\qquad\qquad\qquad D(V, W; x, y)$

$$= \frac{1}{\|z\|} D(u, v; \xi, \eta) \left[I + r^2 \begin{pmatrix} \operatorname{Re} [h(\zeta, t) - \zeta] \\ \operatorname{Im} [h(\zeta, t) - \zeta] \end{pmatrix} (\xi, \eta) \right],$$

where I is the identity transformation. Next, let

$$C = r^2 \begin{pmatrix} \operatorname{Re} [h(\zeta, t) - \zeta] \\ \operatorname{Im} [h(\zeta, t) - \zeta] \end{pmatrix} (\xi, \eta).$$

It is not difficult to show that C has proportional columns, and thus C has rank 1. Hence

$$\det(I + C) = 1 + \operatorname{trace}\{C\}$$

$$= 1 + r^2 \left[\sum_{j=1}^{n} (\operatorname{Re} h_j(\zeta, t) - \xi_j)\xi_j + \sum_{j=1}^{n} (\operatorname{Im} h_j(\zeta, t) - \eta_j)\eta_j \right]$$

$$= 1 + r^2 \mathrm{Re} \, \langle h(\zeta,t) - \zeta, \zeta \rangle = r^2 \mathrm{Re} \, \langle h(\zeta,t), \zeta \rangle$$

$$\geq r^2 \|\zeta\|^2 \frac{1 - c\|\zeta\|}{1 + c\|\zeta\|} \geq \frac{1 - c}{1 + c}.$$

Note that for the above inequalities we have used the left side of (8.5.12). Consequently, in view of (8.5.30) and the above relation, we obtain

$$(8.5.31) \qquad |\det D(V, W; x, y)| = \left| \det \left\{ \frac{1}{\|z\|} D(u, v; \xi, \eta)[I + C] \right\} \right|$$

$$\geq \frac{1}{\|z\|^{2n}} |\det D(u, v; \xi, \eta)| \left(\frac{1 - c}{1 + c} \right) = \frac{1}{\|z\|^{2n}} \left(\frac{1 - c}{1 + c} \right) |\det Df(\zeta,t)|^2.$$

For the last equality we have used the fact that $f(\cdot, t)$ is holomorphic on B, and thus $|J_f(\zeta,t)|^2 = \det D(u, v; \xi, \eta)$.

On the other hand, from (8.5.30) we obtain

$$(8.5.32) \qquad \|D(V, W; x, y)\| \leq \frac{1}{\|z\|} \|D(u, v; \xi, \eta)\| \cdot \|I + C\|$$

$$= \frac{1}{\|z\|} \|Df(\zeta,t)\| \cdot \|I + C\|,$$

where we have used the equality $\|D(u, v; \xi, \eta)\| = \|Df(\zeta,t)\|$. (This equality follows if we take into account the Cauchy-Riemann equations and the holomorphy of $f(\cdot, t)$ on B.)

Next, applying the Schwarz Lemma to the map h, we obtain

$$\|I + C\| \leq 1 + \|C\| \leq 1 + r^2 \|h(\zeta,t) - \zeta\| \|\zeta\| = 1 + r\|h(\zeta,t) - \zeta\|.$$

Moreover, since

$$h(\zeta,t) - \zeta = 2[I - E(\zeta,t)]^{-1} E(\zeta,t)(\zeta)$$

and

$$1 - c \leq \|I - E(\zeta,t)\| \leq 1 + c,$$

we deduce that

$$\|[I - E(\zeta,t)]^{-1}\| \leq [1 - \|E(\zeta,t)\|]^{-1} \leq \frac{1}{1 - c},$$

and thus

$$\|h(\zeta,t) - \zeta\| \leq \frac{2c\|\zeta\|}{1-c} = \frac{2c}{r(1-c)}.$$

Therefore,

$$\|I + C\| \leq 1 + \frac{2c}{1-c} = \frac{1+c}{1-c}.$$

Finally taking into account the above inequality, the relations (8.5.31) and (8.5.32), the quasiregularity of f, and the inequality

$$\frac{\|I - E(\zeta,t)\|^{2n}}{|\det[I - E(\zeta,t)]|^2} \leq \left(\frac{1+c}{1-c}\right)^{2n-2},$$

we obtain

$$\|D(V,W;x,y)\|^{2n} \leq \frac{1}{\|z\|^{2n}}\|Df(\zeta,t)\|^{2n} \cdot \|I + C\|^{2n}$$

$$= \frac{1}{\|z\|^{2n}}e^{2nt}\|Df(\zeta e^{-t})[I - E(\zeta,t)]\|^{2n} \cdot \|I + C\|^{2n}$$

$$\leq \frac{1}{\|z\|^{2n}}e^{2nt}K^2|J_f(\zeta e^{-t})|^2 \cdot \|I - E(\zeta,t)\|^{2n} \cdot \left(\frac{1+c}{1-c}\right)^{2n}$$

$$= \frac{1}{\|z\|^{2n}}K^2\frac{|\det Df(\zeta,t)|^2}{|\det[I - E(\zeta,t)]|^2} \cdot \|I - E(\zeta,t)\|^{2n} \cdot \left(\frac{1+c}{1-c}\right)^{2n}$$

$$\leq \frac{1}{\|z\|^{2n}}K^2|\det Df(\zeta,t)|^2 \cdot \left(\frac{1+c}{1-c}\right)^{4n-2}$$

$$\leq K^2|\det D(V,W;x,y)| \cdot \left(\frac{1+c}{1-c}\right)^{4n-1}.$$

Hence we have proved that F_r has outer dilatation bounded a.e. by a bound independent of r. This completes the proof.

Hamada and Kohr [Ham-Koh12] have recently obtained the following sufficient condition for quasiconformal extension to $\mathbb{C}^n$ of the first element of a g-Loewner chain with $g(\zeta) = (1+\zeta)/(1-\zeta)$, $\zeta \in U$. To this end, we recall that a mapping $f(z,t)$ is a g-Loewner chain with $g(\zeta) = (1+\zeta)/(1-\zeta)$ if and only if $f(z,t)$ is a Loewner chain such that $\{e^{-t}f(z,t)\}_{t\geq 0}$ is a normal family on B. In this case there is a mapping

$$(8.5.33) \qquad h : B \times [0,\infty) \to \mathbb{C}^n$$

such that $h(\cdot, t) \in \mathcal{M}$ for $t \geq 0$, $h(z, \cdot)$ is measurable on $[0, \infty)$ for $z \in B$, and

$$\frac{\partial f}{\partial t}(z, t) = Df(z, t)h(z, t), \quad a.e. \quad t \geq 0, \quad \forall z \in B.$$

We leave the proof of Theorem 8.5.8 for the reader and mention that the idea of the proof is to use similar arguments as in the proof of Theorem 8.5.7. (See Problem 8.5.1.)

Theorem 8.5.8. *Let $f(z, t)$ be a g-Loewner chain with $g(\zeta) = (1+\zeta)/(1-\zeta)$, $\zeta \in U$. Assume the following conditions hold:*

(i) There exist constants $M \geq 1$ and $c \in [0, 1)$ such that

$$\|Df(z, t)\| \leq \frac{Me^t}{(1 - \|z\|)^c}, \quad z \in B, \ t \geq 0;$$

(ii) There exists a constant $c_1 > 0$ such that

$$c_1 \|z\|^2 \leq \text{Re} \ \langle h(z, t), z \rangle, \quad z \in B, \ t \geq 0,$$

where $h(z, t)$ is given by (8.5.33).

(iii) There exists a constant $c_2 > 0$ such that

$$\|h(z, t)\| \leq c_2, \ z \in B, \ t \geq 0;$$

(iv) $f(\cdot, t)$ is quasiregular on B for each $t \geq 0$, and the outer dilatation is bounded independently of t.

Then for each $t \geq 0$, $f(\cdot, t)$ has a continuous and injective extension to $\overline{B}$, again denoted by $f(\cdot, t)$, and $f(z) = f(z, 0)$ extends to a quasiconformal homeomorphism of $\mathbb{R}^{2n}$ onto $\mathbb{R}^{2n}$. The extension F is given by

$$F(z) = \begin{cases} f(z, 0), & \|z\| \leq 1 \\ f\left(\dfrac{z}{\|z\|}, \log \|z\|\right), & \|z\| > 1. \end{cases}$$

Remark 8.5.9. (i) Let $f(z, t)$ be the chain given by (8.5.6) and $f(z) = f(z, 0)$ be a normalized quasiregular holomorphic mapping on B, which satisfies the assumption (8.5.5). Also let $h(z, t)$ be the mapping given by (8.5.10). Then $f(z, t)$ and $h(z, t)$ satisfy the assumptions of Theorem 8.5.8, and hence

f can be extended to a quasiconformal homeomorphism of $\mathbb{R}^{2n}$ onto $\mathbb{R}^{2n}$. We leave the proof for the reader.

(ii) A large class of mappings $h = h(z,t)$ which satisfy the conditions (ii) and (iii) in Theorem 8.5.8 is given by $h(z,t) = [I - E(z,t)]^{-1}[I + E(z,t)](z)$, $z \in B$, $t \geq 0$, where $E(z,t) \in L(\mathbb{C}^n, \mathbb{C}^n)$ for $z \in B$ and $t \geq 0$, $E(\cdot,t)$ is holomorphic on B, $E(0,t) = 0$, $t \geq 0$, and $\|E(z,t)\| \leq c < 1$, $z \in B$, $t \geq 0$.

(iii) Another class of mappings which satisfy the conditions in Theorem 8.5.8 was considered in [Cu3].

Brodskii [Brod] proved the following simple sufficient condition for quasiconformal extension of normalized holomorphic mappings on the unit ball of $\mathbb{C}^n$. We now give the proof of this result by applying Theorem 8.5.8 (see also [Cu3], [Ham-Koh12]). Another application of Theorem 8.5.8, which generalizes both Theorems 8.5.7 and 8.5.10, is given in Problem 8.5.4.

Theorem 8.5.10. *If $f : B \to \mathbb{C}^n$ is a normalized holomorphic mapping such that*

$$\sup_{z \in B} \|Df(z) - I\| < 1, \quad z \in B,$$

then f is univalent and quasiregular on B and extends to a quasiconformal homeomorphism of $\mathbb{R}^{2n}$ onto $\mathbb{R}^{2n}$.

Proof. We imbed f as the first element of the chain

$$f(z,t) = f(ze^{-t}) + (e^t - e^{-t})z = e^t z + \dots, \quad z \in B, t \geq 0,$$

and we shall show that $f(z,t)$ satisfies the conditions in Theorem 8.5.8.

First, we show that $f(z,t)$ is a Loewner chain. (See also Problem 8.4.3.) It is not difficult to see that

$$Df(z,t) = e^t[I - E(z,t)], \quad z \in B, t \geq 0,$$

and

$$\frac{\partial f}{\partial t}(z,t) = e^t[I + E(z,t)](z), \quad z \in B, t \geq 0,$$

where

$$E(z,t) = e^{-2t}[I - Df(ze^{-t})], \quad z \in B, t \geq 0.$$

Using the hypothesis, we deduce that there is a constant $c \in [0,1)$ such that

$$\|Df(z) - I\| \leq c, \quad z \in B,$$

and hence $\|E(z,t)\| \le c\|z\|$, $z \in B$, $t \ge 0$. This implies that $I - E(z,t)$ is an invertible operator, and if $h(z,t) = [I - E(z,t)]^{-1}[I + E(z,t)](z)$ then

$$\frac{\partial f}{\partial t}(z,t) = Df(z,t)h(z,t), \quad z \in B, t \ge 0.$$

Clearly $h(\cdot,t)$ is well defined and is holomorphic on B for each $t \ge 0$. Also $h(z,\cdot)$ is measurable on $[0,\infty)$ for each $z \in B$. Since $\|E(z,t)\| \le c\|z\|$, we deduce as in the proof of Theorem 8.4.1 that $\mathrm{Re}\,\langle h(z,t), z \rangle > 0$ for $z \in B \setminus \{0\}$ and $t \ge 0$, and hence $h(\cdot,t) \in \mathcal{M}$, $t \ge 0$. Using Remark 8.5.9 (ii), we deduce that $h(z,t)$ satisfies the conditions (ii) and (iii) in Theorem 8.5.8.

Moreover, since $\lim\limits_{t\to\infty} e^{-t}f(z,t) = z$ locally uniformly on B, we deduce that $f(z,t)$ is a Loewner chain such that $\{e^{-t}f(z,t)\}_{t\ge 0}$ is a normal family on B.

On the other hand, since $Df(z,t) = e^t[I - E(z,t)]$, we obtain

$$\|Df(z,t)\| \le e^t(1 + c), \; z \in B, \, t \ge 0,$$

so the condition (i) from Theorem 8.5.8 holds.

It remains to show that $f(\cdot,t)$ satisfies the condition (iv) in Theorem 8.5.8 for $t \ge 0$. Indeed, we have

$$\|Df(z,t)\|^n = e^{nt}\|I - E(z,t)\|^n \le e^{nt}(1 + c)^n$$

$$= (1 + c)^n \frac{|\det Df(z,t)|}{|\det[I - E(z,t)]|} \le \left[\frac{1+c}{1-c}\right]^n |\det Df(z,t)|,$$

using the fact that $|\det[I - E(z,t)]| \ge (1 - c)^n$. This completes the proof.

8.5.2 Strongly starlike and strongly spirallike mappings of type α on the unit ball of $\mathbb{C}^n$

Chuaqui [Chu2] proved the following quasiconformal extension result for a quasiregular strongly starlike mapping on B (see Definition 8.3.22). We shall give a proof of this result based on the application of Theorem 8.5.8 (see [Ham-Koh12]). Hamada [Ham4] extended this result to the case of strongly starlike mappings on bounded balanced pseudoconvex domains in $\mathbb{C}^n$ for which the Minkowski function is of class C^1 in $\mathbb{C}^n \setminus \{0\}$. Also Hamada and Kohr [Ham-Koh14] have recently extended Chuaqui's result to the case of strongly starlike mappings on the unit ball of $\mathbb{C}^n$ with an arbitrary norm.

Theorem 8.5.11. *Let $f : B \to \mathbb{C}^n$ be a quasiconformal strongly starlike mapping such that $\|[Df(z)]^{-1}f(z)\|$ is uniformly bounded on B. Then f can be extended to a quasiconformal homeomorphism of $\mathbb{R}^{2n}$ onto $\mathbb{R}^{2n}$.*

Proof. Let $f(z,t) = e^t f(z)$, $z \in B$, $t \geq 0$. Since f is strongly starlike, $f(z,t)$ is a Loewner chain. Of course $\lim\limits_{t\to\infty} e^{-t} f(z,t) = f(z)$ locally uniformly on B and

$$\frac{\partial f}{\partial t}(z,t) = Df(z,t)h(z,t), \quad z \in B, \ t \geq 0,$$

where $h(z,t) = w(z)$ and $w(z) = [Df(z)]^{-1}f(z)$. Also in view of Definition 8.3.22 and the remarks following this definition, there exists $c \in (0,1)$ such that $f \in S_c^0(B)$ and

$$\left| \frac{1}{\|z\|^2} \langle w(z), z \rangle - \frac{1+c^2}{1-c^2} \right| \leq \frac{2c}{1-c^2}, \quad z \in B \setminus \{0\}.$$

Then

$$\|z\|^2 \frac{1-c\|z\|}{1+c\|z\|} \leq \operatorname{Re} \langle w(z), z \rangle \leq \|z\|^2 \frac{1+c\|z\|}{1-c\|z\|}, \quad z \in B,$$

and thus the condition (ii) from Theorem 8.5.8 holds. Also from the above relation, we obtain

$$(8.5.34) \qquad \|z\| \frac{1-c\|z\|}{1+c\|z\|} \leq \|w(z)\|, \quad z \in B.$$

Moreover, from Theorem 8.3.23 and the fact that $f \in S_c^0(B)$, we deduce that

$$(8.5.35) \qquad \frac{\|z\|}{(1+c\|z\|)^2} \leq \|f(z)\| \leq \frac{\|z\|}{(1-c\|z\|)^2}, \quad z \in B,$$

and since $Df(z)w(z) = f(z)$, we obtain from (8.5.34) and (8.5.35) that

$$(8.5.36) \qquad \left\| Df(z) \left(\frac{w(z)}{\|w(z)\|} \right) \right\| = \frac{\|f(z)\|}{\|w(z)\|} \leq \frac{1+c\|z\|}{(1-c\|z\|)^3} \leq \frac{1+c}{(1-c)^3}.$$

Since f is quasiregular, there exists an absolute constant $K \geq 1$ such that

$$(8.5.37) \qquad \|Df(z)\|^n \leq K |J_f(z)|, \quad z \in B.$$

Combining this inequality with (8.5.36), we conclude that $\|Df(z)\|$ is uniformly bounded on B and

$$(8.5.38) \qquad \|Df(z)\| = \sup_{\|u\|=1} \|Df(z)u\| \leq K \frac{1+c}{(1-c)^3}, \quad z \in B.$$

Indeed, since $[Df(z)]^*Df(z)$ is a positive semi-definite matrix, its eigenvalues are real and nonnegative. Let $0 \leq \alpha_1 \leq \ldots \leq \alpha_n$ be these eigenvalues. There is a unitary matrix V such that $W = V^*[Df(z)]^*Df(z)V$ is a diagonal matrix with diagonal components $\alpha_1, \ldots, \alpha_n$. Let $u \in \mathbb{C}^n$ be a unit vector. Then

$$\|Df(z)u\|^2 = \langle Df(z)u, Df(z)u \rangle = \langle WV^*u, V^*u \rangle,$$

and since $v = V^*u$ is also a unit vector, we easily deduce from the above relation that $\|Df(z)u\|^2 \geq \alpha_1$. Thus if $u = w(z)/\|w(z)\|$, $z \in B \setminus \{0\}$, we obtain from (8.5.36) that $\alpha_1 \leq \left[\dfrac{1+c}{(1-c)^3}\right]^2$. Further, in view of (8.5.37) we obtain

$$\alpha_n^n = \|Df(z)\|^{2n} \leq K^2 |\det Df(z)|^2 = K^2 \alpha_1 \cdots \alpha_n.$$

Therefore $\alpha_n \leq K^2 \alpha_1$ and the inequality (8.5.38) now follows. Moreover, the inequality (8.5.38) gives

$$\|Df(z,t)\| = e^t \|Df(z)\| \leq Ke^t \frac{1+c}{(1-c)^3},$$

and hence the condition (i) in Theorem 8.5.8 occurs. Also since f is quasiregular, we obtain

$$\|Df(z,t)\|^n = e^{nt}\|Df(z)\|^n \leq Ke^{nt}|J_f(z)| = K|J_f(z,t)|, \quad z \in B, \ t \geq 0.$$

Therefore $f(\cdot, t)$ satisfies condition (iv) in Theorem 8.5.8 for each $t \geq 0$.

Finally, since $\|w(z)\|$ is uniformly bounded on B, we deduce that the condition (iii) from Theorem 8.5.8 holds. Hence all conditions from Theorem 8.5.8 are satisfied, and thus $f(z) = f(z,0)$ extends to a quasiconformal homeomorphism of $\mathbb{R}^{2n}$ onto $\mathbb{R}^{2n}$. This completes the proof.

Hamada and Kohr [Ham-Koh6] have recently defined the notion of strong spirallikeness of type $\alpha \in (-\pi/2, \pi/2)$ on the unit ball of $\mathbb{C}^n$ (and more generally on bounded balanced pseudoconvex domains for which the Minkowski function is of class C^1 in $\mathbb{C}^n \setminus \{0\}$).

Definition 8.5.12. Let $f : B \to \mathbb{C}^n$ be a normalized locally biholomorphic mapping and $\alpha \in (-\pi/2, \pi/2)$. Also let $a = \tan \alpha$. We say that f is *strongly spirallike of type* α if there is a constant $c < 1$ such that

$$(8.5.39) \qquad \left| \frac{1}{\|z\|^2} \langle h(z), z \rangle - \frac{1+c^2}{1-c^2} \right| \leq \frac{2c}{1-c^2}, \quad z \in B \setminus \{0\},$$

where

(8.5.40) $\qquad h(z) = iaz + (1 - ia)[Df(z)]^{-1}f(z), \quad z \in B.$

Such a strongly spirallike map of type α is of course spirallike of type α, and in view of Theorem 8.2.1,

$$f(z,t) = e^{(1-ia)t}f(e^{iat}z), \quad z \in B, \quad t \geq 0,$$

is a Loewner chain such that

$$\frac{\partial f}{\partial t}(z,t) = Df(z,t)h(z,t), \quad z \in B, \ t \geq 0,$$

where

$$h(z,t) = iaz + (1 - ia)e^{-iat}[Df(e^{iat}z)]^{-1}f(e^{iat}z),$$

for all $z \in B$ and $t \geq 0$. Then we obtain the following result by an argument similar to the proof of Theorem 8.5.11 ([Ham-Koh6]; see also [Ham-Koh12]). We leave the proof for the reader.

Theorem 8.5.13. *Let $f : B \to \mathbb{C}^n$ be a strongly spirallike mapping of type $\alpha \in (-\pi/2, \pi/2)$. Assume f is quasiconformal on B and $\|[Df(z)]^{-1}f(z)\|$ is uniformly bounded on B. Then f can be extended to a quasiconformal homeomorphism of $\mathbb{R}^{2n}$ onto $\mathbb{R}^{2n}$.*

Notes. For more information about quasiconformal extension results for biholomorphic mappings in $\mathbb{C}^n$ which can be embedded in Loewner chains, see [Chu2], [Cu3], [Ham-Koh6], [Ham-Koh12], [Ham-Koh14], [Pfa2].

Problems

8.5.1. Prove Theorem 8.5.8.

Hint. Use the condition (ii) in the hypothesis to prove that there is a constant $d > 0$ such that $\|f(z,s)\| \leq de^s\|z\|$ for $z \in B$ and $s \in [0, \infty)$. For this purpose, fix $z \in B \setminus \{0\}$ and $s \geq 0$, and let $v(t) = v(z,s,t)$ be the solution of the initial value problem

$$\frac{\partial v}{\partial t} = -h(v,t) \quad a.e. \quad t \geq s, \quad v(s) = z.$$

Then prove that $\|e^{t-s}v(z,s,t)\| \le d\|z\|$, $t \ge s$, for some $d > 0$. Letting $t \to \infty$ in this inequality and using the fact that $\lim\limits_{t\to\infty} e^t v(z,s,t) = f(z,s)$ locally uniformly on B, we obtain the claimed conclusion. Next, use the conditions (i)-(iii) to deduce that $f_t(z) = f(z,t)$ has a continuous and injective extension to $\overline{B}$ (again denoted by $f_t(z)$) for each $t \ge 0$, which satisfies Hölder conditions similar to those in (8.5.23) and (8.5.24). We now use the condition (iv) and proceed as in Steps 2-4 in the proof of Theorem 8.5.7 to obtain the desired conclusion.

8.5.2. Let $f(z,t)$ be the chain given by (8.5.6) and $f(z) = f(z,0)$ be a normalized quasiregular holomorphic mapping on B which satisfies the relation (8.5.5). Also let $h(z,t)$ be the mapping given by (8.5.10). Prove that $f(z,t)$ and $h(z,t)$ satisfy the conditions in Theorem 8.5.8, and thus f can be extended to a quasiconformal homeomorphism of $\mathbb{R}^{2n}$ onto $\mathbb{R}^{2n}$. (See also [Ham-Koh12].)

8.5.3. (i) Let $f : B \subset \mathbb{C}^2 \to \mathbb{C}^2$ be given by

$$f(z) = \frac{z}{(1 + ke^{i\alpha}z_1 z_2)^{\frac{1+e^{-i\alpha}}{2}}}, \quad z = (z_1, z_2) \in B,$$

where $\alpha \in (-\pi, \pi)$ and $k \in [0, 1)$. Show that f is strongly spirallike of type $\alpha/2$.

(ii) Let $f_1, \ldots, f_n$ be strongly spirallike functions of type $\alpha \in (-\pi/2, \pi/2)$ on the unit disc U and $f : B \to \mathbb{C}^n$ be given by $f(z) = (f_1(z_1), \ldots, f_n(z_n))$, $z = (z_1, \ldots, z_n) \in B$. Show that f is strongly spirallike of type α on the unit ball B of $\mathbb{C}^n$.

(Hamada and Kohr, 2001 [Ham-Koh6].)

8.5.4. Let $f : B \to \mathbb{C}^n$ be a normalized holomorphic mapping and let $G(z)$ be a nonsingular $n \times n$ matrix which is holomorphic with respect to $z \in B$. Assume $G(0) = I$ and there exist $c \in [0, 1)$ and $K \ge 1$ such that the following conditions hold for all $z \in B$:

$$\left\|[G(z)]^{-1}Df(z) - I\right\| \le c,$$

$$\left\|\|z\|^2\left[[G(z)]^{-1}Df(z) - I\right] + (1 - \|z\|^2)[G(z)]^{-1}DG(z)(z, \cdot)\right\| \le c$$

and

$$\|G(z)\|^n \le K|\det G(z)|.$$

Prove that f is univalent and quasiregular on B and extends to a quasiconformal homeomorphism of $\mathbb{R}^{2n}$ onto $\mathbb{R}^{2n}$.

(Ren and Ma, 1995 [Ren-Ma]. See also [Ham-Koh12].)

Hint. Prove that $f(z,t) = f(ze^{-t}) + (e^t - e^{-t})G(ze^{-t})(z)$, $z \in B$, $t \geq 0$, is a Loewner chain which satisfies the conditions of Theorem 8.5.8 (see also Problem 8.4.5). To this end, we mention that the first and second conditions in the hypothesis imply that

$$(1 - \|z\|^2)\|[G(z)]^{-1}DG(z)(z,\cdot)\| \leq 2c, \quad z \in B.$$

Then a similar argument as in the proof of Lemma 8.5.6 yields that there is a constant $M > 0$ such that

$$\|G(z)\| \leq \frac{M}{(1 - \|z\|)^c}, \quad z \in B.$$

This relation implies the condition (i) in Theorem 8.5.8. On the other hand, the third condition in the hypothesis implies that $f(\cdot, t)$ is quasiregular on B for each $t \geq 0$, and the outer dilatation is bounded independently of t. Next, it suffices to use similar arguments as in the proof of Theorem 8.5.10.

Note that if $G(z) = Df(z)$, $z \in B$, in Problem 8.5.4, we obtain the result of Theorem 8.5.7. In this case, the third relation in the hypothesis is equivalent to the condition that f is quasiregular on B. Also if $G(z) \equiv I$, we obtain the result of Theorem 8.5.10.

8.5.5. Prove Lemmas 8.5.3 and 8.5.4.

8.5.6. Prove Theorem 8.5.13.

Hint. Use similar arguments as in the proof of Theorem 8.5.11.

8.5.7. Verify that the mappings in Remark 8.5.9 (ii) satisfy the conditions (ii) and (iii) in Theorem 8.5.8.

Chapter 9

Bloch constant problems in several complex variables

In this chapter we shall briefly discuss Bloch constant problems in higher dimensions. Let B be the Euclidean unit ball in $\mathbb{C}^n$. We shall be interested in subclasses of $H(B)$, consisting of mappings from B into $\mathbb{C}^n$ for which there is a positive lower bound for the supremum of the radii of the schlicht balls in the image of all mappings in the given subclass. In particular we consider the set of Bloch mappings, although there is not as close a relationship as in one variable between Bloch constant problems for this class of mappings and Bloch constant problems for more general mappings.

9.1 Preliminaries and a generalization of Bonk's distortion theorem

We begin with the following definition:

Definition 9.1.1. Let $f : B \to \mathbb{C}^n$ be a holomorphic mapping and $a \in B$.

(i) A *schlicht ball* of f centered at $f(a)$ is a ball with center $f(a)$ such that f maps biholomorphically a subdomain of B containing a onto this ball.

(ii) Let $r(a, f)$ be the radius of the largest schlicht ball of f centered at $f(a)$. Also let $r(f) = \sup\{r(a, f) : a \in B\}$.

(iii) If $\mathcal{F}$ is a family of holomorphic mappings from B to $\mathbb{C}^n$, the *Bloch constant* for $\mathcal{F}$ is defined by $\mathbf{B}(\mathcal{F}) = \inf\{r(f) : f \in \mathcal{F}\}$.

As we have seen in Section 6.1.7, Bloch's theorem fails in $\mathbb{C}^n$, $n \geq 2$, if we do not require additional assumptions on the mapping besides the usual condition $Df(0) = I$.

The restrictions on mappings which have been considered in order to obtain nontrivial Bloch theorems are of the following principal types: quasiconformality (and similar) assumptions, boundedness assumptions, the class of Bloch mappings, and global geometric assumptions such as starlikeness and convexity.

Quasiconformality assumptions were first considered by Bochner [Boc], then by Takahashi [Tak], Sakaguchi [Sak1], Wu [Wu], Hahn [Hah2], Harris [Harr3], and most recently by Chen and Gauthier [Che-Gau2].

Boundedness assumptions have been considered by Hahn [Hah3], Harris [Harr3], Harris, Reich, and Shoikhet [Harr-Re-Sh], and Chen and Gauthier [Che-Gau2].

Here we shall focus mainly on a generalization of Bonk's distortion theorem (see Theorem 4.2.4) obtained by Liu [LiuX], on a theorem for bounded holomorphic mappings due to Chen and Gauthier [Che-Gau2], and on some results for convex and starlike mappings obtained by Graham and Varolin [Gra-Var2].

Bloch functions and Bloch mappings may be defined as follows on the Euclidean unit ball of $\mathbb{C}^n$:

Definition 9.1.2. (i) Let $f : B \to \mathbb{C}$ be a holomorphic function. Then f is called a *Bloch function* if

$$\|f\| = \sup\{\|D(f \circ \varphi)(0)\| : \varphi \in \text{Aut}(B)\} < \infty.$$

(ii) A holomorphic mapping $g : B \to \mathbb{C}^n$ is called a *Bloch mapping* if

$$\|g\| = \sup\{\|D(g \circ \varphi)(0)\| : \varphi \in \text{Aut}(B)\} < \infty.$$

The quantity $\|f\|$ ($\|g\|$) is called the *Bloch seminorm* of f (g) and is invariant under composition of f (g) with automorphisms of B.

We denote the set of Bloch mappings from B into $\mathbb{C}^n$ by $\mathcal{B}$.

It is clear that a mapping $g \in H(B)$ is a Bloch mapping if and only if each of its component functions is a Bloch function. As in one variable, there are various equivalent characterizations of Bloch functions and Bloch mappings. One of them is given in the following theorem [Tim1], [LiuX] (compare with the one-variable case). Others are given in [Tim1] and in the exercises at the end of this section.

Theorem 9.1.3. *Let* $f : B \to \mathbb{C}^n$ *be a holomorphic mapping. Then* f *is a Bloch mapping if and only if the family*

$$\mathcal{F}_f = \Big\{g : \; g(z) = (f \circ \varphi)(z) - (f \circ \varphi)(0) \; \text{for some} \; \varphi \in \text{Aut}(B)\Big\}$$

is a normal family.

Proof. First, suppose $\mathcal{F}_f$ is a normal family. If $g \in \mathcal{F}_f$ then $g \in H(B)$, $g(0) = 0$, and since $\mathcal{F}_f$ is a normal family we deduce that the set

$$\Big\{\|D(f \circ \varphi)(0)\| : \; \varphi \in \text{Aut}(B)\Big\} = \Big\{\|Dg(0)\| : \; g \in \mathcal{F}_f\Big\}$$

is bounded. Thus f is a Bloch mapping.

Conversely, assume f is a Bloch mapping. Let $a \in B$ and $\varphi \in \text{Aut}(B)$. Also let $\phi_a \in \text{Aut}(B)$ be given by (6.1.9). Then

$$\|D(f \circ \varphi)(a)\| \leq \|D(f \circ \varphi \circ \phi_a)(0)\| \cdot \|[D\phi_a(0)]^{-1}\| \leq \frac{\|f\|}{1 - \|a\|^2},$$

where we have used the fact that $\|[D\phi_a(0)]^{-1}\| = 1/(1 - \|a\|^2)$ (see the relation (10.2.7)). Therefore, the family $\{\|D(f \circ \varphi)(\cdot)\| : \varphi \in \text{Aut}(B)\}$ is locally uniformly bounded, and thus $\mathcal{F}_f$ is a normal family. This completes the proof.

Remark 9.1.4. (i) Using the second part of the proof of Theorem 9.1.3, we deduce that if $f : B \to \mathbb{C}^n$ is a Bloch mapping, then [LiuX]

$$(1 - \|z\|^2)\|Df(z)\| \leq \|f\|, \quad z \in B.$$

(ii) Liu [LiuX] showed that the quantity

$$m(f) = \sup \Big\{(1 - \|z\|^2)\|Df(z)\| : \; z \in B\Big\},$$

is actually equivalent to the Bloch seminorm of f, i.e. there is a positive constant M such that

$$m(f) \leq \|f\| \leq Mm(f), \; \forall \, f \in H(B).$$

Consequently, f is a Bloch mapping if and only if $m(f) < \infty$.

Bloch functions have been studied on bounded homogeneous domains using the covariant derivative in the Bergman metric (see [Fit-Gon2], [Gon3], [Gon-Yu-Zh], [Tim1,2]). Krantz and Ma [Kran-Ma] studied Bloch functions on strongly pseudoconvex domains using the Kobayashi metric.

A somewhat analogous class of mappings may be defined as follows [LiuX]:

Definition 9.1.5. Let $f \in H(B)$ and let

$$(9.1.1) \qquad \|f\|_0 = \sup\left\{|J_{f\circ\varphi}(0)|^{1/n} : \varphi \in \text{Aut}(B)\right\}$$

$$= \sup\left\{(1 - \|z\|^2)^{(n+1)/(2n)}|J_f(z)|^{1/n} : z \in B\right\}.$$

Note that the second equality is a consequence of the relation (6.1.13).

Also let

$$\mathcal{V}B = \{f \in H(B) : \|f\|_0 < \infty\}$$

and

$$\mathcal{V}B_0 = \{f \in \mathcal{V}B : \|f\|_0 = 1 \text{ and } J_f(0) = 1\}.$$

Then $\|\cdot\|_0$ is a seminorm [LiuX] which is invariant under composition with automorphisms of B.

There is a generalization of Bonk's distortion theorem to the class $\mathcal{V}B_0$, due to Liu [LiuX].

Theorem 9.1.6. Let $f \in \mathcal{V}B_0$. Then

$$(9.1.2) \quad |J_f(z)| \geq \text{Re } J_f(z) \geq (1 - \sqrt{n+2}\|z\|)/(1 - \|z\|/\sqrt{n+2})^{n+2}$$

for $\|z\| \leq 2\sqrt{n+2}/(n+3)$. This estimate is sharp.

Proof. Fix $w \in \mathbb{C}^n$, $\|w\| = 1$, and let $g : \overline{U} \to \mathbb{C}$ be given by

$$g(\zeta) = (1 - c_n\psi(\zeta))^{n+1}J_f(\psi(\zeta)w), \quad |\zeta| \leq 1,$$

where

$$\psi(\zeta) = c_n\frac{1-\zeta}{1 - c_n^2\zeta}, \quad c_n = \frac{1}{(n+2)^{1/2}}.$$

It is easy to see that ψ maps the unit disc U onto the disc $E = \{\xi \in \mathbb{C} : |1 - c_n\xi|^2 < 1 - |\xi|^2\} \subset U$, and a short computation yields that $g(1) = 1$. Moreover, since $J_f(0) = 1$ and $\|f\|_0 = 1$, we deduce that

$$|J_f(z)| = 1 + o(\|z\|) \text{ as } \|z\| \to 0,$$

and thus $g'(1) = 1$. On the other hand, using again the fact that $\|f\|_0 = 1$ and $\psi(U) = E$, one deduces that

$$|g(\zeta)| \le (1 - |\psi(\zeta)|^2)^{(n+1)/2}|J_f(\psi(\zeta)w)| \le 1, \quad |\zeta| < 1.$$

Hence $g(U) \subset U$, and in view of Julia's lemma on the unit disc (see Lemma 4.3.1) we deduce that for each $r > 0$, g maps the horodisc

$$\Delta(1, r) = \left\{ \zeta \in \mathbb{C} : \frac{|1 - \zeta|^2}{1 - |\zeta|^2} < r \right\}$$

into itself. The closed horodisc $\overline{\Delta}(1, r)$ is also mapped to itself. Now any point $\zeta \in (-1, 1)$ is a boundary point of exactly one horodisc of the form $\Delta(1, r)$. All other points in the closure of this horodisc have real part larger than ζ. Hence we obtain

$$\operatorname{Re} g(\zeta) \ge \zeta, \quad \zeta \in (-1, 1),$$

and it is clear that this inequality also holds for $\zeta \in [-1, 1]$.

Next, let $\eta = c_n(1 - \zeta)/(1 - c_n^2\zeta)$, $-1 \le \zeta \le 1$. Then $0 \le \eta \le 2\sqrt{n+2}/(n+3)$ and the above inequality is equivalent to

$$\operatorname{Re} J_f(\eta w) \ge (1 - \sqrt{n + 2\eta})/(1 - \eta/\sqrt{n+2})^{n+2}.$$

Finally, letting $z \in \mathbb{C}^n$, $0 < \|z\| \le 2\sqrt{n+2}/(n+3)$, $w = z/\|z\|$, and $\eta = \|z\|$ in the above, we obtain (9.1.2), as desired.

To prove the sharpness of this relation, let $f \in H(B)$ be such that $f(0) = 0$ and

$$Df(z) = \begin{pmatrix} \dfrac{1 - \sqrt{n + 2}z_1}{(1 - z_1/\sqrt{n+2})^{n+2}} & 0 \\ 0 & I_{n-1} \end{pmatrix}, \quad z = (z_1, \dots, z_n) \in B,$$

where I_{n-1} is the $(n-1) \times (n-1)$-identity matrix. Then $J_f(0) = 1$ and it is elementary to see that $\|f\|_0 = 1$. Moreover, for $z = (r, 0, \dots, 0)$, $r \in [0, 2\sqrt{n+2}/(n+3)]$, we have $\|z\| = r$ and $\operatorname{Re} J_f(z) = (1 - \sqrt{n + 2}\|z\|)/(1 - \|z\|/\sqrt{n+2})^{n+2}$. This completes the proof.

Since the relation (9.1.2) is an estimate for the Jacobian determinant of $f(z)$ and not for $Df(z)$, one needs an additional assumption in order to get an estimate for $r(0, f)$. Following Liu [LiuX], we introduce

Definition 9.1.7. Let $1 \leq k < \infty$ and $\mathcal{B}_n(k) = \{f : B \to \mathbb{C}^n : f \in H(B),\ \|f\| \leq k\}$. Also let $\mathcal{B}_{n,0}(k) = \{f \in \mathcal{B}_n(k) : J_f(0) = 1\}$.

In the remainder of this section we shall obtain a lower bound for the Bloch constant $\mathbf{B}(\mathcal{B}_{n,0}(k))$ due to Liu [LiuX], which is actually a generalization of the one-variable estimate of Ahlfors in Corollary 4.2.5. First we need the following definition and lemma.

Definition 9.1.8. Let $f : B \to \mathbb{C}^n$ be a holomorphic mapping and $z_0 \in B$. We say that z_0 is a *critical point* of f if $J_f(z_0) = 0$. In this case, $f(z_0)$ is called a *critical value* of f. A point $w_0 \in \mathbb{C}^n$ is called a *boundary point* of $f(B)$ if there is a sequence $\{z_k\}_{k \in \mathbb{N}}$ in B such that $\{z_k\}_{k \in \mathbb{N}}$ has no limit point in B and the sequence $\{f(z_k)\}_{k \in \mathbb{N}}$ converges to w_0.

(This use of the term "boundary point" occurs only in this section.)

The following lemma is due to Liu [LiuX].

Lemma 9.1.9. *Let $f : B \to \mathbb{C}^n$ be a holomorphic mapping, let Ω be a domain contained in B and $a \in \Omega$. If f maps Ω biholomorphically onto the schlicht ball $B(f(a), r(a, f))$, then either Ω and B have a common boundary point or else there is a critical value $f(z)$ on the boundary of $B(f(a), r(a, f))$ such that the critical point z lies on the boundary of Ω. Consequently, $r(a, f)$ is given by the Euclidean distance from $f(a)$ to a boundary point of $f(B)$ or to a critical value of f.*

Proof. Suppose that Ω and B do not have a common boundary point. In this case $\overline{\Omega} \subset B$ and we can find a family $\{\Omega_k\}_{k \in \mathbb{N}}$ of open subsets of B such that $\Omega_k \supset \overline{\Omega}$, $\Omega_k \supset \overline{\Omega}_{k+1}$, and $\bigcap_{k=1}^{\infty} \Omega_k = \overline{\Omega}$. Using the definition of $B(f(a), r(a, f))$, we deduce the existence of points z_k and w_k in Ω_k such that $z_k \neq w_k$ and $f(z_k) = f(w_k)$, $k \in \mathbb{N}$. Since $\{z_k\}_{k \in \mathbb{N}}$, $\{w_k\}_{k \in \mathbb{N}}$ are bounded sequences, we may suppose that $z_k \to z$ and $w_k \to w$ as $k \to \infty$. Obviously, $z, w \in \partial\Omega$, $f(z) = f(w)$ and $f(z)$ belongs to the boundary of $B(f(a), r(a, f))$.

There are two possibilities: either $z = w$, or $z \neq w$.

If $z = w$, then it is easy to see that $J_f(z) = 0$, and thus $f(z)$ is a critical value of f and z is a critical point of f which lies on the boundary of Ω.

If $z \neq w$, then at least one of the points z and w must be a critical point of f. Otherwise f must be locally univalent near z and w, and combining this fact with the equality $f(z) = f(w)$, yields easily that f cannot be univalent on Ω. However, this contradicts the assumption that Ω is mapped biholomorphically onto $B(f(a), r(a, f))$. This completes the proof.

Now we are able to prove the following result [LiuX]:

Theorem 9.1.10. *Let* $1 \leq k < \infty$ *and let* $f \in \mathcal{B}_n(k) \cap \mathcal{VB}_0$. *Also let*

$$(9.1.3) \qquad\qquad c(k, n)$$

$$= k^{1-n} \int_0^{1/\sqrt{n+2}} [(1 - t^2)^{n-1}(1 - \sqrt{n+2}t)/(1 - t/\sqrt{n+2})^{n+2}]dt$$

$$\geq k^{1-n} \frac{\sqrt{n+2}}{en} \left[\left(1 + \frac{1}{n+1} \right)^{n+1} - 2 \right].$$

Then $r(0, f) \geq c(k, n)$.

Proof. In view of Lemma 9.1.9, $r(0, f)$ is given by the Euclidean distance from $f(0)$ to a point p which is either a boundary point of $f(B)$ or a critical value of f. Let Γ be the straight line segment from $f(0)$ to p. Also let γ be the inverse image of Γ under f. Since there are no critical values of f on Γ except possibly for p, we deduce that γ is a smooth curve from the origin to the boundary of B or a curve from the origin to a critical point z_0 of f in B, smooth except possibly at z_0. In the latter case $\|z_0\| \geq 1/\sqrt{n+2}$ by (9.1.2). Hence in all cases,

$$r(0, f) = \left\| \int_\Gamma dw \right\| = \int_\Gamma \|dw\| = \int_\gamma \|Df(z)dz\|$$

$$= \int_\gamma \left\| Df(z)\frac{dz}{\|dz\|} \right\| \|dz\| \geq \int_\gamma \frac{|J_f(z)|}{\|Df(z)\|^{n-1}} d\|z\|$$

$$\geq \int_0^{1/\sqrt{n+2}} \frac{|J_f(z)|}{\|Df(z)\|^{n-1}} d\|z\|.$$

Here we have used the fact (see Lemma 9.2.2) that $\|Au\| \geq |\det A|/\|A\|^{n-1}$ for any nonzero $n \times n$ matrix A and any unit vector $u \in \mathbb{C}^n$. Combining the above arguments and using the fact that $\|Df(z)\| \leq k/(1 - \|z\|^2)$, $z \in B$, since $f \in \mathcal{B}_n(k)$, we deduce that $r(0, f) \geq c(k, n)$, where $c(k, n)$ is given by (9.1.3). This completes the proof.

Using this result together with a version of Landau's reduction (see Chapter 4) to reduce to the case $\|f\|_0 = 1$, Liu [LiuX] proved the following theorem. We leave this proof for the reader.

Theorem 9.1.11. $\mathbf{B}(\mathcal{B}_{n,0}(k)) \geq c(k,n)$.

Analogous results on classical matrix domains were obtained by FitzGerald and Gong [Fit-Gon2]; see also [Gon-Yu-Zh].

Problems

9.1.1. Let $f : B \to \mathbb{C}$ be a holomorphic function. Show that f is a Bloch function if and only if the family $\{f \circ h \mid h : U \to B,\ h$ holomorphic$\}$ is a family of Bloch functions with uniformly bounded Bloch seminorm.
(Timoney, 1980 [Tim1].)

9.1.2. Let $f : B \to \mathbb{C}$ be a holomorphic function. Show that f is a Bloch function if and only if the directional derivative of f is $O(1/(1 - \|z\|^2))$ in the radial direction and $O(1/(1 - \|z\|^2)^{1/2})$ in the directions orthogonal to the radial direction.
(Timoney, 1980 [Tim1].)

9.1.3. Prove Theorem 9.1.11.

9.1.4. Show that $\|\cdot\|_0$ given by (9.1.1) is a seminorm which is invariant under composition with automorphisms of B.
(Liu, 1992 [LiuX].)

9.2 Bloch constants for bounded and quasiregular holomorphic mappings

In this section we shall discuss a Bloch constant theorem of Chen and Gauthier [Che-Gau2] for bounded holomorphic mappings on the Euclidean unit ball of $\mathbb{C}^n$. It is a generalization of the result of Landau [Lan] given in Problem 4.2.3.

The first step is to generalize Landau's theorem to the case of holomorphic mappings from the unit disc U into $\mathbb{C}^n$. The following lemma is due to Chen and Gauthier [Che-Gau2].

Lemma 9.2.1. *Let $g : U \to \mathbb{C}^n$ be a holomorphic mapping such that $\|g(z)\| < M$, $z \in U$, $g(0) = 0$ and $\|Dg(0)\| = \alpha > 0$. Then*

(i) g is injective on U_{ρ_0} and $Dg(z) \neq 0$ for $z \in U_{\rho_0}$, where

$$\rho_0 = \frac{\alpha}{M + \sqrt{M^2 - \alpha^2}};$$

(ii) For any positive number $\rho \leq \rho_0$, $g(\partial U_\rho)$ lies outside the ball B_R, where

$$R = M\frac{\rho(\alpha - M\rho)}{M - \alpha\rho} \geq M\rho_0\rho.$$

Proof. Let $g = (g_1, \ldots, g_n)$ and let $h : U \to \mathbb{C}$ be given by

$$h(\zeta) = \frac{1}{\alpha}\left[\overline{g_1'(0)}g_1(\zeta) + \ldots + \overline{g_n'(0)}g_n(\zeta)\right], \quad |\zeta| < 1.$$

Then h is a holomorphic function on U, $h(0) = 0$, $h'(0) = \alpha$ and an application of the Schwarz inequality shows that $|h(\zeta)| \leq \|g(\zeta)\| < M$ for $\zeta \in U$. Hence Problem 4.2.3 applies to h and all of the desired conclusions about g follow from this.

We leave the proof of the next lemma for the reader since it suffices to use elementary properties of matrices and eigenvalues. (A proof may be found in [Che-Gau2].)

Lemma 9.2.2. *If $A = (a_{ij})_{1 \leq i,j \leq n} \in L(\mathbb{C}^n, \mathbb{C}^n)$, then for any $w \in \mathbb{C}^n$, $\|w\| = 1$,*

$$|\det A| \leq \|Aw\| \cdot \|A\|^{n-1}.$$

The next result is due to Chen and Gauthier [Che-Gau2].

Lemma 9.2.3. *Let $f : B \to \mathbb{C}^n$ be a holomorphic mapping and let $M > 0$ be such that $\|f(z)\| \leq M$, $z \in B$. Then*

(9.2.1) $$(1 - \|z\|^2)\|Df(z)\| \leq M, \quad z \in B,$$

and

(9.2.2) $$(1 - \|z\|^2)^{\frac{n+1}{2}}|J_f(z)| \leq M^n, \quad z \in B.$$

Proof. Fix $w \in \mathbb{C}^n$ with $\|w\| = 1$. Applying the Cauchy integral formula

$$Df(0)(w) = \frac{1}{2\pi i} \int_{|\zeta|=r} \frac{f(\zeta w)}{\zeta^2}d\zeta, \quad r \in (0,1),$$

and using the inequality $\|f(z)\| \leq M$ for all $z \in B$, we easily deduce that $\|Df(0)(w)\| \leq M$. Since w is arbitrary, we conclude that $\|Df(0)\| \leq M$, and thus we obtain (9.2.1) for the case $z = 0$. Next, fix $a \in B \setminus \{0\}$ and let $\varphi_a(z) = \phi_{-a}(z)$, where $\phi_a \in \mathrm{Aut}(B)$ is given by (6.1.9). Also let $h(z) = f \circ \varphi_a(z)$, $z \in B$. Then $h \in H(B)$ and $h(0) = f(a)$. Moreover,

$$(9.2.3) \quad \|Df(a)\| = \|Dh(0)[D\varphi_a(0)]^{-1}\| \leq \|Dh(0)\| \cdot \|[D\varphi_a(0)]^{-1}\|$$

$$\leq \frac{1}{1 - \|a\|^2} \|Dh(0)\|,$$

making use of the fact that $\|[D\varphi_a(0)]^{-1}\| = 1/(1 - \|a\|^2)$ (see formula (10.2.7)). On the other hand, since $\|h(z)\| \leq M$, $z \in B$, we deduce in view of the first step of the proof that $\|Dh(0)\| \leq M$, and from (9.2.3) and this relation we obtain (9.2.1).

Now we show (9.2.2). Clearly (9.2.2) for the case $z = 0$ is a consequence of (9.2.1). Thus, we have to prove (9.2.2) for $z \in B \setminus \{0\}$. Again fix $a \in B \setminus \{0\}$ and let h be defined as above. Since $\|h(z)\| \leq M$ on B it follows that $|J_h(0)| \leq M^n$. But from (6.1.12) we have

$$|J_h(0)| = |J_f(a)||J_{\varphi_a}(0)| = (1 - \|a\|^2)^{\frac{n+1}{2}} |J_f(a)|,$$

and thus

$$(1 - \|a\|^2)^{\frac{n+1}{2}} |J_f(a)| \leq M^n.$$

This completes the proof.

The following lemma [Che-Gau2] is related to a result of Takahashi [Tak]. We have

Lemma 9.2.4. Let $f : B \to \mathbb{C}^n$ be a holomorphic mapping such that $|J_f(0)| = a > 0$ and $\|f(z)\| < M$, $z \in B$, for some $M > 0$. Then f is univalent on the ball B_{ρ_0}, where $\rho_0 = \dfrac{a}{bM^n}$, and $b \approx 4.2$ is the minimum of the function $(2 - r^2)/(r(1 - r^2))$ for $0 < r < 1$.

Proof. Since $\|f(z)\| < M$, $z \in B$, we can apply (9.2.1) to deduce that

$$\|Df(z) - Df(0)\| \leq M \left[\frac{1}{1 - \|z\|^2} + 1 \right] = \frac{M(2 - \|z\|^2)}{1 - \|z\|^2}, \quad z \in B.$$

Now, the function $g(r) = \dfrac{2 - r^2}{r(1 - r^2)}$, $0 < r < 1$, attains its minimum $b \approx 4.2$ at $r_0 \approx 0.66$. Combining this with the Schwarz lemma for holomorphic mappings (see Lemma 6.1.28), we deduce that

$$\|Df(z) - Df(0)\| \le bM\|z\|, \quad \|z\| \le r_0.$$

On the other hand, from Lemma 9.2.2 and the relation (9.2.1) we obtain

$$(9.2.4) \qquad \|Df(0)(w)\| \ge \frac{a}{\|Df(0)\|^{n-1}} \ge \frac{a}{M^{n-1}}, \quad \|w\| = 1.$$

Moreover, in view of Lemma 9.2.2 we have $|J_f(0)| = a \le M^n$, and thus $\rho_0 \le 1/b$. We also note that $r_0 > 1/b$. Next, let $q_1, q_2 \in B_{\rho_0}$ be two distinct points. Then

$$(9.2.5) \quad \|Df(z) - Df(0)\| \le bM\|z\| < bM\rho_0 = \frac{a}{M^{n-1}}, \quad z \in [q_1, q_2],$$

where $[q_1, q_2]$ is the closed line segment between q_1 and q_2.

From (9.2.4) and (9.2.5), we therefore deduce that

$$\|f(q_2) - f(q_1)\| \ge \left\| \int_{[q_1, q_2]} Df(0)dz \right\| - \int_{[q_1, q_2]} \|Df(z) - Df(0)\| \cdot \|dz\|$$

$$> \frac{a}{M^{n-1}}\|q_1 - q_2\| - \frac{a}{M^{n-1}}\|q_1 - q_2\| = 0.$$

Thus $f(q_1) \ne f(q_2)$, as desired. This completes the proof.

Now we are able to prove the main result of this section, due to Chen and Gauthier [Che-Gau2].

Theorem 9.2.5. *Let $f : B \to \mathbb{C}^n$ be a holomorphic mapping such that $f(0) = 0$, $J_f(0) = a > 0$, and $\|f(z)\| < M$, $z \in B$, for some positive constant M. Let ρ_0 be defined as in Lemma 9.2.4 and let*

$$\rho_1 = \frac{a}{2M^n}, \quad \rho_2 = M\rho_0\rho_1 = \frac{a^2}{2bM^{2n-1}}.$$

Then f maps the ball B_{ρ_0} biholomorphically onto a domain which contains the ball B_{ρ_2}.

Proof. Taking into account Lemma 9.2.4, we deduce that f is univalent on B_{ρ_0}. Next, fix $z \in \partial B_{\rho_0}$ and let $w = z/\|z\| = z/\rho_0 = (w_1, \dots, w_n) \in \partial B$.

We have already noted that $\rho_0 \leq 1/b < 1$. Let $g : U \to \mathbb{C}^n$ be given by $g(\zeta) = f(\zeta w)$, $|\zeta| < 1$. Then g is holomorphic on U, $g(0) = 0$, $\|g(\zeta)\| < M$, $|\zeta| < 1$, and by Lemma 9.2.2 and the relation (9.2.1) we obtain

$$\|Dg(0)\| = \|Df(0)(w)\| \geq \frac{a}{M^{n-1}}.$$

Let $c = \|Dg(0)\|$ and $r_1 = \dfrac{c}{M + \sqrt{M^2 - c^2}}$. Then $r_1 > \dfrac{a}{2M^n} = \rho_1 > \rho_0$. Applying the result of Lemma 9.2.1, we deduce that

$$\|f(z)\| = \|g(\rho_0)\| \geq M\rho_0\rho_1 = \rho_2.$$

Since z was arbitrarily chosen on ∂B_{ρ_0}, we conclude that $f(\partial B_{\rho_0})$ lies outside the ball B_{ρ_2}. This completes the proof.

We conclude this section by discussing some Bloch constant results for quasiregular holomorphic mappings on the unit ball of $\mathbb{C}^n$ (cf. Section 8.5). Such results have been obtained by Bochner [Boc], Takahashi [Tak], Sakaguchi [Sak1], Wu [Wu], Hahn [Hah2], Harris [Harr3], and recently by Chen and Gauthier [Che-Gau2].

The condition of quasiregularity or quasiconformality was the earliest condition considered in studying Bloch constant problems in several complex variables. It can be formulated in various ways. Using the terminology of [Che-Gau2], we introduce the following definitions.

Definition 9.2.6. Let $f : B \to \mathbb{C}^n$ be a holomorphic mapping and $K \geq 1$.
(i) We say that f is a *Wu K-mapping* if

$$\|Df(z)\| \leq K|J_f(z)|^{1/n}, \quad z \in B.$$

Obviously, a Wu K-mapping is also a quasiregular mapping.
Let $\mathcal{F}_{K,n}$ be the set of Wu K-mappings and for $K \geq 1$ and $n \geq 2$, let

$$\mathbf{B}(\mathcal{F}_{K,n}) = \inf \left\{ r(f) : f \in \mathcal{F}_{K,n}, \ J_f(0) = 1 \right\}$$

be the Bloch constant for the n-dimensional Wu K-mappings.
(ii) We say that f is a *Bochner K-mapping* if

$$\|Df(z)\|_2 \leq K|J_f(z)|^{1/n}, \quad z \in B,$$

where $\|A\|_2 = \Big[\sum_{i,j=1}^{n} |a_{ij}|^2 \Big]^{1/2}$ for any $(n \times n)$ complex matrix $A = (a_{ij})_{1 \leq i,j \leq n}$.
Bochner [Boc] proved that for each $K \geq 1$ and $n \geq 2$, the Bloch constant for
Bochner K-mappings f, for which $J_f(0) = 1$, is positive.

(iii) We say that f is a *Takahashi K-mapping* if

$$\max_{\|z\| \leq r} \|Df(z)\|_2 \leq K \max_{\|z\| \leq r} |J_f(z)|^{1/n}, \; r \in [0,1).$$

If f is a Wu K-mapping, then f is also a Takahashi $\widetilde{K}$-mapping with
$\widetilde{K} = \sqrt{n}K$.

(iv) Finally we say that f is a *Hahn K-mapping* if

$$\max_{\|z\|=r} \Lambda(z) \leq K \max_{\|z\|=r} \lambda(z), \; 0 \leq r < 1,$$

where $\lambda^2(z)$ and $\Lambda^2(z)$ are the smallest and the biggest eigenvalues of the
Hermitian matrix $[Df(z)]^*[Df(z)]$, $z \in B$. Then a Wu K-mapping is a Hahn
K^n-mapping.

We mention that Takahashi [Tak] proved that the Bloch constant for
Takahashi K-mappings f, for which $J_f(0) = 1$, is bounded below by $(n -
1)^{n-2}/[12K^{2n-1}]$. Sakaguchi [Sak1] improved this result by showing that the
Bloch constant for Takahashi K-mappings f, for which $J_f(0) = 1$, is at least
$(n - 1)^{n-2}/[8K^{2n-1}]$. Hahn [Hah2] showed that the Bloch constant for Hahn
K-mappings f, for which $J_f(0) = 1$, is at least $K^{1/n}/[4K(2K + 1)]$.

Wu [Wu] proved that $\mathbf{B}(\mathcal{F}_{K,n}) > 0$ for all $K \geq 1$ and $n \geq 2$. The above
result of Sakaguchi [Sak1] applied to Wu K-mappings yields that

$$\mathbf{B}(\mathcal{F}_{K,n}) \geq \frac{1}{8K^{2n-1}n^{3/2}} \left(1 - \frac{1}{n}\right)^{n-2},$$

while Harris [Harr3] proved the estimate

$$\mathbf{B}(\mathcal{F}_{K,n}) \geq \frac{1}{8K^{2n}}.$$

Chen and Gauthier [Che-Gau2] improved the above results, as follows:

$$\mathbf{B}(\mathcal{F}_{K,n}) \geq \frac{1}{10K^{2n-1}} \text{ and } \mathbf{B}(\mathcal{F}_{K,n}) \geq \frac{1}{12K^{n-1}}.$$

Finally we remark that an interesting Bloch constant problem for holomorphic mappings of the unit ball of a Banach space with restricted numerical range has recently been solved by Harris, Reich and Shoikhet [Harr-Re-Sh]; see also [Harr3] for various norm restrictions.

Notes. For more information about Bloch constant problems in several variables, the reader may consult [Che-Gau2], [LiuX], [Fit-Gon2], [Gam-Che], [Gon-Yu-Zh], [Harr3], [Harr-Re-Sh], [Wu].

9.3 Bloch constants for starlike and convex mappings in several complex variables

We conclude this chapter with some results concerning Bloch constants for normalized convex and starlike mappings. For normalized convex mappings, the Bloch constant is known precisely on the unit polydisc in $\mathbb{C}^n$, and on the Euclidean unit ball it is known if one adds the assumption of k-fold symmetry, $k \geq 2$. For normalized starlike mappings of the Euclidean unit ball, the Bloch constant is known precisely in certain cases of k-fold symmetry depending on the dimension, and it can be estimated in the general case.

We begin by recalling the covering theorem of Szegö [Sze] for convex functions in one variable proved in Chapter 4: if $f \in K$ then the image of f contains a disc of radius $\pi/4$, and this result is sharp. It is natural to consider generalizations to several variables. In the case of the unit polydisc P in $\mathbb{C}^n$, this is straightforward, as shown by Graham and Varolin [Gra-Var2]:

Theorem 9.3.1. Let $f : P \to \mathbb{C}^n$ be a convex mapping with $Df(0) = I$. Then $f(P)$ contains a polydisc of radius $\pi/4$, and this result is sharp.

Proof. Using Theorem 6.3.2 and the fact that $Df(0) = I$, we easily deduce that

$$f(z) = (g_1(z_1), \ldots, g_n(z_n)), \quad z = (z_1, \ldots, z_n) \in P,$$

where g_j is a convex function of one variable such that $g_j'(0) = 1$, $j = 1, \ldots, n$.

The image of U under each function g_j contains a disc of radius $\pi/4$ by Theorem 4.3.7, so the image of P contains a polydisc, each of whose radii is $\pi/4$.

In order to prove the sharpness, it suffices to consider the convex mapping $f : P \to \mathbb{C}^n$ given by

$$f(z) = \left(\frac{1}{2} \log \left[\frac{1 + z_1}{1 - z_1} \right], \ldots, \frac{1}{2} \log \left[\frac{1 + z_n}{1 - z_n} \right] \right), \quad z = (z_1, \ldots, z_n) \in P.$$

The corresponding problem for the Euclidean unit ball B in $\mathbb{C}^n$ has not yet been solved (cf. [Gra-Var2]):

Open Problem 9.3.2. *If $f : B \to \mathbb{C}^n$, $n \geq 2$, is a normalized convex mapping then does $f(B)$ contain a ball of radius $\pi/4$?*

However, the answer is positive if f is odd, in view of Theorem 7.2.9 for the case $k = 2$. More generally, we have [Gra-Var2]

Theorem 9.3.3. *The Bloch and Koebe constants for normalized convex mappings of B with k-fold symmetry ($k \geq 2$) coincide and have the value r_k, where $r_k = \int_0^1 \dfrac{dt}{(1 + t^k)^{2/k}}.$*

Proof. It suffices to observe that if Ω is a convex domain in $\mathbb{C}^n$ with k-fold symmetry and $V \subset \Omega$ is a ball, then the convex hull of the union of the balls $e^{2\pi i j/k} V$, $j = 1, \ldots, k$, contains a ball centered at 0 which is at least as large as V.

Another important special case in which the solution of Open Problem 9.3.2 is known is given in Theorem 11.2.9.

We next give an upper bound for the Bloch constant for normalized starlike mappings on the Euclidean unit ball of $\mathbb{C}^n$, $n \geq 2$, which decreases with the dimension and which tends to $1/4$ (the value of the Koebe constant for starlike mappings of the ball B) as $n \to \infty$. This result was obtained by Graham and Varolin [Gra-Var2].

Theorem 9.3.4. *The Bloch constant $\mathbf{B}(S^*(B))$ for normalized starlike mappings on the unit ball B of $\mathbb{C}^n$, $n \geq 2$, satisfies $\mathbf{B}(S^*(B)) < \dfrac{1}{4} \left(\dfrac{\sqrt{n} + 1}{\sqrt{n} - 1} \right)^2.$*

Proof. Let $f(z) = \left(\dfrac{z_1}{(1 - z_1)^2}, \ldots, \dfrac{z_n}{(1 - z_n)^2} \right)$ for $z = (z_1, \ldots, z_n) \in B$. This mapping is normalized starlike and omits the hyperplanes $w_j = -1/4$, $j = 1, \ldots, n$. If there is a ball of radius $r > 1/4$ in the image of f, the center of this ball must be at a distance at least r from each of these hyperplanes. Let $(c_1, \ldots, c_n)$ be the coordinates of the center of such a ball.

Now the Koebe function $k(\zeta) = \dfrac{\zeta}{(1-\zeta)^2}$, $\zeta \in U$, may be written as
$k(\zeta) = -\dfrac{1}{4} + \dfrac{1}{4}\left(\dfrac{1+\zeta}{1-\zeta}\right)^2$. This representation shows that for a given value
of $|\zeta|$, $\left|k(\zeta)+\dfrac{1}{4}\right|$ is maximized when ζ is positive real. Conversely, for a given
value of $\left|k(\zeta)+\dfrac{1}{4}\right|$, $|\zeta|$ is minimized when ζ is positive real. Thus for the
purposes of bounding r above we may assume that $c_1 = \ldots = c_n = r - 1/4$.

If we solve the equation $x/(1-x)^2 = r - 1/4$ and require that $x < n^{-1/2}$,
we deduce that $r < \dfrac{1}{4}\left(\dfrac{\sqrt{n}+1}{\sqrt{n}-1}\right)^2$. This completes the proof.

A similar argument gives an upper bound for the Bloch constant for normalized odd starlike mappings on the Euclidean unit ball of $\mathbb{C}^n$ (see [Gra-Var2]).

Theorem 9.3.5. *The Bloch constant* $\mathbf{B}(S^*_{(2)}(B))$ *for normalized odd starlike mappings on the unit ball B of $\mathbb{C}^n$, $n \geq 2$, satisfies* $\mathbf{B}(S^*_{(2)}(B)) < \dfrac{n+1}{2(n-1)}$.

Proof. Let $f : B \to \mathbb{C}^n$ be given by

$$f(z) = \left(\dfrac{z_1}{1-z_1^2}, \ldots, \dfrac{z_n}{1-z_n^2}\right), \quad z = (z_1, \ldots, z_n) \in B.$$

Then f is odd and starlike, since each of its components is odd and starlike on the unit disc. Now if we consider the function $\zeta \mapsto \dfrac{\zeta}{1-\zeta^2}$ and fix $|\zeta|$, then the modulus of the image point and its distance from the omitted rays are maximized when ζ is real. The image of the above mapping f omits the hyperplanes $w_j = \pm i/2$, $j = 1, \ldots, n$. Suppose there exists a ball of radius $r > 1/2$ in the image of this mapping. For the purpose of estimating r we may assume that the center of this ball is $(c, c, \ldots, c)$, where $c > 0$. Then $r \leq \sqrt{c^2 + 1/4}$. Next, setting $c = x/(1-x^2)$ and requiring $0 < x < n^{-1/2}$, we deduce that $r < (n+1)/[2(n-1)]$. This completes the proof.

We conclude with a result due to Graham and Varolin [Gra-Var2], concerning the Bloch constant for normalized starlike mappings with k-fold symmetry.

Theorem 9.3.6. *The Bloch constant for normalized starlike mappings on the unit ball B of $\mathbb{C}^n$, $n \geq 2$, with k-fold symmetry is given by* $\mathbf{B}(S^*_{(k)}(B)) = 4^{-1/k}$ *when $k = 3$ and $n \geq 4$ and when $k \geq 4$ and $n \geq 2$.*

Proof. Let $F : B \to \mathbb{C}^n$ be given by

$$F(z) = \left(\frac{z_1}{(1 - z_1^k)^{2/k}}, \ldots, \frac{z_n}{(1 - z_n^k)^{2/k}} \right), \quad z = (z_1, \ldots, z_n) \in B.$$

This mapping is k-fold symmetric starlike on B and the image of the unit ball under this mapping contains the ball $B_{\frac{1}{4^{1/k}}}$. The one-variable function $f_k(\zeta) = \dfrac{\zeta}{(1 - \zeta^k)^{2/k}}$ covers a disc of the same radius centered at 0. However, for $k \geq 3$, $r(f_k, w)$ has a local maximum at $w = 0$.

We discuss the cases $k = 3$ and $k > 3$ separately.

If $k = 3$ the endpoints of the three rays omitted by f_3 are located at the points $4^{-1/3} e^{2\pi i j/3}$, $j = 1, 2, 3$, and we must move a distance $4^{-1/3}$ from 0 to find a nonzero w such that $r(f_3, w)$ is as large as $4^{-1/3}$. To minimize $|\zeta|$ such that $f_k(\zeta)$ is at distance $4^{-1/3}$ from 0 we take ζ^3 to be positive, hence we may take $\zeta > 0$. Solving $\dfrac{\zeta}{(1 - \zeta^3)^{2/3}} = 4^{-1/3}$ gives $\zeta = \left(\dfrac{\sqrt{2} - 1}{\sqrt{2} + 1} \right)^{1/3}$.

Requiring that $\zeta < n^{-1/2}$ gives $n < \left(\dfrac{\sqrt{2} + 1}{\sqrt{2} - 1} \right)^{2/3} \approx 3.24$. Hence only $n = 2, 3$ are possible. For all other values of n, the largest ball in the image of F is centered at 0.

If $k > 3$ the entire omitted rays of f_k come into play instead of just the endpoints. A point on the positive real axis which is at distance $a \geq 4^{-1/k}$ from the nearest omitted ray must have distance $a \csc(\pi/k)$ from 0. Solving $f_k(x) = 4^{-1/k} \csc(\pi/k)$ and requiring that $0 < x < n^{-1/2}$ gives

$$n < \left[\frac{1 + \sqrt{1 + (\csc(\pi/k))^k}}{-1 + \sqrt{1 + (\csc(\pi/k))^k}} \right]^{2/k}.$$

The right-hand side has the approximate value 1.6 when $k = 4$ and is a decreasing function of k. Hence there are no values of $n \geq 2$ which satisfy this relation. This completes the proof.

Remark 9.3.7. For the cases $k = 3$, $n = 2, 3$, not covered by Theorem 9.3.6, one can estimate the Bloch constant by using the reasoning in the proof of Theorem 9.3.4. This gives $\mathbf{B}(S_{(3)}^*(B)) \approx 0.8340$ for $n = 2$ and $\mathbf{B}(S_{(3)}^*(B)) \approx 0.6486$ for $n = 3$.

Chapter 10

Linear invariance in several complex variables

In this chapter we are going to demonstrate that the one variable notion of linear-invariant families has a useful generalization to several complex variables. As we have seen in the fifth chapter, linear invariance, introduced by Pommerenke ([Pom1], [Pom2]), has been a powerful tool in extending many ideas of univalent function theory to the study of locally univalent functions on the unit disc. Here we shall present some of the most important applications of this method in geometric function theory of several variables.

In this chapter we shall study growth and distortion results for linear-invariant families on the Euclidean unit ball of $\mathbb{C}^n$ as well as certain examples, open problems, and conjectures which involve the order of a given L.I.F. In fact we shall consider two definitions for the order of a L.I.F. $\mathcal{F}$. The trace order leads to bounds for the growth of the Jacobian determinant of mappings in $\mathcal{F}$, and in fact is determined if sharp bounds for the growth of $|J_f(z)|$, $f \in \mathcal{F}$, are known. The norm order, recently introduced by Pfaltzgraff and Suffridge [Pfa-Su4], contain more information and gives bounds for $\|Df(z)\|$ and an upper bound for $\|f(z)\|$, as well as geometric results such as the radii of convexity, starlikeness, and univalence of a L.I.F. of known order. In one variable, both of these notions of order reduce to the usual one.

We briefly consider L.I.F.'s on the unit polydisc as well.

10.1 Preliminaries concerning the notion of linear invariance in several complex variables

10.1.1 L.I.F.'s and trace order in several complex variables

Let $\mathbb{C}^n$ be the space of n complex variables equipped with the Euclidean structure. As usual, let B be the unit ball of $\mathbb{C}^n$ with respect to the Euclidean norm $\|\cdot\|$.

The underlying set of mappings and transforms which are the subject of the study of linear invariance are the set $\mathcal{LS}(B)$ consisting of normalized locally biholomorphic mappings on the Euclidean unit ball B and the Koebe transform $\Lambda_\phi(f)$ given by

$$(10.1.1) \qquad \Lambda_\phi(f)(z) = [D\phi(0)]^{-1}[Df(\phi(0))]^{-1}(f(\phi(z)) - f(\phi(0)))$$

for $f \in \mathcal{LS}(B)$ and $\phi \in \text{Aut}(B)$. In the case $n = 1$ we denote the set $\mathcal{LS}(U)$ by $\mathcal{LS}$.

It is clear that the Koebe transform has the *group property*, i.e.

$$\Lambda_{\psi \circ \phi} = \Lambda_\phi \circ \Lambda_\psi.$$

In this chapter we shall use an alternative notation for the automorphisms of B in order to conform with source papers. We shall write

$$(10.1.2) \qquad\qquad \varphi_a(z) = T_a\left(\frac{z+a}{1+a^*z}\right), \quad \|a\| < 1,$$

and

$$T_a = \frac{1}{\|a\|^2}\left\{(1 - s_a)aa^* + s_a\|a\|^2 I\right\}, \quad s_a = \sqrt{1 - \|a\|^2}.$$

Since $a^*z = \langle z, a \rangle$, $z \in \mathbb{C}^n$, the operator T_a is the same as in (6.1.10), and we have $\varphi_a(z) = \phi_{-a}(z)$, $z \in B$, where ϕ_a is given by (6.1.9).

The Koebe transforms formed with the automorphisms φ_a, $\|a\| < 1$, play a special role in this subject. Throughout this chapter $f(z; a)$ will denote the Koebe transform formed with φ_a, i.e.

$$(10.1.3) \qquad\qquad f(z; a) = \Lambda_{\varphi_a}(f)(z)$$

$$= [D\varphi_a(0)]^{-1}[Df(\varphi_a(0))]^{-1}(f(\varphi_a(z)) - f(\varphi_a(0))).$$

We introduce the following definition (see [Pfa5]; [Bar-Fit-Gon2]):

Definition 10.1.1. The family $\mathcal{F}$ is called a *linear-invariant family* (L.I.F.) if

(i) $\mathcal{F} \subseteq \mathcal{LS}(B)$ and

(ii) $\Lambda_\phi(f) \in \mathcal{F}$ for all $f \in \mathcal{F}$ and $\phi \in \mathrm{Aut}(B)$.

In the case of one variable, this definition coincides with Definition 5.1.1.

(Rudin [Rud2] used the term $\mathcal{M}$-*invariant family* to mean a family of holomorphic mappings on the Euclidean unit ball B of $\mathbb{C}^n$ that is invariant under the group $\mathrm{Aut}(B)$.)

The following result of Pfaltzgraff and Suffridge [Pfa-Su4] is useful in various situations. Recall that by $\mathcal{U}$ we denote the set of unitary transformations from $\mathbb{C}^n$ to $\mathbb{C}^n$.

Theorem 10.1.2. *Let $\mathcal{F} \subseteq \mathcal{LS}(B)$. Then $\mathcal{F}$ is a linear-invariant family if and only if for each $f \in \mathcal{F}$, the mapping $V^{-1}f(Vz;a) \in \mathcal{F}$ for all $a \in B$ and $V \in \mathcal{U}$.*

Proof. Let $f \in \mathcal{F}$. Taking into account Theorem 6.1.23, we observe that the condition (ii) in Definition 10.1.1 is satisfied if for each $a \in B$ and $V \in \mathcal{U}$, $\Lambda_\varphi(f) \in \mathcal{F}$ for $\varphi = \varphi_a V$, where φ_a is given by (10.1.2). Since $\varphi(0) = \varphi_a(0) = a$ and $[D\varphi(0)]^{-1} = V^{-1}[D\varphi_a(0)]^{-1}$, we deduce that

$$\Lambda_\varphi(f)(z) = [D\varphi(0)]^{-1}[Df(\varphi(0))]^{-1}(f(\varphi(z)) - f(\varphi(0)))$$

$$= V^{-1}[D\varphi_a(0)]^{-1}[Df(a)]^{-1}(f(\varphi_a(Vz)) - f(a)) = V^{-1}f(Vz;a).$$

Therefore $\Lambda_\varphi(f) \in \mathcal{F}$ if and only if $V^{-1}f(Vz;a) \in \mathcal{F}$, as desired. This completes the proof.

Pfaltzgraff [Pfa5] defined the *order* of a L.I.F. in the following way:

Definition 10.1.3. For a linear-invariant family $\mathcal{F}$, let

$$(10.1.4) \qquad \mathrm{ord}\,\mathcal{F} = \sup_{f \in \mathcal{F}} \sup_{\|w\|=1} \left| \mathrm{trace}\left\{ \frac{1}{2} D^2 f(0)(w, \cdot) \right\} \right|$$

$$= \sup_{f \in \mathcal{F}} \sup_{\|w\|=1} \left| \frac{1}{2} \sum_{j=1}^{n} \sum_{l=1}^{n} \frac{\partial^2 f_j}{\partial z_j \partial z_l}(0) w_l \right|$$

be the order of $\mathcal{F}$. We shall also refer to ord $\mathcal{F}$ as the *trace order* of $\mathcal{F}$.

We note that the trace order may be expressed in some alternative forms, using unitary transformations from $\mathbb{C}^n$ to $\mathbb{C}^n$. If $V \in \mathcal{U}$ and $f \in \mathcal{F}$, then also $g \in \mathcal{F}$, where $g(z) = \Lambda_V(f)(z) = V^{-1}f(Vz)$, $z \in B$, and

$$D^2 g(0)(v, \cdot) = V^{-1}[D^2 f(0)(Vv, \cdot)]V, \quad v \in \mathbb{C}^n.$$

Using the fact that the trace is a similarity invariant, one obtains that

$$\text{trace}\{D^2 g(0)(v, \cdot)\} = \text{trace}\{D^2 f(0)(Vv, \cdot)\}$$

$$= \sum_{j=1}^{n}\sum_{l=1}^{n} \frac{\partial^2 f_j}{\partial z_j \partial z_l}(0)(Vv)_l.$$

Next, if v and k are fixed, $1 \leq k \leq n$, we may choose V such that $Vv = e_k$, where e_k is the unit vector in $\mathbb{C}^n$ having the kth-coordinate equal to 1. Thus we obtain

$$\text{trace}\{D^2 g(0)(v, \cdot)\} = \sum_{j=1}^{n} \frac{\partial^2 f_j}{\partial z_j \partial z_k}(0).$$

In view of (10.1.4) and the above equality, we deduce that the order of the L.I.F. $\mathcal{F}$ can be written equivalently as follows (see [Pfa5]):

$$(10.1.5) \qquad \text{ord } \mathcal{F} = \sup\left\{\frac{1}{2}\left|\sum_{j=1}^{n} \frac{\partial^2 f_j}{\partial z_j \partial z_k}(0)\right| : f \in \mathcal{F}\right\}.$$

If we set $k = 1$ and $n = 2$ in (10.1.5), we obtain the expression for ord $\mathcal{F}$ which appears in [Bar-Fit-Gon2]:

$$\text{ord } \mathcal{F} = \sup\left\{\left|d_{2,0}^{(1)} + \frac{1}{2}d_{1,1}^{(2)}\right| : f \in \mathcal{F}\right\},$$

when $f = (f_1, f_2) \in \mathcal{F}$ is normalized by

$$f_j(z) = z_1 + d_{2,0}^{(j)}z_1^2 + d_{1,1}^{(j)}z_1 z_2 + d_{0,2}^{(j)}z_2^2 + \ldots, \, j = 1, 2, \, z = (z_1, z_2) \in B.$$

A similar expression to (10.1.5) was also used by Liu [LiuT1] (cf. [Gon4, p.111]).

Consequently, we have proved the following equalities:

$$(10.1.6) \qquad \operatorname{ord} \mathcal{F} = \sup_{f \in \mathcal{F}} \sup_{\|w\|=1} \left| \operatorname{trace}\left\{ \frac{1}{2} D^2 f(0)(w, \cdot) \right\} \right|$$

$$= \sup_{f \in \mathcal{F}} \sup_{\|w\|=1} \left| \frac{1}{2} \sum_{j=1}^{n} \sum_{l=1}^{n} f_{jl}^{j}(0) w_l \right| = \sup_{f \in \mathcal{F}} \left| \frac{1}{2} \sum_{j=1}^{n} f_{jk}^{j}(0) \right|,$$

for each $k = 1, \ldots, n$, where $f_{jk}^{j}(0) = \dfrac{\partial^2 f_j}{\partial z_j \partial z_k}(0)$.

Remark 10.1.4. In order to define the notion of linear-invariance on a domain in $\mathbb{C}^n$, one needs the domain to have a transitive automorphism group. In this case one can introduce the Koebe transform as in formula (10.1.1). The theory of linear-invariant families was extended to bounded symmetric domains by Gong and Zheng [Gon-Zh1-4] (cf. [Gon4]). They obtained a general distortion theorem involving the order of a L.I.F. Some interesting results for the case of the polydisc were obtained by Pfaltzgraff and Suffridge [Pfa-Su2]. There are some differences from the case of the ball; for instance, it is quite easy to compute the order of the L.I.F. $K(P)$. We shall give some results for the polydisc in this chapter, and we refer the reader to [Gon-Zh1-4] and [Gon4] for the more general case of bounded symmetric domains.

10.1.2 Examples of L.I.F.'s on the Euclidean unit ball of $\mathbb{C}^n$

Next, we give some examples of linear-invariant families (L.I.F.'s) on the Euclidean unit ball of $\mathbb{C}^n$ (see [Pfa5]).

Example 10.1.5. The family $S(B)$ (of normalized biholomorphic mappings of B) is a L.I.F. of infinite order in dimension $n \geq 2$. To see that the order is infinite, let $n = 2$, $l \in \mathbb{N}$, $l \geq 2$, and let

$$f(z) = \left(z_1, \frac{z_2}{(1 - z_1)^l} \right) = (z_1, z_2 + l z_2 z_1 + \ldots), \quad z = (z_1, z_2) \in B.$$

Clearly $f \in S(B)$ and since

$$\sum_{j=1}^{2} \frac{\partial^2 f_j}{\partial z_j \partial z_1}(0) = l \to \infty \quad \text{as} \quad l \to \infty,$$

the conclusion follows (see also [Bar-Fit-Gon2]).

Example 10.1.6. The family $K(B)$ of normalized convex mappings of B is a L.I.F. Its behaviour as a L.I.F. is quite different in higher dimensions than in one. For example, the trace order of $K(B)$ in dimension at least two does not give the minimum possible order of a L.I.F., and its exact value is unknown. Moreover, a L.I.F. of minimum order need not be a subset of $K(B)$.

Example 10.1.7. The set $\mathcal{LS}(B)$ is a L.I.F. of infinite order. The set $\mathcal{U}_n(\alpha) \subseteq \mathcal{LS}(B)$, consisting of the union of all L.I.F.'s in $\mathcal{LS}(B)$ of order not greater than α, is also a L.I.F. on B. This is the n-dimensional generalization of the universal linear-invariant family $\mathcal{U}(\alpha) = \mathcal{U}_1(\alpha)$ (see Example 5.1.6).

Example 10.1.8. Let $\mathcal{G}$ be a nonempty subset of $\mathcal{LS}(B)$ and let $\Lambda[\mathcal{G}]$ denote the *L.I.F. generated by* $\mathcal{G}$, i.e.

$$\Lambda[\mathcal{G}] = \Big\{\Lambda_\phi(g) :\ g \in \mathcal{G},\ \phi \in \mathrm{Aut}(B)\Big\}.$$

To see that $\Lambda[\mathcal{G}]$ is a L.I.F., it suffices to observe that the Koebe transform has the group property. Also it is obvious that $\Lambda[\mathcal{G}] = \mathcal{G}$ if and only if $\mathcal{G}$ is a L.I.F. We shall see that many interesting questions about linear-invariant families arise through this process. For example, if $\mathcal{G} = S^*(B)$ (this set is not a L.I.F. since the Koebe transform translates the point $f(0)$), then $\Lambda[S^*(B)]$ is a L.I.F. whose order is unknown in dimension $n \geq 2$.

Any individual element $f \in \mathcal{LS}(B)$ generates the L.I.F. $\Lambda[\{f\}]$. Its order is called the *order of* f, and is denoted by $\mathrm{ord}\, f$ (see [God-Li-St]).

Example 10.1.9. It is also possible to generate L.I.F.'s in higher dimensions beginning with sets of normalized locally univalent functions on the unit disc. For example, let $f \in \mathcal{LS}$ and $\alpha \geq 0$, and define $F_\alpha : B \to \mathbb{C}^n$ by

$$F_\alpha(z) = \big(f(z_1), z'(f'(z_1))^\alpha\big), \quad z = (z_1, z') \in B,$$

where $z' = (z_2, \ldots, z_n)$ and the branches of the power functions are chosen such that $(f'(z_1))^\alpha|_{z_1=0} = 1$. It is obvious that $F_\alpha \in \mathcal{LS}(B)$. If we begin with a subset $\mathcal{F} \subset \mathcal{LS}$, we can construct the corresponding set $(\mathcal{F})_{n,\alpha} \subset \mathcal{LS}(B)$ given by

(10.1.7) $(\mathcal{F})_{n,\alpha} = \Big\{F_\alpha(z) = \big(f(z_1), z'(f'(z_1))^\alpha\big) :\ f \in \mathcal{F}\Big\}.$

We then obtain a L.I.F. by taking $\Lambda[(\mathcal{F})_{n,\alpha}]$. (We note that even if $\mathcal{F}$ is a L.I.F., it is unlikely that $(\mathcal{F})_{n,\alpha}$ is a L.I.F.)

The set $(\mathcal{F})_{n,1/2}$ will be denoted by $(\mathcal{F})_n$. This set was recently investigated by several authors (see [Pfa5], [Gra-Koh2], [Gra-Ham-Koh-Su], [Lic-St2]). In Section 11.5 we shall study the order of $\Lambda[(\mathcal{F})_{n,\alpha}]$, when $\mathcal{F}$ is a L.I.F. on the unit disc of a given order and $\alpha \in [0, 1/2]$.

An important special case occurs when $\mathcal{F} = K$. In this case $(K)_n$ (and hence $\Lambda[(K)_n]$) is a subset of $K(B)$, as shown recently by Roper and Suffridge [Rop-Su1]. (We shall prove this result in Chapter 11.) We remark that ord $\Lambda[(K)_n] = (n+1)/2$, but as already noted, in several variables the exact value of ord $K(B)$ is unknown.

Example 10.1.10. If $f_j \in \mathcal{LS}$, $j = 1, \ldots, n$, then the mapping

$$(10.1.8) \qquad F(z) = (f_1(z_1), \ldots, f_n(z_n)), \quad z = (z_1, \ldots, z_n) \in B,$$

belongs to $\mathcal{LS}(B)$. Hence if we begin with subsets $\mathcal{F}_j$ of $\mathcal{LS}$, $j = 1, \ldots, n$, we may construct the subset of $\mathcal{LS}(B)$ given by

$$\left\{\mathcal{F}_1, \ldots, \mathcal{F}_n\right\} = \left\{ F(z) = (f_1(z_1), \ldots, f_n(z_n)) : f_k \in \mathcal{F}_k, \, k = 1, \ldots, n \right\}.$$

From this we obtain the L.I.F. $\Lambda[\{\mathcal{F}_1, \ldots, \mathcal{F}_n\}]$ generated by $\{\mathcal{F}_1, \ldots, \mathcal{F}_n\}$.

10.2 Distortion results for linear-invariant families in several complex variables

10.2.1 Distortion results for L.I.F.'s on the Euclidean unit ball of $\mathbb{C}^n$

In this section we are going to obtain a distortion theorem for linear-invariant families on the Euclidean unit ball B of $\mathbb{C}^n$. This theorem gives upper and lower bounds for the growth of the Jacobian determinant of mappings in a L.I.F. in terms of the trace order. Later, in Section 10.4, we shall see that there is a distortion theorem for $\|Df(z)\|$ in terms of the norm order. Both of these results are generalizations of Theorem 5.1.8.

We need to deduce some computational results concerning the automorphisms of B. We have seen in Theorem 6.1.23 that up to multiplication by a unitary transformation, the set $\mathrm{Aut}(B)$ consists of all mappings φ_a given by

(10.1.2). If T_a is the linear operator defined by (6.1.10), then it is not difficult to check the following relations (see [Pfa5], [Pfa-Su4]):

(10.2.1) $$T_a(a) = a, \quad a^*T_a(\cdot) = a^*(\cdot),$$

(10.2.2) $$T_a^{-1} = \frac{1}{s_a^2}(I - aa^*)T_a, \quad T_a^2 = aa^* + s_a^2 I.$$

Also straightforward computations yield the following formulas:

(10.2.3) $$(I - aa^*)^{-1} = I + \frac{1}{s_a^2}aa^*,$$

(10.2.4) $$D\varphi_a(z) = T_a\left\{\frac{I(\cdot)}{1 + a^*z} - \frac{z + a}{(1 + a^*z)^2}a^*(\cdot)\right\}$$

and

(10.2.5) $$D^2\varphi_a(z)(v, \cdot) = T_a\left\{\frac{-[va^*(\cdot) + (a^*v)I(\cdot)]}{(1 + a^*z)^2}\right.$$
$$\left. + \frac{2(z + a)(a^*v)a^*(\cdot)}{(1 + a^*z)^3}\right\}.$$

Hence, combining (10.2.4), (10.2.5), (10.2.1) and (10.2.2), we obtain

(10.2.6) $$D\varphi_a(0) = T_a(I - aa^*),$$

(10.2.7) $$[D\varphi_a(0)]^{-1} = \frac{1}{s_a^2}T_a$$

and

(10.2.8) $$[D\varphi_a(0)]^{-1}D^2\varphi_a(0)(v, \cdot) = -(a^*v)I - va^*.$$

We give the details of the proof of (10.2.8) and we leave the proofs of the other relations for the reader:

$$[D\varphi_a(0)]^{-1}D^2\varphi_a(0)(v, \cdot)$$
$$= \frac{1}{s_a^2}(aa^* + s_a^2 I)(-va^* - (a^*v)I + 2a(a^*v)a^*)$$
$$= \frac{1}{s_a^2}\left[-(a^*v)aa^* - s_a^2 va^* - (a^*v)aa^* - s_a^2(a^*v)I\right.$$
$$\left. + 2(a^*a)(a^*v)aa^* + 2s_a^2(a^*v)aa^*\right]$$

$$= \frac{1}{s_a^2}\Big[-2(a^*v)aa^* + 2\|a\|^2(a^*v)aa^* \Big]$$

$$+2(a^*v)aa^* - va^* - (a^*v)I$$

$$= \frac{2}{s_a^2}(-1 + \|a\|^2)(a^*v)aa^* + 2(a^*v)aa^* - va^* - (a^*v)I$$

$$= -(a^*v)I - va^*.$$

The following lemma, due to Pfaltzgraff [Pfa5], is very useful and plays a key role in the proof of the distortion theorem for L.I.F.'s on B.

Lemma 10.2.1. *If $f \in \mathcal{LS}(B)$, $\varphi \in \mathrm{Aut}(B)$, $v \in \mathbb{C}^n$, and $g(z) = \Lambda_\varphi(f)(z)$, then*

(10.2.9)
$$\mathrm{trace}\{D^2 g(0)(v, \cdot)\} = \mathrm{trace}\{[D\varphi(0)]^{-1}D^2\varphi(0)(v, \cdot)\}$$

$$+\mathrm{trace}\{[Df(\varphi(0))]^{-1}D^2 f(\varphi(0))(D\varphi(0)v, \cdot)\}.$$

Proof. By differentiating the formula

$$g(z) = [D\varphi(0)]^{-1}[Df(\varphi(0))]^{-1}(f(\varphi(z)) - f(\varphi(0))),$$

we obtain

$$Dg(z) = [D\varphi(0)]^{-1}[Df(\varphi(0))]^{-1}Df(\varphi(z))D\varphi(z)$$

and

$$D^2 g(z)(v, \cdot) = [D\varphi(0)]^{-1}[Df(\varphi(0))]^{-1}\Big\{Df(\varphi(z))D^2\varphi(z)(v, \cdot)$$

$$+D^2 f(\varphi(z))(D\varphi(z)v, \cdot)D\varphi(z)\Big\}.$$

Setting $z = 0$ in the above gives

$$D^2 g(0)(v, \cdot) = [D\varphi(0)]^{-1}[Df(\varphi(0))]^{-1}\Big\{Df(\varphi(0))D^2\varphi(0)(v, \cdot)$$

$$+D^2 f(\varphi(0))(D\varphi(0)v, \cdot)D\varphi(0)\Big\}.$$

Taking the trace in both sides of this relation and using the fact that the trace is a similarity invariant gives (10.2.9).

Remark 10.2.2. The formula (10.2.9) is a local result and hence a similar formula is valid on the polydisc.

Another fact needed to prove the distortion theorem for L.I.F.'s on B is the *trace formula* for differentiating the determinant of an $n \times n$ matrix-valued holomorphic function of one complex variable (see [Golb]). If $A : \zeta \mapsto A(\zeta)$ is such a function, then

$$\frac{d}{d\zeta} \det(A(\zeta)) = \text{trace}\left\{A^*(\zeta)\frac{d}{d\zeta}A(\zeta)\right\},$$

where $A^*(\zeta)$ is the adjoint matrix of $A(\zeta)$. Moreover, if A is nonsingular, then we can divide both sides of the above equality by $\det(A(\zeta))$ to obtain

$$\frac{d}{d\zeta} \log(\det A(\zeta)) = \text{trace}\left\{A^{-1}(\zeta)\frac{d}{d\zeta}A(\zeta)\right\}.$$

Further, if f is a normalized locally biholomorphic mapping on B and $\|w\| = 1$, then applying the above equality to $A(\rho) = Df(\rho w)$, $\rho \in U$, we obtain (cf. [Pfa2]):

(10.2.10) $$\frac{d}{d\rho} \log[J_f(\rho w)] = \text{trace}\left\{[Df(\rho w)]^{-1}D^2 f(\rho w)(w, \cdot)\right\}.$$

In particular, this result is true when $0 \le \rho < 1$.

We are now able to prove the distortion theorem for the Jacobian determinant on the unit ball in $\mathbb{C}^n$. The first result in this direction was a theorem for L.I.F.'s on the unit ball in $\mathbb{C}^2$ due to Barnard, FitzGerald, and Gong [Bar-Fit-Gon2]. The estimate (10.2.12) was obtained by Liu [LiuT1] and Pfaltzgraff [Pfa5]. Another interesting proof was recently given by Gong and Yu [Gon-Yu]. The estimate (10.2.11) was given in a preprint of Liu [LiuT1] (cf. [Gon4]), and is also contained in a more general distortion theorem of Gong and Zheng [Gon-Zh1, Theorem 2] (cf. [Gon4, Theorem 5.4.1]). (The work of Gong and Zheng is carried out on bounded symmetric domains in $\mathbb{C}^n$ and measures the growth of $|J_f(z)|$ relative to the Bergman kernel function. Originally they considered only L.I.F.'s consisting of biholomorphic maps; locally biholomorphic maps are considered in [Gon4].) The proof below uses similar arguments as in [Pfa5, Theorem 5.1].

Theorem 10.2.3. *Let $\mathcal{F} \subset LS(B)$ be a L.I.F. with* $\text{ord}\,\mathcal{F} = \alpha < \infty$ *and* $f \in \mathcal{F}$. *Then*

(10.2.11) $$\left|\log\left[J_f(z)(1 - \|z\|^2)^{\frac{n+1}{2}}\right]\right| \le \alpha\log\left(\frac{1 + \|z\|}{1 - \|z\|}\right), \quad z \in B,$$

where the branch of the logarithm on the left is chosen to have the value 0 when $z = 0$. Consequently,

$$(10.2.12) \quad \frac{(1 - \|z\|)^{\alpha - \left(\frac{n+1}{2}\right)}}{(1 + \|z\|)^{\alpha + \left(\frac{n+1}{2}\right)}} \leq |J_f(z)| \leq \frac{(1 + \|z\|)^{\alpha - \left(\frac{n+1}{2}\right)}}{(1 - \|z\|)^{\alpha + \left(\frac{n+1}{2}\right)}}, \quad z \in B,$$

and

$$(10.2.13) \quad |\arg J_f(z)| \leq \alpha \log \left(\frac{1 + \|z\|}{1 - \|z\|}\right), \quad z \in B.$$

Proof. Let $\varphi = \varphi_a \in \text{Aut}(B)$ be given by (10.1.2) with $a \in B \setminus \{0\}$, and let $g(z) = \Lambda_\varphi(f)(z)$. From (10.2.8) we have

$$[D\varphi(0)]^{-1} D^2 \varphi(0)(v, \cdot) = -(a^* v)I - va^*, \quad v \in \mathbb{C}^n,$$

and hence in view of (10.2.9) we deduce that

$$\text{trace}\{D^2 g(0)(v, \cdot)\} = \text{trace}\{-(a^* v)I - va^*\}$$

$$+ \text{trace}\{[Df(a)]^{-1} D^2 f(a)(D\varphi_a(0)v, \cdot)\}.$$

Letting

$$v = [D\varphi_a(0)]^{-1} a = \frac{1}{s_a^2} T_a(a) = \frac{1}{s_a^2} a$$

in the above, we obtain

$$\text{trace}\left\{D^2 g(0)\left(\frac{1}{s_a^2} a, \cdot\right)\right\} = -\frac{1}{s_a^2} \text{trace}\{(a^* a)I + aa^*\}$$

$$+ \text{trace}\{[Df(a)]^{-1} D^2 f(a)(a, \cdot)\}$$

$$= -\frac{1}{s_a^2}(n + 1)\|a\|^2 + \text{trace}\{[Df(a)]^{-1} D^2 f(a)(a, \cdot)\}.$$

Now, fix $w \in B \setminus \{0\}$ and set $a = \rho w$, $\rho \in (0, 1]$, in the preceding formula. Applying the relation (10.2.10) to the term $\text{trace}\{[Df(\rho w)]^{-1} D^2 f(\rho w)(\rho w, \cdot)\}$ we obtain

$$\text{trace}\left\{D^2 g(0)\left(\frac{\rho w}{1 - \rho^2 \|w\|^2}, \cdot\right)\right\} = -(n + 1)\frac{\rho^2 \|w\|^2}{1 - \rho^2 \|w\|^2}$$

$$+ \rho \frac{d}{d\rho} \log(J_f(\rho w)),$$

or equivalently,

$$\frac{n+1}{2}\frac{d}{d\rho}\log(1-\rho^2\|w\|^2) + \frac{d}{d\rho}\log(J_f(\rho w))$$

$$= \text{trace}\left\{\frac{1}{2}D^2g(0)\left(\frac{w}{\|w\|}, \cdot\right)\right\}\frac{d}{d\rho}\log\left(\frac{1+\rho\|w\|}{1-\rho\|w\|}\right).$$

Therefore, we deduce that

$$(10.2.14) \qquad \frac{d}{d\rho}\log\left[J_f(\rho w)(1-\rho^2\|w\|^2)^{\frac{n+1}{2}}\right]$$

$$= \text{trace}\left\{\frac{1}{2}D^2g(0)\left(\frac{w}{\|w\|}, \cdot\right)\right\}\frac{d}{d\rho}\log\left(\frac{1+\rho\|w\|}{1-\rho\|w\|}\right).$$

Since ord $\mathcal{F} = \alpha$, we have

$$\left|\text{trace}\left\{\frac{1}{2}D^2g(0)\left(\frac{w}{\|w\|}, \cdot\right)\right\}\right| \le \alpha,$$

and hence from (10.2.14) we obtain

$$\left|\frac{d}{d\rho}\log\left[J_f(\rho w)(1-\rho^2\|w\|^2)^{\frac{n+1}{2}}\right]\right| \le \alpha\frac{d}{d\rho}\log\left(\frac{1+\rho\|w\|}{1-\rho\|w\|}\right).$$

If we integrate both sides of this inequality with respect to ρ over the interval $[0, 1]$ we obtain (10.2.11), as desired.

Finally, it suffices to observe that (10.2.11) implies both relations (10.2.12) and (10.2.13), by taking real and imaginary parts. This completes the proof.

We remark that equality in (10.2.11) and (10.2.12) is achieved by the mapping

$$(10.2.15) \qquad F_{n,\alpha}(z) = \left(f_{2\alpha/(n+1)}(z_1), z'\sqrt{f'_{2\alpha/(n+1)}(z_1)}\right),$$

for $z = (z_1, z') \in B$, where $z' = (z_2, \ldots, z_n)$ and

$$(10.2.16) \qquad f_\beta(z_1) = \frac{1}{2\beta}\left\{\left(\frac{1+z_1}{1-z_1}\right)^\beta - 1\right\}, \qquad \beta > 0.$$

Note that the mapping $F_{n,\alpha}$ is convex on B for $\alpha = (n+1)/2$ (in this case $F_{n,1/2}(z) = z/(1-z_1)$, $z = (z_1, \ldots, z_n) \in B$), biholomorphic on B for $(n+1)/2 \le \alpha \le n+1$, and locally biholomorphic on B for $\alpha > n+1$.

Remark 10.2.4. In one variable we have seen that equality in the distortion result (5.1.8) holds if and only if f is a rotation of the generalized Koebe function, i.e. f is given by (5.1.9). In several variables, the situation is different and there are many mappings for which equality in (10.2.12) holds. To see this, we consider the following example from [God-Li-St]. Let $n \geq 2$ and let $f_1, \ldots, f_n$ be holomorphic functions on U such that $f_k(0) = 1$, $k = 1, \ldots, n$, and

$$\prod_{k=1}^{n} f_k(\zeta) = \frac{(1+\zeta)^{\alpha - \left(\frac{n+1}{2}\right)}}{(1-\zeta)^{\alpha + \left(\frac{n+1}{2}\right)}}, \quad \zeta \in U.$$

For j fixed between 1 and n, let $G_j : B \to \mathbb{C}^n$ be the mapping whose jth component is $\int_0^{z_j} f_j(\zeta)d\zeta$, and whose other components are given by $z_k f_k(z_j)$, $k \neq j$. It is easy to see that $G_j \in LS(B)$, and a short computation yields

$$J_{G_j}(z) = \prod_{k=1}^{n} f_k(z_j), \quad z = (z_1, \ldots, z_n) \in B.$$

For each G_j, equality occurs in (10.2.12) when z_j is real and $z_k = 0$, $k \neq j$.

Among many important consequences of Theorem 10.2.3, we mention the following generalization to higher dimensions of Theorem 5.1.12. This result was obtained in [Pfa5].

Theorem 10.2.5. Let $\mathcal{F} \subset LS(B)$ be a L.I.F. with ord $\mathcal{F} = \alpha$. Then $\alpha \geq (n+1)/2$.

Proof. Suppose $\alpha < (n+1)/2$. From the lower estimate in (10.2.12) we conclude that $|J_f(z)| \to \infty$ as $\|z\| \to 1$, for all $f \in \mathcal{F}$. On the other hand, since each mapping $f \in \mathcal{F}$ is locally biholomorphic on B, $J_f(z) \neq 0$ for $z \in B$. Therefore we obtain a contradiction with the minimum principle for holomorphic functions. Hence we must have ord $\mathcal{F} \geq (n+1)/2$. This completes the proof.

Theorem 10.2.3 also leads to a characterization of the order of a L.I.F. $\mathcal{F}$ in terms of upper and lower bounds for the growth of the Jacobian determinant of the mappings in $\mathcal{F}$. This result was proved by Gong and Zheng in the context of bounded symmetric domains [Gon-Zh3, Lemma 2.1], [Gon4, Lemma 5.4.2]. Also Godula, Liczberski, and Starkov [God-Li-St] gave a proof for the case of the ball in which they showed that the order could be characterized using only the upper bound for the growth of the Jacobian determinant. We shall use similar arguments as in [God-Li-St, Theorem 1].

Theorem 10.2.6. *Let $\mathcal{F} \subset LS(B)$ be a L.I.F. such that* ord $\mathcal{F} = \alpha < \infty$. *Then α is the smallest positive number such that the estimate (10.2.12) holds for all $f \in \mathcal{F}$ and $z \in B$.*

Proof. Let $\beta \geq (n+1)/2$ be such that

$$\frac{(1 - \|z\|)^{\beta - \left(\frac{n+1}{2}\right)}}{(1 + \|z\|)^{\beta + \left(\frac{n+1}{2}\right)}} \leq |J_f(z)| \leq \frac{(1 + \|z\|)^{\beta - \left(\frac{n+1}{2}\right)}}{(1 - \|z\|)^{\beta + \left(\frac{n+1}{2}\right)}},$$

for all $f \in \mathcal{F}$ and $z \in B$. These inequalities are equivalent to the following:

$$\left| \log \left[|J_f(z)|(1 - \|z\|^2)^{\frac{n+1}{2}} \right] \right| \leq \beta \log \left(\frac{1 + \|z\|}{1 - \|z\|} \right).$$

We have to show that $\beta \geq \alpha$. For this purpose, let $g \in \mathcal{F}$. In view of the above distortion result, we have

$$\left| \log \left[|J_g(z)|(1 - \|z\|^2)^{\frac{n+1}{2}} \right] \right| \leq \beta \log \left(\frac{1 + \|z\|}{1 - \|z\|} \right)$$

for $z \in B$. Next, let $z = \rho w$ with w fixed, $\|w\| = 1$, and $0 < \rho < 1$. Then we deduce that

$$\left| \log \left[|J_g(\rho w)|(1 - \rho^2)^{\frac{n+1}{2}} \right] \right| \leq \beta \log \left(\frac{1 + \rho}{1 - \rho} \right),$$

that is,

$$-\beta \log \left(\frac{1 + \rho}{1 - \rho} \right) \leq \mathrm{Re} \, \log \left[J_g(\rho w)(1 - \rho^2)^{\frac{n+1}{2}} \right] \leq \beta \log \left(\frac{1 + \rho}{1 - \rho} \right).$$

Dividing these inequalities by ρ and then letting $\rho \searrow 0$, and using the fact that $J_g(0) = 1$, we obtain

$$-\beta \leq \frac{1}{2} \mathrm{Re} \left\{ \frac{d}{d\rho} \log J_g(\rho w) \right\} \Big|_{\rho = 0} \leq \beta.$$

Taking into account (10.2.10), this relation is equivalent to

$$\left| \frac{1}{2} \mathrm{Re} \left\{ \mathrm{trace} \left\{ D^2 g(0)(w, \cdot) \right\} \right\} \right| \leq \beta.$$

Since $g \in \mathcal{F}$ and w, $\|w\| = 1$, were arbitrarily chosen, we deduce that

$$\sup_{g \in \mathcal{F}} \sup_{\|w\|=1} \left| \frac{1}{2} \mathrm{Re} \left\{ \mathrm{trace} \left\{ D^2 g(0)(w, \cdot) \right\} \right\} \right| \leq \beta.$$

Finally it suffices to remark that

$$\sup_{g \in \mathcal{F}} \sup_{\|w\|=1} \left| \frac{1}{2} \mathrm{trace} \left\{ D^2 g(0)(w, \cdot) \right\} \right|$$

$$= \sup_{g \in \mathcal{F}} \sup_{\|w\|=1} \left| \frac{1}{2} \mathrm{Re} \left\{ \mathrm{trace} \left\{ D^2 g(0)(w, \cdot) \right\} \right\} \right|,$$

since we are free to multiply w by a complex scalar of modulus 1. Consequently, $\alpha \leq \beta$, as desired. This completes the proof.

Remark 10.2.7. Let $\mathcal{F}$ be a L.I.F. of finite order. From Theorem 10.2.6 one may deduce another equivalent definition of the order of $\mathcal{F}$ (cf. [God-Li-St], [Gon-Zh3],[Gon4]):

$$(10.2.17) \qquad \mathrm{ord}\, \mathcal{F} = \inf \left\{ \alpha : \frac{(1 - \|z\|)^{\alpha - \left(\frac{n+1}{2}\right)}}{(1 + \|z\|)^{\alpha + \left(\frac{n+1}{2}\right)}} \leq |J_f(z)| \right.$$

$$\left. \leq \frac{(1 + \|z\|)^{\alpha - \left(\frac{n+1}{2}\right)}}{(1 - \|z\|)^{\alpha + \left(\frac{n+1}{2}\right)}}, \ f \in \mathcal{F}, \ z \in B \right\}.$$

This latter definition of the order of L.I.F.'s has an interesting consequence, due to Godula, Liczberski and Starkov [God-Li-St].

Corollary 10.2.8. *Let $f_1, f_2 \in \mathcal{LS}(B)$ be such that $J_{f_1}(z) = J_{f_2}(z)$, $z \in B$. Then $\mathrm{ord}\, f_1 = \mathrm{ord}\, f_2$.*

Proof. Let $\phi \in \mathrm{Aut}(B)$ and $g_i = \Lambda_\phi(f_i)$, $i = 1, 2$. It is obvious that

$$J_{g_i}(z) = \frac{J_{f_i}(\phi(z)) J_\phi(z)}{J_{f_i}(\phi(0)) J_\phi(0)}, \quad z \in B, \quad i = 1, 2,$$

and hence $J_{g_1}(z) = J_{g_2}(z)$, $z \in B$. Taking into account (10.2.17), the conclusion follows.

10.2.2   Distortion results for L.I.F.'s on the unit polydisc of $\mathbb{C}^n$

In this section we obtain a distortion theorem for L.I.F.'s on the unit polydisc P of $\mathbb{C}^n$, and thus we consider $\mathbb{C}^n$ with the maximum norm $\|\cdot\|_\infty$. For the proof we use similar arguments as in the proof of Theorem 10.2.3, including the fact that Lemma 10.2.1 is valid on the polydisc. The distortion estimate (10.2.19) was obtained in [Pfa-Su2], and (10.2.18) as well as (10.2.19) can be obtained from the work of Gong and Zheng [Gon-Zh1] (cf. [Gon4, Theorem 5.4.1]), if we interpret their result for the polydisc.

We let $\mathcal{LS}(P)$ denote the set of normalized locally biholomorphic mappings from P into $\mathbb{C}^n$. The Koebe transform and the notions of linear-invariance and order of a L.I.F. on P are completely analogous to the case of the unit ball. The automorphisms of the polydisc are given in Theorem 6.1.25.

Theorem 10.2.9. *Let $\mathcal{F} \subset \mathcal{LS}(P)$ be a L.I.F. with* $\operatorname{ord} \mathcal{F} = \alpha < \infty$. *If* $f \in \mathcal{F}$, *then*

$$(10.2.18) \qquad \left| \log \left[J_f(z) \prod_{j=1}^{n} (1 - |z_j|^2) \right] \right| \leq \alpha \log \left(\frac{1 + \|z\|_\infty}{1 - \|z\|_\infty} \right),$$

for all $z = (z_1, \ldots, z_n) \in P$, *where the branch of the logarithm on the left is chosen to have the value 0 when $z = 0$. Consequently,*

$$(10.2.19) \qquad \frac{(1 - \|z\|_\infty)^{\alpha-n}}{(1 + \|z\|_\infty)^{\alpha+n}} \leq |J_f(z)| \leq \left(\frac{1 + \|z\|_\infty}{1 - \|z\|_\infty} \right)^\alpha \prod_{j=1}^{n} (1 - |z_j|^2)^{-1},$$

for all $z \in P$.

Proof. Let $a = (a_1, \ldots, a_n) \in P \setminus \{0\}$ and let $\psi \in \operatorname{Aut}(P)$ be given by

$$\psi(z) = \psi_{-a}(z) = \left(\frac{z_1 + a_1}{1 + \bar{a}_1 z_1}, \ldots, \frac{z_n + a_n}{1 + \bar{a}_n z_n} \right), \quad z = (z_1, \ldots, z_n) \in P,$$

in the notation of Theorem 6.1.25. Let $g = \Lambda_\psi(f)$. Using the trace formula (10.2.9) and taking into account (6.1.14) and (6.1.15), we obtain

$$\operatorname{trace}\{D^2 g(0)(v, \cdot)\} = -2a^* v + \operatorname{trace}\{[Df(a)]^{-1} D^2 f(a)(D\psi(0)v, \cdot)\},$$

for all $v \in \mathbb{C}^n$. Setting

$$v = [D\psi(0)]^{-1} a = \left(\ldots, \frac{a_j}{1 - |a_j|^2}, \ldots \right)$$

in the above gives

$$\text{trace}\Big\{[Df(a)]^{-1}D^2f(a)(a,\cdot)\Big\} - 2\sum_{j=1}^{n}\frac{|a_j|^2}{1-|a_j|^2}$$

$$= \text{trace}\Big\{D^2g(0)\Big(\Big(\cdots,\frac{a_j}{1-|a_j|^2},\cdots\Big),\cdot\Big)\Big\}.$$

Further, if we fix $w \in B \setminus \{0\}$ and set $a = \rho w$, $0 < \rho \leq 1$, in the above, and use similar arguments as in the proof of Theorem 10.2.3, we deduce the relation

$$\frac{d}{d\rho}\log\Big[J_f(\rho w)\prod_{j=1}^{n}(1-\rho^2|w_j|^2)\Big]$$

$$= \text{trace}\Big\{\frac{1}{2}D^2g(0)(\lambda,\cdot)\Big\}\frac{d}{d\rho}\log\Big(\frac{1+\rho\|w\|_\infty}{1-\rho\|w\|_\infty}\Big),$$

where $\lambda = (\lambda_1,\ldots,\lambda_n)$ and

$$\lambda_j = \frac{1-\rho^2\|w\|_\infty^2}{1-\rho^2|w_j|^2}\cdot\frac{w_j}{\|w\|_\infty}, \quad j = 1,\ldots,n.$$

Since $g \in \mathcal{F}$, $\|\lambda\|_\infty = 1$ and $\text{ord}\,\mathcal{F} = \alpha$, it follows that

$$\Big|\text{trace}\Big\{\frac{1}{2}D^2g(0)(\lambda,\cdot)\Big\}\Big| \leq \alpha,$$

and hence

$$\Big|\frac{d}{d\rho}\log\Big[J_f(\rho w)\prod_{j=1}^{n}(1-\rho^2|w_j|^2)\Big]\Big| \leq \alpha\frac{d}{d\rho}\log\Big(\frac{1+\rho\|w\|_\infty}{1-\rho\|w\|_\infty}\Big).$$

Integrating both sides of the above inequality in ρ between 0 and 1, we obtain (10.2.18). Also, taking the real part inside the left-hand side of (10.2.18) and using the fact that the minimum of $|J_f(z)|$ on the closed polydisc $\|z\|_\infty \leq r$, $r < 1$, occurs on the distinguished boundary, we easily deduce (10.2.19). This completes the proof.

Note that equality in (10.2.18) and (10.2.19) is achieved by the mapping

$$F_\alpha(z) = (f_{\alpha/n}(z_1),\ldots,f_{\alpha/n}(z_n)), \quad z = (z_1,\ldots,z_n) \in P,$$

where f_β is given by (10.2.16). We remark that the mapping F_α is in $K(P)$ for $\alpha = n$, is biholomorphic on P for $n \leq \alpha \leq 2n$, and is locally biholomorphic on P for $\alpha > n$.

The lower bound for the distortion in (10.2.19) and the minimum principle for holomorphic functions give a lower bound for the order of a L.I.F. on the unit polydisc (see [Pfa-Su2]):

Theorem 10.2.10. *If $\mathcal{F} \subset LS(P)$ is a L.I.F., then* ord $\mathcal{F} \geq n$.

Moreover, similar reasoning as in the proof of Theorem 10.2.6 yields the following result (cf. [Gon4, Lemma 5.4.2]; [Ham-Koh11]):

Theorem 10.2.11. *Let $\mathcal{F}$ be a L.I.F. on P such that* ord $\mathcal{F} = \alpha < \infty$. *Then α is the smallest positive number for which (10.2.19) holds for all $f \in \mathcal{F}$ and $z \in P$.*

Proof. Let $\beta \geq n$ be such that

$$\frac{(1 - \|z\|_\infty)^{\beta-n}}{(1 + \|z\|_\infty)^{\beta+n}} \leq |J_f(z)| \leq \left(\frac{1 + \|z\|_\infty}{1 - \|z\|_\infty}\right)^\beta \prod_{j=1}^{n}(1 - |z_j|^2)^{-1},$$

for all $f \in \mathcal{F}$ and $z = (z_1, \ldots, z_n) \in P$. We have to show that $\beta \geq \alpha$. For this purpose, let $f \in \mathcal{F}$. It is not difficult to see that the above inequalities imply the relation

$$\left| \log\left[|J_f(z)| \prod_{j=1}^{n}(1 - |z_j|^2)\right]\right| \leq \beta \log\left(\frac{1 + \|z\|_\infty}{1 - \|z\|_\infty}\right),$$

for all $z = (z_1, \ldots, z_n) \in P$.

Next, setting $z = \rho w$ with $\|w\|_\infty = 1$ and $\rho \in (0, 1)$, we obtain

$$-\beta \log\left(\frac{1 + \rho}{1 - \rho}\right) \leq \text{Re } \log\left[J_f(\rho w) \prod_{j=1}^{n}(1 - \rho^2|w_j|^2)\right] \leq \beta \log\left(\frac{1 + \rho}{1 - \rho}\right).$$

Dividing these inequalities by ρ and letting $\rho \searrow 0$, and using the fact that $J_f(0) = 1$, we obtain

$$-\beta \leq \frac{1}{2}\text{Re}\left\{\frac{d}{d\rho} \log J_f(\rho w)\right\}\bigg|_{\rho=0} \leq \beta.$$

With similar reasoning as in the proof of Theorem 10.2.6, we deduce that

$$\sup_{f \in \mathcal{F}} \sup_{\|w\|_\infty=1} \left|\frac{1}{2}\text{Re}\left\{\text{trace}\left\{D^2 f(0)(w, \cdot)\right\}\right\}\right| \leq \beta.$$

Since

$$\sup_{f\in\mathcal{F}}\sup_{\|w\|_\infty=1}\left|\frac{1}{2}\text{trace}\left\{D^2f(0)(w,\cdot)\right\}\right|$$

$$= \sup_{f\in\mathcal{F}}\sup_{\|w\|_\infty=1}\left|\frac{1}{2}\text{Re}\left\{\text{trace}\left\{D^2f(0)(w,\cdot)\right\}\right\}\right|,$$

we conclude that $\text{ord}\,\mathcal{F} = \alpha \leq \beta$, as desired. This completes the proof.

Therefore, as in the case of the Euclidean unit ball, we can give an alternative definition of the order of a L.I.F. on the unit polydisc in terms of the growth of the Jacobian determinant (see [Gon4, Lemma 5.4.2] and [Ham-Koh11]. Compare with Remark 10.2.7):

Remark 10.2.12. Let $\mathcal{F}$ be a L.I.F. of finite order on P. Then

$$\text{ord}\,\mathcal{F} = \inf\left\{\alpha : \frac{(1-\|z\|_\infty)^{\alpha-n}}{(1+\|z\|_\infty)^{\alpha+n}} \leq |J_f(z)|\right.$$

$$\left.\leq \left(\frac{1+\|z\|_\infty}{1-\|z\|_\infty}\right)^\alpha \prod_{j=1}^n(1-|z_j|^2)^{-1},\, f\in\mathcal{F},\, z=(z_1,\ldots,z_n)\in P\right\}.$$

Moreover, there is also an analog of Corollary 10.2.8 on the unit polydisc. We have [Ham-Koh11]

Corollary 10.2.13. Let $f_1, f_2 \in \mathcal{LS}(P)$ be such that $J_{f_1}(z) = J_{f_2}(z)$, $z \in P$. Then $\text{ord}\,f_1 = \text{ord}\,f_2$, where $\text{ord}\,f_j = \text{ord}\,\Lambda[\{f_j\}]$, $j=1,2$.

Problems

10.2.1. Let $\mathcal{F} \subset \mathcal{LS}$ be a L.I.F. on the unit disc with $\text{ord}\,\mathcal{F} = \alpha < \infty$ and let $(\mathcal{F})_n = (\mathcal{F})_{n,1/2}$ be the set of locally univalent mappings of the Euclidean unit ball B generated as in Example 10.1.8. Show that

$$\text{ord}\,\Lambda[(\mathcal{F})_n] = \frac{\alpha(n+1)}{2}.$$

(Pfaltzgraff, [Pfa5,6], Liczberski and Starkov, [Lic-St2].)

10.2.2. Prove Theorem 10.2.10. Also prove Corollary 10.2.13.

10.2.3. Let $f \in \mathcal{U}_n(\alpha)$ and $M(r, J_f) = \max_{\|z\|=r}|J_f(z)|$, $0 \leq r < 1$, where $\|\cdot\|$ is the Euclidean norm of $\mathbb{C}^n$. Show that

$$M(r, J_f)\frac{(1-r)^{\alpha+\left(\frac{n+1}{2}\right)}}{(1+r)^{\alpha-\left(\frac{n+1}{2}\right)}}$$

is a non-increasing function of $r \in [0,1)$. Show that for each $v \in \mathbb{C}^n$, $\|v\| = 1$,

$$|J_f(rv)|\frac{(1-r)^{\alpha+\left(\frac{n+1}{2}\right)}}{(1+r)^{\alpha-\left(\frac{n+1}{2}\right)}}$$

is also a non-increasing function of $r \in [0,1)$.
(Liczberski and Starkov, 2000 [Lic-St1].)

   10.2.4. Verify the relations (10.2.1)-(10.2.7).

   10.2.5. Show that the mapping $F_{n,\alpha}$ given by (10.2.15) is biholomorphic on the Euclidean unit ball B of $\mathbb{C}^n$ for $(n+1)/2 \leq \alpha \leq n+1$, and locally biholomorphic on B for $\alpha > n+1$.

10.3 Examples of L.I.F.'s of minimum order on the Euclidean unit ball and the unit polydisc of $\mathbb{C}^n$

10.3.1 Examples of L.I.F.'s of minimum order on the Euclidean unit ball of $\mathbb{C}^n$

   In this section we shall prove two important and unexpected results, due to Pfaltzgraff and Suffridge [Pfa-Su2]:

   • *The Cayley transform does not give bounds for the growth of the Jacobian determinant of all normalized convex mappings of the Euclidean unit ball of* $\mathbb{C}^n$, $n \geq 2$.

   • *The n-dimensional analog of Corollary 5.2.5 (i.e. $\mathcal{F}$ has minimum order if and only if $\mathcal{F} \subseteq K$) does not hold on the Euclidean unit ball of $\mathbb{C}^n$, $n \geq 2$. Indeed, $K(B)$ does not have minimum order.*

   At this time, the order of $K(B)$ is unknown when $n \geq 2$. The first estimates for ord $K(B)$ in several variables were obtained by Barnard, FitzGerald and Gong [Bar-Fit-Gon2] in dimension two.

   Theorem 10.3.1. *Let $n = 2$ and let $B \subset \mathbb{C}^2$ denote the Euclidean unit ball of $\mathbb{C}^2$. Then*

(10.3.1) $$\frac{3}{2} \leq \text{ord}\, K(B) < 1.761.$$

   Proof. It is obvious that ord $K(B) \geq 3/2$, in view of Theorem 10.2.5. Thus we have only to show the upper bound in (10.3.1). For this purpose, let

$f \in K(B)$. From Theorem 6.3.4 we have

$$(10.3.2) \qquad 1 - \text{Re } \langle [Df(z)]^{-1} D^2 f(z)(v,v), z \rangle > 0$$

for $z \in B$ and $v \in \mathbb{C}^2$, $\|v\| = 1$ and $\text{Re } \langle z, v \rangle = 0$.

Since f is normalized, f can be written as follows:

$$f_j(z) = z_1 + d_{2,0}^{(j)} z_1^2 + d_{1,1}^{(j)} z_1 z_2 + d_{0,2}^{(j)} z_2^2 + \dots, \qquad z \in B, \ j = 1, 2.$$

Short computations combined with (10.3.2) yield that

$$(10.3.3) \quad \text{Re } \left\{ 1 - z^* \begin{pmatrix} 2d_{2,0}^{(1)} & d_{1,1}^{(1)} & d_{1,1}^{(1)} & 2d_{0,2}^{(1)} \\ 2d_{2,0}^{(2)} & d_{1,1}^{(2)} & d_{1,1}^{(2)} & 2d_{0,2}^{(2)} \end{pmatrix} \begin{pmatrix} v_1^2 \\ v_1 v_2 \\ v_1 v_2 \\ v_2^2 \end{pmatrix} + \dots \right\} > 0.$$

Fix $z \in B \setminus \{0\}$. Let c_1 and c_2 be positive numbers and define $v = (v_1, v_2)$ by

$$v_1 = \frac{ic_1 z_1}{\sqrt{c_1^2 |z_1|^2 + c_2^2 |z_2|^2}}, \quad v_2 = \frac{ic_2 z_2}{\sqrt{c_1^2 |z_1|^2 + c_2^2 |z_2|^2}}.$$

Then $\|v\| = 1$, $\text{Re } \langle z, v \rangle = 0$ and in view of (10.3.3), we obtain

$$\text{Re } \left\{ 1 + \frac{1}{c_1^2 |z_1|^2 + c_2^2 |z_2|^2} z^* \begin{pmatrix} 2d_{2,0}^{(1)} & d_{1,1}^{(1)} & d_{1,1}^{(1)} & 2d_{0,2}^{(1)} \\ 2d_{2,0}^{(2)} & d_{1,1}^{(2)} & d_{1,1}^{(2)} & 2d_{0,2}^{(2)} \end{pmatrix} \begin{pmatrix} c_1^2 z_1^2 \\ c_1 c_2 z_1 z_2 \\ c_1 c_2 z_1 z_2 \\ c_2^2 z_2^2 \end{pmatrix} + \dots \right\} > 0.$$

We may write z in the form $z = \zeta w$ where $\zeta \in \mathbb{C}$, $0 < |\zeta| < 1$, and $w = (w_1, w_2) \in \mathbb{C}^2$ with $\|w\| = 1$. Then the above inequality becomes

$$\text{Re } \left\{ 1 + \frac{1}{c_1^2 |w_1|^2 + c_2^2 |w_2|^2} \zeta w^* \begin{pmatrix} 2d_{2,0}^{(1)} & d_{1,1}^{(1)} & d_{1,1}^{(1)} & 2d_{0,2}^{(1)} \\ 2d_{2,0}^{(2)} & d_{1,1}^{(2)} & d_{1,1}^{(2)} & 2d_{0,2}^{(2)} \end{pmatrix} \begin{pmatrix} c_1^2 w_1^2 \\ c_1 c_2 w_1 w_2 \\ c_1 c_2 w_1 w_2 \\ c_2^2 w_2^2 \end{pmatrix} + \dots \right\} > 0.$$

Clearly the left-hand side of the above inequality can be written as $\text{Re } h(\zeta)$, where h is a holomorphic function on U such that $h(0) = 1$. Since $\text{Re } h(\zeta) > 0$, $\zeta \in U$, the coefficient of ζ in the power series expansion of h about the origin may be expressed as

$$\frac{(\overline{w}_1, \overline{w}_2)}{c_1^2 |w_1|^2 + c_2^2 |w_2|^2} \begin{pmatrix} 2c_1^2 w_1^2 d_{2,0}^{(1)} + 2c_1 c_2 w_1 w_2 d_{1,1}^{(1)} + 2c_2^2 w_2^2 d_{0,2}^{(1)} \\ 2c_1^2 w_1^2 d_{2,0}^{(2)} + 2c_1 c_2 w_1 w_2 d_{1,1}^{(2)} + 2c_2^2 w_2^2 d_{0,2}^{(2)} \end{pmatrix}$$

$$= 2 \int_0^{2\pi} e^{-it} d\mu(t),$$

where μ is a non-decreasing function on $[0, 2\pi]$ such that $\mu(2\pi) - \mu(0) = 1$ (see the proof of Theorem 2.1.5). Hence

(10.3.4)
$$c_1^2 w_1 |w_1|^2 d_{2,0}^{(1)} + c_1 c_2 |w_1|^2 w_2 d_{1,1}^{(1)} + c_2^2 \overline{w}_1 w_2^2 d_{0,2}^{(1)}$$

$$+ c_1^2 w_1^2 \overline{w}_2 d_{2,0}^{(2)} + c_1 c_2 w_1 |w_2|^2 d_{1,1}^{(2)} + c_2^2 w_2 |w_2|^2 d_{0,2}^{(2)}$$

$$= (c_1^2 |w_1|^2 + c_2^2 |w_2|^2) \int_0^{2\pi} e^{-it} d\mu(t).$$

Let $w_1 = |w_1| e^{i\theta_1}$ and $w_2 = |w_2| e^{i\theta_2}$, where $\theta_1, \theta_2 \in [0, 2\pi]$. Multiplying both sides of equation (10.3.4) by $\overline{w}_1$ and then integrating with respect to $d\theta_1/2\pi$, we deduce that

$$\left| c_1^2 |w_1|^4 d_{2,0}^{(1)} + c_1 c_2 |w_1|^2 |w_2|^2 d_{1,1}^{(2)} \right| \leq (c_1^2 |w_1|^2 + c_2^2 |w_2|^2) |w_1|,$$

and hence

(10.3.5)
$$\left| d_{2,0}^{(1)} + \frac{c_2}{c_1} \left| \frac{w_2}{w_1} \right|^2 d_{1,1}^{(2)} \right| \leq \frac{1}{|w_1|} \left[1 + \left(\frac{c_2}{c_1} \right)^2 \left| \frac{w_2}{w_1} \right|^2 \right].$$

Let $(c_2/c_1)|w_2/w_1|^2 = \alpha$ and $|w_1| = x$. Thus (10.3.5) is equivalent to

$$|d_{2,0}^{(1)} + \alpha d_{1,1}^{(2)}| \leq \frac{1}{x} + \alpha^2 \frac{x}{1 - x^2}.$$

Since

$$\operatorname{ord} K(B) = \sup \left\{ \frac{1}{2} \left| \sum_{j=1}^{2} \frac{\partial^2 f_j}{\partial z_j \partial z_1}(0) \right| : f \in K(B) \right\}$$

$$= \sup \left\{ \left| d_{2,0}^{(1)} + \frac{1}{2} d_{1,1}^{(2)} \right| : f \in K(B) \right\},$$

we only need to consider the case $\alpha = 1/2$ in the estimate for $|d_{2,0}^{(1)} + \alpha d_{1,1}^{(2)}|$.

In this situation we obtain

(10.3.6)
$$\left| d_{2,0}^{(1)} + \frac{1}{2} d_{1,1}^{(2)} \right| \leq \frac{1}{x} + \frac{1}{4} \frac{x}{1 - x^2}, \quad 0 < x < 1.$$

It is easy to see that the function on the right-hand side of (10.3.6) has a minimum at $x = \sqrt{\left(9 - \sqrt{33}\right)/6}$ and that this minimum is strictly less than 1.761. Thus from (10.3.6) we obtain that

$$\left| d_{2,0}^{(1)} + \frac{1}{2} d_{1,1}^{(2)} \right| \leq \min_{0 < x < 1} \left\{ \frac{1}{x} + \frac{1}{4} \frac{x}{1 - x^2} \right\} < 1.761.$$

Since f is arbitrary, the conclusion now follows. This completes the proof.

Remark 10.3.2. Liu [LiuT1] showed that if B is the Euclidean unit ball of $\mathbb{C}^n$, then the upper estimate

$$\operatorname{ord} K(B) \leq \frac{n+1}{2} + \frac{\sqrt{2}-1}{2}(n-1)$$

holds (cf. [Gon4]). Therefore, we know that

$$\frac{n+1}{2} \leq \operatorname{ord} K(B) \leq \frac{n+1}{2} + \frac{\sqrt{2}-1}{2}(n-1),$$

when B is the Euclidean unit ball of $\mathbb{C}^n$, $n \geq 2$. Barnard, FitzGerald and Gong [Bar-Fit-Gon2] conjectured that in the case of n complex variables, $n \geq 2$, $\operatorname{ord} K(B) = (n+1)/2$.

However, this conjecture is false, as shown by Pfaltzgraff and Suffridge [Pfa-Su2]:

Theorem 10.3.3. $\operatorname{ord} K(B) > (n+1)/2$ for $n \geq 2$.

Proof.

First step. Let $F : B \to \mathbb{C}^n$ be given by

$$(10.3.7) \quad F(z) = ((1 + (\sqrt{2}/2)z_n)z_1, \ldots, (1 + (\sqrt{2}/2)z_n)z_{n-1}, z_n).$$

We first show that F is convex on the unit ball B. It suffices to prove this assertion in the case $n = 2$, since only minor modifications are needed for $n > 2$ (see also Example 6.3.14 and [Gon4]).

Since

$$F(z) = \left(z_1 + \frac{\sqrt{2}}{2} z_1 z_2, z_2 \right),$$

we have

$$DF(z) = \begin{pmatrix} 1 + \dfrac{\sqrt{2}}{2} z_2 & \dfrac{\sqrt{2}}{2} z_1 \\ 0 & 1 \end{pmatrix},$$

$$[DF(z)]^{-1} = \begin{pmatrix} \left(1 + \dfrac{\sqrt{2}}{2} z_2\right)^{-1} & -\dfrac{z_1}{\sqrt{2}}\left(1 + \dfrac{\sqrt{2}}{2} z_2\right)^{-1} \\ 0 & 1 \end{pmatrix}.$$

Consequently F is a normalized locally biholomorphic mapping on B and

$$D^2 F(z)(v,v) = \left(\sqrt{2} v_1 v_2, 0\right), \quad v = (v_1, v_2) \in \mathbb{C}^2.$$

Moreover, for $z \in B$, $v \in \mathbb{C}^2$, $\|v\| = 1$ and $\mathrm{Re}\,\langle z, v\rangle = 0$, we obtain

$$\mathrm{Re}\,\langle [DF(z)]^{-1} D^2 F(z)(v,v), z\rangle = \mathrm{Re}\left[\frac{\sqrt{2}\bar{z}_1 v_1 v_2}{1 + \dfrac{z_2}{\sqrt{2}}}\right] \le 2|v_1||v_2| \frac{\dfrac{|z_1|}{\sqrt{2}}}{1 - \dfrac{|z_2|}{\sqrt{2}}}.$$

Now, let $h : [0, 1/\sqrt{2}] \to [0, \infty)$ be given by

$$h(y) = \frac{\sqrt{\dfrac{1}{2} - y^2}}{1 - y}, \quad 0 \le y \le 1/\sqrt{2}.$$

Then it is elementary to see that h assumes its maximum value 1 when $y = 1/2$, and hence we deduce that

$$\mathrm{Re}\,\langle [DF(z)]^{-1} D^2 F(z)(v,v), z\rangle \le 2|v_1||v_2| \frac{\dfrac{|z_1|}{\sqrt{2}}}{1 - \dfrac{|z_2|}{\sqrt{2}}} \le \max_{0 \le y \le 1/\sqrt{2}} h(y) = 1.$$

In view of Theorem 6.3.4, we conclude that F is a convex mapping of B.

Second step. We now prove that $\mathrm{ord}\, K(B) > (n+1)/2$ for $n \ge 2$. For this purpose, let $x \in (-1, 1)$ and $\phi \in \mathrm{Aut}(B)$ be given by

$$\phi(z) = \left(\frac{\sqrt{1 - x^2}}{1 - x z_n} z_1, \ldots, \frac{\sqrt{1 - x^2}}{1 - x z_n} z_{n-1}, \frac{z_n - x}{1 - x z_n}\right).$$

Also let F be given by (10.3.7) and $c = \sqrt{2}/2$. Straightforward computation shows that the Koebe transform $\Lambda_\phi(F)$ is given by

$$\Lambda_\phi(F)(z) = \left(\frac{1 + \dfrac{c - x}{1 - cx} z_n}{(1 - x z_n)^2} z_1, \ldots, \frac{1 + \dfrac{c - x}{1 - cx} z_n}{(1 - x z_n)^2} z_{n-1}, \frac{z_n}{1 - x z_n}\right).$$

Let $G = \Lambda_\phi(F)$. Then $G \in K(B)$ and it is easy to deduce that

$$\frac{\partial^2 G_j}{\partial z_j \partial z_n}(0) = \begin{cases} 2x + \dfrac{c-x}{1-cx}, & 1 \leq j \leq n-1 \\ 2x, & j = n. \end{cases}$$

Hence

$$\frac{1}{2}\left| \sum_{j=1}^{n} \frac{\partial^2 G_j}{\partial z_j \partial z_n}(0) \right| = \left| nx + \frac{n-1}{2} \cdot \frac{c-x}{1-cx} \right|.$$

Taking into account (10.1.6) as well as the above equality, one concludes that

$$\text{ord}\, K(B) \geq \sup_{-1 < x < 1} \left| nx + \frac{n-1}{2} \cdot \frac{c-x}{1-cx} \right|.$$

Let

$$g(x) = \left| nx + \frac{n-1}{2} \cdot \frac{c-x}{1-cx} \right|, \quad -1 < x < 1.$$

Straightforward computation yields that g has a maximum value at $x_n = \sqrt{2} - \sqrt{2(n-1)}/(2\sqrt{n})$. Then

$$\text{ord}\, K(B) \geq g(x_n) = \frac{3n-1}{\sqrt{2}} - \sqrt{2n(n-1)} > \frac{n+1}{2} \quad \text{for} \quad n \geq 2.$$

Consequently, $\text{ord}\, K(B) > (n+1)/2$ for $n \geq 2$, as claimed. This completes the proof.

Note that for $n = 2$, $g(x_2) = 5\sqrt{2}/2 - 2 \approx 1.535 > 3/2$ and for $n = 3$, $g(x_3) = 4\sqrt{2} - 2\sqrt{3} \approx 2.19 > 2$.

Open Problem 10.3.4. *Find* $\text{ord}\, K(B)$ *when* B *is the Euclidean unit ball of* $\mathbb{C}^n$, $n \geq 2$.

The following example of Pfaltzgraff and Suffridge [Pfa-Su2] shows that the Cayley transform $f(z) = \dfrac{z}{1-z_1}$, $z \in B$, does not give bounds for the growth of the Jacobian determinant of all normalized convex mappings of B when $n \geq 2$. In fact, sharp bounds for $|J_f(z)|$, $f \in K(B)$, are not known.

Example 10.3.5. The inequality

$$(10.3.8) \qquad \frac{1}{(1+\|z\|)^{n+1}} \leq |J_f(z)| \leq \frac{1}{(1-\|z\|)^{n+1}}$$

is false for some values of $z \in B$ and some $f \in K(B)$, when B is the Euclidean unit ball of $\mathbb{C}^n$ with $n \geq 2$.

Proof. Let F and ϕ be as in the proof of Theorem 10.3.3. Also let $G(z) = \Lambda_\phi(F)(z)$. We prove that for the mapping G, the distortion estimate (10.3.8) does not hold for certain z sufficiently near 0. We have

$$G(z) = \left(\frac{1 + \frac{\frac{1}{\sqrt{2}} - x}{1 - \frac{1}{\sqrt{2}}x} z_n}{(1 - x z_n)^2} z_1, \ldots, \frac{1 + \frac{\frac{1}{\sqrt{2}} - x}{1 - \frac{1}{\sqrt{2}}x} z_n}{(1 - x z_n)^2} z_{n-1}, \frac{z_n}{1 - x z_n} \right).$$

A simple computation yields that

$$J_G(z) = \left(1 + \frac{\frac{1}{\sqrt{2}} - x}{1 - \frac{1}{\sqrt{2}}x} z_n \right)^{n-1} \cdot \frac{1}{(1 - x z_n)^{2n}}.$$

Let $z_j = 0$ for $1 \le j \le n - 1$ and $z_n = t$, $-1 < t < 1$. For $n \ge 3$ and $x = 1/\sqrt{2}$, we obtain

$$\frac{d}{dt} \left[J_G(z) - \frac{1}{(1-t)^{n+1}} \right]_{t=0} = \frac{d}{dt} \left[\frac{1}{\left(1 - \frac{t}{\sqrt{2}}\right)^{2n}} - \frac{1}{(1-t)^{n+1}} \right]_{t=0}$$

$$= n(\sqrt{2} - 1) - 1 > 0.$$

Consequently, when $n \ge 3$, we have $|J_G(z)| > \frac{1}{(1 - \|z\|)^{n+1}}$ for t sufficiently near 0 and positive, and $|J_G(z)| < \frac{1}{(1 + \|z\|)^{n+1}}$ for t sufficiently near 0 and negative.

When $n = 2$, choose $x = \sqrt{2} - 1/2$. Then

$$\frac{d}{dt} \left[J_G(z) - \frac{1}{(1-t)^3} \right] = \frac{d}{dt} \left[\frac{1 + \frac{\frac{1}{\sqrt{2}} - x}{1 - \frac{1}{\sqrt{2}}x} t}{(1 - xt)^4} - \frac{1}{(1-t)^3} \right]_{t=0}$$

$$= 4x - 3 + \frac{\frac{1}{\sqrt{2}} - x}{1 - \frac{x}{\sqrt{2}}} = 5\sqrt{2} - 7 > 0.$$

Therefore, again we deduce that $|J_G(z)| > \dfrac{1}{(1 - \|z\|)^3}$ for t near·0 and positive, while $|J_G(z)| < \dfrac{1}{(1 + \|z\|)^3}$ for t near 0 and negative.

Open Problem 10.3.6. *Find the sharp bounds for $|J_f(z)|$ when f is a normalized convex mapping of the Euclidean unit ball in $\mathbb{C}^n$, $n \geq 2$.*

We now recall that if $\mathcal{F}$ is a L.I.F. on the unit disc, then in view of Corollary 5.2.5,

$$\operatorname{ord} \mathcal{F} = 1 \iff \mathcal{F} \subseteq K.$$

In several complex variables the analogous property (i.e. $\mathcal{F}$ has minimal order if and only if $\mathcal{F} \subset K(B)$) is not true, as shown recently by Pfaltzgraff and Suffridge [Pfa-Su2,3]. They constructed some examples of L.I.F.'s $\mathcal{F}$ such that $\operatorname{ord} \mathcal{F} = (n+1)/2$, but $\mathcal{F} \not\subset K(B)$ for $n \geq 2$.

Also recently Graham and Kohr [Gra-Koh2] gave another such example. Their construction involves the operators

$$(10.3.9) \quad \Psi_{n,\alpha}(f)(z) = F_\alpha(z) = \left(f(z_1), \left(\frac{f(z_1)}{z_1} \right)^\alpha z' \right), \quad z \in B,$$

where $z' = (z_2, \ldots, z_n)$, $\alpha \in [0,1]$ and f is a normalized locally univalent function on the unit disc U such that $f(\zeta) \neq 0$ for $\zeta \in U \setminus \{0\}$. We choose the branch of the power function such that

$$\left(\frac{f(z_1)}{z_1} \right)^\alpha \bigg|_{z_1=0} = 1.$$

We remark that these operators preserve some interesting geometric properties. For example, in [Gra-Koh2] it is shown that $\Psi_{n,\alpha}(S^*) \subset S^*(B)$ and also $\Psi_{n,\alpha}(S) \subset S^1(B)$, where $S^1(B)$ is the set of normalized univalent mappings of B which can be embedded in Loewner chains (see Section 8.3). However, the operator $\Psi_{n,\alpha}$ does not preserve convexity when $n \geq 2$, and the example of Graham and Kohr is based on this fact. To see that convexity is not preserved, we note the following

Example 10.3.7. Let $\alpha \in [0,1]$, $n = 2$, and $f : U \to \mathbb{C}$ be given by

$$f(\zeta) = \frac{1}{2} \log \left(\frac{1+\zeta}{1-\zeta} \right), \quad \zeta \in U.$$

Then f is convex, but $\Psi_{2,\alpha}(f)$ is not convex on the Euclidean unit ball in $\mathbb{C}^2$.

Proof. Let

$$u = \frac{1}{2} \log\left(\frac{1+z_1}{1-z_1}\right)$$

and

$$v = z_2 \left[\frac{1}{2z_1} \log\left(\frac{1+z_1}{1-z_1}\right)\right]^{\alpha}.$$

If $\Psi_{2,\alpha}(f)(B)$ is a convex domain, then so is its intersection with the plane $\mathrm{Im}\, u = 0$, $\mathrm{Im}\, v = 0$. This intersection contains the entire real u-axis and precisely the interval $(-1,1)$ of the real v-axis. In order to show that convexity is not satisfied, it suffices to show that if $z_1 \to 1$ along the real axis then we are constrained to have $|v| \to 0$.

If $z_1 \to 1$ along the real axis then $\mathrm{Re}\, u \to \infty$ and $\mathrm{Im}\, u = 0$, and

$$|v|^2 = |z_2|^2 \left[\frac{1}{2z_1} \log\left(\frac{1+z_1}{1-z_1}\right)\right]^{2\alpha}.$$

Let $z_2 = \varepsilon > 0$, with ε small and let $z_1 = \sqrt{1-\varepsilon^2}$. Then it is elementary to show that for $\alpha \in (0,1]$, we have

$$|v|^2 = \varepsilon^2 \left[\frac{1}{2\sqrt{1-\varepsilon^2}} \log\left(\frac{1+\sqrt{1-\varepsilon^2}}{1-\sqrt{1-\varepsilon^2}}\right)\right]^{2\alpha}$$

$$= \left[\frac{\varepsilon^{1/\alpha}}{2\sqrt{1-\varepsilon^2}} \left(\log\left(1+\sqrt{1-\varepsilon^2}\right) - \log\left(1-\sqrt{1-\varepsilon^2}\right)\right)\right]^{2\alpha}$$

$$= \left[\frac{\varepsilon^{1/\alpha}}{\sqrt{1-\varepsilon^2}} \log\left(1+\sqrt{1-\varepsilon^2}\right) - \frac{\varepsilon^{1/\alpha}}{\sqrt{1-\varepsilon^2}} \log \varepsilon\right]^{2\alpha}.$$

Now it is easy to check that $|v|^2 \to 0$ as $\varepsilon \to 0$.

On the other hand, if $\alpha = 0$ then

$$|v|^2 = \varepsilon^2 \to 0 \quad \text{as} \quad \varepsilon \to 0.$$

Hence $|v^2| \to 0$ as $\varepsilon \to 0$, for $\alpha \in [0,1]$. This completes the proof.

Using the idea in Example 10.1.8, we are now able to prove the following result [Gra-Koh2]. Another example of a L.I.F. with minimum order will be given in Chapter 11. We shall use similar arguments as in the proof of [Lic-St2, Theorem 5].

Theorem 10.3.8. *The L.I.F.* $\Lambda[\Psi_{n,0}(K)]$ *has minimum order, that is* $\operatorname{ord}\Lambda[\Psi_{n,0}(K)] = (n+1)/2$. *However,* $\Lambda[\Psi_{n,0}(K)]$ *is not a subset of* $K(B)$ *for* $n \geq 2$.

Proof. The fact that $\Lambda[\Psi_{n,0}(K)] \not\subseteq K(B)$ for $n \geq 2$ is a direct consequence of Example 10.3.7. Thus we have only to show that $\operatorname{ord}\Lambda[\Psi_{n,0}(K)] = (n+1)/2$. For this purpose, let $\mathcal{G} = \Lambda[\Psi_{n,0}(K)]$ and let $\beta = \operatorname{ord}\mathcal{G}$. From Theorem 10.2.6 we know that β is the smallest positive number for which the inequality

$$(10.3.10) \qquad \frac{(1-\|z\|)^{\beta-\frac{n+1}{2}}}{(1+\|z\|)^{\beta+\frac{n+1}{2}}} \leq |J_G(z)| \leq \frac{(1+\|z\|)^{\beta-\frac{n+1}{2}}}{(1-\|z\|)^{\beta+\frac{n+1}{2}}}$$

holds for all $G \in \mathcal{G}$ and $z \in B$.

Let $G \in \mathcal{G}$. Then $G(z) = \Lambda_\phi(F)(z)$ for some $F \in \Psi_{n,0}(K)$ and $\phi \in \operatorname{Aut}(B)$. Taking into account Theorem 6.1.23, we may write $\phi(z) = Vh_a(z)$ for some $V \in \mathcal{U}$ and $a = (a_1, \ldots, a_n) \in B$, where $\mathcal{U}$ is the set of unitary transformations in $\mathbb{C}^n$,

$$h_a(z) = -\phi_a(z) = T_a\left(\frac{a-z}{1-a^*z}\right), \ z \in B,$$

and ϕ_a, T_a and s_a are given by (6.1.9), (6.1.10) and (6.1.11), respectively.

Straightforward computation yields the relation

$$|J_G(z_1, 0, \ldots, 0)| = \left| \frac{J_F(\phi(z_1, 0, \ldots, 0))J_\phi(z_1, 0, \ldots, 0)}{J_F(\phi(0))J_\phi(0)} \right|$$

$$= \left| \frac{J_F(\phi(z_1, 0, \ldots, 0))}{J_F(\phi(0))} \right| \cdot \frac{1}{|1 - z_1\bar{a}_1|^{n+1}}, \quad |z_1| < 1.$$

Here we have used the fact that

$$|J_\phi(z)| = |J_{\phi_a}(z)| = \left[\frac{s_a}{|1-a^*z|}\right]^{n+1}, \ z \in B,$$

by (6.1.12).

Now since $F \in \Psi_{n,0}(K)$, there exists a function $f \in K$ such that

$$F(z) = (f(z_1), z'), \quad z = (z_1, z') \in B.$$

It follows that

$$|J_F(z)| = |f'(z_1)|, \quad z = (z_1, z') \in B,$$

and thus

$$(10.3.11) \quad |J_G(z_1, 0, \ldots, 0)| = \left| \frac{f'((\phi(z_1, 0, \ldots, 0))_1)}{f'((\phi(0))_1)} \right| \cdot \frac{1}{|1 - z_1\bar{a}_1|^{n+1}}.$$

Further, let $b = (b_1, \ldots, b_n) = Va$ and let $v = (v_1, \ldots, v_n)$ denote the first column of the unitary matrix V. Then $\|b\| = \|a\|$ and a short computation shows that

$$\phi(z_1, 0, \ldots, 0) = \frac{b - sz_1v - (1-s)z_1\|b\|^{-2}(b^*v)b}{1 - z_1b^*v}$$

$$= h_b(z_1v), \quad |z_1| < 1,$$

where $s = s_a = s_b$. If we let

$$c = sv_1 + (1-s)b_1\|b\|^{-2}b^*v$$

and

$$\psi(z_1) = (\phi(z_1, 0, \ldots, 0))_1,$$

then we have

$$\psi(z_1) = \frac{b_1 - z_1c}{1 - z_1b^*v} = \frac{b_1 - z_1c}{1 - z_1\bar{a}_1}.$$

There are two possibilities: either $b_1\bar{a}_1 \neq c$ or $b_1\bar{a}_1 = c$. In the first case, ψ is a univalent function on the unit disc U and $\psi(U)$ is a disc contained in U. Taking into account (10.3.11), we obtain

$$(10.3.12) \qquad |J_G(z_1, 0, \ldots, 0)| = \left| \frac{f'(\psi(z_1))}{f'(\psi(0))} \right| \cdot \frac{1}{|1 - z_1\bar{a}_1|^{n+1}}$$

$$= \left| \frac{f'(\psi(z_1))}{f'(b_1)(1 - z_1\bar{a}_1)^2} \right| \cdot \frac{1}{|1 - z_1\bar{a}_1|^{n-1}}.$$

Setting

$$g(z_1) = \Lambda_\psi(f)(z_1) = \frac{f(\psi(z_1)) - f(b_1)}{f'(b_1)(b_1\bar{a}_1 - c)}, \quad |z_1| < 1,$$

we deduce that $g \in K$. To show this, it suffices to observe that if f is a convex univalent function on U and D is any disc such that $D \subseteq U$, then $f(D)$ is a convex domain. (This follows from the fact that $f(U_r)$ is convex for $0 < r < 1$.) In particular, $f(\psi(U))$ is a convex domain. We also note that g is normalized.

Consequently, the distortion result

$$\frac{1}{(1+|z_1|)^2} \leq |g'(z_1)| \leq \frac{1}{(1-|z_1|)^2}$$

holds on U (see (2.2.7)). Thus, if we let $\gamma = (n+1)/2$ and use the above relation and the equalities (10.3.12), we deduce that

$$(10.3.13) \qquad \frac{(1-|z_1|)^{\gamma - \frac{n+1}{2}}}{(1+|z_1|)^{\gamma + \frac{n+1}{2}}} \leq |J_G(z_1, 0, \ldots, 0)| \leq \frac{(1+|z_1|)^{\gamma - \frac{n+1}{2}}}{(1-|z_1|)^{\gamma + \frac{n+1}{2}}},$$

for all $z_1 \in U$.

In the second case, if $b_1 \bar{a}_1 = c$, then $\psi(z_1) = b_1$, $|z_1| < 1$, and

$$|J_G(z_1, 0, \ldots, 0)| = \frac{1}{|1 - z_1 \bar{a}_1|^{n+1}}.$$

Now, it is obvious that if $\gamma = (n+1)/2$, we have

$$\frac{(1-|z_1|)^{\gamma - \frac{n+1}{2}}}{(1+|z_1|)^{\gamma + \frac{n+1}{2}}} \leq \frac{1}{|1 - z_1 \bar{a}_1|^{n+1}} \leq \frac{(1+|z_1|)^{\gamma - \frac{n+1}{2}}}{(1-|z_1|)^{\gamma + \frac{n+1}{2}}}, \quad |z_1| < 1.$$

Therefore, again the relation (10.3.13) holds.

On the other hand, since the equality $|J_{\Lambda_W(G)}(z)| = |J_G(Wz)|$ holds for all $W \in \mathcal{U}$ and $z \in B$, we conclude from (10.3.13) that

$$\frac{(1-\|z\|)^{\gamma - \frac{n+1}{2}}}{(1+\|z\|)^{\gamma + \frac{n+1}{2}}} \leq |J_G(z)| \leq \frac{(1+\|z\|)^{\gamma - \frac{n+1}{2}}}{(1-\|z\|)^{\gamma + \frac{n+1}{2}}}, \quad z \in B.$$

Since G was arbitrarily chosen, it follows from (10.3.10) and the above estimate that $\beta = \operatorname{ord} \mathcal{G} \leq \gamma$. However, Theorem 10.2.5 gives $\operatorname{ord} \mathcal{G} \geq \gamma$, and so we must have $\operatorname{ord} \mathcal{G} = \gamma$.

Another interesting example of a L.I.F. of minimum order on the unit ball of $\mathbb{C}^n$ is the following due to Pfaltzgraff and Suffridge [Pfa-Su3]. Using Corollary 10.2.8, we can give a simpler proof than the original of Pfaltzgraff and Suffridge. This proof was obtained by Godula, Liczberski and Starkov [God-Li-St]. We have

Theorem 10.3.9. *Let* $\mathcal{F} = \{F \in \mathcal{LS}(B) : J_F(z) \equiv 1\}$ *and let* $\Lambda[\mathcal{F}]$ *be the L.I.F. generated by* $\mathcal{F}$. *Then* $\operatorname{ord} \Lambda[\mathcal{F}] = (n+1)/2$.

Proof. Let $f_0(z) = z$, $z \in B$. Then $\operatorname{ord} f_0 = (n+1)/2$, and in view of Corollary 10.2.8, one deduces that $\operatorname{ord} f = \operatorname{ord} f_0$, for all $f \in \mathcal{F}$. This completes the proof.

Note that Theorems 10.3.8 and 10.3.9 are particular cases of the following result ([Ham-Koh11]; compare with [Lic-St2]). We leave the proof for the reader, since it suffices to use arguments similar to those in the proof of Theorem 10.3.8.

Theorem 10.3.10. *Let* $\mathcal{F}$ *be a nonempty subset of* $\mathcal{LS}(B)$ *such that* $F \in \mathcal{F}$ *if and only if there is* $f \in K$ *for which* $J_F(z) = f'(z_1)$, $z = (z_1, z') \in B$. *Then* $\operatorname{ord} \Lambda[\mathcal{F}] = (n+1)/2$.

We mention that other examples of L.I.F.'s of minimum order on the Euclidean unit ball of $\mathbb{C}^n$, $n \geq 2$, have recently been obtained in [Pfa-Su2,3] and [Gra-Ham-Koh-Su].

10.3.2   Examples of L.I.F.'s of minimum order on the unit polydisc of $\mathbb{C}^n$

We next consider some examples of L.I.F.'s on the unit polydisc P of $\mathbb{C}^n$, $n \geq 2$. Again we find that there are L.I.F.'s that have minimum order and which are not subsets of $K(P)$. However, as shown by Pfaltzgraff and Suffridge [Pfa-Su2], in contrast to the situation for the Euclidean unit ball it is very easy to obtain the exact value of the trace order of $K(P)$, and $\operatorname{ord} K(P)$ is minimal.

Theorem 10.3.11. $\operatorname{ord} K(P) = n$.

Proof. Let $f \in K(P)$. In view of Theorem 6.3.2, there exist $f_j \in K$, $j = 1, \ldots, n$, such that

$$f(z) = (f_1(z_1), \ldots, f_n(z_n)), \; z = (z_1, \ldots, z_n) \in P.$$

It follows that $D^2 f(0)(w, \cdot)$ has the diagonal matrix

$$D^2 f(0)(w, \cdot) = \begin{pmatrix} f_1''(0)w_1 & \cdots & 0 \\ \cdots & \cdots & \cdots \\ 0 & \cdots & f_n''(0)w_n \end{pmatrix},$$

and hence

$$\text{trace}\left\{\frac{1}{2}D^2f(0)(w,\cdot)\right\} = \frac{1}{2}\sum_{j=1}^{n} f_j''(0)w_j.$$

Now, since $f_j \in K$, we have $|f_j''(0)/2| \le 1$, and therefore

$$\left|\text{trace}\left\{\frac{1}{2}D^2f(0)(w,\cdot)\right\}\right| \le \sum_{j=1}^{n}|w_j|\left|\frac{f_j''(0)}{2}\right| \le \sum_{j=1}^{n}|w_j| \le n\|w\|_\infty.$$

Consequently, $\text{ord}\,K(P) \le n$. On the other hand, in view of Theorem 10.2.10, $\text{ord}\,K(P) \ge n$, and hence the conclusion follows.

The next result shows that, as in the case of the ball, there exist L.I.F.'s of minimum order on the unit polydisc P of $\mathbb{C}^n$, $n \ge 2$, which are not subsets of $K(P)$. (See [Pfa-Su2] and compare with Theorem 10.3.8. and Corollary 5.2.5.)

Theorem 10.3.12. *For $n \ge 2$ there exist L.I.F.'s $\mathcal{F} \subset \mathcal{LS}(P)$ such that $\text{ord}\,\mathcal{F} = n$, but $\mathcal{F}$ is not contained in $K(P)$.*

Proof. Consider any mapping $F : P \to \mathbb{C}^n$ of the form

(10.3.14) $$F(z) = (\widetilde{z}, z_n + g(\widetilde{z})), \quad z = (\widetilde{z}, z_n) \in P,$$

where $\widetilde{z} = (z_1, \ldots, z_{n-1})$ and g is a nonconstant holomorphic function from $\|\widetilde{z}\|_\infty < 1$ into $\mathbb{C}$ such that $g(0) = 0$ and $Dg(0) = 0$. These mappings are not convex on P since they do not satisfy the criterion given in Theorem 6.3.2. However, we show that $\text{ord}\,\Lambda[\{F\}] = n$. For this purpose, we consider the set of mappings

$$\{F_\phi(z) = F(\phi(z)) : \phi \in \text{Aut}(P)\}.$$

From (10.3.14) we obtain

$$F_\phi(z) = (\phi_1(z_1), \ldots, \phi_{n-1}(z_{n-1}), \phi_n(z_n) + g(\widetilde{\phi(z)})).$$

Short computations yield that

$$\text{trace}\left\{\frac{1}{2}[DF_\phi(0)]^{-1}D^2F_\phi(0)(w,\cdot)\right\} = \frac{1}{2}\sum_{j=1}^{n}\frac{\phi_j''(0)}{\phi_j'(0)}w_j.$$

Now, suppose that $f \in \Lambda[\{F\}]$ and V is a unitary transformation which preserves the polydisc, i.e. V is the composition of a diagonal unitary transformation and a permutation matrix. Then if $h = \Lambda_V(f)$, we have

$$\text{trace}\{D^2h(0)(v,\cdot)\} = \text{trace}\{D^2f(0)(Vv,\cdot)\}.$$

It therefore suffices to consider automorphisms ϕ of P which have the form $\phi_j(z_j) = e^{i\theta_j}(z_j - a_j)/(1 - \bar{a}_j z_j)$ for $|z_j| < 1$, where $|a_j| < 1$ and $\theta_j \in \mathbb{R}$, $j = 1, \ldots, n$ (see Theorem 6.1.25). In this case we deduce that

$$\frac{1}{2}\left|\sum_{j=1}^{n}\frac{\phi_j''(0)}{\phi_j'(0)}w_j\right| = \left|\sum_{j=1}^{n}\bar{a}_j w_j\right| \leq n\|w\|_\infty.$$

Therefore

$$\left|\text{trace}\left\{\frac{1}{2}[DF_\phi(0)]^{-1}D^2 F_\phi(0)(w, \cdot)\right\}\right| \leq n, \quad \|w\|_\infty = 1,$$

and hence $\text{ord}\,\Lambda[\{F\}] \leq n$. Finally, applying Theorem 10.2.10 we deduce that $\text{ord}\,\Lambda[\{F\}] = n$, as claimed. This completes the proof.

We note that the example constructed in the proof of Theorem 10.3.12 is a particular case of the result below [Ham-Koh11]. (Compare with Theorem 10.3.9.)

Theorem 10.3.13. *Let* $\mathcal{F} = \{f \in \mathcal{LS}(P) : J_f(z) \equiv 1\}$. *Then* $\text{ord}\,\Lambda[\mathcal{F}] = n$, *but* $\Lambda[\mathcal{F}] \not\subseteq K(P)$ *when* $n \geq 2$.

Proof. Let $f_0(z) = z$, $z \in P$. It is not difficult to check that $\text{ord}\,f_0 = n$. Since each $f \in \mathcal{F}$ has the property that $J_f(z) \equiv 1$, it follows from Corollary 10.2.13 that $\text{ord}\,f = \text{ord}\,f_0$, and thus $\text{ord}\,\Lambda[\mathcal{F}] = n$, as desired. The fact that $\Lambda[\mathcal{F}] \not\subseteq K(P)$ when $n \geq 2$ is obvious, since the map F given by (10.3.14) belongs to $\mathcal{F}$ but is not convex on P.

Problems

10.3.1. Let $\alpha \in [0, 1]$ and let $\Psi_{n,\alpha}$ denote the operator defined by (10.3.9). Show that

(i) $\Psi_{n,\alpha}(S^*) \subseteq S^*(B)$.

(ii) $\text{ord}\,\Lambda[\Psi_{n,0}(S)] = \text{ord}\,\Lambda[\Psi_{n,0}(S^*)] = (n+3)/2$.

(Graham and Kohr, 2002 [Gra-Koh2].)

10.3.2. Complete the details in the proof of Theorem 10.3.13.

10.3.3. Let P be the unit polydisc of $\mathbb{C}^n$, $a = (a_1, \ldots, a_n) \in P$ and let $\psi_a \in \text{Aut}(P)$ be given by (6.1.14)-(6.1.15). Show that a subset $\mathcal{F} \subset \mathcal{LS}(P)$

is a linear-invariant family if and only if for each $f \in \mathcal{F}$, the mapping $V^{-1}\Lambda_{\psi_a}(f)(V(\cdot)) \in \mathcal{F}$ for all $a \in P$ and all unitary transformations V of $\mathbb{C}^n$ which preserve P.

10.4 Norm order of linear-invariant families in several complex variables

In this section we are going to study a new notion of order, recently introduced by Pfaltzgraff and Suffridge [Pfa-Su4], for a linear-invariant family of locally biholomorphic mappings of the Euclidean unit ball in $\mathbb{C}^n$. The reader will see throughout this section that the norm order has a much broader range of applicability to the study of geometric properties of locally biholomorphic mappings than does the trace order. For example, using the norm order we may deduce results about the radius of convexity, starlikeness and univalence of mappings in a L.I.F. In the case of one variable both the norm order and the trace order reduce to the usual order of a L.I.F. The basic source for this section is [Pfa-Su4].

We shall keep the notation from the previous sections of this chapter. In particular we recall the notation $f(z; a)$ (see (10.1.3)) for the Koebe transform of a mapping f with $\varphi_a \in \text{Aut}(B)$ given by (10.1.2).

First we give a lemma of Pfaltzgraff and Suffridge [Pfa-Su4] which is useful in the computation of the norm order.

Lemma 10.4.1. Let $\mathcal{F}$ be a L.I.F. and $f \in \mathcal{F}$. Let $a \in B$, $V \in \mathcal{U}$, and let $g(z) = f(z; a)$ and $G(z) = f(Vz; a)$, $z \in B$. Then

$$(10.4.1) \qquad \sup\{\|D^2 G(0)(v, v)\| : \|v\| = 1, \ V \in \mathcal{U}\}$$

$$= \sup\{\|D^2 g(0)(\gamma, \gamma) : \|\gamma\| = 1\}.$$

Proof. Since $G(z) = g(Vz)$ for $z \in B$, it follows that

$$DG(z) = Dg(Vz)V \text{ and } D^2 G(0)(\cdot, \cdot) = D^2 g(0)(V(\cdot), V(\cdot)).$$

Hence

$$D^2 G(0)(v, v) = D^2 g(0)(\gamma, \gamma), \quad \gamma = Vv,$$

and (10.4.1) follows, because it is obvious that $\|\gamma\| = \|v\| = 1$. This completes the proof.

Recall that if $\widetilde{A}_k : \prod_{j=1}^{k} \mathbb{C}^n \to \mathbb{C}^n$ is a symmetric k-linear mapping, then by Theorem 7.2.19, we have

$$(10.4.2) \quad \|\widetilde{A}_k\| = \sup_{\substack{\|z^{(j)}\|=1 \\ 1 \le j \le k}} \|\widetilde{A}_k(z^{(1)}, \ldots, z^{(k)})\| = \sup_{\|z\|=1} \|\widetilde{A}_k(z, \ldots, z)\|.$$

We are now able to define the *norm order* of a L.I.F. on the Euclidean unit ball of $\mathbb{C}^n$. This notion has recently been introduced by Pfaltzgraff and Suffridge [Pfa-Su4].

Definition 10.4.2. If $\mathcal{F}$ is a L.I.F., we define the *norm order* of $\mathcal{F}$ by

$$\|\mathrm{ord}\| \, \mathcal{F} = \sup \left\{ \frac{1}{2}\|D^2 f(0)\| : f \in \mathcal{F} \right\}.$$

In view of Lemma 10.4.1 and (10.4.2) we have

$$\|\mathrm{ord}\| \, \mathcal{F} = \sup \left\{ \frac{1}{2}\|D^2 f(0; a)(v, v)\| : f \in \mathcal{F}, \ \|v\| = 1, \ \|a\| < 1 \right\}.$$

Next we compute the second order Fréchet derivative of $f(z; a)$ at $z = 0$. For this purpose, it suffices to use formulas (10.2.6), (10.2.7), (10.2.8), and the equalities

$$Df(z; a) = [D\varphi_a(0)]^{-1}[Df(a)]^{-1}Df(\varphi_a(z))D\varphi_a(z),$$

and

$$D^2 f(z; a)(\cdot, \cdot) = [D\varphi_a(0)]^{-1}[Df(a)]^{-1}\Big\{ D^2 f(\varphi_a(z))(D\varphi_a(z)(\cdot), D\varphi_a(z)(\cdot))$$

$$+ Df(\varphi_a(z))D^2\varphi_a(z)(\cdot, \cdot) \Big\}.$$

Evaluating at $z = 0$, we deduce for $v \in \mathbb{C}^n$ that

$$(10.4.3) \qquad\qquad \frac{1}{2}D^2 f(0; a)(v, v) = -(a^*v)v +$$

$$+ \frac{1}{2s_a^2}T_a[Df(a)]^{-1}D^2 f(a)(D\varphi_a(0)v, D\varphi_a(0)v),$$

where T_a and s_a are defined by (6.1.10) and (6.1.11).

The next result of Pfaltzgraff and Suffridge [Pfa-Su4] gives the minimum possible value of the norm order of a L.I.F. on B. The corresponding result for the trace order is given in Theorem 10.2.5. In Theorem 10.5.1 we shall see that the lower bound is assumed by the L.I.F. of normalized convex mappings on B.

Theorem 10.4.3. $\|\mathrm{ord}\|\,\mathcal{F} \geq 1$ *for every L.I.F. $\mathcal{F}$.*

Proof. Let $f \in \mathcal{F}$ and $a \in B \setminus \{0\}$, and let $v = a/\|a\|$ in (10.4.3). Since $D\varphi_a(0)a = s_a^2 a$ by (10.2.6), we obtain

$$\frac{1}{2}D^2 f(0;a)\left(\frac{a}{\|a\|},\frac{a}{\|a\|}\right) = -a + \frac{1}{2s_a^2}T_a[Df(a)]^{-1}D^2 f(a)\left(\frac{s_a^2 a}{\|a\|},\frac{s_a^2 a}{\|a\|}\right)$$

$$= -a + \frac{s_a^2}{2\|a\|^2}T_a[Df(a)]^{-1}D^2 f(a)(a,a).$$

Further, let

$$g(\zeta) = \frac{\zeta s_a^2}{2\|a\|^2}\langle T_a[Df(\zeta a)]^{-1}D^2 f(\zeta a)(a,a),a\rangle, \quad |\zeta| < \frac{1}{\|a\|}.$$

Then g is a holomorphic function on the disc $|\zeta| < 1/\|a\|$ and since $g(0) = 0$, for every r with $r < 1/\|a\|$ there is a point ζ, $|\zeta| = r$, such that $\mathrm{Re}\,g(\zeta) \leq 0$. Since $f(z;a) \in \mathcal{F}$, it follows that

$$\|\mathrm{ord}\|\,\mathcal{F} \geq \left\|\frac{1}{2}D^2 f(0;a)\left(\frac{a}{\|a\|},\frac{a}{\|a\|}\right)\right\|$$

$$\geq \left|\frac{1}{2}\left\langle D^2 f(0;a)\left(\frac{a}{\|a\|},\frac{a}{\|a\|}\right),a\right\rangle\right|.$$

We now replace a by ζa where ζ is chosen so that $|\zeta| = 1$ and $\mathrm{Re}\,g(\zeta) \leq 0$. Then

$$\|\mathrm{ord}\|\,\mathcal{F} \geq \left|\frac{1}{2}\left\langle D^2 f(0;\zeta a)\left(\frac{\zeta a}{\|a\|},\frac{\zeta a}{\|a\|}\right),\zeta a\right\rangle\right|$$

$$= \left|-\|a\|^2 + \frac{\zeta s_a^2}{2}\left\langle T_{\zeta a}[Df(\zeta a)]^{-1}D^2 f(\zeta a)\left(\frac{a}{\|a\|},\frac{a}{\|a\|}\right),a\right\rangle\right|$$

$$= |\|a\|^2 - g(\zeta)| \geq \|a\|^2 - \mathrm{Re}\,g(\zeta)| = \|a\|^2 + |\mathrm{Re}\,g(\zeta)|,$$

where we have used the fact that $T_{\zeta a} = T_a$ and $\mathrm{Re}\,g(\zeta) \leq 0$. Hence $\|\mathrm{ord}\|\,\mathcal{F} \geq \|a\|^2$ for all $a \in B \setminus \{0\}$, which implies $\|\mathrm{ord}\|\,\mathcal{F} \geq 1$, as desired.

The following distortion theorem, due to Pfaltzgraff and Suffridge [Pfa-Su4], is a generalization to higher dimensions of the distortion estimate (5.1.8). It involves estimating $\|Df(z)\|$, which is considerably more difficult than estimating $|J_f(z)|$. Since the proof is rather long, we leave it for the reader. However, it is a very interesting proof.

Theorem 10.4.4. *If $\mathcal{F}$ is a L.I.F. with $\|\text{ord}\|\,\mathcal{F} = \alpha < \infty$, then*

$$\frac{(1 - \|z\|)^{\alpha-1}}{(1 + \|z\|)^{\alpha+1}} \leq \inf_{g \in \mathcal{F}} \left\| Dg(z)\left(\frac{z}{\|z\|}\right) \right\|$$

$$\leq \left\| Df(z)\left(\frac{z}{\|z\|}\right) \right\| \leq \|Df(z)\| \leq \frac{(1 + \|z\|)^{\alpha-1}}{(1 - \|z\|)^{\alpha+1}},$$

for all $f \in \mathcal{F}$ and $z \in B$ (with $z \neq 0$ where $z/\|z\|$ occurs).

An upper estimate for the growth of mappings in a L.I.F. follows from Theorem 10.4.4., as shown by Pfaltzgraff and Suffridge [Pfa-Su4]. Only an upper bound is obtained, since a general L.I.F. can include maps with zeros in $B \setminus \{0\}$.

Theorem 10.4.5. *If $\mathcal{F}$ is a L.I.F. with $\|\text{ord}\|\,\mathcal{F} = \alpha < \infty$, then*

$$\|f(z)\| \leq \frac{1}{2\alpha}\left[\left(\frac{1 + \|z\|}{1 - \|z\|}\right)^{\alpha} - 1\right],$$

for all $f \in \mathcal{F}$ and $z \in B$. Equality is achieved by the mapping

(10.4.4) $$F(z) = (f_\alpha(z_1), z'h(z_1)), \quad z = (z_1, z') \in B,$$

where $z' = (z_2, \ldots, z_n)$, $f_\alpha(z_1) = \dfrac{1}{2\alpha}\left[\left(\dfrac{1 + z_1}{1 - z_1}\right)^{\alpha} - 1\right]$ and $h(z_1) = \dfrac{1}{1 - z_1}$.

Proof. Let $z \in B \setminus \{0\}$ and $f \in \mathcal{F}$. Assume that $f(z) \neq 0$ and let $G : [0, 1] \to \mathbb{R}$ be given by

$$G(t) = \text{Re}\left\langle f(tz), \frac{f(z)}{\|f(z)\|}\right\rangle, \quad 0 \leq t \leq 1.$$

Using the upper bound for $\|Df(\cdot)\|$ from Theorem 10.4.4, we obtain the estimate

$$\|f(z)\| = G(1) = \int_0^1 G'(t)dt = \int_0^1 \text{Re}\left\langle Df(tz)z, \frac{f(z)}{\|f(z)\|}\right\rangle dt$$

$$\leq \int_0^1 \|Df(tz)\| \cdot \|z\| dt \leq \int_0^1 \frac{(1+t\|z\|)^{\alpha-1}}{(1-t\|z\|)^{\alpha+1}} \|z\| dt$$

$$= \frac{1}{2\alpha} \left\{ \left(\frac{1+\|z\|}{1-\|z\|}\right)^\alpha - 1 \right\}.$$

This completes the proof.

It is of course possible to obtain upper and lower bounds for the Jacobian determinant of maps in a L.I.F. of given norm order. To do so, it suffices to note that ord $\mathcal{F} \leq n\|\text{ord}\| \mathcal{F}$ (since $|\text{trace}\{A\}| \leq n\|A\|$ for any $n \times n$ matrix A), and to apply Theorem 10.2.3.

Corollary 10.4.6. Let $\mathcal{F}$ be a L.I.F. with $\|\text{ord}\| \mathcal{F} = \alpha < \infty$. If $F \in \mathcal{F}$ then

$$\frac{(1-\|z\|)^{n\alpha-\left(\frac{n+1}{2}\right)}}{(1+\|z\|)^{n\alpha+\left(\frac{n+1}{2}\right)}} \leq |J_F(z)| \leq \frac{(1+\|z\|)^{n\alpha-\left(\frac{n+1}{2}\right)}}{(1-\|z\|)^{n\alpha+\left(\frac{n+1}{2}\right)}}, \quad z \in B.$$

Problems

10.4.1. Let $\mathcal{F}$ be a L.I.F. and let $cl(\mathcal{F})$ be the closure of $\mathcal{F}$ with respect to the usual topology of local uniform convergence. Prove that $cl(\mathcal{F})$ is also a L.I.F. on B and $\|\text{ord}\| cl(\mathcal{F}) = \|\text{ord}\| \mathcal{F}$.
(Pfaltzgraff and Suffridge, 2000 [Pfa-Su4].)

10.4.2. Let $\alpha \geq 1$ and let B denote the Euclidean unit ball of $\mathbb{C}^n$. Also let $F : B \to \mathbb{C}^n$ be given by (10.4.4). Let $\Lambda[\{F\}]$ denote the L.I.F. generated by $\{F\}$. Prove that $\|\text{ord}\| cl(\Lambda[\{F\}]) = \alpha$.
(Pfaltzgraff and Suffridge, 2000 [Pfa-Su4].)

10.4.3. Prove that if $\mathcal{F}$ is a L.I.F. with $\|\text{ord}\| \mathcal{F} = \alpha < \infty$ and if $f \in \mathcal{F}$ then

$$\frac{(1-\|z\|)^{(2n-1)\alpha+(n-3)/2}}{(1+\|z\|)^{(2n-1)\alpha-(n-3)/2}} \|w\| \leq \|Df(z)w\|, \quad z \in B, w \in \mathbb{C}^n.$$

(Pfaltzgraff and Suffridge, 2000 [Pfa-Su4].)

10.5　Norm order and univalence on the Euclidean unit ball of $\mathbb{C}^n$

In this section we determine the norm order of the linear-invariant family $K(B)$, and we investigate some radius problems which are analogous to results in one variable obtained in Chapter 5 (radius of convexity, starlikeness, and univalence). These results indicate that the norm order is much more closely related to geometric properties of the mappings in a given L.I.F. on the Euclidean unit ball in $\mathbb{C}^n$ than the trace order. We begin by showing that, in contrast to the situation for the trace order of the L.I.F. $K(B)$ (see Open Problem 10.3.4), it is possible to give a complete characterization of $\|ord\| K(B)$. This result was obtained recently by Pfaltzgraff and Suffridge [Pfa-Su4].

Theorem 10.5.1. $\|ord\| K(B) = 1$.

Proof. It suffices to apply Theorem 7.2.16.

The next result gives a lower bound for the radius of convexity of a L.I.F. in terms of its norm order [Pfa-Su4]:

Theorem 10.5.2. Let $\mathcal{F}$ be a L.I.F. with $\|ord\| \mathcal{F} = \alpha < \infty$ and let $f \in \mathcal{F}$. Then f maps the ball $B_{1/(2\alpha)}$ biholomorphically onto a convex domain.

Proof. Let $f \in \mathcal{F}$ and $g(z) = f(z;a)$, $z \in B$, $a \in B \setminus \{0\}$. In view of (10.4.3), we have

$$D^2 g(0)(v,v) = -2\langle v, a\rangle v + \frac{1}{s_a^2} T_a [Df(a)]^{-1} D^2 f(a)(D\varphi_a(0)v, D\varphi_a(0)v).$$

We will show that the sufficient condition for convexity in Theorem 6.3.4 holds if $\|a\| < 1/(2\alpha)$. For this purpose, let $w \in \mathbb{C}^n$, $\|w\| = 1$, be such that $\text{Re}\langle w, a\rangle = 0$. Also let $v = [D\varphi_a(0)]^{-1} w = (1/s_a^2) T_a(w)$ in the above formula. Then

$$D^2 g(0)(v,v) = -\frac{2}{s_a^4}\langle T_a(w), a\rangle T_a(w) + \frac{1}{s_a^2} T_a[Df(a)]^{-1} D^2 f(a)(w,w)$$

$$= -\frac{2}{s_a^4}\langle w, a\rangle T_a(w) + \frac{1}{s_a^2} T_a[Df(a)]^{-1} D^2 f(a)(w,w).$$

On the other hand, a straightforward computation yields that

$$\langle D^2 g(0)(v,v), a\rangle = -\frac{2}{s_a^4}\langle w, a\rangle\langle T_a(w), a\rangle + \frac{1}{s_a^2}\langle T_a[Df(a)]^{-1} D^2 f(a)(w,w), a\rangle$$

$$= -\frac{2}{s_a^4}\langle w, a\rangle^2 + \frac{1}{s_a^2}\langle [Df(a)]^{-1}D^2f(a)(w,w), a\rangle$$

$$= \frac{1}{s_a^2}\left\{\langle [Df(a)]^{-1}D^2f(a)(w,w), a\rangle + \frac{2}{s_a^2}|\langle w, a\rangle|^2\right\},$$

where we have used the fact that $\operatorname{Re}\langle w, a\rangle = 0$, and thus $\langle w, a\rangle^2 = -|\langle w, a\rangle|^2$.

Since $\|\operatorname{ord}\|\mathcal{F} = \alpha$, it follows that $\|D^2g(0)(v,v)\| \leq 2\alpha\|v\|^2$. Also since

$$\|v\|^2 = \left\|\frac{1}{s_a^2}T_a(w)\right\|^2 = \frac{1}{s_a^4}(|\langle a, w\rangle|^2 + s_a^2),$$

we obtain in view of the above arguments that

$$\left|\left\{1 - \langle [Df(a)]^{-1}D^2f(a)(w,w), a\rangle\right\} - \left\{1 + \frac{2}{s_a^2}|\langle a, w\rangle|^2\right\}\right|$$

$$= s_a^2|\langle D^2g(0)(v,v), a\rangle| \leq s_a^2\|D^2g(0)(v,v)\| \cdot \|a\|$$

$$\leq \frac{2\alpha}{s_a^2}(|\langle a, w\rangle|^2 + s_a^2)\|a\|.$$

Therefore, we deduce that

$$1 - \operatorname{Re}\langle [Df(a)]^{-1}D^2f(a)(w,w), a\rangle$$

$$\geq 1 + \frac{2}{s_a^2}|\langle a, w\rangle|^2 - \frac{2\alpha}{s_a^2}(|\langle a, w\rangle|^2 + s_a^2)\|a\|$$

$$= (1 - 2\alpha\|a\|) + \frac{2}{s_a^2}|\langle a, w\rangle|^2(1 - \alpha\|a\|) > 0$$

for $\|a\| < 1/(2\alpha)$. From Theorem 6.3.4 we conclude that f maps the ball $B_{1/(2\alpha)}$ biholomorphically onto a convex domain. This completes the proof.

Pfaltzgraff and Suffridge [Pfa-Su4] formulated the following conjecture for the radius of convexity of a L.I.F. of given norm order. If this conjecture is true, it would be a generalization of Theorem 5.2.3. We note that in dimension $n \geq 2$, there is in general no such connection between the trace order of a L.I.F. on B and its radius of convexity. (Theorem 10.5.6 below gives a weaker result in the direction of the conjecture.)

Conjecture 10.5.3. If $\mathcal{F}$ is a L.I.F. on the Euclidean unit ball B of $\mathbb{C}^n$, $n \geq 2$, with $\|\operatorname{ord}\|\mathcal{F} = \alpha < \infty$, then the radius of convexity of $\mathcal{F}$ is $\alpha - \sqrt{\alpha^2 - 1}$.

A covering theorem for L.I.F.'s of given norm order may be deduced from Theorem 10.5.2 [Pfa-Su4]:

Theorem 10.5.4. *If $\mathcal{F}$ is a L.I.F. with $\|\text{ord}\| \, \mathcal{F} = \alpha < \infty$, then for each $f \in \mathcal{F}$ the image $f(B)$ contains the ball $B_{1/(4\alpha)}$.*

Proof. Since $f \in \mathcal{F}$, $g(z) = 2\alpha f\left(\dfrac{z}{2\alpha}\right)$ is a normalized convex mapping of the unit ball B, by Theorem 10.5.2. In view of Theorem 7.2.2, $g(B)$ contains the ball $B_{1/2}$, and hence $f(B)$ contains the ball $B_{1/(4\alpha)}$. This completes the proof.

A lower bound for the radius of starlikeness of a L.I.F. in terms of its norm order was also obtained by Pfaltzgraff and Suffridge [Pfa-Su4] (compare with Corollary 5.2.6):

Theorem 10.5.5. *If $\mathcal{F}$ is a L.I.F. with $\|\text{ord}\| \, \mathcal{F} = \alpha < \infty$, then each $f \in \mathcal{F}$ maps the ball $B_{\rho(\alpha)}$, where $\rho(\alpha) = 4\alpha/(1 + 4\alpha^2)$, biholomorphically onto a starlike domain.*

Proof. Let $z_0 \in B$ with $\|z_0\| < 4\alpha/(1 + 4\alpha^2)$ and choose $a \in B$ such that $z_0 = 2a/(1 + \|a\|^2)$. Then $\|a\| < 1/(2\alpha)$. If $f \in \mathcal{F}$ then the Koebe transform $f(z; a)$ belongs to $\mathcal{F}$ too. Taking into account Theorem 10.5.2, one deduces that $f(z; a)$ maps the ball $B_{1/(2\alpha)}$ biholomorphically onto a convex domain. Because $\|a\| < 1/(2\alpha)$, one concludes that the mapping $f(z; a) - f(-a; a)$ is a starlike mapping on the ball $\|z\| < 1/(2\alpha)$. Therefore, using Theorem 6.2.2, one obtains that

$$\text{Re} \, \langle [Df(z; a)]^{-1}(f(z; a) - f(-a; a)), z \rangle > 0, \quad \|z\| < \frac{1}{2\alpha}.$$

Setting $z = a$ in the above, one deduces that

$$0 < \text{Re} \, \langle [Df(a; a)]^{-1}(f(a; a) - f(-a; a)), a \rangle$$

$$= \frac{(1 + \|a\|^2)^3}{2(1 - \|a\|^2)} \text{Re} \, \langle [Df(z_0)]^{-1} f(z_0), z_0 \rangle.$$

Since z_0 is arbitrary, we conclude that f is starlike on the ball $B_{\rho(\alpha)}$ with $\rho(\alpha) = 4\alpha/(1 + 4\alpha^2)$.

As a further application of the norm order, we show that it is possible to solve a radius problem involving quasi-convexity of type B. We recall (Definition 6.3.20) that if $f \in \mathcal{LS}(B)$, then f is quasi-convex of type B (or b-quasiconvex) if

$$\text{Re } \langle z + [Df(z)]^{-1}D^2f(z)(z,z), z \rangle > 0, \quad z \in B \setminus \{0\}.$$

The following result was obtained in [Pfa-Su4] (compare with Theorem 5.2.3):

Theorem 10.5.6. *If $\mathcal{F}$ is a L.I.F. with $\|\text{ord}\| \mathcal{F} = \alpha < \infty$ and if $f \in \mathcal{F}$, then f is b-quasiconvex on the ball $B_{\rho(\alpha)}$, where $\rho(\alpha) = \alpha - \sqrt{\alpha^2 - 1}$. Consequently, if $\|\text{ord}\| \mathcal{F} = 1$ then $\mathcal{F}$ is a subset of the b-quasiconvex maps of B.*

Proof. Let $g(z) = f(z;a)$ be the Koebe transform of f with the automorphism φ_a, $a \in B \setminus \{0\}$, given by (10.1.2). Then

$$D^2g(0)(v,v) = -2\langle v,a \rangle v + \frac{1}{s_a^2}T_a[Df(a)]^{-1}D^2f(a)(D\varphi_a(0)v, D\varphi_a(0)v),$$

by (10.4.3). Since

$$\left\| \frac{1}{2}D^2g(0)(v,v) \right\| \leq \alpha, \quad v \in \mathbb{C}^n, \|v\| = 1,$$

we have

$$(10.5.1) \quad \left\| -\langle v,a \rangle v + \frac{s_a^2}{2}T_a[Df(a)]^{-1}D^2f(a)(T_a^{-1}v, T_a^{-1}v) \right\| \leq \alpha,$$

using the fact that $D\varphi_a(0) = s_a^2 T_a^{-1}$, by (10.2.7).

Let $v = a/\|a\|$. Since $T_a^{-1}(a) = T_a(a) = a$, we deduce from (10.5.1) that

$$\left\| \frac{1 - \|a\|^2}{2\|a\|^2}T_a[Df(a)]^{-1}D^2f(a)(a,a) - a \right\| \leq \alpha,$$

and hence

$$\text{Re } \langle T_a[Df(a)]^{-1}D^2f(a)(a,a), a \rangle - \frac{2\|a\|^4}{1 - \|a\|^2} \geq -\frac{2\alpha\|a\|^3}{1 - \|a\|^2}.$$

Using the fact that $T_a^*(a) = a$, the above relation is equivalent to

$$\text{Re } \langle a + [Df(a)]^{-1}D^2f(a)(a,a), a \rangle \geq \frac{\|a\|^2(\|a\|^2 - 2\alpha\|a\| + 1)}{1 - \|a\|^2}.$$

and the last quantity is positive for $\|a\| < \alpha - \sqrt{\alpha^2 - 1}$. This completes the proof.

We close this section with another result due to Pfaltzgraff and Suffridge [Pfa-Su4] which relates the radius of univalence r_1 of a L.I.F. of finite norm order with its radius of nonvanishing r_0 (compare with Theorem 5.2.7 in the case $n = 1$).

More precisely, let r_0 denote the largest number such that no mapping in the L.I.F. $\mathcal{F}$ has zeros on $B_{r_0} \setminus \{0\}$. Also let r_1 denote the largest number such that every mapping in the L.I.F. $\mathcal{F}$ is univalent on B_{r_1}. Then we have the following connection between r_0 and r_1:

Theorem 10.5.7. *If $\mathcal{F}$ is a L.I.F. of finite norm order, then $r_1 = \dfrac{r_0}{1 + \sqrt{1 - r_0^2}}$.*

Proof. Let $f \in \mathcal{F}$, $r \leq r_0/(1 + \sqrt{1 - r_0^2})$ and $z', z'' \in B_r$ be such that $z' \neq z''$. Also let $f(z; a)$ be the Koebe transform of f with φ_a given by (10.1.3), i.e.

$$f(z; a) = [D\varphi_a(0)]^{-1}[Df(\varphi_a(0))]^{-1}(f(\varphi_a(z)) - f(\varphi_a(0))),$$

where φ_a is the biholomorphic automorphism of B given by (10.1.2). Since $f \in \mathcal{F}$, $f(z; a) \in \mathcal{F}$ too. Choosing $a = z'$ and

$$z = \varphi_{z'}^{-1}(z'') = \varphi_{-z'}(z'') = T_{z'}\left(\frac{z'' - z'}{1 - \langle z'', z' \rangle}\right)$$

in the above, we obtain that

$$(10.5.2) \qquad f\left(T_{z'}\left(\frac{z'' - z'}{1 - \langle z'', z' \rangle}\right); z'\right)$$

$$= \frac{1}{1 - \|z'\|^2} T_{z'}[Df(z')]^{-1}(f(z'') - f(z')).$$

On the other hand, it is easy to verify the relation

$$1 - \|\varphi_a(z)\|^2 = \frac{(1 - \|z\|^2)(1 - \|a\|^2)}{|1 + \langle z, a \rangle|^2}, \quad z, a \in B.$$

(Also see Lemma 6.1.24.)

If we choose $a = -z'$ and $z = z''$ in this equality, and use the fact that $\|z'\| < r$, $\|z''\| < r$, we easily obtain

$$1 - \left\|T_{z'}\left(\frac{z'' - z'}{1 - \langle z'', z'\rangle}\right)\right\|^2 > \frac{(1 - r^2)^2}{(1 + r^2)^2},$$

and therefore

$$\left\|T_{z'}\left(\frac{z'' - z'}{1 - \langle z'', z'\rangle}\right)\right\| < \frac{2r}{1 + r^2} \leq r_0.$$

Combining this inequality with (10.5.2), we conclude that $f(z'') \neq f(z')$. Consequently f is univalent on B_r, and hence $r_1 \geq r_0/(1 + \sqrt{1 - r_0^2})$.

Conversely, let $z_0 \in B$ be such that $0 < \|z_0\| < 2r_1/(1 + r_1^2)$ and choose $a \in B$ such that $z_0 = 2a/(1 + \|a\|^2)$. Then it is obvious that $0 < \|a\| < r_1$, and short computations yield the relations

$$f(a; a) = [D\varphi_a(0)]^{-1}[Df(a)]^{-1}(f(z_0) - f(a))$$

and

$$f(-a; a) = -[D\varphi_a(0)]^{-1}[Df(a)]^{-1}f(a).$$

Hence

$$f(z_0) = Df(a)D\varphi_a(0)(f(a; a) - f(-a; a)).$$

Since $0 < \|a\| < r_1$, we have $f(a; a) \neq f(-a; a)$, which implies that $f(z_0) \neq 0$. Hence $r_0 \geq 2r_1/(1 + r_1^2)$, or equivalently $r_1 \leq r_0/(1 + \sqrt{1 - r_0^2})$. This completes the proof.

Problems

10.5.1. Show that if $\mathcal{F}$ is a L.I.F. with $\|\mathrm{ord}\|\,\mathcal{F} = \alpha < \infty$, then $\mathcal{F}$ is a normal family.

(Pfaltzgraff and Suffridge, 2000 [Pfa-Su4].)

10.5.2. Does the set of quasi-convex maps of type B of the Euclidean unit ball in $\mathbb{C}^n$, $n \geq 2$, form a linear-invariant family?

10.6 Linear-invariant families in complex Hilbert spaces

We end this chapter with a few ideas concerning the notion of linear invariance in complex Hilbert spaces.

Recently Hamada and Kohr [Ham-Koh8] generalized the notion of norm order to the case of a L.I.F. $\mathcal{F}$ on the unit ball B of a complex Hilbert space X. As in finite dimensions, we have the following

Definition 10.6.1. The family $\mathcal{F}$ is called a linear-invariant family if

(i) $\mathcal{F} \subseteq \mathcal{LS}(B) = \{f : B \to X \mid f$ is normalized and locally biholomorphic$\}$,

(ii) $\Lambda_\phi(f) \in \mathcal{F}$ whenever $f \in \mathcal{F}$ and $\phi \in \mathrm{Aut}(B)$.

(The Koebe transform $\Lambda_\phi(f)$ of f is defined in a similar manner as in the finite dimensional case.)

Definition 10.6.2. The norm order of the L.I.F. $\mathcal{F}$ is

$$\|\mathrm{ord}\| \, \mathcal{F} = \sup\left\{\frac{1}{2}\|D^2 f(0)\| : f \in \mathcal{F}\right\}.$$

An essential fact is that Hörmander's result in Theorem 7.2.19 can be applied in the case of complex Hilbert spaces, and thus

$$\|\mathrm{ord}\| \, \mathcal{F} = \sup\left\{\frac{1}{2}\|D^2 f(0)(w, w)\| : f \in \mathcal{F}, \ \|w\| = 1\right\}.$$

Using this observation and the fact that the set $\mathrm{Aut}(B)$ of biholomorphic automorphisms of B has similar behaviour as in the finite dimensional case (see Section 6.1.5), Hamada and Kohr [Ham-Koh8] obtained the following generalizations of some of the results contained in the last two sections of this chapter. We leave the proofs for the reader.

Theorem 10.6.3. Let $\mathcal{F}$ be a L.I.F. on a complex Hilbert space X. Then $\|\mathrm{ord}\| \, \mathcal{F} \geq 1$. Also $\|\mathrm{ord}\| \, K(B) = 1$.

For the following result we need the definition of quasi-convexity of type B (or b-quasiconvexity) in the case of complex Hilbert spaces. This definition is the same as in the case $X = \mathbb{C}^n$. Then we have

Theorem 10.6.4. Let $\mathcal{F}$ be a L.I.F. on a complex Hilbert space X with $\|\mathrm{ord}\| \, \mathcal{F} = \alpha < \infty$. If $f \in \mathcal{F}$ then f is b-quasiconvex on the ball $B_{r(\alpha)}$, where $r(\alpha) = \alpha - \sqrt{\alpha^2 - 1}$.

Theorem 10.6.5. *Let $\mathcal{F} \subset S(B)$ be a L.I.F. on a complex Hilbert space X with $\|\text{ord}\|\, \mathcal{F} = \alpha < \infty$, and $f \in \mathcal{F}$. Assume that for each $r \in (0,1)$ there exists $M = M(r) > 0$ such that*

$$\|[Df(z)]^{-1}(f(z) - f(u))\| \leq M(r), \; z, u \in B_r.$$

Then f maps the ball $B_{1/(2\alpha)}$ biholomorphically onto a convex domain in X. Moreover, f maps the ball $B_{\rho(\alpha)}$ with $\rho(\alpha) = 4\alpha/(1 + 4\alpha^2)$ biholomorphically onto a starlike domain in X.

In addition, certain results involving the norm order of a L.I.F. on the unit polydisc of $\mathbb{C}^n$ have recently been obtained by Hamada and Kohr [Ham-Koh13].

Problems

10.6.1. Prove Theorem 10.6.3.
10.6.2. Prove Theorem 10.6.4.
10.6.3. Prove Theorem 10.6.5.

Chapter 11

Univalent mappings and the Roper-Suffridge extension operator

The aim of this chapter is to study an operator, introduced by Roper and Suffridge, that provides a way of extending a (locally) univalent function $f \in H(U)$ to a (locally) biholomorphic mapping F on the Euclidean unit ball B of $\mathbb{C}^n$. This operator has the remarkable property that if f is a convex function on U then F is a convex mapping on B. We shall investigate its behaviour on other classes of functions as well. In particular, if f is starlike on U then F is starlike on B, and if f is a univalent Bloch function on U then F is a Bloch mapping on B. Growth and covering theorems for the extended mappings may also be obtained, including covering theorems of Bloch type.

There are some interesting connections with the theory of Loewner chains, which we shall study in the context of a generalization of the Roper-Suffridge extension operator, denoted by $\Phi_{n,\alpha}$, $\alpha \in [0, 1/2]$. We shall prove that every mapping $F_\alpha \in \Phi_{n,\alpha}(S)$ may be embedded in a Loewner chain, and in fact has parametric representation. We shall also consider radius problems associated with the operator $\Phi_{n,\alpha}$. Related conjectures and open problems will be discussed. Finally we study the concept of linear-invariant families as it relates to families generated by the operator $\Phi_{n,\alpha}$, and we obtain generalizations of

recent results of several mathematicians concerning the order of the families
so generated.

11.1   Convex, starlike and Bloch mappings and the Roper-Suffridge extension operator

Let $\mathbb{C}^n$ be the Euclidean space of n-complex variables with the usual inner
product $\langle \cdot, \cdot \rangle$ and the induced norm $\| \cdot \|$. As usual, let B be the Euclidean unit
ball of $\mathbb{C}^n$. Let $z' = (z_2, \ldots, z_n)$ so that $z = (z_1, z')$.

The *Roper-Suffridge extension operator* is defined for normalized locally
univalent functions on the unit disc U by

$$(11.1.1) \qquad \Phi_n(f)(z) = F(z) = \left(f(z_1), \sqrt{f'(z_1)}z' \right).$$

We choose the branch of the square root such that $\sqrt{f'(0)} = 1$. Note that
if $f \in S$ then $\Phi_n(f) \in S(B)$.

This operator was introduced in [Rop-Su1] as a means of constructing a
convex mapping on the Euclidean unit ball in $\mathbb{C}^n$ given an arbitrary convex
function on the unit disc. It is surprisingly difficult to do this.

If $f_1, \ldots, f_n$ are normalized starlike functions on the unit disc U then it is
very easy to show that

$$F(z) = (f_1(z_1), \ldots, f_n(z_n)), \quad z = (z_1, \ldots, z_n) \in B,$$

is a normalized starlike mapping on B (see also Problem 6.2.5). However, if
$f_1, \ldots, f_n$ are convex functions then F need not be convex on the Euclidean
unit ball B of $\mathbb{C}^n$, $n \geq 2$. The following well known example illustrates this
phenomenon (see also Problem 6.3.3):

Example 11.1.1. The mapping

$$F(z) = \left(\frac{z_1}{1 - z_1}, \ldots, \frac{z_n}{1 - z_n} \right), \quad z = (z_1, \ldots, z_n) \in B,$$

is not convex on the Euclidean unit ball B of $\mathbb{C}^n$, $n \geq 2$, even though $f(\zeta) = \zeta/(1 - \zeta)$ is a convex function on U.

Proof. The following proof is due to Gong, Wang and Yu [Gon-Wa-Yu3]: Suppose F is convex on B. Then obviously

$$\frac{1}{n}\sum_{k=1}^{n} F(re_k) \in F(B),$$

for all $r \in (0,1)$, where e_k is the kth-standard basis vector in $\mathbb{C}^n$. Consequently,

$$F^{-1}\Big(\frac{1}{n}\sum_{k=1}^{n} F(re_k)\Big) \in B.$$

However, a short computation yields that

$$F^{-1}\Big(\frac{1}{n}\sum_{k=1}^{n} F(re_k)\Big) = \Big(\frac{r}{r+n(1-r)}, \ldots, \frac{r}{r+n(1-r)}\Big)$$

and if r is sufficiently close to 1, then $\dfrac{r}{r+n(1-r)} > \dfrac{1}{\sqrt{n}}$, which implies that

$$F^{-1}\Big(\frac{1}{n}\sum_{k=1}^{n} F(re_k)\Big) \notin B.$$

This is a contradiction.

As observed by Roper and Suffridge [Rop-Su1], similar arguments show that there is no convex mapping of the form

$$(11.1.2) \qquad F(z) = \Big(\frac{z_1}{1-z_1}, g(z_2)\Big), \quad z = (z_1, z_2) \in B.$$

In fact, prior to the work of Roper and Suffridge, there was no general way of constructing convex mappings of B out of convex functions on U. Subsequently their extension operator was used in [Gra-Var2] to show that certain constants in covering theorems for convex mappings are sharp (see Theorem 7.2.9), and by Pfaltzgraff [Pfa5] in studying linear-invariant families (see Sections 10.1 and 10.2).

We shall present a simplified proof of the theorem of Roper and Suffridge which is due to Graham and Kohr [Gra-Koh1]. Another interesting and quite different proof has recently been given by Gong and Liu [Gon-Liu3].

Theorem 11.1.2. *Let $f \in K$ and let $F : B \to \mathbb{C}^n$ be defined by (11.1.1).*
Then F is convex.

Proof. We shall give the proof when $n = 2$; this case contains all the
essential features of the general case.

It is clear that F is locally biholomorphic on B, since $J_F(z) = (f'(z_1))^{3/2} \neq$
0 for $z = (z_1, z_2) \in B$. Therefore, using the criterion for convexity given in
Theorem 6.3.4, we have to prove that

$$1 - \operatorname{Re} \langle [DF(z)]^{-1} D^2 F(z)(u, u), z \rangle > 0,$$

for all $z = (z_1, z_2) \in B$ and $u = (u_1, u_2) \in \mathbb{C}^2$ with $\|u\| = 1$ and $\operatorname{Re} \langle z, u \rangle = 0$.

We may assume that $z = (z_1, z_2) \in B$, $z_2 \neq 0$, because the case $z = (z_1, 0)$
is easily handled. Indeed, if $z = (z_1, 0)$, $|z_1| < 1$ and $u = (u_1, u_2) \in \mathbb{C}^2$ are
such that $\|u\| = 1$ and $\operatorname{Re} \langle z, u \rangle = 0$, a short computation yields

$$1 - \operatorname{Re} \langle [DF(z)]^{-1} D^2 F(z)(u, u), z \rangle = 1 - \operatorname{Re} \left[\frac{u_1^2 \bar{z}_1 f''(z_1)}{f'(z_1)} \right]$$

$$= |u_1|^2 \operatorname{Re} \left[1 + \frac{z_1 f''(z_1)}{f'(z_1)} \right] + 1 - |u_1|^2 > 0.$$

Here we have used the fact that $z_1 \bar{u}_1 + \bar{z}_1 u_1 = 0$ and $f \in K$.

It is obvious that F is holomorphic at all points $z = (z_1, z_2) \in \overline{B}$ such that
$z_2 \neq 0$.

Now we can write $z = (z_1, z_2) \in \overline{B} \setminus \{0\}$, $z_2 \neq 0$, and u as $z = \alpha V$ and
$u = (\alpha/|\alpha|)W$ respectively, where $\alpha \in \mathbb{C}$, $0 < |\alpha| \leq 1$, $\|V\| = \|W\| = 1$,
$V = (V_1, V_2)$, $V_2 \neq 0$, and $\operatorname{Re} \langle V, W \rangle = 0$. Hence we obtain

$$1 - \operatorname{Re} \langle [DF(z)]^{-1} D^2 F(z)(u, u), z \rangle$$

$$= 1 - \operatorname{Re} \left\{ \alpha \langle [DF(\alpha V)]^{-1} D^2 F(\alpha V)(W, W), V \rangle \right\}.$$

The expression $1 - \operatorname{Re} \left\{ \alpha \langle [DF(\alpha V)]^{-1} D^2 F(\alpha V)(W, W), V \rangle \right\}$ is the real
part of an analytic function of α and thus is harmonic. Applying the minimum
principle for harmonic functions, this function can only attain its minimum
when $|\alpha| = 1$. Therefore it suffices to prove that

$$(11.1.3) \qquad 1 - \operatorname{Re} \langle [DF(z)]^{-1} D^2 F(z)(u, u), z \rangle \geq 0,$$

for all $z = (z_1, z_2) \in \mathbb{C}^2$, $|z_1|^2 + |z_2|^2 = 1$, $z \neq (z_1, 0)$, $u = (u_1, u_2) \in \mathbb{C}^2$, $|u_1|^2 + |u_2|^2 = 1$ and Re $\langle z, u \rangle = 0$.

After short computations we deduce that the relation (11.1.3) is equivalent to

$$(11.1.4) \qquad 1 \geq \mathrm{Re} \left\{ u_1(u_1\bar{z}_1 + u_2\bar{z}_2) \frac{f''(z_1)}{f'(z_1)} + \frac{u_1^2|z_2|^2}{2}\{f; z_1\} \right\},$$

for all $z_1, z_2 \in \mathbb{C}$, $|z_1|^2 + |z_2|^2 = 1$, $z_2 \neq 0$, $u_1, u_2 \in \mathbb{C}$, $|u_1|^2 + |u_2|^2 = 1$ and Re $\{u_1\bar{z}_1 + u_2\bar{z}_2\} = 0$, where $\{f; z_1\}$ denotes the Schwarzian derivative of f at z_1.

We consider the following three cases:

(i) $u_1 = 0$, $|u_2| = 1$.

(ii) $u_2 = 0$, $|u_1| = 1$.

(iii) $0 < |u_1| < 1$, $0 < |u_2| < 1$, $|u_1|^2 + |u_2|^2 = 1$.

In fact the first case is obvious, so we need to consider only the second and third possibilities.

(ii) Assume $u_2 = 0$, $|u_1| = 1$. In this case the relation (11.1.4) becomes

$$1 \geq \mathrm{Re} \left\{ u_1^2\bar{z}_1 \frac{f''(z_1)}{f'(z_1)} + \frac{u_1^2|z_2|^2}{2}\{f; z_1\} \right\},$$

for $z_1, z_2 \in \mathbb{C}$, $z_2 \neq 0$, $|z_1|^2 + |z_2|^2 = 1$, $u_1 \in \mathbb{C}$, $|u_1| = 1$ and Re $\{u_1\bar{z}_1\} = 0$, i.e. $u_1\bar{z}_1 + \bar{u}_1 z_1 = 0$.

Therefore we have to show that

$$(11.1.5) \qquad 1 \geq -\mathrm{Re} \left\{ \frac{z_1 f''(z_1)}{f'(z_1)} \right\} + \frac{1 - |z_1|^2}{2} \mathrm{Re} \left\{ u_1^2\{f; z_1\} \right\},$$

for $z_1 \in \mathbb{C}$, $|z_1| < 1$ and $u_1 \in \mathbb{C}$, $|u_1| = 1$, Re $\{u_1\bar{z}_1\} = 0$.

We need to make use of an estimate for the Schwarzian derivative of a convex function which we considered in Chapter 2 (see (2.2.16) and the preceding remarks): if $f \in K$ then

$$(11.1.6) \quad (1 - |z_1|^2)^2|\{f; z_1\}| \leq 2 \left(1 - \left| \frac{1}{2}(1 - |z_1|^2)\frac{f''(z_1)}{f'(z_1)} - \bar{z}_1 \right|^2 \right).$$

Then we obtain

$$-\mathrm{Re} \left\{ \frac{z_1 f''(z_1)}{f'(z_1)} \right\} + \frac{1 - |z_1|^2}{2} \mathrm{Re} \left\{ u_1^2\{f; z_1\} \right\} - 1$$

$$\leq -\frac{2}{1-|z_1|^2}\text{Re}\left\{z_1\left(\frac{1-|z_1|^2}{2}\frac{f''(z_1)}{f'(z_1)}-\bar{z}_1\right)\right\}$$

$$-\frac{1+|z_1|^2}{1-|z_1|^2}+\frac{1}{1-|z_1|^2}\left(1-\left|\frac{1-|z_1|^2}{2}\frac{f''(z_1)}{f'(z_1)}-\bar{z}_1\right|^2\right)$$

$$\leq -\frac{1}{1-|z_1|^2}\left\{|z_1|^2-2|z_1|\left|\frac{1-|z_1|^2}{2}\frac{f''(z_1)}{f'(z_1)}-\bar{z}_1\right|+\left|\frac{1-|z_1|^2}{2}\frac{f''(z_1)}{f'(z_1)}-\bar{z}_1\right|^2\right\}$$

$$=-\frac{1}{1-|z_1|^2}\left(|z_1|-\left|\frac{1-|z_1|^2}{2}\frac{f''(z_1)}{f'(z_1)}-\bar{z}_1\right|\right)^2\leq 0.$$

Thus the assertion (11.1.5) is proved.

(iii) Assume that $0 < |u_1| < 1$, $0 < |u_2| < 1$ and $|u_1|^2 + |u_2|^2 = 1$. In fact we can suppose that $0 < u_1 < 1$, $0 < u_2 < 1$. For if the statement (11.1.4) fails for some choice of f, z, and u, then choose θ_1 and θ_2 such that $u_1 e^{i\theta_1} > 0$, $u_2 e^{i\theta_2} > 0$, and consider the rotation of f given by $f_{-\theta_1}(z_1) = e^{i\theta_1} f(e^{-i\theta_1} z_1)$. Then since

$$\frac{f''_{-\theta_1}(z_1 e^{i\theta_1})}{f'_{-\theta_1}(z_1 e^{i\theta_1})} = e^{-i\theta_1}\frac{f''(z_1)}{f'(z_1)}$$

and

$$\{f_{-\theta_1}; z_1 e^{i\theta_1}\} = e^{-2i\theta_1}\{f; z_1\},$$

we see that (11.1.4) also fails for $f_{-\theta_1}$, $(z_1 e^{i\theta_1}, z_2 e^{i\theta_2})$, and $(u_1 e^{i\theta_1}, u_2 e^{i\theta_2})$.

Hence we have to prove that

$$1 \geq u_1 \text{Re}\left\{(u_1\bar{z}_1 + u_2\bar{z}_2)\frac{f''(z_1)}{f'(z_1)}\right\} + \frac{u_1^2|z_2|^2}{2}\text{Re}\{f; z_1\},$$

for all $u_1, u_2 \in (0,1)$, $u_1^2 + u_2^2 = 1$ and $z_1, z_2 \in \mathbb{C}$, $z_2 \neq 0$, $|z_1|^2 + |z_2|^2 = 1$, and

$$u_1\text{Re}\, z_1 + u_2\text{Re}\, z_2 = 0.$$

Let $z_j = x_j + iy_j$ for $j = 1, 2$. Then $u_1 x_1 + u_2 x_2 = 0$, and since $|z_1|^2 + |z_2|^2 = 1$, we have

$$x_1^2 + y_1^2 + \frac{u_1^2 x_1^2}{1-u_1^2} + y_2^2 = 1,$$

and hence

(11.1.7) $$1 - x_1^2 - y_1^2 = u_1^2 - u_1^2 y_1^2 + u_2^2 y_2^2.$$

Also we can suppose that $u_1 y_1 + u_2 y_2 \geq 0$, since the other case can be treated in a similar manner.

Using these remarks, we need to show that

$$1 \geq u_1(u_1 y_1 + u_2 y_2)\mathrm{Im}\left[\frac{f''(z_1)}{f'(z_1)}\right] + \frac{u_1^2(1 - |z_1|^2)}{2}\mathrm{Re}\{f; z_1\}.$$

Taking into account the relations (11.1.6) and (11.1.7), we obtain

$$u_1(u_1 y_1 + u_2 y_2)\mathrm{Im}\left[\frac{f''(z_1)}{f'(z_1)}\right] + \frac{u_1^2(1 - |z_1|^2)}{2}\mathrm{Re}\{f; z_1\}$$

$$\leq \frac{2u_1}{1 - |z_1|^2}(u_1 y_1 + u_2 y_2)\mathrm{Im}\left\{\frac{1 - |z_1|^2}{2}\frac{f''(z_1)}{f'(z_1)} - \bar{z}_1\right\}$$

$$-\frac{2u_1 y_1}{1 - |z_1|^2}(u_1 y_1 + u_2 y_2) + \frac{u_1^2}{1 - |z_1|^2}\left\{1 - \left|\frac{1 - |z_1|^2}{2}\frac{f''(z_1)}{f'(z_1)} - \bar{z}_1\right|^2\right\}$$

$$\leq \frac{1}{1 - |z_1|^2}\left\{u_1^2 - u_1^2\left|\frac{1 - |z_1|^2}{2}\frac{f''(z_1)}{f'(z_1)} - \bar{z}_1\right|^2\right\}$$

$$+\frac{2u_1(u_1 y_1 + u_2 y_2)}{1 - |z_1|^2}\left|\frac{1 - |z_1|^2}{2}\frac{f''(z_1)}{f'(z_1)} - \bar{z}_1\right| - \frac{2u_1^2 y_1^2 + 2u_1 u_2 y_1 y_2}{1 - |z_1|^2} \leq 1.$$

Here we have used the fact that the last inequality is equivalent to

$$\left[u_1\left|\frac{1 - |z_1|^2}{2}\frac{f''(z_1)}{f'(z_1)} - \bar{z}_1\right| - (u_1 y_1 + u_2 y_2)\right]^2 \geq 0.$$

This completes the proof.

It seems to be difficult to perturb either the Roper-Suffridge operator or the domain without losing the convexity-preserving property. We shall give two illustrations of this. One is an example of Graham, Hamada, Kohr, and Suffridge [Gra-Ham-Koh-Su], and the other is a theorem of Roper and Suffridge [Rop-Su1].

Example 11.1.3. Let $n = 2$ and let $F : B \subset \mathbb{C}^2 \to \mathbb{C}^2$ be given by

$$F(z) = \left(\frac{z_1}{1 - z_1}, \frac{z_2 g(z_1)}{1 - z_1}\right), \quad z = (z_1, z_2) \in B,$$

where g is a non-vanishing analytic function on U such that $g(0) = 1$. We show that the mapping F is only convex for $g(z_1) \equiv 1$.

Proof. Observe that the line $L = \{(it, 0) : t \in \mathbb{R}\} \subset F(B)$. Since $F(B)$ is convex, an elementary argument shows that for every $W \in F(B)$, the line $W + L \subset F(B)$.

Let

$$u = \frac{z_1}{1 - z_1} \quad \text{and} \quad v = \frac{z_2 g(z_1)}{1 - z_1}.$$

Then Re $u > -1/2$ and

$$|z_2|^2 < 1 - |z_1|^2 = 1 - \frac{|u|^2}{|1 + u|^2} = \frac{1 + 2\text{Re } u}{|1 + u|^2},$$

so the mapping F is of the form

$$(u, \rho e^{i\varphi}\sqrt{1 + 2\text{Re } u}\, h(u)), \quad \rho < 1, \quad h(u) = g\left(\frac{u}{1 + u}\right).$$

It can be seen from the nature of the mapping F and the remark at the beginning of the proof that

$$(11.1.8) \quad (u, v) \in F(B) \Leftrightarrow (u, |v|) \in F(B) \Leftrightarrow (u + it, |v|) \in F(B),$$

for all $t \in \mathbb{R}$.

For each u, let

$$M(u) = \sup\{|v| : (u, v) \in F(B)\} = \sqrt{1 + 2\text{Re } u}\,|h(u)|.$$

Then (11.1.8) implies that $M(u)$ is independent of Im u, so that $|h(u)|$ is constant on the lines Re $u = $ constant $> -1/2$.

By the Schwarz reflection principle, we may reflect the function h with the domain restricted to the right half-plane across the unit circle (it is the unit circle since $h(0) = 1$) by $h(-\bar{u}) = 1/\overline{h(u)}$. This extended function is entire because $h(u) \neq 0$. Write $u = \sigma + i\tau$ and observe that $h(u) = Re^{i\phi}$ where R is independent of τ. Using the Cauchy-Riemann equations, it is easy to see that ϕ is independent of σ and in fact $h(u) = e^{au}$ for some real a. Using the convexity of the mapping, it follows that the set (u, v) such that $u > -1/2$ and $0 < v < \sqrt{1 + 2ue^{au}} = k(u)$ is convex. By elementary calculus, since $k''(u) > 0$ for large u when $a \neq 0$, the above set cannot be convex unless $a = 0$. This completes the proof.

Theorem 11.1.4. *Let $f \in K$, let $B(p)$ be the unit ball in $\mathbb{C}^2$ with respect to the p-norm, $1 \leq p \leq \infty$, and let $F : B(p) \to \mathbb{C}^2$ be given by*

$$F(z_1, z_2) = \left(f(z_1), z_2 \sqrt{f'(z_1)} \right), \quad (z_1, z_2) \in B(p).$$

Then F is convex if and only if $p = 2$.

Proof. The case $p = \infty$ follows from Theorem 6.3.2. For the case $1 \leq p < \infty$ we will use a similar argument as in [Rop-Su1]. For this purpose, let $f(\zeta) = \frac{1}{2} \log \left(\frac{1+\zeta}{1-\zeta} \right)$, $\zeta \in U$. Then f is convex on U.

Also let

$$u = \frac{1}{2} \log \left(\frac{1 + z_1}{1 - z_1} \right) \quad \text{and} \quad v = \frac{z_2}{(1 - z_1^2)^{1/2}}.$$

If $F(B(p))$ is convex then so is its intersection with the plane Im $u = 0$, Im $v = 0$. This intersection contains the entire real u-axis and precisely the interval $(-1, 1)$ of the real v-axis. In order to show that convexity is not satisfied, it suffices to show that if $z_1 \to 1$ along the real axis then we are constrained to have $|v| \to 0$ if $p < 2$, while it is possible to have $|v| \to \infty$ if $p > 2$. It suffices to consider the images of the points $((1 - \varepsilon^p)^{1/p}, \pm \varepsilon)$ for small $\varepsilon > 0$. We have

$$|v|^2 = \frac{\varepsilon^2}{1 - (1 - \varepsilon^p)^{2/p}}$$

$$= \frac{\varepsilon^2}{1 - \left(1 - \frac{2}{p} \varepsilon^p + \frac{2}{p} \left(\frac{2}{p} - 1 \right) \varepsilon^{2p} + O(\varepsilon^{3p}) \right)}$$

$$= \frac{\varepsilon^2}{\frac{2}{p} \varepsilon^p \left(1 + \left(1 - \frac{2}{p} \right) \varepsilon^p \right) + O(\varepsilon^{3p})}$$

$$= \frac{p \varepsilon^{2-p}}{2 \left(1 + \left(1 - \frac{2}{p} \right) \varepsilon^p \right) + O(\varepsilon^{2p})}.$$

Now it is easy to see that if $p = 2$ then $|v|^2 = 1$, if $p < 2$ then $|v|^2 \to 0$ as $\varepsilon \to 0$, and if $p > 2$ then $|v|^2 \to \infty$ as $\varepsilon \to 0$. This completes the proof.

Theorem 11.1.2 can be extended to the case of complex Hilbert spaces without any essential new difficulties (see [Rop-Su1]). Let X be a complex

Hilbert space and let B denote the unit ball of X. In view of Theorem 6.3.8 (ii), we obtain

Theorem 11.1.5. *Let $f \in K$ and $u \in X$ with $\|u\| = 1$. Let $F_u : B \to X$ be given by*

$$(11.1.9) \quad F_u(z) = f(\langle z, u \rangle) u + \sqrt{f'(\langle z, u \rangle)}(z - \langle z, u \rangle u), \quad z \in B.$$

Then F_u is convex.

Proof. It is not hard to show that F_u is biholomorphic. Moreover, using the convexity of f and Theorem 2.2.8, we deduce that for each $r \in (0,1)$ there exists $M = M(r) > 0$ such that

$$\|[DF_u(z)]^{-1}(F_u(z) - F_u(v))\| \le M(r), \quad \|z\| \le r, \quad \|v\| \le r.$$

Finally, similar reasoning as in Theorem 11.1.2 yields that

$$\|v\|^2 - \mathrm{Re}\, \langle [DF_u(z)]^{-1} D^2 F_u(z)(v,v), z \rangle > 0,$$

for $z \in B$ and $v \in X \setminus \{0\}$, $\mathrm{Re}\, \langle z, v \rangle = 0$. Hence Theorem 6.3.8 (ii) implies that F_u is convex. This completes the proof.

We now turn to other properties of the Roper-Suffridge extension operator. First we show that the operator Φ_n preserves starlikeness. This result was obtained in [Gra-Koh1].

Theorem 11.1.6. *Let $f \in S^*$ and $F : B \to \mathbb{C}^n$ be defined by (11.1.1). Then F is starlike.*

Proof. We give the proof in the case $n = 2$; the general case is entirely similar. Using the criterion for starlikeness in Theorem 6.2.2, we have to prove that

$$\mathrm{Re}\, \langle [DF(z)]^{-1} F(z), z \rangle \ge 0,$$

for all $z = (z_1, z_2) \in B$, $z_2 \neq 0$, since the case $z = (z_1, 0)$ is obvious. Taking into account the minimum principle for harmonic functions, it suffices to prove that

$$\mathrm{Re}\, \langle [DF(z)]^{-1} F(z), z \rangle \ge 0,$$

for all $z = (z_1, z_2) \in \mathbb{C}^2$, $|z_1|^2 + |z_2|^2 = 1$, $z_2 \neq 0$.

After short computations, we obtain

$$\langle [DF(z)]^{-1}F(z), z \rangle = |z_1|^2 \frac{f(z_1)}{z_1 f'(z_1)} + |z_2|^2 - \frac{|z_2|^2 f''(z_1)f(z_1)}{2(f'(z_1))^2}.$$

Hence we need to show that

$$(11.1.10) \ \mathrm{Re} \left\{ |z_1|^2 \frac{f(z_1)}{z_1 f'(z_1)} + 1 - |z_1|^2 - \frac{1-|z_1|^2}{2} \frac{f''(z_1)f(z_1)}{(f'(z_1))^2} \right\} \geq 0$$

for $|z_1| < 1$. Now, the function p defined by $p(z_1) = \dfrac{f(z_1)}{z_1 f'(z_1)}$, $z_1 \in U$, satisfies $p \in H(U)$, $p(0) = 1$ and $\mathrm{Re}\, p(z_1) > 0$, i.e. $p \in \mathcal{P}$. Therefore from the Herglotz formula (2.1.3), we have

$$p(z_1) = \int_0^{2\pi} \frac{1 + z_1 e^{-i\theta}}{1 - z_1 e^{-i\theta}} d\mu(\theta), \quad z_1 \in U,$$

where μ is a non-decreasing function on $[0, 2\pi]$ with $\mu(2\pi) - \mu(0) = 1$.

Also, since

$$\frac{f''(z_1)f(z_1)}{(f'(z_1))^2} = 1 - z_1 p'(z_1) - p(z_1),$$

it follows from (11.1.10) that we have to prove the relation

$$(11.1.11) \ (1 + |z_1|^2)\mathrm{Re}\, p(z_1) + 1 - |z_1|^2 + (1 - |z_1|^2)\mathrm{Re}\, [z_1 p'(z_1)] \geq 0$$

for $z_1 \in U$. Using the fact that

$$z_1 p'(z_1) = 2 \int_0^{2\pi} \frac{z_1 e^{-i\theta}}{(1 - z_1 e^{-i\theta})^2} d\mu(\theta),$$

it suffices to observe that the relation (11.1.11) is equivalent to

$$(11.1.12) \qquad \mathrm{Re} \int_0^{2\pi} \frac{1 - |z_1|^2 z_1^2 e^{-2i\theta}}{(1 - z_1 e^{-i\theta})^2} d\mu(\theta) \geq 0, \quad z_1 \in U.$$

Since μ is non-decreasing on $[0, 2\pi]$ and

$$\mathrm{Re}\, \frac{1 - |w|^2 w^2}{(1 - w)^2} > 0, \quad |w| < 1,$$

the inequality (11.1.12) follows. This completes the proof.

In Theorem 11.1.4 we have seen that the operator Φ_n, regarded as an extension operator from U to $B(p)$, does not preserve convexity if $p \neq 2$. It is natural to see what happens in the case of starlikeness. In this situation, we can prove that Theorem 11.1.6 does not hold in the unit polydisc P of $\mathbb{C}^2$ [Gra-Koh1].

Remark 11.1.7. Let $f \in S^*$ and let $F : P \subset \mathbb{C}^2 \to \mathbb{C}^2$ be given by $F(z_1, z_2) = \left(f(z_1), z_2\sqrt{f'(z_1)} \right)$. Then F need not be starlike on P.

Proof. Since

$$[DF(z)]^{-1}F(z) = \left(\frac{f(z_1)}{f'(z_1)}, -\frac{z_2}{2}\frac{f''(z_1)f(z_1)}{(f'(z_1))^2} + z_2 \right), \ z = (z_1, z_2) \in P,$$

F will be starlike if and only if the following relations are satisfied on the unit disc (see Theorem 6.2.3):

(i) $\operatorname{Re}\left[\dfrac{f(z_1)}{z_1 f'(z_1)} \right] > 0$,

(ii) $1 - \dfrac{1}{2}\operatorname{Re}\left[\dfrac{f''(z_1)f(z_1)}{(f'(z_1))^2} \right] > 0$.

The first inequality is obvious since $f \in S^*$. However, the second is not always true. To see this, it suffices to consider $f(z_1) = \dfrac{z_1}{(1 - z_1)^2}$, $|z_1| < 1$.

Then $f \in S^*$, but (ii) is equivalent to $\operatorname{Re}\dfrac{1}{(1 + z_1)^2} > 0$, $|z_1| < 1$. Since this relation is not satisfied everywhere in the unit disc, F is not starlike on P.

As in the case of convex maps, Theorem 11.1.6 may be extended to complex Hilbert spaces (see [Gra-Koh1]). To do so, it suffices to apply Theorem 6.2.6 and similar reasoning as in the proof of Theorem 11.1.6. Let X be a complex Hilbert space and let B denote the unit ball of X. We have

Theorem 11.1.8. Let $f \in S^*$ and $u \in X$, $\|u\| = 1$. Define $F_u : B \to X$ by (11.1.9). Then F_u is starlike.

Since we have proved that $\Phi_n(S^*) \subseteq S^*(B)$ and $\Phi_n(K) \subseteq K(B)$, when B is the Euclidean unit ball of $\mathbb{C}^n$, it would be interesting to give an answer to the following:

Conjecture 11.1.9. If $f : U \to \mathbb{C}$ is a normalized close-to-convex function with respect to a normalized starlike function g on U (i.e. $\operatorname{Re}\left[\dfrac{\zeta f'(\zeta)}{g(\zeta)} \right] > 0$, $\zeta \in U$), then $\Phi_n(f)$ is close-to-starlike with respect to $\Phi_n(g)$ on the Euclidean unit ball of $\mathbb{C}^n$, $n \geq 2$.

Finally we consider the extension of Bloch functions using the Roper-Suffridge operator. Let $\mathcal{B}$ denote the set of Bloch functions on U, and let $\mathcal{B}_1$ denote the subset of $\mathcal{B}$ consisting of functions which are normalized and satisfy

(11.1.13)
$$\sup_{|\zeta|<1}(1 - |\zeta|^2)|f'(\zeta)| = f'(0) = 1.$$

(See Chapter 4.)

We have the following result obtained in [Gra-Koh1]:

Theorem 11.1.10. *If $f \in S \cap \mathcal{B}_1$ then $F = \Phi_n(f)$ is a Bloch mapping.*

Proof. It suffices to prove that the quantity $m(F)$ given by

$$m(F) = \sup_{z \in B}(1 - \|z\|^2)\|DF(z)\|$$

is finite (see Remark 9.1.4). For this purpose, we shall again restrict our attention to the case $n = 2$.

Short computations yield that

$$DF(z)u = \left(u_1 f'(z_1), \frac{z_2 f''(z_1) u_1}{2\sqrt{f'(z_1)}} + \sqrt{f'(z_1)}u_2\right),$$

for all $z = (z_1, z_2) \in B$ and $u = (u_1, u_2) \in \mathbb{C}^2$. Since $f \in S$, it follows that

$$\left|\frac{1 - |z_1|^2}{2} \cdot \frac{f''(z_1)}{f'(z_1)} - \bar{z}_1\right| \le 2, \quad z_1 \in U.$$

Hence we obtain

$$\|DF(z)u\|^2 = |u_1|^2|f'(z_1)|^2 + |f'(z_1)|\left|\frac{z_2 f''(z_1)u_1}{2f'(z_1)} + u_2\right|^2$$

$$\le |f'(z_1)|^2 + |f'(z_1)|\left[\frac{1 - |z_1|^2}{4}\left|\frac{f''(z_1)}{f'(z_1)}\right|^2 + 1 + \left|\frac{f''(z_1)}{f'(z_1)}\right|\right]$$

$$\le |f'(z_1)|^2 + |f'(z_1)|\frac{6|z_1| + 9}{1 - |z_1|^2},$$

for all $z = (z_1, z_2) \in B$ and $u \in \mathbb{C}^2$, $\|u\| = 1$.

From (11.1.13) we have

$$|f'(z_1)| \le \frac{1}{1 - |z_1|^2}, \quad z_1 \in U,$$

and hence

$$(1 - \|z\|^2)^2 \|DF(z)\|^2$$

$$\leq (1 - |z_1|^2)^2 |f'(z_1)|^2 + (1 - |z_1|^2)|f'(z_1)|(6|z_1| + 9) \leq 16,$$

for all $z \in B$. Therefore $m(F) \leq 4$ and we conclude that F is a Bloch mapping, as claimed. This completes the proof.

Problems

11.1.1. Show that there is no convex mapping of the form (11.1.2) on the Euclidean unit ball of $\mathbb{C}^2$.

11.2 Growth and covering theorems associated with the Roper-Suffridge extension operator

In this section we shall prove that if f belongs to a class of univalent functions in the unit disc which satisfy a growth theorem and a distortion theorem, then the extension $F = \Phi_n(f)$ of f satisfies a growth theorem and consequently a covering theorem on the Euclidean unit ball of $\mathbb{C}^n$. We shall also prove a Bloch-type covering theorem for convex mappings obtained using the operator Φ_n.

For this purpose, we need to extend the definition of the Roper-Suffridge operator slightly. If f is not normalized, so that there is not a canonical choice of the branch of the square root $\sqrt{f'(z_1)}$, there are two possible ways to define $\Phi_n(f)$. Both of them have the same range, however, and their growth properties are identical. Without specifying a rule for choosing between them, we shall sometimes write $\Phi_n(f)$ in discussing results which are valid for both choices of the branch of the square root.

Other properties of Φ_n which are easily verified include the following:

• If $f(z_1) = z_1/(1 - z_1)$ then $F = \Phi_n(f)$ is the normalized Cayley transform, i.e. $F(z) = z/(1 - z_1)$ for $z = (z_1, \ldots, z_n) \in B$.

• *If f is an automorphism of U then $F = \Phi_n(f)$ (i.e. either choice of F) is an automorphism of B.*

- *If f is odd or more generally k-fold symmetric, then so is $F = \Phi_n(f)$.*

Another property of the Roper-Suffridge operator which shows that this operator is to some extent canonical among possible extension operators is the following [Gra6]:

Lemma 11.2.1. *If $f : U \to \mathbb{C}$ is a locally univalent function and T is an automorphism of U then $\Phi_n(f)$ and $\Phi_n(f \circ T)$ have the same range.*

Remark 11.2.2. For specific f and T, we may choose the branches of the square roots such that $\Phi_n(f \circ T) = \Phi_n(f) \circ \Phi_n(T)$. However it is not possible to choose the branch of the square root in a systematic way for all f so that this formula is always valid.

Proof of Lemma 11.2.1. Using the definition of Φ_n, we have

$$(11.2.1) \qquad \Phi_n(f \circ T)(z) = \left((f \circ T)(z_1), \sqrt{(f \circ T)'(z_1)}z' \right)$$

$$= \left(f(T(z_1)), \pm\sqrt{f'(T(z_1))}\sqrt{T'(z_1)}z' \right).$$

Whether the correct choice of sign (for a particular f and T) is plus or minus is of no consequence, for in either case $\left(T(z_1), \pm\sqrt{T'(z_1)}z' \right)$ gives an automorphism of B. Hence the right-hand side of (11.2.1) has the form $\left(f(\zeta_1), \sqrt{f'(\zeta_1)}\zeta' \right)$ for some $\zeta \in B$, and $\zeta = (\zeta_1, \zeta')$ ranges over B as z does.

We next show that, under conditions which are quite commonly encountered in practice, if $\mathcal{F}$ is a subset of S whose members satisfy a growth theorem and a distortion theorem, then the mappings in $\Phi_n(\mathcal{F})$ satisfy a similar growth theorem. As we have seen in Section 6.1.7, this is in contrast to the behaviour of the full class $S(B)$, and additional assumptions are needed to obtain positive results (see Chapter 7).

The following result was obtained by Graham [Gra6]:

Theorem 11.2.3. *Suppose $\mathcal{F}$ is a subset of S such that all $f \in \mathcal{F}$ satisfy*

$$(11.2.2) \qquad \varphi(|z_1|) \le |f(z_1)| \le \psi(|z_1|), \quad |z_1| < 1$$

$$(11.2.3) \qquad \varphi'(|z_1|) \le |f'(z_1)| \le \psi'(|z_1|), \quad |z_1| < 1,$$

where

$$(11.2.4) \qquad \varphi, \psi \text{ are twice differentiable on } [0, 1),$$

$$(11.2.5) \qquad \varphi(r) \le r, \quad \varphi'(r) \ge 0, \quad \varphi''(r) \le 0 \quad \text{on} \quad [0, 1)$$

(11.2.6) $\psi(r) \geq r, \quad \psi'(r) \geq 0, \quad \psi''(r) \geq 0 \quad \text{on} \quad [0,1).$

Then all mappings $F \in \Phi_n(\mathcal{F})$ satisfy

(11.2.7) $\varphi(\|z\|) \leq \|F(z)\| \leq \psi(\|z\|), \quad z \in B.$

Furthermore if for some $f \in \mathcal{F}$ the lower (respectively upper) bound in (11.2.2) is sharp at $z_1 \in U$, then the lower (respectively upper) bound in (11.2.7) is sharp for $\Phi_n(f)$ at $(z_1, 0, \ldots, 0)$.

For the proof of this result we need the following lemma [Gra6]:

Lemma 11.2.4. *Suppose φ and ψ are functions which satisfy the conditions (11.2.4)-(11.2.6). Then for fixed $r \in [0,1)$,*

the minimum of $(\varphi(t))^2 + (r^2 - t^2)\varphi'(t)$ for $t \in [0,r]$ occurs when $t = r$;

the maximum of $(\psi(t))^2 + (r^2 - t^2)\psi'(t)$ for $t \in [0,r]$ occurs when $t = r$.

Proof. It is easy to check the sign of the first derivative in $[0,r]$.

Proof of Theorem 11.2.3. Let $\|z\| = r < 1$. Taking into account the result of Lemma 11.2.4, it is not difficult to obtain the sharp upper and lower bounds for

$$|f(z_1)|^2 + \|z'\|^2 |f'(z_1)| = |f(z_1)|^2 + (r^2 - |z_1|^2)|f'(z_1)|.$$

As particular cases of Theorem 11.2.3, we obtain the following growth results [Gra6]:

Corollary 11.2.5. *If $f \in S$ and $r \in [0,1)$ then*

$$\frac{r}{(1+r)^2} \leq \|\Phi_n(f)(z)\| \leq \frac{r}{(1-r)^2}, \quad \|z\| = r.$$

If $f \in S$ is k-fold symmetric then

$$\frac{r}{(1+r^k)^{2/k}} \leq \|\Phi_n(f)(z)\| \leq \frac{r}{(1-r^k)^{2/k}}, \quad \|z\| = r.$$

If $f \in K$ then

(11.2.8) $\dfrac{r}{1+r} \leq \|\Phi_n(f)(z)\| \leq \dfrac{r}{1-r}, \quad \|z\| = r.$

If $f \in K$ and $f''(0) = 0, \ldots, f^{(k)}(0) = 0$ then

$$\int_0^r \frac{dt}{(1+t^k)^{2/k}} \leq \|\Phi_n(f)(z)\| \leq \int_0^r \frac{dt}{(1-t^k)^{2/k}}, \quad \|z\| = r.$$

If $f \in S \cap B_1$ then

$$(11.2.9) \qquad \frac{1}{2}\left(1 - \exp\left(-\frac{2r}{1-r}\right)\right) \le \|\Phi_n(f)(z)\| \le \frac{1}{2}\log\left(\frac{1+r}{1-r}\right)$$

for $\|z\| = r$.

Remark 11.2.6. The bounds in (11.2.8) are clear because we have already proved that $\Phi_n(K) \subset K(B)$. All of the above estimates are sharp except for the lower estimate in (11.2.9). (We mention that the lower estimate in (11.2.9) is not sharp in one variable for functions in $S \cap B_1$. The corresponding distortion estimate $|f'(\zeta)| \ge \varphi'(r)$, $|\zeta| = r < 1$, is true for all normalized locally univalent Bloch functions with Bloch seminorm 1, and is sharp in this case. See [Liu-Min] and also Chapter 4.)

A covering theorem of Koebe type follows from Theorem 11.2.3 by using a standard argument [Gra6].

Theorem 11.2.7. *Suppose that the family $\mathcal{F} \subset S$ and the functions φ and ψ satisfy the hypothesis of Theorem 11.2.3. Then for all $f \in \mathcal{F}$, the image of $\Phi_n(f)$ contains the ball B_ρ, where $\rho = \lim_{r \nearrow 1} \varphi(r)$.*

Proof. The existence of ρ follows from the fact that φ is a bounded increasing function on $[0,1)$. On the other hand, since $\Phi_n(f)$ is an open mapping, we may conclude that $\Phi_n(f)(B) \supseteq B_\rho$, as claimed.

Some particular cases of Theorem 11.2.7 are as follows [Gra6]:

Corollary 11.2.8. *(i) If $f \in S$ then $\Phi_n(f)(B) \supseteq B_{1/4}$.*

(ii) If $f \in S$ and f is k-fold symmetric then $\Phi_n(f)(B) \supseteq B_{4^{-1/k}}$.

(iii) If $f \in K$ then $\Phi_n(f)(B) \supseteq B_{1/2}$.

(iv) If $f \in K$ and $f''(0) = 0$ then $\Phi_n(f)(B) \supseteq B_{\pi/4}$.

(v) If $f \in K$ and $f''(0) = 0, \ldots, f^{(k)}(0) = 0$, then $\Phi_n(f)(B) \supseteq B_{r_k}$, where

$$r_k = \int_0^1 \frac{dt}{(1+t^k)^{2/k}}.$$

(vi) If $f \in S \cap B_1$ then $\Phi_n(f)(B) \supseteq B_{1/2}$.

All results are sharp except for the last.

Finally we obtain covering theorems of Bloch type for the families $\Phi_n(S)$ and $\Phi_n(K)$. These results were obtained by Graham [Gra6].

Theorem 11.2.9. *(i) The image of every $F \in \Phi_n(K)$ contains a ball of radius $\pi/4$. This result is sharp.*

(ii) The image of every $F \in \Phi_n(S)$ contains a ball of radius 1/2.

Proof. These results may be obtained from Corollary 11.2.8 (iv) and (vi) respectively, if we show that, in addition to the stated assumptions, it is possible to assume that $f \in S \cap \mathcal{B}_1$ (which implies $f''(0) = 0$ (see Remark 4.2.3)). To do so we apply Landau's reduction in the z_1 variable and use Lemma 11.2.1. For this purpose, suppose $f \in S$ and f is not in $\mathcal{B}_1$. By dilating we may assume that f is holomorphic on $\overline{U}$, and hence $(1 - |z_1|^2)|f'(z_1)|$ has an interior maximum on U, say at a. Since $f \notin \mathcal{B}_1$, it follows that $a \neq 0$. Let T be a disc automorphism such that $T(0) = a$ and let $(f \circ T)'(0) = \alpha$. Since $a \neq 0$, we must have $|\alpha| > 1$. Let

$$g_T(f)(z_1) = \frac{(f \circ T)(z_1) - (f \circ T)(0)}{\alpha}$$

and consider the mapping $\Phi_n(g_T(f))$ given by

$$\Phi_n(g_T(f))(z) = \left(g_T(f)(z_1), \sqrt{(g_T(f))'(z_1)}\, z' \right) = A(\Phi_n(f \circ T)(z) - b),$$

where $A = \operatorname{diag}(\alpha^{-1}, \alpha^{-1/2}, \ldots, \alpha^{-1/2})$, the branch of $\alpha^{-1/2}$ is determined by the choice of branch of $\sqrt{(f \circ T)'(z_1)}$ so that we have

$$\sqrt{(g_T(f))'(z_1)} = \alpha^{-1/2} \sqrt{(f \circ T)'(z_1)},$$

$b = ((f \circ T)(0), 0, \ldots, 0)$, and the action of A on $\Phi_n(f \circ T)(z) - b$ is determined by writing the latter as a column vector. From this we see that if the image of $\Phi_n(g_T(f))$ contains the ball

(11.2.10) $$|w_1 - c_1|^2 + \|w' - c'\|^2 < \rho^2,$$

then the image of $\Phi_n(f \circ T)$ or of $\Phi_n(f)$ contains the ellipsoid

$$|\alpha|^{-2}|w_1 - c_1 - (f \circ T)(0)|^2 + |\alpha|^{-1}\|w' - c'\|^2 < \rho^2.$$

This ellipsoid contains the ball

$$|w_1 - c_1 - (f \circ T)(0)|^2 + \|w' - c'\|^2 < \rho^2 |\alpha|,$$

which is larger than the ball (11.2.10). Further, note that $g_T(f)$ is convex whenever f is, so the same is true for $\Phi_n(g_T(f))$. Thus given any $f \in S$ (resp.

K), we can produce a function $g \in S \cap B_1$ (resp. $g \in K \cap B_1$), such that the largest ball in $\Phi_n(g)(B)$ has smaller radius than that of the largest ball in $\Phi_n(f)(B)$. Consequently, we may indeed reduce to the case $f \in S \cap B_1$ and apply the statements (iv) and (vi) from Corollary 11.2.8.

The sharpness of part (i) follows by considering the mapping $\Phi_n(f)$, where

$$f(z_1) = \frac{1}{2} \log \left(\frac{1 + z_1}{1 - z_1} \right), \quad |z_1| < 1.$$

Remark 11.2.10. In Chapter 9 (and originally in [Gra-Var2]), the question of whether the image of any convex mapping of B contains a ball of radius $\pi/4$ was considered, and an affirmative answer was obtained for the case of odd mappings. We have just proved that the same result holds for convex mappings in the family $\Phi_n(K)$. However, the general case of the problem remains open.

11.3 Loewner chains and the operator $\Phi_{n,\alpha}$

In this section we study the embeddability in Loewner chains of mappings constructed using the Roper-Suffridge extension operator. In fact we shall consider simultaneously the family of extension operators given by

$$(11.3.1) \quad \Phi_{n,\alpha}(f)(z) = F_\alpha(z) = (f(z_1), (f'(z_1))^\alpha z'), \quad z = (z_1, z') \in B,$$

where $\alpha \in [0, 1/2]$ and f is a normalized locally univalent function on U. We choose the branch of the power function such that $(f'(z_1))^\alpha|_{z_1=0} = 1$. Of course when $\alpha = 1/2$ we obtain the Roper-Suffridge extension operator Φ_n.

We remark that all of the $\Phi_{n,\alpha}$ fall into the general class of operators of the form $\Phi_g(f)(z) = (f(z_1), z' g(z_1))$, where f is a normalized locally univalent function on U, and g is a non-vanishing holomorphic function on U such that $g(0) = 1$. If f is univalent on U and g is analytic and non-zero on U, then the mapping $\Phi_g(f)$ is univalent on B. Among such operators, the Roper-Suffridge operator has some canonical properties as we have seen, but it is of interest to determine to what extent operators of the more general type have useful properties. At a minimum, some connection between the functions f and g

appears to be necessary if we wish the extended mapping to satisfy some geometric condition.

We shall obtain a number of extension results for the operators $\Phi_{n,\alpha}$, $\alpha \in [0, 1/2]$, on the unit ball of $\mathbb{C}^n$ with the Euclidean structure. Our main result is that if $f \in S$ then $\Phi_{n,\alpha}(f)$ can be embedded in a Loewner chain, and moreover $\Phi_{n,\alpha}(f) \in S^0(B)$, where $S^0(B)$ consists of those normalized univalent mappings on B which have parametric representation. In particular, our proof shows that if $f \in S^*$ then $\Phi_{n,\alpha}(f) \in S^*(B)$. On the other hand, Example 11.1.3 implies that convexity is preserved only when $\alpha = 1/2$. Thus the dependence of extension operators from S to $S(B)$ on parameters appears to be an interesting subject.

We begin with the main result of this section, obtained recently by Graham, Kohr and Kohr [Gra-Koh-Koh1]. We give a different proof from that originally obtained in [Gra-Koh-Koh1]. This proof is based on the application of Theorem 8.1.5 instead of Theorem 8.1.6, and gives the conclusion that F_α has parametric representation on B.

Theorem 11.3.1. *Suppose that $f \in S$ and $\alpha \in [0, 1/2]$. Then $F_\alpha = \Phi_{n,\alpha}(f) \in S^0(B)$.*

Proof. It suffices to give the proof in the case $n = 2$. Since $f \in S$, there exists a Loewner chain $f(z_1, t) = e^t z_1 + a_2(t) z_1^2 + \dots$ such that $f(z_1) = f(z_1, 0)$, $z_1 \in U$ (see Theorem 3.1.8). Let $v = v(z_1, s, t)$ be the transition function associated to the Loewner chain $f(z_1, t)$. Then $v'(0, s, t) = e^{s-t}$, $0 \leq s \leq t < \infty$, by the normalization of $f(z_1, t)$, and $f(z_1, s) = f(v(z_1, s, t), t)$ for $z_1 \in U$ and $0 \leq s \leq t < \infty$.

In view of Theorem 3.1.12, there exists a function $p(z_1, t)$ such that $p(\cdot, t) \in \mathcal{P}$ for each $t \geq 0$, $p(z_1, t)$ is measurable in $t \in [0, \infty)$ for each $z_1 \in U$, and for almost all $t \geq 0$,

$$\frac{\partial f}{\partial t}(z_1, t) = z_1 f'(z_1, t) p(z_1, t), \quad \forall z_1 \in U.$$

Moreover, $v(t) = v(z_1, s, t)$ satisfies the initial value problem

$$(11.3.2) \qquad \frac{\partial v}{\partial t} = -v p(v, t), \quad a.e. \quad t \geq s, \quad v(s) = z_1,$$

for all $s \geq 0$ and $z_1 \in U$, and

(11.3.3)
$$\lim_{t \to \infty} e^t v(z_1, s, t) = f(z_1, s)$$

locally uniformly on U.

Now, let $h = h(z,t) : B \times [0, \infty) \to \mathbb{C}^n$ be given by

$$h(z,t) = \left(z_1 p(z_1, t), z_2 \left[1 - \alpha + \alpha p(z_1, t) + \alpha z_1 p'(z_1, t) \right] \right)$$

for $z = (z_1, z_2) \in B$ and $t \geq 0$.

Clearly $h(\cdot, t) \in H(B)$, $h(0, t) = 0$, $Dh(0, t) = I$, and we shall show that $h(\cdot, t) \in \mathcal{M}$. Indeed, we have

(11.3.4)
$$\text{Re } \langle h(z,t), z \rangle = |z_1|^2 \text{Re } p(z_1, t) + (1 - \alpha)|z_2|^2$$

$$+ \alpha|z_2|^2 \text{Re } p(z_1, t) + \alpha|z_2|^2 \text{Re } [z_1 p'(z_1, t)], \quad z \in B, \ t \geq 0.$$

We may assume that $z = (z_1, z_2)$, $z_2 \neq 0$, because it is obvious that (11.3.4) is positive when $z_2 = 0$ and $z_1 \in U \setminus \{0\}$.

Applying the minimum principle for harmonic functions, it suffices to show that

$$\text{Re } \langle h(z,t), z \rangle \geq 0, \ z = (z_1, z_2) \in \mathbb{C}^2, \ |z_1|^2 + |z_2|^2 = 1, \ z \neq (z_1, 0), \ t \geq 0.$$

Since $p(0, t) = 1$ and $\text{Re } p(z_1, t) > 0$, $z_1 \in U$, $t \geq 0$, the estimate (2.1.6) gives

$$|p'(z_1, t)| \leq \frac{2}{1 - |z_1|^2} \text{Re } p(z_1, t), \quad z_1 \in U, \ t \geq 0.$$

Consequently,
$$\text{Re } [z_1 p'(z_1, t)] \geq -\frac{2|z_1|}{1 - |z_1|^2} \text{Re } p(z_1, t),$$

and using (11.3.4) and the fact that $\alpha \in [0, 1/2]$, we deduce that

$$\text{Re } \langle h(z,t), z \rangle$$

$$\geq (1 - \alpha)(1 - |z_1|^2) + \text{Re } p(z_1, t)[(1 - \alpha)|z_1|^2 - 2\alpha|z_1| + \alpha] > 0$$

for $z = (z_1, z_2) \in \mathbb{C}^2$, $|z_1|^2 + |z_2|^2 = 1$, $z_2 \neq 0$, and $t \geq 0$. Thus $h(\cdot, t) \in \mathcal{M}$ for $t \geq 0$, and hence $h(\cdot, t)$ satisfies the assumption (i) of Theorem 8.1.3. Obviously the measurability condition (ii) in Theorem 8.1.3 is also satisfied.

Next, for $z = (z_1, z_2) \in B$ and $s \geq 0$, let $V(t) = V(z, s, t)$, $t \geq s$, be given by

(11.3.5) $\quad V(t) = V(z, s, t) = \left(v(z_1, s, t), z_2 e^{(1-\alpha)(s-t)} (v'(z_1, s, t))^\alpha \right).$

The branch of the power function is chosen such that $(v'(z_1, s, t))^\alpha|_{z_1=0} = e^{\alpha(s-t)}$ for each $t \in [s, \infty)$. Then $V(\cdot, s, t) \in H(B)$, $V(0, s, t) = 0$ and $DV(0, s, t) = e^{s-t} I$.

From (11.3.3), (11.3.5) and Weierstrass' theorem we deduce that the limit

$$F_\alpha(z, s) = \lim_{t \to \infty} e^t V(z, s, t)$$

exists locally uniformly on B for each $s \geq 0$, and

(11.3.6) $\qquad F_\alpha(z, s) = \left(f(z_1, s), z_2 e^{(1-\alpha)s} (f'(z_1, s))^\alpha \right)$

for $z = (z_1, z_2) \in B$ and $s \geq 0$. The branch of the power function $(f'(z_1, s))^\alpha$ is chosen such that $(f'(z_1, s))^\alpha|_{z_1=0} = e^{\alpha s}$.

Moreover, from (11.3.2) we obtain for almost all $t \geq s$,

$$\frac{\partial^2 v(z_1, s, t)}{\partial z_1 \partial t} = \frac{\partial^2 v(z_1, s, t)}{\partial t \partial z_1}$$

$$= -v'(z_1, s, t) p(v(z_1, s, t), t) - v(z_1, s, t) p'(v(z_1, s, t), t) v'(z_1, s, t).$$

Here we have made use of the fact that $v(z_1, s, t)$ is a Lipschitz continuous function of t locally uniformly with respect to $z_1 \in U$, and Vitali's theorem, to conclude that the order of differentiation can be changed. Then after straightforward computations, we obtain for almost all $t \geq s$,

$$\frac{\partial V}{\partial t}(z, s, t) = -\left(v(z_1, s, t) p(v(z_1, s, t), t), z_2 e^{(1-\alpha)(s-t)} (v'(z_1, s, t))^\alpha \times \right.$$

$$\left. \left[1 - \alpha + \alpha p(v(z_1, s, t), t) + \alpha v(z_1, s, t) p'(v(z_1, s, t), t) \right] \right)$$

$$= -\left(V_1(z, s, t) p(V_1(z, s, t), t), V_2(z, s, t) \times \right.$$

$$\left. \left[1 - \alpha + \alpha p(V_1(z, s, t), t) + \alpha V_1(z, s, t) p'(V_1(z, s, t), t) \right] \right).$$

Therefore $V(t) = V(z, s, t)$ is the solution of the initial value problem

$$\frac{\partial V}{\partial t}(z, s, t) = -h(V(z, s, t), t), \quad a.e. \quad t \geq s, \quad V(s) = z.$$

Since h satisfies the assumptions (i) and (ii) of Theorem 8.1.3, we conclude from Theorem 8.1.5 that the mapping $F_\alpha(z, t)$ given by (11.3.6) is a Loewner chain. The conditions of Definition 8.3.2 (for $g(\zeta) = (1 + \zeta)/(1 - \zeta)$, $|\zeta| < 1$) are therefore satisfied, and we have $F_\alpha(z) = F_\alpha(z, 0) \in S^0(B)$, as claimed. This completes the proof.

In the previous section we have seen that the Roper-Suffridge extension operator Φ_n has the following properties: $\Phi_n(K) \subseteq K(B)$ and $\Phi_n(S^*) \subseteq S^*(B)$. We have now added another one: $\Phi_n(S) \subseteq S^0(B)$.

As consequences of Theorem 11.3.1, we can also prove that the operator $\Phi_{n,\alpha}$, $\alpha \in [0, 1/2]$, preserves starlikeness and spirallikeness of type γ with $|\gamma| < \pi/2$. In particular, for $\alpha = 1/2$ we obtain another proof that the Roper-Suffridge operator Φ_n preserves starlikeness. The following results were obtained in [Gra-Koh-Koh1]:

Corollary 11.3.2. *Let $f \in S^*$ and $\alpha \in [0, 1/2]$. Then $F_\alpha = \Phi_{n,\alpha}(f) \in S^*(B)$.*

Proof. Since $f \in S^*$, $f(z_1, t) = e^t f(z_1)$ is a Loewner chain. Let $F_\alpha(z, t)$ be given by (11.3.6). Theorem 11.3.1 implies that $F_\alpha(z, t)$ is a Loewner chain, and since $F_\alpha(z, t) = e^t F_\alpha(z)$ we deduce from Corollary 8.2.3 that F_α is starlike. This completes the proof.

Corollary 11.3.3. *Assume f is a normalized spirallike function of type γ, where $\gamma \in \mathbb{R}$, $|\gamma| < \pi/2$, and let $F_\alpha = \Phi_{n,\alpha}(f)$, with $\alpha \in [0, 1/2]$. Then F_α is a spirallike mapping of type γ.*

Proof. Since f is spirallike of type γ, $f(z_1, t) = e^{(1-ia)t} f(e^{iat} z_1)$ is a Loewner chain, where $a = \tan \gamma$ (see Corollary 3.2.9). A short computation using (11.3.6) yields that $F_\alpha(z, t) = e^{(1-ia)t} F_\alpha(e^{iat} z)$. From Theorem 11.3.1 one concludes that $F_\alpha(z, t)$ is a Loewner chain, and Theorem 8.2.1 therefore yields that F_α is spirallike of type γ. This completes the proof.

As we have already noted, Example 11.1.3 shows that the only case in which $\Phi_{n,\alpha}$ maps K into $K(B)$ is the case $\alpha = 1/2$. Another example which leads to the same conclusion is the following. We leave the proof for the reader.

Example 11.3.4. Let $n = 2$ and

$$f(z_1) = \frac{1}{2} \log \left(\frac{1 + z_1}{1 - z_1} \right), \quad z_1 \in U.$$

Then f is convex, but $\Phi_{n,\alpha}(f)$ is not convex, for $\alpha \in [0, 1/2)$.

Remark 11.3.5. It seems the value $\alpha = 1/2$ plays a unique role in preserving convexity under the operator $\Phi_{n,\alpha}$, $\alpha \in [0, 1/2]$, on the Euclidean unit ball. However, Gong and Liu [Gon-Liu3] have recently proved that if we consider the operator $\Phi_{n,\alpha}$ on the domain

$$\Omega_{n,1/\alpha} = \left\{ z = (z_1, \ldots, z_n) \in \mathbb{C}^n : |z_1|^2 + \sum_{k=2}^{n} |z_k|^{1/\alpha} < 1 \right\}$$

when $\alpha \in (0, 1]$, and $\Omega_{n,1/\alpha}$ is the unit polydisc of $\mathbb{C}^n$ when $\alpha = 0$, then $\Phi_{n,\alpha}$ preserves both starlikeness and convexity on $\Omega_{n,1/\alpha}$ for $\alpha \in [0, 1]$. Their method of proof is completely different from that which we have used in the case of the unit ball B and is of interest for the reader.

11.4 Radius problems and the operator $\Phi_{n,\alpha}$

In this section we continue the study of the operator $\Phi_{n,\alpha}$ with $\alpha \in [0, 1/2]$, and we obtain the radius of starlikeness of $\Phi_{n,\alpha}(S)$ and the radius of convexity of $\Phi_n(S)$ (i.e. of $\Phi_{n,1/2}(S)$). We recall that $\mathcal{LS}(B)$ denotes the set of normalized locally biholomorphic mappings on B. Also if $\mathcal{F}$ is a non-empty subset of $\mathcal{LS}(B)$, then $r^*(\mathcal{F})$ is the largest number such that every mapping in the set $\mathcal{F}$ is starlike on $B_{r^*(\mathcal{F})}$. The quantity $r^*(\mathcal{F})$ is called the *radius of starlikeness* of $\mathcal{F}$. Also $r_c(\mathcal{F})$ is the largest number such that every mapping in the set $\mathcal{F}$ is convex on $B_{r_c(\mathcal{F})}$. This quantity is called the *radius of convexity* of $\mathcal{F}$.

Radius problems in several complex variables were considered by Shi [Shi], who showed that the radius of convexity of $S^*(B)$ is strictly positive on the Euclidean unit ball B of $\mathbb{C}^n$, $n \geq 2$. He also showed that there exists a positive radius of convexity for the set of bounded normalized biholomorphic mappings on B.

Remark 11.4.1. It is obvious that if $f : U \to \mathbb{C}$ is a normalized locally univalent function, and if for some $\alpha \in [0, 1/2]$ and some $r \in (0, 1)$, $\Phi_{n,\alpha}(f)$ is univalent on B_r, then f must also be univalent on U_r. Also if $\Phi_{n,\alpha}(f)$ is starlike on B_r (resp. convex on B_r) then f will likewise be starlike on U_r (resp. convex on U_r).

The following result is due to Graham, Kohr and Kohr [Gra-Koh-Koh1].

Theorem 11.4.2. $r^*(\Phi_{n,\alpha}(S)) = \tanh \dfrac{\pi}{4}$, for all $\alpha \in [0, 1/2]$.

Proof. Using Corollary 3.2.3, we have $r^*(S) = \tanh \dfrac{\pi}{4}$. Let $f \in S$. Then

$$\text{Re}\left[\frac{z_1 f'(z_1)}{f(z_1)}\right] > 0, \quad |z_1| < \tanh \frac{\pi}{4},$$

and this quantity can be negative if $|z_1| > \tanh \dfrac{\pi}{4}$.

Now let $F_\alpha = \Phi_{n,\alpha}(f)$. Taking into account Corollary 11.3.2 and Remark 11.4.1, we deduce that F_α is starlike on B_r with $r = \tanh \dfrac{\pi}{4}$, and further that F_α may fail to be starlike in any ball B_{r_1} with $r_1 > r$. Therefore $r = \tanh \dfrac{\pi}{4}$ is the largest radius for which each $F_\alpha \in \Phi_{n,\alpha}(S)$ is starlike on B_r. This completes the proof.

Since $\Phi_{n,\alpha}(S) \subseteq S^0(B)$ for $\alpha \in [0, 1/2]$, we must have $r^*(S^0(B)) \leq r^*(\Phi_{n,\alpha}(S)) = \tanh \dfrac{\pi}{4}$ for $n \geq 2$. Hence Theorem 11.4.2 leads to the following [Gra-Koh-Koh1]:

Conjecture 11.4.3. $r^*(S^0(B)) = \tanh \dfrac{\pi}{4}$ in dimension $n \geq 2$.

With similar reasoning as in the proof of Theorem 11.4.2, we deduce the following result concerning the radius of convexity of $\Phi_n(S)$ [Gra-Koh-Koh1]:

Theorem 11.4.4. $r_c(\Phi_n(S)) = r_c(\Phi_n(S^*)) = 2 - \sqrt{3}$.

Proof. Let $F \in \Phi_n(S)$ (or $F \in \Phi_n(S^*)$). Then $F = \Phi_n(f)$ for some $f \in S$ (or $f \in S^*$). By Theorem 2.2.22, the radius of convexity for S (or for S^*) is $r = 2 - \sqrt{3}$. Hence

$$\text{Re}\left[\frac{z_1 f''(z_1)}{f'(z_1)} + 1\right] > 0, \quad |z_1| < r,$$

and this quantity can be negative if $|z_1| > r$.

Now if $g \in K(U_\rho)$, $0 < \rho \leq 1$, then $g_\rho \in K(U)$, where $g_\rho(\zeta) = \dfrac{1}{\rho} g(\rho\zeta)$, $\zeta \in U$. Hence from Theorem 11.1.2 we deduce that $\Phi_n(g_\rho) \in K(B)$, which implies that $\Phi_n(g)$ is convex on B_ρ because

$$\Phi_n(g_\rho)(z) = \frac{1}{\rho} \Phi_n(g)(\rho z), \quad z \in B.$$

Using this observation, we conclude that $F = \Phi_n(f)$ is convex on B_r where $r = 2 - \sqrt{3}$. Furthermore, taking into account Remark 11.4.1, we deduce that F may fail to be convex in any ball B_{r_1} with $r_1 > r$. Therefore $r_c(\Phi_n(S)) = r_c(\Phi_n(S^*)) = 2 - \sqrt{3}$. This completes the proof.

Since $\Phi_n(S) \subseteq S^0(B)$, $\Phi_n(S^*) \subseteq S^*(B)$ and $S^*(B) \subset S^0(B)$, we conclude from Theorem 11.4.4 that

$$r_c(S^0(B)) \leq r_c(S^*(B)) \leq 2 - \sqrt{3}.$$

However, an example provided by Suffridge shows that in $\mathbb{C}^n$, $n \geq 2$, the radius of convexity of $S^*(B)$ is strictly less than $2 - \sqrt{3}$. Therefore, it remains an open problem to find this radius in several complex variables.

Example 11.4.5. Let $n = 2$ and let $f : B \subset \mathbb{C}^2 \to \mathbb{C}^2$ be given by

$$f(z) = (z_1 + az_2^2, z_2), \quad z = (z_1, z_2) \in B,$$

where $a \in \mathbb{C}$, $|a| = 3\sqrt{3}/2$. Then f is starlike on B and is convex on B_r where $r = 1/(3\sqrt{3})$. However f is not convex in any ball centered at zero and of radius greater than r.

Proof. Since $|a| = 3\sqrt{3}/2$, Problem 6.2.1 implies that f is starlike on B. With a similar argument as in the proof of Example 6.3.13, we can deduce that f is convex on B_r where $r = 1/(3\sqrt{3})$.

Indeed, using the criterion of convexity in Theorem 6.3.10 we have to show that

$$\mathrm{Re}\,\langle [Df(z)]^{-1}(f(z) - f(u)), z \rangle \geq 0, \quad \|u\| \leq \|z\| < r.$$

Straightforward computations yield (see the proof of Example 6.3.13)

$$\mathrm{Re}\,\langle [Df(z)]^{-1}(f(z) - f(u)), z \rangle$$

$$\geq (\|z\|^2 - \mathrm{Re}\,\langle z, u \rangle)(1 - 2a|z_1|) + |a|\|z_1\||z_1 - u_1|^2 \geq 0,$$

for all $z = (z_1, z_2) \in B_r$, $u = (u_1, u_2) \in B_r$, $\|u\| \leq \|z\|$, when $|a| = 3\sqrt{3}/2$ and $r = 1/(3\sqrt{3})$.

On the other hand, f is not convex in any ball B_{r_1} with $r_1 > 1/(3\sqrt{3})$. To see this, let $z = (z_1, z_2) \in B$ and $u = (u_1, u_2) \in B$, where $z_1 = u_1$, $z_2 = -u_2 \in \mathbb{R} \setminus \{0\}$, $|z_1| > 1/(3\sqrt{3})$ and $\mathrm{Re}\,\{a\bar{z}_1\} > 1/2$. Hence $\|z\| > 1/(3\sqrt{3})$ and

$$\mathrm{Re}\,\langle [Df(z)]^{-1}(f(z) - f(u)), z \rangle$$

$$= \|z\|^2 - \operatorname{Re}\langle z, u \rangle - \operatorname{Re}\{a\bar{z}_1(z_2 - u_2)^2\}$$

$$= \|z\|^2 - \operatorname{Re}\{z_1\bar{z}_1 - z_2\bar{z}_2\} - 4z_2^2\operatorname{Re}\{a\bar{z}_1\}$$

$$= 2z_2^2\{1 - 2\operatorname{Re}\{a\bar{z}_1\}\} < 0.$$

Thus f is not convex in any ball of radius greater than $r = 1/(3\sqrt{3})$ and centered at 0. This completes the proof.

Open Problem 11.4.6. *Find $r_c(S^*(B))$ and $r_c(S^0(B))$ in dimension $n \geq$* 2.

In connection with this problem, we make the following observation:

Remark 11.4.7. There is no positive radius of convexity for the class of normalized starlike mappings on the unit polydisc P of $\mathbb{C}^n$ with $n \geq 2$.

Proof. Let $F : P \to \mathbb{C}^n$ be given by

$$F(z) = \left(\frac{z_1}{(1 - z_1)^2}, \ldots, \frac{z_n}{(1 - z_1)^2}\right), \quad z = (z_1, \ldots, z_n) \in P.$$

It is not difficult to deduce that F is starlike. On the other hand, it is obvious that F violates the decomposition result in Theorem 6.3.2, and hence F is not convex on P_r for any $r \in (0, 1]$. This completes the proof.

11.5 Linear-invariant families and the operator $\Phi_{n,\alpha}$

In this section we use the operator $\Phi_{n,\alpha}$ given by (11.3.1) to obtain linear-invariant families on the Euclidean unit ball B of $\mathbb{C}^n$ and to study the order of these L.I.F.'s. To make the dimension explicit, in this section we shall denote the Euclidean unit ball in $\mathbb{C}^k$ by B^k. Also by $\mathcal{LS}(B^k)$ we shall denote the set of normalized locally biholomorphic mappings from B^k into $\mathbb{C}^k$.

As we have already seen in Theorem 6.1.23, the automorphisms of B^n, up to multiplication by unitary transformations, are the mappings

$$h_a(z) = T_a\left(\frac{a - z}{1 - a^*z}\right), \quad z \in B^n, \ a \in B^n,$$

where $T_0 = I$ and for $a \neq 0$, T_a is the linear operator given by

$$T_a = \frac{1}{\|a\|^2}\{(1 - s_a)aa^* + s_a\|a\|^2 I\},$$

and

$$s_a = \sqrt{1 - \|a\|^2}.$$

(In fact, $h_a(z) = -\phi_a(z)$, where $\phi_a(z)$ is given by (6.1.9).)

In other words, $\text{Aut}(B^n) = \{V h_a : a \in B^n, V \in \mathcal{U}\}$, where $\mathcal{U}$ denotes the set of unitary transformations in $\mathbb{C}^n$.

The following lemmas will be useful in this section. The proof of Lemma 11.5.1 is contained in the first part of the proof of [Pfa-Su4, Theorem 3.3], and Lemma 11.5.2 has recently been obtained in [Gra-Ham-Koh-Su].

Lemma 11.5.1. *Let $\mathcal{F} \subset LS(B^n)$ and $\Lambda[\mathcal{F}]$ be the L.I.F. generated by $\mathcal{F}$ on B^n. Let $a \in U$ and $b \in B^{n-1}$. Then*

$$\text{ord } \Lambda[\mathcal{F}] = \sup \left\{ \left| \text{trace}\left\{ \frac{1}{2} D^2 \Lambda_{h_b} \Lambda_{h_a}(F)(0)(\gamma, \cdot) \right\} \right| \right.$$
$$\left. : |a| < 1, \|b\| < 1, \|\gamma\| = 1, F \in \mathcal{F} \right\},$$

where $h_a = h_{ae_1}$ and $h_b = h_{(0,b)}$.

Proof. We first observe that $\{h_a \circ h_b : a \in U, b \in B^{n-1}\}$ is a family of automorphisms φ of B^n such that $\varphi(0) = \left(a, \sqrt{1 - |a|^2} b \right)$. Since this includes all points of B^n as a and b vary, we conclude that $\text{Aut}(B^n)$ consists of the composition of all unitary mappings with members of this special family. Since the trace is invariant under similarity, it follows that it is sufficient to consider automorphisms of the above type. Since $\Lambda_{h_b} \circ \Lambda_{h_a} = \Lambda_{h_a \circ h_b}$, the lemma now follows.

Lemma 11.5.2. *Assume $f, g : U \to \mathbb{C}$ are holomorphic functions such that f is locally univalent, $f(0) = 0$ and $f'(0) = 1 = g(0)$. Define $F : B^n \to \mathbb{C}^n$ by $F(z) = (f(z_1), g(z_1)z'), z = (z_1, z') \in B^n$. With $G(z) = \Lambda_{h_b}(F)(z)$ we have*

$$\sup \left\{ \left| \text{trace}\left\{ D^2 G(0)(\gamma, \cdot) \right\} \right| : \|b\| < 1, \|\gamma\| = 1 \right\}$$
$$= \max \left\{ n + 1, \sup \left\{ \left| \text{trace}\left\{ D^2 F(0)(\gamma, \cdot) \right\} \right| : \|\gamma\| = 1 \right\} \right\}.$$

Proof. Without loss of generality, we may assume that the coordinates are chosen so that $b = xe_2$ where $0 \le x < 1$. We write $z = (z_1, z_2, v)$ where $v \in B^{n-2}$ and $\|z\| < 1$. Of course if $n = 2$, v will not appear. Then

$$h_b(z) = \frac{\sqrt{1 - x^2}}{1 - xz_2}(-z_1, 0, -v) + \frac{x - z_2}{1 - xz_2}e_2 = \sum_{j=1}^{n} h_b^{(j)}(z)e_j.$$

Since

$$DF(h_b(0))Dh_b(0) = \begin{pmatrix} -\sqrt{1-x^2} & 0 & 0 \\ -x\sqrt{1-x^2}g'(0) & -(1-x^2) & 0 \\ 0 & 0 & -\sqrt{1-x^2}I \end{pmatrix},$$

it follows that $G(z) = H(F(h_b(z)) - xe_2)$, where

$$H = \begin{pmatrix} -\dfrac{1}{\sqrt{1-x^2}} & 0 & 0 \\ \dfrac{xg'(0)}{1-x^2} & -\dfrac{1}{1-x^2} & 0 \\ 0 & 0 & -\dfrac{1}{\sqrt{1-x^2}}I \end{pmatrix}.$$

Because of the form of G, the trace of $D^2G(0)(\gamma, \cdot)$ is

$$\frac{\partial^2 G_1}{\partial z_1^2}(0)\gamma_1 + \frac{\partial^2 G_1}{\partial z_1 \partial z_2}(0)\gamma_2 + \frac{\partial^2 G_2}{\partial z_1 \partial z_2}(0)\gamma_1 + \frac{\partial^2 G_2}{\partial z_2^2}(0)\gamma_2$$

$$+ \sum_{k=3}^{n} \left(\frac{\partial^2 G_k}{\partial z_1 \partial z_k}(0)\gamma_1 + \frac{\partial^2 G_k}{\partial z_2 \partial z_k}(0)\gamma_2 \right),$$

and the entries on the diagonal of $D^2G(0)(\gamma, \cdot)$ are

$$-\sqrt{1-x^2}f''(0)\gamma_1 + x\gamma_2, -\sqrt{1-x^2}g'(0)\gamma_1 + 2x\gamma_2,$$

$$-\sqrt{1-x^2}g'(0)\gamma_1 + x\gamma_2, \ldots, -\sqrt{1-x^2}g'(0)\gamma_1 + x\gamma_2.$$

The trace is therefore

$$-\sqrt{1-x^2}\Big(f''(0) + (n-1)g'(0)\Big)\gamma_1 + (n+1)x\gamma_2.$$

By elementary calculus, the supremum of this quantity with $0 \le x < 1$ and $\|\gamma\| = 1$ is

$$\max\Big\{n+1, |f''(0) + (n-1)g'(0)|\Big\}.$$

Since

$$\text{trace}\Big\{ D^2 F(0)(\gamma, \cdot)\Big\} = \Big(f''(0) + (n-1)g'(0)\Big)\gamma_1,$$

the lemma now follows.

The main result of this section is a theorem of Graham, Hamada, Kohr and Suffridge [Gra-Ham-Koh-Su] concerning the order of L.I.F.'s generated using the operator $\Phi_{n,\alpha}$.

Theorem 11.5.3. *Let $\mathcal{F}$ be a L.I.F. on U such that* ord $\mathcal{F} = \delta < \infty$ *and let $\alpha \in [0, 1/2]$. Then* ord $\Lambda[\Phi_{n,\alpha}(\mathcal{F})] = \eta$, *where*

$$\eta = (1 + (n-1)\alpha)\delta + \frac{(n-1)(1-2\alpha)}{2}.$$

Proof. Let $f \in \mathcal{F}$ and set $G = \Phi_{n,\alpha}(f)$. Using Lemmas 11.5.1 and 11.5.2, it follows that

$$\text{ord } \Lambda[\Phi_{n,\alpha}(\mathcal{F})]$$

$$= \sup \left\{ \left| \frac{\text{trace}\{D^2\Lambda_{h_a}(G)(0)(\gamma, \cdot)\}}{2} \right| : a \in U, \|\gamma\| = 1, f \in \mathcal{F} \right\}.$$

Letting

$$\tilde{f}(z_1; a) = \frac{f\left(\dfrac{a - z_1}{1 - \bar{a}z_1}\right) - f(a)}{-(1 - |a|^2)f'(a)},$$

we have

$$\Lambda_{h_a}(G)(z) = \left(\tilde{f}(z_1; a), \left(\frac{f'\left(\dfrac{a - z_1}{1 - \bar{a}z_1}\right)}{f'(a)} \right)^{\alpha} \frac{1}{1 - \bar{a}z_1}z' \right).$$

Now the diagonal of $D^2\Lambda_{h_a}(G)(0)(\gamma, \cdot)$ has

$$\left(-\frac{(1 - |a|^2)f''(a)}{f'(a)} + 2\bar{a} \right)\gamma_1$$

as its first entry and

$$\left(-\alpha\frac{(1 - |a|^2)f''(a)}{f'(a)} + \bar{a} \right)\gamma_1$$

in the remaining positions. The trace is therefore

$$\left(\frac{-(1-|a|^2)f''(a)}{f'(a)} + 2\bar{a}\right)\gamma_1(1+(n-1)\alpha) + (1-2\alpha)(n-1)\bar{a}\gamma_1.$$

Since we may replace f by a function $g \in \mathcal{F}$ such that

$$g''(0) = \frac{-(1-|a|^2)f''(a)}{f'(a)} + 2\bar{a}$$

(i.e. $g(z_1) = \tilde{f}(z_1; a)$), it is clear that we want to find

$$\sup_{g \in \mathcal{F}, |a| < 1} \left|\frac{g''(0)}{2}(1+(n-1)\alpha) + (1-2\alpha)\frac{n-1}{2}\bar{a}\right|.$$

This is evidently

$$\sup_{g \in \mathcal{F}, |a| < 1} \left(\frac{|g''(0)|}{2}(1+(n-1)\alpha) + (1-2\alpha)\frac{n-1}{2}|a|\right)$$

$$= (1+(n-1)\alpha)\delta + \frac{(n-1)(1-2\alpha)}{2},$$

and the proof is complete.

There are some interesting particular cases of the above theorem. When $\alpha = 1/2$, we obtain the following result for the Roper-Suffridge operator (see [Pfa5,6], [Lic-St1]):

Corollary 11.5.4. *Let $\mathcal{F}$ be a L.I.F. on U such that $\operatorname{ord}\mathcal{F} = \delta < \infty$. Then $\operatorname{ord}\Lambda[\Phi_n(\mathcal{F})] = \delta(n+1)/2$.*

On the other hand, if $\mathcal{F} = K$ in Theorem 11.5.3, we obtain another consequence due to Graham, Hamada, Kohr and Suffridge [Gra-Ham-Koh-Su], which illustrates an interesting phenomenon in several complex variables.

Corollary 11.5.5. *Let $\alpha \in [0, 1/2]$. Then $\operatorname{ord}\Lambda[\Phi_{n,\alpha}(K)] = (n+1)/2$.*

Proof. It suffices to apply Theorem 11.5.3 and then to use the fact that $\operatorname{ord}K = 1$.

Remark 11.5.6. We have seen in Example 11.3.4 that the operator $\Phi_{n,\alpha}$, $\alpha \in [0, 1/2]$, preserves convexity on the unit ball of $\mathbb{C}^n$, $n \geq 2$, if and only if $\alpha = 1/2$. Therefore Corollary 11.5.5 shows that in several complex variables there exist L.I.F.'s of minimum order which are not subsets of $K(B^n)$. (Compare with Theorem 10.3.8.) Indeed, ord $\Lambda[\Phi_{n,\alpha}(K)] = (n+1)/2$, but for $\alpha \neq 1/2$ and $n \geq 2$, $\Lambda[\Phi_{n,\alpha}(K)] \not\subset K(B^n)$.

Finally we shall give simple applications to radius problems for families of mappings generated using the operator $\Phi_{n,\alpha}$. The result below is a generalization of Theorem 11.4.4 and has recently been obtained in [Gra-Ham-Koh-Su].

Theorem 11.5.7. *Let $\mathcal{F}$ be a L.I.F. on U such that* ord $\mathcal{F} = \gamma < \infty$. *Then* $r_c(\Phi_n(\mathcal{F})) = \gamma - \sqrt{\gamma^2 - 1}$. *In particular, $r_c(\Phi_n(K)) = 1$ and $r_c(\Phi_n(S)) = 2 - \sqrt{3}$.*

Proof. Let $F \in \Phi_n(\mathcal{F})$. Then $F = \Phi_n(f)$ for some $f \in \mathcal{F}$. In view of Theorem 5.2.3, we have $f \in K(U_r)$ with $r = \gamma - \sqrt{\gamma^2 - 1}$, and in fact this number is the radius of convexity of $\mathcal{F}$. Since $\Phi_n(f) \in K(B_r^n)$ by Theorem 11.1.2, it follows that $r_c(\Phi_n(\mathcal{F})) = r$. This completes the proof.

Recall that in dimension $n > 1$, there is in general no such connection between the order of a L.I.F. $\mathcal{M}$ of B^n and its radius of convexity (see Chapter 10 and also [Pfa-Su2], [Pfa-Su4]).

We finish this section with a consequence of Corollary 11.3.2 obtained in [Gra-Ham-Koh-Su].

Theorem 11.5.8. *Let $\mathcal{F}$ be a L.I.F. on U such that* ord $\mathcal{F} = \gamma < \infty$. *Also let $\alpha \in [0, 1/2]$. Then $\Phi_{n,\alpha}(\mathcal{F}) \subseteq S^*(B_r^n)$, where $r = 1/\gamma$.*

Proof. Let $F_\alpha \in \Phi_{n,\alpha}(\mathcal{F})$. Then $F_\alpha = \Phi_{n,\alpha}(f)$ for some $f \in \mathcal{F}$. Since ord $\mathcal{F} = \gamma$, we deduce from Corollary 5.2.6 that $f \in S^*(U_r)$ with $r = 1/\gamma$. We then conclude from Corollary 11.3.2 that $F_\alpha \in S^*(B_r^n)$, as desired.

Problems

11.5.1. Let $\alpha \geq 0$ and $\beta \geq 0$. Also let $\Psi_{n,\alpha,\beta}$ denote the operator

$$\Psi_{n,\alpha,\beta}(f)(z) = \left(f(z_1), z' \left(\frac{f(z_1)}{z_1} \right)^\alpha (f'(z_1))^\beta \right),$$

where f is a locally univalent function on U, normalized by $f(0) = f'(0) - 1 = 0$ and such that $f(z_1) \neq 0$ for $z_1 \in U \setminus \{0\}$. The branches of the power functions are chosen such that $\left(\frac{f(z_1)}{z_1} \right)^\alpha \Big|_{z_1=0} = 1$ and $(f'(z_1))^\beta |_{z_1=0} = 1$. Prove the following assertions:

(i) $\Psi_{n,\alpha,\beta}(S) \subseteq S^0(B)$ for $\alpha \in [0,1]$, $\beta \in [0,1/2]$ and $\alpha + \beta \leq 1$.

(ii) $\Psi_{n,\alpha,\beta}(S^*) \subseteq S^*(B)$ for $\alpha \in [0,1]$, $\beta \in [0,1/2]$ and $\alpha + \beta \leq 1$.

(iii) $\Psi_{n,\alpha,\beta}(K) \subset K(B)$ for $n \geq 2$, if and only if $(\alpha, \beta) = (0, 1/2)$.

(Graham, Hamada, Kohr, Suffridge, 2002 [Gra-Ham-Koh-Su].)

11.5.2. Show that $r^*(\Psi_{n,\alpha,\beta}(S)) = \tanh(\pi/4)$ for all $\alpha \in [0,1]$ and $\beta \in [0,1/2]$ with $\alpha + \beta \leq 1$, where $\Psi_{n,\alpha,\beta}$ is the operator defined in Problem 11.5.1. (Graham, Hamada, Kohr, Suffridge, 2002 [Gra-Ham-Koh-Su].)

Bibliography

[Aha] D. Aharonov, *A necessary and sufficient condition for univalence of a meromorphic function*, Duke Math. J., **36**(1969), 599-604.

[Ahl1] L.V. Ahlfors, *An extension of Schwarz's lemma*, Trans. Amer. Math. Soc., **43**(1938), 359-364.

[Ahl2] L.V. Ahlfors, Conformal Invariants. Topics in Geometric Function Theory, Mc Graw-Hill, New York, 1973.

[Ahl3] L.V. Ahlfors, *Sufficient conditions for quasiconformal extension*, Ann. Math. Studies, **79**(1974), 23-29.

[Ahl4] L.V. Ahlfors, Lectures on Quasiconformal Mappings, Wadsworth & Brooks/Cole, Monterey, CA, 1987.

[Ahl-Gru] L.V. Ahlfors, H. Grunsky, *Über die Blochsche Konstante*, Math. Z., **42**(1937), 671-673.

[Ai-Sh] L. Aizenberg, D. Shoikhet, *Boundary behavior of semigroups of holomorphic mappings on the unit ball in* $\mathbb{C}^n$, Complex Variables, **47**(2002), 109-121.

[Aks-Ka] L.A. Aksent'ev, A.V. Kazantsev, *A new property of the Nehari class and its application*, Izv. Vyssh. Uchebn. Zaved. Mat. 8(1989), 69-72 (in Russian); Soviet Math. (Iz. VUZ), **33**(1989), 94-99.

[Al] J.W. Alexander, *Functions which map the interior of the unit circle upon simple regions*, Ann. of Math., **17**(1915-1916), 12-22.

477

[Ale]   I.A. Aleksandrov, Parametric Extensions in the Theory of Univa-
               lent Functions, Izdat. "Nauka", Moscow, 1976 (in Russian).

[Al-Moc]   H.Al-Amiri, P.T. Mocanu, *Spirallike nonanalytic functions*, Proc.
               Amer. Math. Soc., **82**(1981), 61-65.

[An-Lem]   E. Andersén, L. Lempert, *On the group of holomorphic automor-
               phisms of* $\mathbb{C}^n$, Invent. Math., **110**(1992), 371-388.

[And]   J.M. Anderson, *Bloch functions: the basic theory*, Operators and
               Function Theory (Lancaster, 1984), 1-17; NATO Adv. Sci. Inst.
               Ser. C Math. Phys. Sci., **153**, Reidel, Dordrecht, 1985.

[And-Cl-Pom]   J.M. Anderson, J.G. Clunie, C. Pommerenke, *On Bloch func-
               tions and normal functions*, J. Reine Angew. Math., **270**(1974),
               12-37.

[And-Hin]   J.M. Anderson, A. Hinkkanen, *Univalence criteria and quasicon-
               formal extensions*, Trans. Amer. Math. Soc., **324**(1991), 823-842.

[Andr]   C. Andreian Cazacu, Theory of Functions of Several Complex Vari-
               ables, Ed. Acad. Rom., Bucharest, 1965 (in Romanian).

[Andr-Con-Jur] C. Andreian Cazacu, C. Constantinescu, M. Jurchescu, Mod-
               ern Problems of the Theory of Functions, Ed. Acad. Rom.,
               Bucharest, 1965 (in Romanian).

[Avk-Aks1]   F.G. Avkhadiev, L.A. Aksent'ev, *Sufficient conditions for the uni-
               valence of analytic functions*, Dokl. Akad. Nauk SSSR, **198**(1971),
               743-746.

[Avk-Aks2]   F.G. Avkhadiev, L.A. Aksent'ev, *The main results on sufficient
               conditions for an analytic function to be schlicht*, Uspehi Mat.
               Nauk, **30**(4)(1975), 3-60 (in Russian); Russian Math. Surveys,
               30(4)(1975), 1-63.

[Bae]   A. Baernstein, *Univalence and bounded mean oscillation*, Michigan
               Math. J., **23**(1976), 217-223.

[Bae-Dra-Dur-Mar] A. Baernstein, D. Drasin, P. Duren, A. Marden (Editors), The Bieberbach Conjecture. Proceedings of the symposium on the occasion of the proof of the Bieberbach conjecture held at Purdue University, March 11-14, 1985. American Math. Soc., Providence, RI 1986.

[Bar-Fit-Gon1] R.W. Barnard, C.H. FitzGerald, S. Gong, *The growth and 1/4-theorems for starlike mappings in* $\mathbb{C}^n$, Pacif. J. Math., **150**(1991), 13-22.

[Bar-Fit-Gon2] R.W. Barnard, C.H. FitzGerald, S. Gong, *A distortion theorem for biholomorphic mappings in* $\mathbb{C}^2$, Trans. Amer. Math. Soc., **344**(1994), 907-924.

[Baz1] I.E. Bazilevich, *Sur les théorèmes de Koebe-Bieberbach*, Mat. Sb., **1**(43)(1936), 283-292.

[Baz2] I.E. Bazilevich, *On a case of integrability by quadratures of the equation of Loewner-Kufarev*, Mat. Sb., **37**(1955), 471-476.

[Baz3] I.E. Bazilevich, *On the orthogonal systems of functions associated with solutions of the Loewner differential equation*, Zap. Nauchn. Sem. Leningrad. Otdel. Mat. Inst. Steklov, **125**(1983), 24-35.

[Bec1] J. Becker, *Löwnersche differentialgleichung und quasikonform fortsetzbare schlichte funktionen*, J. Reine Angew. Math., **255**(1972), 23-43.

[Bec2] J. Becker, *Löwnersche differentialgleichung und Schlichtheitskriterien*, Math. Ann., **202**(1973), 321-335.

[Bec3] J. Becker, *Über die Lösungsstruktur einer differentialgleichung in der konformen Abbildung*, J. Reine Angew. Math., **285**(1976), 66-74.

[Bec4] J. Becker, *Conformal mappings with quasiconformal extensions*, Aspects of Contemporary Complex Analysis (D. Brannan and J. Clunie, Editors), 37-77, Academic Press, London-New York, 1980.

[Bec-Pom] J. Becker, C. Pommerenke, *Schlichtheitskriterien und Jordangebiete*, J. Reine Angew. Math., **354**(1984), 74-94.

[Bel-Hum] E. Beller, J.A. Hummel, *On the univalent Bloch constant*, Complex Variables Theory Appl., 4(1985), 243-252.

[Ber]   S. Bergman, The Kernel Function and Conformal Mapping, 2nd Edition, American Mathematical Society, Providence, R.I. 1970.

[Bern]   S.D. Bernardi, Bibliography of Schlicht Functions, Mariner Publishing Co., Inc., Tampa, Florida, 1982.

[Bet1]   T. Betker, *Löwner chains and Hardy spaces*, Bull. London Math. Soc., **23**(1991), 367-371.

[Bet2]   T. Betker, *Univalence criteria and Löwner chains*, Bull. London. Math. Soc., **23**(1991), 563-567.

[Bet3]   T. Betker, *Löwner chains and quasiconformal extensions*, Complex Variables, **20**(1992), 107-111.

[Bie1]   L. Bieberbach, *Über die Koeffizienten derjenigen Potenzreihen, welche eine schlichte Abbildung des Einheitskreiss vermitteln.* S.-B. Preuss. Akad. Wiss. 1916, 940-955.

[Bie2]   L. Bieberbach, *Über einige Extremalprobleme im Gebiete der Konformen Abbildung*, Math. Ann. **77**(1916), 153-172.

[Bie3]   L. Bieberbach, Lehrbuch der Funktionentheorie. Band I: Elemente der Funktionentheorie (B.G. Teubner, Leipzig 1921); Band II: Moderne Funktionentheorie (Zweite Auflage, B.G. Teubner, Leipzig 1931). (Reprinted by Johnson Reprint Corp. New York, 1968.)

[Biel-Lew] A. Bielecki, Z.Lewandowski, *Sur un théorème concernant les fonctions univalentes linéairement accessibles de M. Biernacki*, Ann. Polon. Math. **12**(1962), 61-63.

[Bier]   M. Biernacki, *Sur la représentation conforme des domaines linéairement accessibles*, Prace Mat.-Fiz., **44**(1936), 293-314.

[Bla]   C. Blatter, *Ein Verzerrungssatz für schlichte Funktionen*, Comment. Math. Helv., **53**(1978), 651-659.

[Blo1]   A. Bloch, *Les théorèmes de Valiron sur les fonctions entières et la théorie de l'uniformisation*, Comptes Rendus Acad. Sci. Paris, **178**(1924), 2051-2052.

[Blo2]   A. Bloch, *Les théorèmes de Valiron sur les fonctions entières et la théorie de l'uniformisation*, Ann. Fac. Sci. Univ. Toulouse, Ser. **3**, **17**(1925), 1-22.

[Boc]   S. Bochner, *Bloch's theorem for real variables*, Bull. Amer. Math. Soc., **52**(1946), 715-719.

[Boc-Sic]   J. Bochnak, J. Siciak, *Analytic functions in topological vector spaces*, Studia Math., **39**(1971), 77-112.

[Boj-Iw]   B. Bojarski, T. Iwaniec, *Analytical foundations of the theory of quasiconformal mappings in $\mathbb{R}^n$*, Ann. Acad. Sci. Fenn. Ser. A I Math., **8**(1983), 257-324.

[Bon1]   M. Bonk, Extremal Probleme bei Bloch-Funktionen, Dissertation, Technische Universität Braunschweig, 1988.

[Bon2]   M. Bonk, *On Bloch's constant*, Proc. Amer. Math. Soc., **110** (1990), 889-894.

[Bon3]   M. Bonk, *Distortion estimates for Bloch functions*, Bull. London Math. Soc., **23**(1991), 454-456.

[Bon-Min-Yan1] M. Bonk, D. Minda, H. Yanagihara, *Distortion theorems for locally univalent Bloch functions*, J. Analyse Math., **69**(1996), 73-95.

[Bon-Min-Yan2] M. Bonk, D. Minda, H. Yanagihara, *Distortion theorems for Bloch functions*, Pacif. J. Math., **179**(1997), 241-262.

[Bran]   D.A. Brannan, *The Löwner differential equation*, Aspects of Con-
            temporary Complex Analysis (D. Brannan and J. Clunie, Editors),
            79-95, Academic Press, London-New York, 1980.

[Bran-Clu] D.A. Brannan, J. Clunie (Editors), Aspects of Contemporary Com-
            plex Analysis, Academic Press, London-New York, 1980.

[Bri]   J.L. Brickman, *Φ-like analytic functions I*, Bull. Amer. Math. Soc.,
            79(1973), 555-558.

[Brod]   A.A. Brodskii, *Quasiconformal extension of biholomorphic map-
            pings*, Theory of Mappings and Approximation of Functions (G.
            Suvorov, Ed.), pp. 30-34, "Naukova Dumka", Kiev, 1983.

[Brow]   J.E. Brown, *Images of discs under convex and starlike functions*,
            Math. Z., **202**(1989), 457-462.

[Cam]   D.M. Campbell, *Applications and proof of a uniqueness theorem
            for linearly invariant families of finite order*, Rocky Mountain J.
            Math., **4**(1974), 621-634.

[Cam-Cim-Pfa] D.M. Campbell, J.A. Cima, J.A. Pfaltzgraff, *Linear spaces
            and linear invariant families of locally univalent functions*, Manu-
            scripta Math., **4**(1971), 1-30.

[Cam-Pf]   D.M. Campbell, J.A. Pfaltzgraff, *Boundary behaviour and linear
            invariant families*, J. Anal. Math., **29**(1976), 67-92.

[Cam-Zi]   D.M. Campbell, M.R. Ziegler, *The argument of the derivative
            of linear invariant families of finite order and the radius of
            close-to-convexity*, Ann. Univ. Mariae Curie-Sklodowska, Sect A.,
            28(1974), 5-22.

[Car]   P. Caraman, Homeomorfisme Cvasiconforme *n*-Dimensionale, Ed.
            Acad. Rom., Bucharest, 1968.

[Cara1]   C. Carathéodory, *Über den Variabilitätsbereich der Koeffizienten
            von Potenzreihen, die gegebene werte nicht annehmen*, Math. Ann.,
            64(1907), 95-115.

[Cara2]   C. Carathéodory, *Unterschungen über die konformen Abbildungen von festen und veränderlichen Gebieten*, Math. Ann. **72**(1912), 107-144.

[Cara3]   C. Carathéodory, Conformal Representation, Reprint of the 1952 second edition, Dover Publications, Inc., Mineola, NY, 1988.

[Carn]   K. Carne, *The Schwarzian derivative for conformal maps*, J. Reine Angew. Math., **408**(1990), 10-33.

[Cart1]   H. Cartan, *Les transformations analytiques des domaines cerclés les unes dans les autres*, Comptes Rendus Acad. Sci. Paris, **190**(1930), 718-720.

[Cart2]   H. Cartan, *Sur la possibilité d'étendre aux fonctions de plusieurs variables complexes la théorie des fonctions univalentes*, 129-155, Note added to P. Montel, Leçons sur les Fonctions Univalentes ou Multivalentes, Gauthier-Villars, Paris, 1933.

[Cart3]   H. Cartan, Calcul Différentiel. Formes Différentielles, Hermann, Paris, 1967, Russian transl. MIR, Moscou, 1971.

[Căl]   G. Călugăreanu, *Sur la condition nécessaire et suffisante pour l'univalence d'une fonction holomorphe dans un cercle*, C.R. Acad. Sci. Paris, **193**(1931), 1150-1153.

[Cha]   B. Chabat, Introduction à l'Analyse Complexe, I-II, Ed. MIR, Moscou, 1990.

[CheH]   H. Chen, *On the Bloch constant*. In: Approximation, Complex Analysis, and Potential Theory (N. Arakelian and P.M. Gauthier, Eds.), 129-161, Kluwer Acad. Publ., Dordrecht, 2001.

[CheHB]   H.B. Chen, *A growth theorem for biholomorphic convex mappings in the unit ball of a complex Banach space and in a bounded pseudoconvex domain*, Chin. Ann. Math. Ser. A., **21**(2000), 595-600.

[CheZ]   Z. Chen, *Criteria for starlikeness and univalency of maps on a class of bounded strictly balanced domains in* $\mathbb{C}^n$, Chin. Ann. Math. Ser. A, **16**(1995), 230-237.

[Che-Gau1] H. Chen, P.M. Gauthier, *On Bloch's constant*, J. Anal. Math., **69**(1996), 275-291.

[Che-Gau2] H. Chen, P.M. Gauthier, *Bloch constants in several variables*, Trans. Amer. Math. Soc., **353**(2001), 1371-1386.

[Che-Ren]   H.B. Chen, F. Ren, *Univalence of holomorphic mappings and growth theorems for close-to starlike mappings in finitely dimensional Banach spaces*, Acta Math. Sinica (N.S.), **10**(1994), Special Issue, 207-214.

[Cher]   S.S. Chern, *On holomorphic mappings of hermitian manifolds of the same dimension.* In: Entire Functions and Related Parts of Analysis, Proc. Symp. Pure Math. **21**(1968), 157-170.

[Chu1]   M. Chuaqui, *A unified approach to univalence criteria*, Proc. Amer. Math. Soc., **123**(1995), 441-453.

[Chu2]   M. Chuaqui, *Applications of subordination chains to starlike mappings in* $\mathbb{C}^n$, Pacif. J. Math., **168**(1995), 33-48.

[Chu-Pom]   M. Chuaqui, C. Pommerenke, *Characteristic properties of Nehari functions*, Pacif. J. Math., **188**(1999), 83-94.

[Chu-Os1]   M. Chuaqui, B. Osgood, *Sharp distortion theorems associated with the Schwarzian derivative*, J. London Math. Soc., **48**(1993), 289-298.

[Chu-Os2]   M. Chuaqui, B. Osgood, *General univalence criteria in the unit disk: extensions and extremal functions*, Ann. Acad. Sci. Fenn. Math., **23**(1998), 101-132.

[Chu-Os-Pom] M. Chuaqui, B. Osgood, C. Pommerenke, *John domains, quasidisks, and the Nehari class*, J. Reine Angew. Math., **471**(1996), 77-114.

[Cim] J.A. Cima, *The basic properties of Bloch functions*, Internat. J. Math. & Math. Sci. 2(1979), 369-413.

[Cim-Wog] J.A. Cima, W.R. Wogen, *Extreme points of the unit ball of the Bloch space B_0*, Michigan Math. J., 25(1978), 213-222.

[Clu-Pom] J. Clunie, C. Pommerenke, *On the coefficients of close-to-convex univalent functions*, J. London Math. Soc., 41(1966), 161-165.

[Coi-Roc-Wei] R.R. Coifman, R. Rochberg, G. Weiss, *Factorization theorems for Hardy spaces in several variables*, Ann. Math., 103(1976), 611-635.

[Con] J.B. Conway, Functions of One Complex Variable II, Springer-Verlag, New York, 1995.

[Cri1] M. Cristea, *Certain sufficient conditions of univalency*, Studii şi Cercetări Matematice (Mathematical Report), 44(1992), 37-42.

[Cri2] M. Cristea, *Certain conditions for global univalence*, Mathematica (Cluj), 36(1994), 137-142.

[Cri3] M. Cristea, Topological Theory of Analytic Functions, Ed. Univ. Bucharest, 1999 (in Romanian).

[Cu1] P. Curt, *A generalization in n-dimensional complex space of Ahlfors and Becker's criterion for univalence*, Studia Univ. Babeş-Bolyai (Mathematica), 39(1994), 31-38.

[Cu2] P. Curt, *A Marx-Strohhäcker theorem in several complex variables*, Mathematica (Cluj), 39(62)(1997), 59-70.

[Cu3] P. Curt, Special Chapters of Geometric Function Theory of Several Complex Variables, Editura Albastră, Cluj-Napoca, 2001 (in Romanian).

[Cu-Koh1] P. Curt, G. Kohr, *Properties of subordination chains and transition mappings in several complex variables*, submitted.

[Cu-Koh2] P. Curt, G. Kohr, *Subordination chains and Loewner differential equation in several complex variables*, submitted.

[Cu-Pas] P. Curt, N. Pascu, *Loewner chains and univalence criteria for holomorphic mappings in* $\mathbb{C}^n$, Bull. Malaysian Math. Soc., **18**(1995), 45-48.

[DA] J.P. D'Angelo, Several Complex Variables and the Geometry of Real Hypersurfaces, CRC Press, Boca Raton, 1993.

[DeB] L. de Branges, *A proof of the Bieberbach conjecture*, Acta Math., **154**(1985), 137-152.

[Die1] J. Dieudonné, *Sur les fonctions univalentes*, C.R. Acad. Sci. Paris, **192**(1931), 1148-1150.

[Die2] J. Dieudonné, *Recherches sur quelques problèmes relatifs aux polynômes et aux fonctions bornées d'une variable complexe*, Ann. Ecole Nor. Sup., **48**(1931), 247-358.

[Die3] J. Dieudonné, Foundations of Modern Analysis, Academic Press, New York and London, 1960.

[Din1] S. Dineen, The Schwarz Lemma, Clarendon Press, Oxford, 1989.

[Din2] S. Dineen, Complex Analysis on Infinite Dimensional Spaces, Springer-Verlag, Berlin-New York, 1999.

[Din-Tim-Vig] S. Dineen, R. Timoney, J.P. Vigué, *Pseudodistances invariantes sur les domaines d'un espace localement convexe*, Ann. Scuola Norm. Sup. Pisa Cl. Sci, **12**(1985), 515-529.

[Don-Zha] D. Dong, W. Zhang, *The growth and 1/4-theorems for starlike mappings in a Banach space*, Chin. Sci. Bull., **37**(1992), 1062-1064.

[Dun-Sch] N. Dunford, J. Schwartz, *Linear Operators*, I, John Wiley and Sons, Inc., New York, 1958.

[Dur] P.L. Duren, Univalent Functions, Springer-Verlag, New York, 1983.

[Dur-Rud] P.L. Duren, W. Rudin, *Distortion in several variables*, Complex Variables, **5**(1986), 323-326.

[Dur-Sh-Sh] P.L. Duren, H.S. Shapiro, A.L. Shields, *Singular measures and domains not of Smirnov type*, Duke Math. J., **33**(1966), 247-254.

[Ep1] C.L. Epstein, *The hyperbolic Gauss map and quasiconformal reflections*, J. Reine Angew. Math., **372**(1986), 96-135.

[Ep2] C.L. Epstein, *Univalence criteria and surfaces in hyperbolic space*, J. Reine Angew. Math., **380**(1987), 196-214.

[Er] A. Eremenko, *Bloch radius, normal families and quasiregular mappings*, Proc. Amer. Math. Soc., **128**(2000), 557-560.

[Es-Ke] M. Essén, F. Keogh, *The Schwarzian derivative and estimates of functions analytic in the unit disk*, Math. Proc. Cambridge Philos. Soc., **78**(1975), 501-511.

[Fit1] C.H. FitzGerald, *Quadratic inequalities and coefficient estimates for schlicht functions*, Arch. Rational Mech. Anal., **46**(1972), 356-368.

[Fit2] C.H. FitzGerald, *Geometric function theory in one and several complex variables: parallels and problems*, Complex analysis and its applications (C.C. Yang, G.C. Wen, K.Y. Li and Y.M. Chiang, Ed.), 14-25, Longman Scientific and Technical, Harlow, 1994.

[Fit-Gon1] C.H. FitzGerald, S. Gong, *The Schwarzian derivative in several complex variables*, Science in China, Ser. A, **36**(1993), 513-523.

[Fit-Gon2] C.H. FitzGerald, S. Gong, *The Bloch theorem in several complex variables*, J. Geom. Anal., **4**(1994), 35-58.

[Fit-Pom] C.H. FitzGerald, C. Pommerenke, *The de Branges theorem on univalent functions*, Trans. Amer. Math. Soc., **290**(1985), 683-690.

[Fit-Th] C.H. FitzGerald, C. Thomas, *Some bounds on convex mappings in several complex variables*, Pacif. J. Math., **165**(1994), 295-320.

[Fla]   H. Flanders, *The Schwarzian derivative in several complex variables*, Science in China, Ser. A, **36**(1993), 513-523.

[Fra-Ve]   T. Franzoni, E. Vesentini, Holomorphic Maps and Invariant Distances, North-Holland, Amsterdam 1980.

[Gal-Nik]   D. Gale, H. Nikaido, *The Jacobian matrix and global univalence of mappings*, Math. Ann., **159**(1965), 81-93.

[Gam-Che]   J. Gamaliel, H. Chen, *On the Bloch constant for K-quasiconformal mappings in several complex variables*, Acta Math. Sinica (Engl. Ser.), **17**(2001), 237-242.

[Gar-Sch]   P.R. Garabedian, M. Schiffer, *A proof of the Bieberbach conjecture for the fourth coefficient*, J. Rational Mech. Anal. **4**(1955), 427-465.

[Gau]   P.M. Gauthier, *Covering properties of holomorphic mappings*, Complex Geometric Analysis in Pohang (1997), 211-218, Contemp. Math., **222**, Amer. Math. Soc., Providence, RI, 1999.

[Geh]   F.W. Gehring, *Univalent functions and the Schwarzian derivative*, Comment. Math. Helv., **52**(1977), 561-572.

[Geh-Pom]   F.W. Gehring, C. Pommerenke, *On the Nehari univalence criterion and quasicircles*, Comment. Math. Helv., **59**(1984), 226-242.

[God-Li-St] J. Godula, P. Liczberski, V. Starkov, *Order of linearly invariant family of mappings in $\mathbb{C}^n$*, Complex Variables, **42**(2000), 89-96.

[God-St]   J. Godula, V. Starkov, *Linearly invariant families of holomorphic functions in the polydisc*, Banach Center Publ., **37**(1996), 115-127.

[Golb]   M.A. Golberg, *The derivative of a determinant*, Amer. Math. Monthly, **79**(1972), 1124-1126.

[Gol1]   G.M. Goluzin, *On the theory of univalent conformal mappings*, Mat. Sbornik N.S., **42**(1935), 169-190 (in Russian).

[Gol2]   G.M. Goluzin, *On distortion theorems in the theory of conformal mappings*, Mat. Sb., **1(43)**(1936), 127-135 (in Russian).

[Gol3]   G.M. Goluzin, *Some estimates for coefficients of univalent functions*, Mat. Sb., **3(45)**(1938), 321-330 (in Russian).

[Gol4]   G.M. Goluzin, Geometric Theory of Functions of a Complex Variable, Moscow, 1952; English Transl., Amer. Math. Soc., Providence, R.I., 1969.

[Gon1]   S. Gong, *Contributions to the theory of schlicht functions I, Distortion theorem*, Scientia Sinica, 4(1955), 229-249; *II, The coefficient problem*, Scientia Sinica, 4(1955), 359-373.

[Gon2]   S. Gong, *Biholomorphic mappings in several complex variables*, Contemp. Math., **142**(1993), 15-48.

[Gon3]   S. Gong, *The Bloch constant of locally biholomorphic mappings on bounded symmetric domains*, Chin. Ann. Math. Ser. B, **17**(1996), 271-278.

[Gon4]   S. Gong, Convex and Starlike Mappings in Several Complex Variables, Kluwer Acad. Publ., Dordrecht, 1998.

[Gon5]   S. Gong, The Bieberbach Conjecture, Amer. Math. Soc. Intern. Press, Providence, R.I., 1999.

[Gon-Liu1] S. Gong, T. Liu, *The growth theorem of biholomorphic convex mappings on B_p*, Chin. Quart. J. Math., **6**(1991), 78-82 (in Chinese).

[Gon-Liu2] S. Gong, T. Liu, *Distortion theorems for biholomorphic convex mappings on bounded convex circular domains*, Proc. of the Fifth International Colloquium on Complex Analysis, (1997), 73-80; Chin. Ann. Math. Ser. B, **20**(1999), 297-304.

[Gon-Liu3] S. Gong, T. Liu, *On Roper-Suffridge extension operator*, J. Analyse Math., to appear.

[Gon-Wa-Yu1] S. Gong, S. Wang, Q. Yu, *The growth and 1/4-theorems for starlike mappings on B_p*, Chin. Ann. Math. Ser. B, **11**(1990), 100-104.

[Gon-Wa-Yu2] S. Gong, S. Wang, Q. Yu, *A necessary and sufficient condition that biholomorphic mappings are starlike on Reinhardt domains*, Chin. Ann. Math. Ser. B, **13**(1992), 95-104.

[Gon-Wa-Yu3] S. Gong, S. Wang, Q. Yu, *Biholomorphic convex mappings of ball in $\mathbb{C}^n$*, Pacif. J. Math., **161**(1993), 287-306.

[Gon-Wa-Yu4] S. Gong, S. Wang, Q. Yu, *The growth theorem for biholomorphic mappings in several complex variables*, Chin. Ann. Math., **14B**(1993), 93-104.

[Gon-Wa-Yu5] S. Gong, S. Wang, Q. Yu, *Necessary and sufficient conditions for holomorphic mappings to be starlike on bounded starlike circular domains*, Acta Math. Sinica, **42**(1999), 13-16.

[Gon-Yu] S. Gong, Q. Yu, *The distortion theorems of linear invariant family on the unit ball*, Asian J. Math., **4**(2000), 795-815.

[Gon-Yu-Zh] S. Gong, Q. Yu, X. Zheng, Bloch Constant and Schwarzian Derivative (in Chinese), Shanghai Scientific and Technical Publishers, Shanghai, 1997.

[Gon-Zh1] S. Gong, X. Zheng, *Distortion theorem for biholomorphic mappings in transitive domains I*, Intern. Symp. in memory of L.K. Hua, Vol.II, 111-121, Springer-Verlag, New York, 1991.

[Gon-Zh2] S. Gong, X. Zheng, *Distortion theorem for biholomorphic mappings in transitive domains II*, Chin. Ann. Math., **13B**(1992), 471-484.

[Gon-Zh3] S. Gong, X. Zheng, *Distortion theorem for biholomorphic mappings in transitive domains III*, Chin. Ann. Math., **14B**(1993), 367-386.

[Gon-Zh4] S. Gong, X. Zheng, *Distortion theorem for biholomorphic mappings in transitive domains IV*, Chin. Ann. Math., **16B**(1995), 203-212.

[Goo1]   A.W. Goodman, Univalent Functions, I-II, Mariner Publ. Co., Tampa Florida, 1983.

[Goo2]   A.W.Goodman, *On uniformly starlike functions*, J. Math. Anal. Appl., **155**(1991), 364-370.

[Goo3]   A.W. Goodman, *On uniformly convex functions*, Ann. Polon. Math., **56**(1991), 87-92.

[Gra1]   I. Graham, *Boundary behavior of the Carathéodory and Kobayashi metrics on strongly pseudoconvex domains in* $\mathbb{C}^n$, Trans. Amer. Math. Soc., **270**(1975), 219-240.

[Gra2]   I. Graham, *Distortion theorems for holomorphic maps between convex domains in* $\mathbb{C}^n$, Complex Variables **15**(1990), 37-42.

[Gra3]   I. Graham, *Holomorphic mappings into convex domains*, Complex Analysis (Wuppertal 1990), (K. Diederich, Ed.), 127-133, Vieweg, Braunschweig 1991.

[Gra4]   I. Graham, *Sharp constants for the Koebe theorem and for estimates of intrinsic metrics on convex domains*, Proc. Symp. Pure Math., **52** Part 2(1991), 233-238.

[Gra5]   I. Graham, *On the Bloch constant for convex maps of the unit ball in* $\mathbb{C}^n$, Geometric Complex Analysis, (J. Noguchi, Ed.), 215-218, World Scientific, Singapore, 1996.

[Gra6]   I. Graham, *Growth and covering theorems associated with the Roper-Suffridge extension operator*, Proc. Amer. Math. Soc., **127** (1999), 3215-3220.

[Gra-Ham-Koh] I. Graham, H. Hamada, G. Kohr, *Parametric representation of univalent mappings in several complex variables*, Canadian J. Math., **54**(2002), 324-351.

[Gra-Ham-Koh-Su] I. Graham, H. Hamada, G. Kohr, T.J. Suffridge, *Extension operators for locally univalent mappings*, Michigan Math. J., **50**(2002), 37-55.

[Gra-Koh1] I. Graham, G. Kohr, *Univalent mappings associated with the Roper-Suffridge extension operator*, J. Analyse Math., **81**(2000), 331-342.

[Gra-Koh2] I. Graham, G. Kohr, *An extension theorem and subclasses of univalent mappings in several complex variables*, Complex Variables, **47**(2002), 59-72.

[Gra-Koh-Koh1] I. Graham, G. Kohr, M. Kohr, *Loewner chains and the Roper-Suffridge extension operator*, J. Math. Anal. Appl., **247**(2000), 448-465.

[Gra-Koh-Koh2] I. Graham, G. Kohr, M. Kohr, *Loewner chains and parametric representation in several complex variables*, J. Math. Anal. Appl., to appear.

[Gra-Min] I. Graham, D. Minda, *A Schwarz lemma for multivalued functions and distortion theorems for Bloch functions with branch points*, Trans. Amer. Math. Soc., **351**(1999), 4741-4752.

[Gra-Var1] I. Graham, D. Varolin, *On translations of the images of analytic maps*, Complex Variables, **24**(1994), 205-208.

[Gra-Var2] I. Graham, D. Varolin, *Bloch constants in one and several variables*, Pacif. J. Math. **174**(1996), 347-357.

[Gra-Wu] I. Graham, H. Wu, *Some remarks on the intrinsic measures of Eisenman*, Trans. Amer. Math. Soc., **288**(1985), 625-660.

[Gre-Kra] R.E. Greene, S.G. Krantz, Function Theory of One Complex Variable, Second Edition, American Mathematical Society, Providence, RI, 2002.

[Gre-Wu] R.E. Greene, H. Wu, *Bloch's theorem for meromorphic functions*, Math. Z., **116**(1970), 247-257.

[Gro] T.H. Gronwall, *Some remarks on conformal representation*, Ann. Math., **16**(1914/15), 72-76.

[Grö]   H. Grötzsch, *Über die Verzerrung bei schlichten nicht konformen Abbildungen und über eine damit zusammenhängende Erweiterung des Picardschen Satzes*, Ber. Verh. Sächs. Akad. Wiss. Leipzig, **80**(1928), 503-507.

[Gru1]   H. Grunsky, *Neue Abschätzungen zur konformen Abbildung ein- und mehrfach zusammenhängender Bereiche*, Schr. Math. Sem. und Inst. Angew. Math. Univ. Berlin, **11**(1932), 95-140.

[Gru2]   H. Grunsky, *Zwei Bemerkungen zur konformen Abbildung*, Jahresber. Deutsch. Math. Verein, **43**(1933), 140-143.

[Gug]   H.W. Guggenheimer, Differential Geometry, Mc Graw-Hill, New York, 1963.

[Gun]   R.C. Gunning, Introduction to Holomorphic Functions of Several Variables, Vol.I, Wadsworth & Brooks/Cole, Monterey, CA., 1990.

[Gun-Ros]   R.C. Gunning, H. Rossi, Analytic Functions of Several Complex Variables, Prentice-Hall, Inc., Englewood Cliffs, N.J., 1965.

[Gur]   K. Gurganus, *Φ-like holomorphic functions in $\mathbb{C}^n$ and Banach spaces*, Trans. Amer. Math. Soc., **205**(1975), 389-406.

[Hah1]   K.T. Hahn, *Subordination principle and distortion theorems on holomorphic mappings in the space $\mathbb{C}^n$*, Trans. Amer. Math. Soc., **162**(1971), 327-336.

[Hah2]   K.T. Hahn, *Higher dimensional generalizations of the Bloch constant and their lower bounds*, Trans. Amer. Math. Soc., **179**(1973), 263-274.

[Hah3]   K.T. Hahn, *Holomorphic mappings of the hyperbolic space into the complex Euclidean space and the Bloch theorem*, Canad. J. Math., **27**(1975), 446-458.

[Hal-MG]   D.J. Hallenbeck, T.H. MacGregor, Linear Problems and Convexity Techniques in Geometric Function Theory, Pitman, Boston, 1984.

[Ham1]   H. Hamada, *A Schwarz lemma on complex ellipsoids*, Ann. Polon. Math., **67**(1997), 269-275.

[Ham2]   H. Hamada, *The growth theorem of convex mappings on the unit ball of* $\mathbb{C}^n$, Proc. Amer. Math. Soc., **127**(1999), 1075-1077.

[Ham3]   H. Hamada, *Univalent holomorphic mappings on a complex manifold with a* C^1 *exhaustion function*, Manuscripta Math., **99**(1999), 359-369.

[Ham4]   H. Hamada, *Starlike mappings on bounded balanced domains with* C^1*-plurisubharmonic defining functions*, Pacif. J. Math., **194**(2000), 359-371.

[Ham5]   H. Hamada, *Univalence and quasiconformal extension of holomorphic maps on balanced pseudoconvex domains*, preprint.

[Ham-Koh1] H. Hamada, G. Kohr, *Spirallike mappings on bounded balanced pseudoconvex domains in* $\mathbb{C}^n$, Zesz. Nauk. Politech. Rzeszow (Matematyka), **22**(1998), 9-21.

[Ham-Koh2] H. Hamada, G. Kohr, *Convex mappings in several complex variables*, Glasnik Matem., **34(54)**(1999), 203-210.

[Ham-Koh3] H. Hamada, G. Kohr, *Spirallike non-holomorphic mappings on balanced pseudoconvex domains*, Complex Variables, **41**(2000), 253-265.

[Ham-Koh4] H. Hamada, G. Kohr, *Subordination chains and the growth theorem of spirallike mappings*, Mathematica (Cluj), **42(65)**(2000), 153-161.

[Ham-Koh5] H. Hamada, G. Kohr, *Subordination chains and univalence of holomorphic mappings on bounded balanced pseudoconvex domains*, Ann. Univ. Mariae Curie Sklodowska, Sect. A, **55**(2001), 61-80.

[Ham-Koh6] H. Hamada, G. Kohr, *The growth theorem and quasiconformal extension of strongly spirallike mappings of type* α, Complex Variables, **44**(2001), 281-297.

[Ham-Koh7] H. Hamada, G. Kohr, *Some necessary and sufficient conditions of convexity on bounded balanced pseudoconvex domains in* $\mathbb{C}^n$, Complex Variables, **45**(2001), 101-115.

[Ham-Koh8] H. Hamada, G. Kohr, *Linear invariance of locally biholomorphic mappings in Hilbert spaces*, Complex Variables, **47**(2002), 277-289.

[Ham-Koh9] H. Hamada, G. Kohr, *Growth and distortion results for convex mappings in infinite dimensional spaces*, Complex Variables, **47**(2002), 291-301.

[Ham-Koh10] H. Hamada, G. Kohr, *Φ-like and convex mappings in infinite dimensional spaces*, Rev. Roum. Math. Pures Appl., to appear.

[Ham-Koh11] H. Hamada, G. Kohr, *Order of linear invariant families on the unit ball and polydisc of* $\mathbb{C}^n$, Revue Roum. Math. Pures Appl., to appear.

[Ham-Koh12] H. Hamada, G. Kohr, *Loewner chains and quasiconformal extension of holomorphic mappings*, Ann. Polon. Math., to appear.

[Ham-Koh13] H. Hamada, G. Kohr, *Linear invariant families on the unit polydisc*, Mathematica (Cluj), to appear.

[Ham-Koh14] H. Hamada, G. Kohr, *Simple criteria for strongly starlikeness and quasiconformal extension*, submitted.

[Ham-Koh-Koh] H. Hamada, G. Kohr, M. Kohr, *Strongly starlike mappings of order alpha on bounded balanced pseudoconvex domains in* $\mathbb{C}^n$, Rev. Roum. Math. Pures Appl., **44**(1999), 583-594.

[Ham-Koh-Lic1] H. Hamada, G. Kohr, P. Liczberski, *Φ-like holomorphic mappings on balanced pseudoconvex domains*, Complex Variables, **39**(1999), 279-290.

[Ham-Koh-Lic2] H. Hamada, G. Kohr, P. Liczberski, *Starlike mappings of order* α *on the unit ball in complex Banach spaces*, Glasnik Matematiki, **36**(**56**)(2001), 39-48.

[Har1]   R. Harmelin, *Invariant operators and univalent functions*, Trans.
           Amer. Math. Soc., **272**(1982), 721-731.

[Har2]   R. Harmelin, *On the derivatives of the Schwarzian derivative of
           a univalent function and their symmetric generating function*, J.
           London Math. Soc., **27**(1983), 489-499.

[Har3]   R. Harmelin, *Locally convex functions and the Schwarzian deriva-
           tive*, Israel J. Math., **67**(1989), 367-379.

[Harr1]   L.A. Harris, *Schwarz's lemma in normed linear spaces*, Proc. Nat.
           Acad. Sci. U.S.A., **62**(1969), 1014-1017.

[Harr2]   L.A. Harris, *The numerical range of holomorphic functions in Ba-
           nach spaces*, Amer. J. Math., **93**(1971), 1005-1019.

[Harr3]   L.A. Harris, *On the size of balls covered by analytic transforma-
           tions*, Monatshefte für Math., **83**(1977), 9-23.

[Harr-Re-Sh] L.A. Harris, S. Reich, D. Shoikhet, *Dissipative holomorphic
           functions, Bloch radii, and the Schwarz lemma*, J. Anal. Math.,
           82(2000), 221-232.

[Hay]   W.K. Hayman, Multivalent Functions (second edition), Cambridge
           Univ. Press, 1994.

[Hay-Hum] W.K. Hayman, J.A. Hummel, *Coefficients of powers of univalent
           functions*, Complex Variables Theory Appl., **7**(1986), 51-70.

[Hay-Su]   T.L. Hayden, T.J. Suffridge (Editors), Proceedings on Infinite Di-
           mensional Holomorphy, Lecture Notes in Math., **364**, Springer-
           Verlag, New York, 1974.

[Hea-Su]   L.F. Heath, T.J. Suffridge, *Starlike, convex, close-to-convex, spiral-
           like and Φ-like maps in a commutative Banach algebra with iden-
           tity*, Trans. Amer. Math. Soc., **250**(1979), 195-212.

[Hei]   M. Heins, *On a class of conformal metrics*, Nagoya Math. J.,
           21(1962), 1-60.

[Hel]   S. Helgason, Differential Geometry, Lie Groups, and Symmetric Spaces, American Mathematical Society, Providence, R.I., 2001.

[Hen]   P. Henrici, Applied and Computational Complex Analysis, III, Wiley Classical Library, J. Wiley & Sons, New York, 1993.

[Hen-Le]   G.M. Henkin, J. Leiterer, Theory of Functions on Complex Manifolds, Birkhäuser, Boston, 1984.

[Hen-Sch] W. Hengartner, G. Schober, *On schlicht mappings to domains convex in one direction*, Comment. Math. Helv., **45**(1970), 303-314.

[Her]   G. Herglotz, *Über Potenzreihen mit positivem, reelen Teil im Einheitskreis*, S.-B. Sächs. Akad. Wiss. Leipzig Math.-Natur. Kl., **63**(1911), 501-511.

[Herv]   M. Hervé, Analyticity in Infinite Dimensional Spaces, Walter de Gruyter & Co., Berlin-New York, 1989.

[Hil1]   E. Hille, *Remarks on a paper by Zeev Nehari*, Bull. Amer. Math. Soc., **55**(1949), 552-553.

[Hil2]   E. Hille, Analytic Function Theory, vol.II, Ginn and Company, Boston, 1962.

[Hil-Phi] E. Hille, R.S. Phillips, Functional Analysis and Semigroups, Amer. Math. Soc. Coll. Publ., **31**, Providence, R.I., 1957.

[Hor]   D. Horowitz, *A refinement for coefficient estimates of univalent functions*, Proc. Amer. Math. Soc., **54**(1976), 176-178.

[Hör1]   L. Hörmander, *On a theorem of Grace*, Math. Scand., **2**(1954), 55-64.

[Hör2]   L. Hörmander, An Introduction to Complex Analysis in Several Variables, Second Edition, North-Holland, Amsterdam, 1973.

[Hum]   J.A. Hummel, *The coefficient regions of starlike functions*, Pacif. J. Math., **7**(1957), 1381-1389.

[Iw-Ma]   T. Iwaniec, G. Martin, Geometric Function Theory and Non-Linear Analysis, Oxford Mathematical Monographs, Clarendon Press, Oxford, 2001.

[Jac]   I.S. Jack, *Functions starlike and convex of order* α, J. London Math. Soc., **3**(1971), 469-474.

[Jan]   E. Janiec, *Some sufficient conditions for univalence of holomorphic functions*, Demonstratio Math., **22**(1989), 717-727.

[Jar-Pf]   M. Jarnicki, P. Pflug, Invariant Distances and Metrics in Complex Analysis, Walter de Gruyter & Co., Berlin-New York, 1993.

[Jen1]   J.A. Jenkins, Univalent Functions and Conformal Mapping, 2nd Ed. Springer-Verlag, Berlin-New York, 1965.

[Jen2]   J.A. Jenkins, *On weighted distortion in conformal mapping II*, Bull. London Math. Soc., **30**(1998), 151-158.

[Jen3]   J.A. Jenkins, *On the schlicht Bloch constant II*, Indiana Univ. Math. J. **47**(1998), 1059-1063.

[Kan-Lec] S. Kanas, A. Lecko, *Univalence criteria connected with arithmetic and geometric means II*, Zesz. Nauk. Politech. Rzeszow, Mat., **20**(1996), 49-57.

[Kan-Wi]   S. Kanas, A. Wisniowska, *Conic regions and k-uniform convexity*, J. Comput. Appl. Math., **105**(1999), 327-336.

[Kap]   W. Kaplan, *Close-to-convex schlicht functions*, Michigan Math. J., **1**(1952), 169-185.

[Kas]   S.A. Kas'yanyuk, *On the method of structural formulae and the principle of correspondence of boundaries under conformal mappings*, Dop. Akad. Nauk Ukrain, RSR 1959, 14-17.

[Kat1]   T. Kato, *Estimation of iterated matrices with application to the von Neumann condition*, Numer. Math., **2**(1960), 22-29.

[Kat2]   T. Kato, *Nonlinear semigroups and evolution equations*, J. Math.
            Soc. Japan, **19**(1967), 508-520.

[Kau-Kau]   L. Kaup, B. Kaup, Holomorphic Functions of Several Variables,
            Walter de Gruyter & Co., Berlin-New York, 1983.

[Ker-Ste]   N. Kerzman, E.M. Stein, *The Cauchy kernel, the Szegö kernel, and
            the Riemann mapping function*, Math. Ann., **236**(1978), 85-93.

[Kik]   K. Kikuchi, *Starlike and convex mappings in several complex vari-
            ables*, Pacif. J. Math., **44**(1973), 569-580.

[Kim-Min]   S.A.Kim, D. Minda, *Two point distortion theorems for univalent
            functions*, Pacif. J. Math., **163**(1994), 137-157.

[Kir]   W.E. Kirwan, *Extremal properties of slit conformal mappings*, As-
            pects of Contemporary Complex Analysis (D. Brannan and J. Clu-
            nie, Editors), 439-449, Academic Press, London-New York 1980.

[Klo]   R. Klouth, *Estimation for generalized Schwarz derivatives in lin-
            ear invariant families of functions*, Arch. Math. (Basel), **36**(1981),
            455-462 (in German).

[Kob]   S. Kobayashi, Hyperbolic Manifolds and Holomorphic Mappings,
            Marcel Dekker Inc, New York, 1970.

[Koe]   P. Koebe, *Über die Uniformisierung beliebiger analytischer kurven*,
            Nachr. Akad. Wiss. Göttingen, Math. Phys. Kl., 1907, 191-210.

[Koep1]   W. Koepf, *Close-to-convex functions and linear-invariant families*,
            Ann. Acad. Sci. Fenn., Ser. A I Math., **8**(1983), 349-355.

[Koep2]   W. Koepf, *Convex functions and the Nehari univalence criterion*,
            Complex Analysis, Joensu, 1987, 214-218. Lecture Notes in Math.,
            1351, Springer-Verlag, Berlin-New York 1988.

[Koep3]   W. Koepf, *On close-to-convex functions and linearly accessible do-
            mains*, Complex Variables, **11**(1989), 269-279.

[Koh1]   G. Kohr, *On some partial differential inequalities for holomorphic mappings in* $\mathbb{C}^n$, Complex Variables, **31**(1996), 131-140.

[Koh2]   G. Kohr, *Certain partial differential inequalities and applications for holomorphic mappings defined on the unit ball of* $\mathbb{C}^n$, Ann. Univ. Mariae Curie Skl., Sect. A, **50**(1996), 87-94.

[Koh3]   G. Kohr, *On some conditions of spirallikeness in* $\mathbb{C}^n$, Complex Variables, **32**(1997), 79-88.

[Koh4]   G. Kohr, *Some sufficient conditions of starlikeness for mappings of* C^1 *class*, Complex Variables, **36**(1998), 1-9.

[Koh5]   G. Kohr, *On some alpha convex mappings on the unit ball of* $\mathbb{C}^n$, Demonstratio Math., **31**(1998), 209-222.

[Koh6]   G. Kohr, *On some best bounds for coefficients of subclasses of biholomorphic mappings in* $\mathbb{C}^n$, Complex Variables, **36**(1998), 261-284.

[Koh7]   G. Kohr, *On starlikeness and strongly-starlikeness of order alpha in* $\mathbb{C}^n$, Mathematica (Cluj), **40(63)**(1998), 95-109.

[Koh8]   G. Kohr, *On some distortion results for convex mappings in* $\mathbb{C}^n$, Complex Variables, **39**(1999), 161-175.

[Koh9]   G. Kohr, *Using the method of Löwner chains to introduce some subclasses of biholomorphic mappings in* $\mathbb{C}^n$, Rev. Roum. Math. Pures Appl., **46**(2001), 743-760.

[Koh10]   G. Kohr, *Biholomorphic mappings and parametric representation in several complex variables*, submitted.

[Koh-Koh]   G. Kohr, M. Kohr, *Partial differential subordinations for holomorphic mappings of several complex variables*, Studia Univ. Babeş-Bolyai (Mathematica), **40**(1995), 45-62.

[Koh-Lic1] G. Kohr, P. Liczberski, *On some sufficient conditions for univalence in* $\mathbb{C}^n$, Demonstratio Math., **29**(1996), 407-412.

[Koh-Lic2] G. Kohr, P. Liczberski, Univalent Mappings of Several Complex Variables, Cluj University Press, Cluj-Napoca, Romania, 1998.

[Koh-Lic3] G. Kohr, P. Liczberski, *On strongly starlikeness of order alpha in several complex variables*, Glasnik Matem., **33(53)**(1998), 185-198.

[Kra]   I. Kra, *Deformations of Fuchsian groups* II, Duke Math. J., **38** (1971), 499-508.

[Kran]   S.G. Krantz, Function Theory of Several Complex Variables, Reprint of the 1992 Edition, AMS Chelsea Publishing, Providence, R.I., 2001.

[Kran-Ma] S.G. Krantz, D. Ma, *Bloch functions on strongly pseudoconvex domains*, Indiana Univ. Math. J., **37**(1988), 145-163.

[Krau]   W. Kraus, *Über den Zusammenhang einiger Charakteristiken eines einfach zusammenhängenden Bereiches mit der Kreisabbildung*, Mitt. Math. Sem Giessen **21**(1932), 1-28.

[Krz1]   J. Krzyz, *On the maximum modulus of univalent functions*, Bull. Acad. Polon. Sci., Cl. III, **3**(1955), 203-206.

[Krz2]   J. Krzyz, *The radius of close-to-convexity within the family of univalent functions*, Bull. Acad. Polon. Sci., **10**(1962), 201-204.

[Krz3]   J. Krzyz, *Some remarks on close-to-convex functions*, Bull. Acad. Polon. Sci., **12**(1964), 25-28.

[Krz4]   J. Krzyz, *Convolution and quasiconformal extension*, Comment. Math. Helv., **51**(1976), 99-104.

[Krz-Rea] J. Krzyz, M.O. Reade, *The radius of univalence of certain analytic functions*, Michigan Math. J., **11**(1964), 157-159.

[Kub-Por] E. Kubicka, T. Poreda, *On the parametric representation of starlike maps of the unit ball in $\mathbb{C}^n$ into $\mathbb{C}^n$*, Demonstratio Math., **21**(1988), 345-355.

[Kuf1]   P.P. Kufarev, *On one parameter families of analytic functions*, Mat. Sb., **13(55)**(1943), 87-118 (in Russian).

[Kuf2]   P.P. Kufarev, *A remark on integrals of the Loewner equation*, Dokl. Akad. Nauk SSSR, **57**(1947), 655-656 (in Russian).

[Kuf3]   P.P. Kufarev, *A theorem on solutions of a differential equation*, Uchen. Zap. Tomsk. Gos. Univ., **5**(1947), 20-21 (in Russian).

[Kuf4]   P.P. Kufarev, *On the theory of univalent functions*, Dokl. Akad. Nauk SSSR, **57**(1947), 751-754 (in Russian).

[Kuf5]   P.P. Kufarev, *On a method for investigation of extremal problems in the theory of univalent functions*, Dokl. Akad. Nauk SSSR, **107**(1956), 633-635 (in Russian).

[Küh]   R. Kühnau (Editor), Handbook of Complex Analysis, Volume 1: Geometric Function Theory, Elsevier Science, Amsterdam, 2002.

[Lad-Lak]   G.E. Ladas, V. Laksmikantham, Differential Equations in Abstract Spaces, Academic Press, New York, 1972.

[Lan]   E. Landau, *Über die Blochsche Konstante und zwei verwandte Weltkonstanten*, Math. Z., **30**(1929), 608-634.

[Leb]   N. Lebedev, The Area Principle in the Theory of Univalent Functions, Izdat. "Nauka", Moscow, 1975 (in Russian).

[Lec1]   A. Lecko, *Some subclasses of close-to-convex functions*, Ann. Polon. Math., **58**(1993), 53-64.

[Lec2]   A. Lecko, *On coefficient inequalities in the Carathéodory class of functions*, Ann. Polon. Math., **75**(2000), 59-67.

[Leh]   O. Lehto, Univalent Functions and Teichmüller Spaces, Springer-Verlag, Berlin-New York, 1987.

[Leh-Vir1] O. Lehto, K.I. Virtanen, *Boundary behaviour and normal meromorphic functions*, Acta Math., **97**(1957), 47-65.

[Leh-Vir2] O. Lehto, K.I. Virtanen, Quasiconformal Mappings in the Plane, Springer-Verlag, New York-Heidelberg-Berlin, 1973.

[Lem1] L. Lempert, *La métrique de Kobayashi et la représentation des domaines sur la boule*, Bull. Soc. Math. France, **109**(1981), 427-474.

[Lem2] L. Lempert, *Holomorphic retracts and intrinsic metrics in convex domains*, Analysis Math., **8**(1982), 257-261.

[Leu] Y.J. Leung, *Notes on Loewner differential equations*, Contemporary Mathematics, **38**(1985), 1-11.

[Lew1] Z. Lewandowski, *Sur l'identité de certaines classes de fonctions univalentes*, I, II, Ann. Univ. Mariae Curie Sklodowska, **12**(1958), 131-146; **14**(1960), 19-46.

[Lew2] Z. Lewandowski, *Some remarks on univalence criteria*, Ann. Univ. Mariae Curie-Sklodowska, **36/37**(1982/1983), 87-95.

[Lew-St] Z. Lewandowski, J. Stankiewicz, *Some sufficient conditions for univalence*, Zeszyty Nauk. Politech. Rzeszow. Mat. Fiz., **1**(1984), 11-16.

[Lic1] P. Liczberski, *On the subordination of holomorphic mappings in* $\mathbb{C}^n$, Demonstratio Math., **19**(1986), 293-301.

[Lic2] P. Liczberski, *Some remarks on the subordination of holomorphic mappings from the unit ball in* $\mathbb{C}^n$ *into* $\mathbb{C}^n$, Zesz. Nauk. Politech. Lodz, **22**(1991), 31-42.

[Lic3] P. Liczberski, *Some remarks on biholomorphic mappings in* $\mathbb{C}^n$, Complex Variables, **28**(1996), 371-373.

[Lic-St1] P. Liczberski, V. Starkov, *Regularity theorems for linearly invariant families of holomorphic mappings in* $\mathbb{C}^n$, Ann. Univ. Mariae Curie-Sklodowska, Sect. A, **54**(2000), 61-73.

[Lic-St2]   P. Liczberski, V. Starkov, *Linearly invariant families of holomorphic mappings of a ball. The dimension reduction method*, Siberian Math. J., **42**(2001), 715-730.

[Lic-St3]   P. Liczberski, V. Starkov, *Distortion theorems for biholomorphic convex mappings in* $\mathbb{C}^n$, J. Math. Anal. Appl., **274**(2002), 495-504.

[LiuH1]   H. Liu, *On spirallike mappings in several complex variables*, Chin. Quart. J. Math., **14**(1999), 62-72.

[LiuH2]   H. Liu, *Class of starlike mappings, its extensions and subclasses in several complex variables*, Doctoral dissertation, Univ. Sci. Tech. China, 1999.

[LiuT1]   T. Liu, *The distortion theorem for biholomorphic mappings in* $\mathbb{C}^n$, preprint, 1989.

[LiuT2]   T. Liu, *The growth theorems and covering theorems for biholomorphic mappings on classical domains*, Doctoral Thesis, Univ. Sci.Tech. China, 1989.

[Liu-Ren1] T. Liu, G. Ren, *Distortion theorem of convex mappings on classical domains*, Proc. of the Fifth International Colloquium on Complex Analysis, (1997), 205-210.

[Liu-Ren2] T. Liu, G. Ren, *The growth theorem for starlike mappings on bounded starlike circular domains*, Chin. Ann. Math. Ser. B, **19**(1998), 401-408.

[Liu-Ren3] T. Liu, G. Ren, *Decomposition theorem of normalized biholomorphic convex mappings*, J. Reine Angew. Math., **496**(1998), 1-13.

[Liu-Ren4] T. Liu, G. Ren, *Growth theorem of convex mappings on bounded convex circular domains*, Science in China, Ser. A, **41**(1998), 123-130.

[LiuX]   X. Liu, *Bloch functions of several complex variables*, Pacif. J. Math., **152**(1992), 347-363.

[Liu-Min]   X. Liu, D. Minda, *Distortion theorems for Bloch functions*, Trans.
            Amer. Math. Soc., **333**(1992), 325-338.

[Lö1]   K. Löwner, *Untersuchungen über die Verzerrung bei konformen
            Abbildungen des Einheitskreises $|z| < 1$, die durch Funktionen
            mit nichtverschwindender Ableitung geliefertwerden*, S.-B. Sächs.
            Akad. Wiss., **69**(1917), 89-106.

[Lö2]   K. Löwner, *Untersuchungen über schlichte konforme Abbildungen
            des Einheitskreises, I*, Math. Ann., **89**(1923), 103-121.

[Ma-Min1]   W. Ma, D. Minda, *Linear invariance and uniform local univalence*,
            Complex Variables, **16**(1991), 9-19.

[Ma-Min2]   W. Ma, D. Minda, *Euclidean linear invariance and uniform local
            convexity*, J. Austr. Math. Soc. Ser. A, **52**(1992), 401-418.

[Ma-Min3]   W. Ma, D. Minda, *Uniformly convex functions*, Ann. Polon. Math.,
            57(1992), 165-175.

[Ma-Min4]   W. Ma, D. Minda, *Hyperbolically convex functions*, Ann. Polon.
            Math., **60**(1994), 81-100.

[Ma-Min5]   W. Ma, D. Minda, *Hyperbolic linear invariance and hyperbolic k-
            convexity*, J. Austr. Math. Soc., Ser. A, **58**(1995), 73-93.

[Ma-Min6]   W. Ma, D. Minda, *Two point distortion theorems for bounded uni-
            valent functions*, Ann. Acad. Sci. Fenn. Math., **22**(1997), 425-444.

[Ma-Min7]   W. Ma, D. Minda, *Two point distortion for univalent functions*, J.
            Comput. Appl. Math. **105**(1999), 385-392.

[Mac1]   T.H. MacGregor, *A covering theorem for convex functions*, Proc.
            Amer. Math. Soc., **15**(1964), 310.

[Mac2]   T.H. MacGregor, *Translation of the image domains of analytic
            functions*, Proc. Amer. Math. Soc., **16**(1965), 1280-1286.

[Mac3]   T.H. MacGregor, *Majorization by univalent functions*, Duke Math.
            J., **34**(1967), 95-102.

[Mac4]   T.H. MacGregor, *Geometric problems in complex analysis*, Amer.
            Math. Monthly, **79**(1972), 447-468.

[Mard-Ric] A. Marden, S. Rickman, *Holomorphic mappings of bounded dis-
            tortion*, Proc. Amer. Math. Soc., **46**(1974), 226-228.

[Mart-Ric-Väi1] O. Martio, S. Rickman, J. Väisälä, *Definitions for quasireg-
            ular mappings*, Ann. Acad. Sci. Fenn. Ser. A I, **448**(1969), 1-40.

[Mart-Ric-Väi2] O. Martio, S. Rickman, J. Väisälä, *Topological and metric
            properties of quasiregular mappings*, Ann. Acad. Sci. Fenn. Ser. A
            I, **488**(1971), 1-31.

[Marx]   A. Marx, *Untersuchungen über schlichte Abbildungen*, Math. Ann.,
            107(1932), 40-67.

[Mat]   T. Matsuno, *Star-like theorems and convex-like theorems in the
            complex vector space*, Sci. Rep. Tokyo Kyoiku Daigaku, Sect.A,
            5(1955), 88-95.

[Mej-Pom1] D. Mejia, C. Pommerenke, *On hyperbolically convex functions*, J.
            Geom. Anal., **10**(2000), 361-374.

[Mej-Pom2] D. Mejia, C. Pommerenke, *On spherically convex univalent func-
            tions*, Michigan Math. J., **47**(2000), 163-172.

[Mer-Rob-Sco] E.P. Merkes, M.S. Robertson, W.T. Scott, *On products of star-
            like functions*, Proc. Amer. Math. Soc., **13**(1962), 960-964.

[Mia-Wes]   J. Miazga, A. Weselowski, *A univalence criterion and the
            Schwarzian derivative*, Demonstratio Math., **21**(1988), 761-766.

[Mili1]   I.M. Milin, *Estimation of coefficients of univalent functions*, Soviet
            Math. Dokl., **6**(1965), 196-198.

[Mili2]   I.M. Milin, Univalent Functions and Orthonormal Systems, English transl., Amer. Math. Soc., Providence, R.I., 1977.

[Mill]   S.S. Miller, *Distortion properties of alpha-starlike functions*, Proc. Amer. Math. Soc., **38**(1973), 311-318.

[Mill-Moc1] S.S. Miller, P.T. Mocanu, *Second order differential inequalities in the complex plane*, J. Math. Anal. Appl., **65**(1978), 289-305.

[Mill-Moc2] S.S. Miller, P.T. Mocanu, Differential Subordinations. Theory and Applications, Marcel Dekker Inc., New York, 2000.

[Mill-Moc-Rea1] S.S. Miller, P.T. Mocanu, M.O. Reade, *All α-convex functions are starlike*, Rev. Roum. Math. Pures Appl., **17**(1972), 1395-1397.

[Mill-Moc-Rea2] S.S. Miller, P.T. Mocanu, M.O. Reade, *All α-convex functions are univalent and starlike*, Proc. Amer. Math. Soc., **37**(1973), 553-554.

[Mill-Moc-Rea3] S.S. Miller, P.T. Mocanu, M.O. Reade, *Bazilevich functions and generalized convexity*, Rev. Roum. Math. Pures Appl., **19**(1974), 213-224.

[Min1]   D. Minda, *Bloch constants*, J. Anal. Math., **41**(1982), 54-84.

[Min2]   D. Minda, *Marden constants for Bloch and normal functions*, J. Anal. Math., **42**(1982/83), 117-127.

[Min3]   D. Minda, *Lower bounds for the hyperbolic metric in convex regions*, Rocky Mountain J. Math. **13**(1983), 61-69.

[Min4]   D. Minda, *The Schwarzian derivative and univalence criteria*, Contemp. Math., **38**(1985), 43-52.

[Min5]   D. Minda, *The Bloch and Marden constants*, Computational Methods and Function Theory, 131-142, Lecture Notes in Math., **1435**, Springer-Verlag, New York 1990.

[Moc1]   P.T. Mocanu, *Une propriété de convexité généralisée dans la théorie de la représentation conforme*, Mathematica (Cluj), **11(34)**(1969), 127-133.

[Moc2]   P.T. Mocanu, *Starlikeness and convexity for nonanalytic functions in the unit disc*, Mathematica (Cluj), **22(45)**(1980), 77-83.

[Moc-Bu-Să] P.T. Mocanu, T. Bulboacă, G. Sălăgean, Geometric Theory of Univalent Functions, Casa Cărţii de Ştiinţă, Cluj-Napoca, 1999 (in Romanian).

[Moc-Koh-Koh] P.T. Mocanu, G. Kohr, M. Kohr, *Two simple sufficient conditions for convexity*, Studia Univ. Babeş-Bolyai (Mathematica), **37**(1992), 23-33.

[Moc-Rea] P.T. Mocanu, M.O. Reade, *On generalized convexity in conformal mappings*, Rev. Roum. Math. Pures Appl., **16**(1971), 1541-1544.

[Mol-Mor] R. Molzon, P. Mortensen, *Univalence of holomorphic mappings*, Pacif. J. Math., **180**(1997), 125-133.

[Mon]   P. Montel, Leçons sur les Fonctions Univalentes ou Multivalentes, Paris, Gauthier-Villars, 1933.

[Mu-Su]   J.R. Muir, T.J. Suffridge, *Unbounded convex mappings of the ball in $\mathbb{C}^n$*, Proc. Amer. Math. Soc., **129**(2001), 3389-3393.

[Muj]   J. Mujica, Complex Analysis in Banach Spaces. Holomorphic Functions and Domains of Holomorphy in Finite and Infinite Dimensions, North-Holland, Amsterdam, 1986.

[Nac]   L. Nachbin, Topology on Spaces of Holomorphic Mappings, Springer-Verlag, New York, 1969.

[Nar]   R. Narasimhan, Several Complex Variables, The University of Chicago Press, 1971.

[Nat]   I.P. Natanson, Theory of Functions of a Real Variable, Revised Edition, Volume I, Ungar, New York, 1961.

[Neh1]   Z. Nehari, *The Schwarzian derivative and schlicht functions*, Bull.
           Amer. Math., **55**(1949), 545-551.

[Neh2]   Z. Nehari, Conformal Mapping, McGraw-Hill, New York, 1952.

[Neh3]   Z. Nehari, *A property of convex conformal maps*, J. Analyse Math.,
           30(1976), 390-393.

[Neh4]   Z. Nehari, *Univalence criteria depending on the Schwarzian deriv-
           ative*, Illinois J. Math., **23**(1979), 345-351.

[Nev1]   R. Nevanlinna, *Über die schlichten Abbildungen des Einheit-
           skreises*, Översikt Finska Vetenskaps-Soc. Förh., **62A**(1920), 1-14.

[Nev2]   R. Nevanlinna, *Über die konforme Abbildung von Sterngebieten*,
           Översikt av Finska Vetenskaps-Soc.Förh., **63A**(1920-1921), 1-21.

[Nog-Och] J. Noguchi, T. Ochiai, Geometric Function Theory in Several Com-
           plex Variables, American Mathematical Society, Providence, RI,
           1990.

[Nos]   K. Noshiro, *On the theory of schlicht functions*, J. Fac. Sci.
           Hokkaido Univ., **2**(1934-35), 129-155.

[On]   I. Ono, *Analytic vector functions of several complex variables*, J.
           Math. Soc. Japan, **8**(1956), 216-246.

[Os1]   B. Osgood, *Some properties of f''/f' and the Poincaré metric*,
           Indiana Univ. Math. J., **31**(1982), 449-461.

[Os2]   B. Osgood, *Old and new on the Schwarzian derivative*, In: Qua-
           siconformal Mappings and Analysis (Ann Arbor 1995), 275-308,
           Springer-Verlag, New York 1998.

[Os-St1]   B. Osgood, D. Stowe, *A generalization of Nehari's univalence cri-
           terion*, Comment. Math. Helv., **65**(1990), 234-242.

[Os-St2]   B. Osgood, D. Stowe, *The Schwarzian derivative and conformal
           mappings of Riemannian manifolds*, Duke Math. J., **67**(1992), 57-
           99.

[Ov]   M. Overholt, *The extreme points of the set of Schwarzians of uni-*
            valent functions, Complex Variables Theory Appl., **11**(1989), 197-
            202.

[Ove]   H. Ovesea, *A generalization of Ruscheweyh's univalence criterion*,
            J. Math. Anal. Appl., **258**(2001), 102-109.

[Oz]   S. Ozaki, *On the theory of multivalent functions*, Sci. Rep. Tokyo
            Bunrika Daigaku, A, **40**(1935), 167-188.

[Par]   G.B. Park, *On geometric properties of the linear invariant families*
            of holomorphic functions, Honam. Math. J., **9**(1987), 57-69.

[Pas]   N.N. Pascu, *Loewner chains and univalence criteria*, Mathematica
            (Cluj), **37(60)**(1995), 215-217.

[Pes1]   E. Peschl, *Über die Verwendung von Differentialinvarianten bei*
            gewissen Funktionenfamilien und die Übertragung einer darauf
            gegründeten Methode auf partialle Differentialgleichungen von el-
            liptischen Typus, Ann. Acad. Sci. Fenn. A I Math., **336/6**(1963),
            2-22.

[Pes2]   E. Peschl, *Über unverzweigte konforme Abbildungen*, Österreich
            Akad. Wiss. Math.-Natur. Kl. S.-B. II **185**(1976), 55-78.

[Pfa1]   J.A. Pfaltzgraff, *Subordination chains and univalence of holomor-*
            phic mappings in $\mathbb{C}^n$, Math. Ann., **210**(1974), 55-68.

[Pfa2]   J.A. Pfaltzgraff, *Subordination chains and quasiconformal exten-*
            sion of holomorphic maps in $\mathbb{C}^n$, Ann. Acad. Sci. Fenn, Ser. A,
            1(1975), 13-25.

[Pfa3]   J.A. Pfaltzgraff, *Loewner theory in* $\mathbb{C}^n$, Abstract of papers pre-
            sented to AMS, **11(66)**(1990), 46.

[Pfa4]   J.A. Pfaltzgraff, *k-Quasiconformal extension criteria in the disk*,
            Complex Variables, **21**(1993), 293-301.

[Pfa5]   J.A. Pfaltzgraff, *Distortion of locally biholomorphic maps of the n-ball*, Complex Variables, **33**(1997), 239-253.

[Pfa6]   J.A. Pfaltzgraff, *Distortion of locally biholomorphic maps of the n-ball, erratum*, Complex Variables, **45**(2001), 197-200.

[Pfa-Re-Um] J.A. Pfaltzgraff, M.O. Reade, T. Umezawa, *Sufficient conditions for univalence*, Ann. Fac. Sci. Univ. Nat. Zaire (Kinshasa) Sect. Math.-Phys., **2**(1976), 211-218.

[Pfa-Su1]   J.A. Pfaltzgraff, T.J. Suffridge, *Close-to-starlike holomorphic functions of several variables*, Pacif. J. Math., **57**(1975), 271-279.

[Pfa-Su2]   J.A. Pfaltzgraff, T.J. Suffridge, *Linear invariance, order and convex maps in $\mathbb{C}^n$*, Complex Variables, **40**(1999), 35-50.

[Pfa-Su3]   J.A. Pfaltzgraff, T.J. Suffridge, *An extension theorem and linear invariant families generated by starlike maps*, Ann. Univ. Mariae Curie Sklodowska, Sect.A, **53**(1999), 193-207.

[Pfa-Su4]   J.A. Pfaltzgraff, T.J. Suffridge, *Norm order and geometric properties of holomorphic mappings in $\mathbb{C}^n$*, J. Analyse Math., **82**(2000), 285-313.

[Pic]   G. Pick, *Über die konforme Abbildung eines Kreises auf ein schlichtes und zugleich beschränktes Gebiet*, S.-B. Kaiserl Akad. Wiss. Wien, **126**(1917), 247-263.

[Poi]   H. Poincaré, *Les fonctions analytiques de deux variables et la représentation conforme*, Rend. Circ. Mat. Palermo, **23**(1907), 185-220.

[Pok]   V.V. Pokornyi, *On some sufficient conditions for univalence*, Dokl. Akad. Nauk SSSR, **79**(1951), 743-746 (in Russian).

[Pol]   E.A. Poletsky, *Holomorphic quasiregular mappings*, Proc. Amer. Math. Soc., **92**(1985), 235-241.

[Pol-Sh]   E.A. Poletskii, B.V. Shabat, Invariant Metrics, Several Complex
             Variables III, (G.N. Khenkin, Ed.), 63-112, Springer-Verlag, New-
             York, 1989.

[Poly-Sze]   G. Polya, G. Szegö, Problems and Theorems in Analysis I,
             Springer-Verlag, Berlin-New York, 1972.

[Pom1]   C. Pommerenke, *Linear-invariante Familien analytischer Funktio-
             nen*, I, Math. Ann., **155**(1964), 108-154.

[Pom2]   C. Pommerenke, *Linear-invariante Familien analytischer Funktio-
             nen*, II, Math. Ann., **156**(1964), 226-262.

[Pom3]   C. Pommerenke, *Über die Subordination analytischer Funktionen*,
             J. Reine Angew Math., **218**(1965), 159-173.

[Pom4]   C. Pommerenke, *On Bloch functions*, J. London Math. Soc.,
             2(1970), 689-695.

[Pom5]   C. Pommerenke, Univalent Functions, Vandenhoeck & Ruprecht,
             Göttingen, 1975.

[Pom6]   C. Pommerenke, *Schlichte Funktionen und analytische Funktio-
             nen von beschränkter mittlerer Oszillation*, Comment. Math. Helv.
             52(1977), 591-602.

[Pom7]   C. Pommerenke, *On univalent functions, Bloch functions, and
             VMOA*, Math. Ann. **236**(1978), 199-208.

[Pom8]   C. Pommerenke, *On the Becker univalence criterion*, Ann. Univ.
             Mariae Curie-Sklodowska, Ser. A., **36/37**(1982/1983), 123-124.

[Pom9]   C. Pommerenke, *On the Epstein univalence criterion*, Results in
             Math., **10**(1986), 143-146.

[Pom10]   C. Pommerenke, Boundary Behaviour of Conformal Maps,
             Springer-Verlag, Berlin-New York, 1992.

[Pom11]   C. Pommerenke, *On Bloch functions and conformal mapping*, Complex Variables, **21**(1993), 287-292.

[Por1]   T. Poreda, *On the univalent holomorphic maps of the unit polydisc in $\mathbb{C}^n$ which have the parametric representation, I - the geometrical properties*, Ann. Univ. Mariae Curie Sklodowska, Sect A, **41**(1987), 105-113.

[Por2]   T. Poreda, *On the univalent holomorphic maps of the unit polydisc in $\mathbb{C}^n$ which have the parametric representation, II - necessary and sufficient conditions*, Ann. Univ. Mariae Curie Sklodowska, Sect A, **41**(1987), 114-121.

[Por3]   T. Poreda, *On the univalent subordination chains of holomorphic mappings in Banach spaces*, Commentationes Math., **28**(1989), 295-304.

[Por4]   T. Poreda, *On generalized differential equations in Banach spaces*, Dissertationes Mathematicae, **310**(1991), 1-50.

[Por5]   T. Poreda, *On the geometrical properties of the starlike maps in Banach spaces*, Problemy Mat., **12**(1993), 59-71.

[Por-Sz]   T. Poreda, A. Szadkowska, *On the holomorphic solutions of certain differential equations of first order for the mappings of the unit ball of $\mathbb{C}^n$ into $\mathbb{C}^n$*, Demonstratio Mathematica, **22**(1989), 983-996.

[Pra]   H. Prawitz, *Über die Mittelwerte analytische Funktionen*, Ark. Mat. Ast. Fys., **20A**(1927), 1-12.

[Rad]   H. Rademacher, *On the Bloch-Landau constant*, Amer. J. Math., **65**(1943), 387-390.

[Rah]   B.N. Rahmanov, *On the theory of univalent functions*, Dokl. Akad. Nauk. SSSR (N.S.), **78**(1951), 209-211; **91**(1953), 729-732; **97**(1954), 973-976 (in Russian).

[Ran]   M. Range, Holomorphic Functions and Integral Representations in
           Several Complex Variables, Springer-Verlag, New York, 1986.

[Răd]   D. Răducanu, *On univalence of holomorphic mappings in* $\mathbb{C}^n$,
           Demonstratio Mathematica, **34**(2001), 789-794.

[Rea]   M. O. Reade, *On close-to-convex univalent functions*, Mich. Math.
           J., **3**(1955-56), 59-62.

[Rem]   R. Remmert, Classical Topics in Complex Function Theory, Sprin-
           ger-Verlag, New York, 1998.

[Ren-Ma]   F. Ren, J. Ma, *Quasiconformal extension of biholomorphic map-
           pings of several complex variables*, J. Fudan Univ. Natur. Sci.,
           34(1995), 545-556.

[Res]   Y.G. Reshetnyak, Space Mappings with Bounded Distortion,
           Transl. of Math. Monographs, **73**, Amer. Math. Soc. Providence,
           R.I., 1989.

[Ric]   S. Rickman, Quasiregular Mappings, Springer-Verlag, New York
           1993.

[Robe1]   M.S. Robertson, *On the theory of univalent functions*, Ann. Math.,
           37(1936), 374-408.

[Robe2]   M.S. Robertson, *A remark on the odd schlicht functions*, Bull.
           Amer. Math. Soc., **42**(1936), 366-370.

[Robe3]   M.S. Robertson, *Applications of the subordination principle to uni-
           valent functions*, Pacif. J. Math., **11**(1961), 315-324.

[Robe4]   M.S. Robertson, *Quasi-subordination and coefficient conjectures*,
           Bull. Amer. Math. Soc., **76**(1970), 1-9.

[Robi]   R.M. Robinson, *Univalent majorants*, Trans.Amer. Math. Soc.,
           61(1947), 1-35.

[Rog1]   W. Rogosinski, *On subordinate functions*, Proc. Cambridge Phil. Soc., **35**(1939), 1-26.

[Rog2]   W. Rogosinski, *On the coefficients of subordinate functions*, Proc. London Math. Soc., **48**(1943), 48-82.

[Røn1]   F. Rønning, *Uniformly convex functions and a corresponding class of starlike functions*, Proc. Amer. Math. Soc., **118**(1993), 189-196.

[Røn2]   F. Rønning, *On uniform starlikeness and related properties of univalent functions*, Complex Variables, **24**(1994), 233-239.

[Rop-Su1] K. Roper, T.J. Suffridge, *Convex mappings on the unit ball of* $\mathbb{C}^n$, J. Anal. Math., **65**(1995), 333-347.

[Rop-Su2] K. Roper, T.J. Suffridge, *Convexity properties of holomorphic mappings in* $\mathbb{C}^n$, Trans. Amer. Math. Soc., **351**(1999), 1803-1833.

[Ros]   P.C. Rosenbloom, *Conformal mapping of nearly circular domains and Loewner's differential equation*, Inequalities III (O. Shisha, Ed.), 301-310, Academic Press, New York, 1972.

[Ros-Rov] M. Rosenblum, J. Rovnyak, Topics in Hardy Classes and Univalent Functions, Birkhäuser Verlag, Boston, 1994.

[Ros-Rud] J.P. Rosay, W. Rudin, *Holomorphic maps from* $\mathbb{C}^n$ *to* $\mathbb{C}^n$, Trans. Amer. Math. Soc., **310**(1988), 47-86.

[Rot]   O. Roth, *A remark on the Loewner differential equation*, Computational Methods and Function Theory 1997 (Nicosia), 461-469, World Scientific, Singapore, 1999.

[Rov]   J. Rovnyak, *A vector extension of Loewner's differential equation*, Linear Operators in Function Spaces (Timişoara 1988), Oper. Theory Adv. Appl., 1990, 301-308.

[Roy]   H.L. Royden, Real Analysis, (Third Edition), MacMillan, N.Y., 1988.

[Rud1]   W. Rudin, Function Theory in the Polydisc, W.A. Benjamin Inc.,
          New York, 1969.

[Rud2]   W. Rudin, Function Theory in the Unit Ball of $\mathbb{C}^n$, Springer-
          Verlag, New York, 1980.

[Rud3]   W. Rudin, Real and Complex Analysis, third edition, McGraw-
          Hill, New York, 1987.

[Rus1]   S. Ruscheweyh, *An extension of Becker's univalence condition*,
          Math. Ann., **220**(1976), 285-290.

[Rus2]   S. Ruschweyh, Convolutions in Geometric Function Theory, Les
          Presses de l'Université de Montreal, 1982.

[Rus-She] S. Ruscheweyh, T. Sheil-Small, *Hadamard products of schlicht
          functions and the Pólya-Schoenberg conjecture*, Comment. Math.
          Helv., **48**(1973), 119-135.

[Rus-Wir1] S. Ruscheweyh, K.J. Wirths, *On extreme Bloch functions with
          prescribed critical points*, Math. Z., **180**(1982), 91-105.

[Rus-Wir2] S. Ruscheweyh, K.J. Wirths, *Extreme Bloch functions with many
          critical points*, Analysis, 4(1984), 237-247.

[Sak1]   K. Sakaguchi, *On Bloch's theorem for several complex variables*,
          Sci. Rep. Tokyo Kyoiku Daigaku, Sect. A, **5**(1956), 149-154.

[Sak2]   K. Sakaguchi, *On a certain univalent mapping*, J. Math. Soc.
          Japan, **11**(1959), 72-75.

[Sak3]   K. Sakaguchi, *A note on p-valent functions*, J. Math. Soc. Japan,
          14(1962), 312-321.

[Sak-Fuk] K. Sakaguchi, S. Fukui, *On alpha-starlike functions and related
          functions*, Bull. Nara Univ. of Education, **28**(1979), 5-12.

[Saks]   S. Saks, Theory of the Integral, Warsaw, 1937.

[Scha-Spe] A.C. Schaeffer, D.C. Spencer, Coefficient Regions for Schlicht Functions, Amer. Math. Soc. Colloq. Publ., **35**(1950).

[Sch]   M. Schiffer, *Sur un principe nouveau pour l'évaluation des fonctions holomorphes*, Bull. Soc. Math. France, **64**(1936), 231-240.

[Sch-Schm] M. Schiffer, H.G. Schmidt, *A new set of coefficient inequalities for univalent functions*, Arch. Rational Mech. Anal., **42**(1971), 346-368.

[Schif]   J.L. Schiff, Normal Families, Springer-Verlag, New York, 1993.

[Schip]   E. Schippers, *Distortion theorems for higher order Schwarzian derivatives of univalent functions*, Proc. Amer. Math. Soc., **128**(2000), 3241-3249.

[Scho]   G. Schober, Univalent Functions-Selected Topics, Lecture Notes in Math., **478**, Springer-Verlag, New York, 1975.

[Schw]   B. Schwarz, *On two univalence criteria of Nehari*, Illinois J. Math., **27**(1983), 346-351.

[She1]   T. Sheil-Small, *On convex univalent functions*, J. London Math. Soc., **1**(1969), 483-492.

[She2]   T. Sheil-Small, *On linearly accessible univalent functions*, J. London Math. Soc., **6**(1973), 385-398.

[Shi]   J.H. Shi, *On the bound of convexity of univalent maps of the ball*, Chin. Sci. Bull., **27**(1982), 473-476.

[Shi-Wil] A.L. Shields, D.L. Williams, *Bounded projections, duality and multipliers in spaces of analytic functions*, Trans. Amer. Math. Soc., **162**(1971), 287-302.

[Sho]   D. Shoikhet, Semigroups in Geometrical Function Theory, Kluwer Acad. Publ., Dordrecht, 2001.

[Spa]   L. Spaček, *Contribution à la théorie des fonctions univalentes*, Časopis Pěst. Mat., **62**(1932), 12-19 (in Russian).

[Sta]   J. Stankiewicz, *Geometric interpretations of some subclasses of univalent functions*, Zesz. Nauk. Politech. Rzeszow Mat.-Fiz., **18**(1993), 43-49.

[Str]   E. Strohhäcker, *Beitrage zur Theorie der schlichten Funktionen*, Math. Z., **37**(1933), 356-380.

[Stu]   E. Study, *Vorlesungen über ausgewählte Gegenstände der Geometrie*, 2. Heft, Teubner, Leipzig and Berlin, 1913.

[Su1]   T.J. Suffridge, *Some remarks on convex maps of the unit disc*, Duke Math. J., **37**(1970), 775-777.

[Su2]   T.J. Suffridge, *The principle of subordination applied to functions of several variables*, Pacif. J. Math., **33**(1970), 241-248.

[Su3]   T.J. Suffridge, *Starlike and convex maps in Banach spaces*, Pacif. J. Math., **46**(1973), 575-589.

[Su4]   T.J. Suffridge, *Starlikeness, convexity and other geometric properties of holomorphic maps in higher dimensions*, Lecture Notes in Math., **599**, 146-159, Springer-Verlag, New York, 1976.

[Su5]   T.J. Suffridge, *Biholomorphic mappings of the ball onto convex domains*, Abstract of papers presented to AMS, **11**(**66**)(1990), 46.

[Su6]   T.J. Suffridge, *Holomorphic mappings of domains in $\mathbb{C}^n$ onto convex domains*, submitted.

[Sze]   G. Szegö, *Über eine Extremalaufgabe aus der Theorie der Schlichten Abbildungen*, Sitzungsberichte der Berliner Mathematische Geselleschaft **22**(1923), 38-47. [Gabor Szegö: Collected Papers, Ed. by Richard Askey, Birkhäuser Verlag, Boston-Basel-Stuttgart, 1982, Vol.I, 607-618.]

[Tak]   S. Takahashi, *Univalent mappings in several complex variables*, Ann. Math., **53**(1951), 464-471.

[Tam]   H. Tamanoi, *Higher Schwarzian operators and combinatorics of the Schwarzian derivative*, Math. Ann., **305**(1996), 127-151.

[Tho-Wh] E. Thorp, R. Whitley, *The strong maximum modulus theorem for analytic functions into a Banach space*, Proc. Amer. Math. Soc., **18**(1967), 640-646.

[Tim1]   R.M. Timoney, *Bloch functions in several complex variables*, I, Bull. London Math. Soc., **12**(1980), 241-267.

[Tim2]   R.M. Timoney, *Bloch functions in several complex variables*, II, J. Reine Angew. Math., **319**(1980), 1-22.

[Tit]   E.C. Titchmarsh, The Theory of Functions, 2nd. ed, Oxford Univ. Press, 1976.

[Tri]   S.Y. Trimble, *A coefficient inequality for convex univalent functions*, Proc. Amer. Math. Soc., **48**(1975), 266-267.

[Tsu]   M. Tsuji, Potential Theory in Modern Function Theory, Maruzen Co., Tokyo, 1959.

[Väi]   J. Väisälä, Lectures on n-Dimensional Quasiconformal Mappings, Lectures Notes in Math., **229**, Springer-Verlag, New York, 1971.

[Val]   F.A. Valentine, Convex Sets, McGraw Hill, New York, 1964.

[Vuo]   M. Vuorinen, Conformal Geometry and Quasiregular Mappings, Lecture Notes in Math., **1319**, Springer-Verlag, New York 1988.

[War]   S.E. Warschawski, *On the higher derivatives at the boundary in conformal mapping*, Trans. Amer. Math. Soc., **38**(1935), 310-340.

[Wei]   L. Weinstein, *The Bieberbach conjecture*, International Mathematics Research Notices, Duke Math. J., **64**(1991), 61-64.

[Wir1]   K.J. Wirths, *Über holomorphe Funktionen, die einer Wachstums-
           beschränkung unterliegen*, Arch. Math., **30**(1978), 606-612.

[Wir2]   K.J. Wirths, *On holomorphic functions satisfying* $|f(z)|(1-|z|^2) \leq$
           1 *in the disc*, Proc. Amer. Math. Soc., **85**(1982), 19-23.

[Wir3]   K.J. Wirths, *Bounds for the Taylor coefficients of locally univalent
           Bloch functions*, Complex Variables Theory Appl., **41**(2000), 45-
           61.

[Wol]   J. Wolff, *L'intégrale d'une fonction holomorphe et à partie réelle
           positive dans un demi plan est univalente*, C.R. Acad. Sci. Paris,
           198(1934), 1209-1210.

[Wu]   H. Wu, *Normal families of holomorphic mappings*, Acta Math.,
           119(1967), 193-233.

[Ya]   S. Yamashita, *Norm estimates for functions starlike or convex of
           order alpha*, Hokkaido Math. J., **28**(1999), 217-230.

[Yan1]   H. Yanagihara, *Sharp distortion estimate for locally schlicht Bloch
           functions*, Bull. London Math. Soc., **26**(1994), 539-542.

[Yan2]   H. Yanagihara, *On the locally univalent Bloch constant*, J. Anal.
           Math., **65**(1995), 1-17.

[ZhaM]   M. Zhang, *Ein Überdeckungssatz für konvexe Gebiete*, Acad. Sinica
           Science Record, **5**(1952), 17-21.

[ZhaS]   S. Zhang, *On the schlicht Bloch constant* (Chinese), Acta Scientar-
           ium Naturalium Universitas Pekinensis, **25**(1989), 537-540.

[Zhu]   K. Zhu, Operator Theory in Function Spaces, Marcel Dekker Inc.,
           New York, 1990.

List of Symbols

Symbols in Chapter 1

$\mathbb{C}$	the complex plane
$\mathbb{R}$	the field of real numbers
Re z	the real part of z
Im z	the imaginary part of z
$U(z_0, r)$	the open disc centered at z_0 and of radius r
U_r	the open disc centered at 0 and of radius r
U	the open unit disc
Δ	the exterior of the closed unit disc
$H(G)$	the set of holomorphic functions on an open subset G of $\mathbb{C}$
$H_u(D)$	the set of univalent functions on a domain D of $\mathbb{C}$
S	the class of normalized univalent functions on U
Σ	the class of univalent functions φ on Δ with a simple pole at ∞ and normalized so that $\varphi(\zeta) = \zeta + \alpha_0 + \alpha_1/\zeta + \cdots$
$M_\infty(r, f)$	$= \max\limits_{|z|=r} |f(z)|$ for $f \in H(U)$ and $0 < r < 1$
$d_h(a, b)$	the distance function induced by the hyperbolic metric
$D_1 f(z)$	$= (1 - |z|^2) f'(z)$

Symbols in Chapter 2

$\mathcal{P}$	the Carathéodory class
$\prec$	subordination
$\mathcal{V}$	the class of Schwarz functions
S^*	the class of normalized starlike functions on U

521

K	the class of normalized convex functions on U
$S^*(\alpha)$	the class of normalized starlike functions of order α
$K(\beta)$	the class of normalized convex functions of order β
M_α	the class of α-convex functions
C	the class of normalized close-to-convex functions
$\widehat{S}_\alpha$	the class of normalized spirallike functions of type α
$\{f;z\}$	the Schwarzian derivative of $f \in H(U)$ at $z \in U$
$\widehat{\Omega}$	the convex hull of a set $\Omega \subset \mathbb{C}$
$r^*(\mathcal{F})$	the radius of starlikeness of $\mathcal{F}$
$r_c(\mathcal{F})$	the radius of convexity of $\mathcal{F}$

Symbols in Chapter 3

$G_n \to G$	kernel convergence
$f(z,t)$	subordination chain, Loewner chain
$v(z,s,t)$	transition functions for a Loewner chain
$f'(z,t)$	$= \frac{\partial f}{\partial z}(z,t)$
$f_t(z)$	$= f(z,t)$
$p(z,t)$	a function which belongs to $\mathcal{P}$ as a function of $z \in U$ and is measurable in $t \in [0,\infty)$
$\widehat{\mathbb{C}}$	the extended complex plane

Symbols in Chapter 4

$\mathrm{Aut}(U)$	the set of automorphisms of U
$\mathcal{B}$	the set of Bloch functions
$\mathcal{B}_1$	the set of normalized locally univalent Bloch functions of Bloch seminorm 1
$\mathbf{B}$	the Bloch constant
$\mathbf{B}_0$	the locally univalent Bloch constant
$\Delta(1,r)$	horodisc internally tangent to ∂U at 1
$\overline{\Delta}(1,r)$	closure of $\Delta(1,r)$ relative to U
$\|\cdot\|$	the Bloch seminorm

$\| \cdot \|_B$	the Bloch norm
f^*	the spherical derivative of f
$r(a, f)$	the radius of the largest schlicht disc centered at $f(a)$
L	the Landau constant
A	the univalent Bloch constant

Symbols in Chapter 5

$\mathcal{LS}$	the set of normalized locally univalent functions on U
$\Lambda_\phi(f)$	the Koebe transform of f with respect to $\phi \in \mathrm{Aut}(U)$
$f(z; \zeta)$	the Koebe transform of f with respect to the disc automorphism $\phi(z) = (z + \zeta)/(1 + \overline{\zeta}z)$
L.I.F.	linear-invariant family
ord $\mathcal{F}$	the order of a L.I.F. $\mathcal{F}$
$\rho(z, f)$	the hyperbolic radius of the largest hyperbolic disc in U centered at z in which f is univalent
$\mathcal{U}(\alpha)$	the universal L.I.F. of order α
$\Lambda[\mathcal{G}]$	the L.I.F. generated by a nonempty subset $\mathcal{G}$ of $\mathcal{LS}$

Symbols in Chapter 6

$\mathbb{C}^n$	the space of n-complex variables
$\langle \cdot, \cdot \rangle$	the Euclidean inner product in $\mathbb{C}^n$
$B(a, r)$	open ball centered at a and of radius r
B	the Euclidean unit ball of $\mathbb{C}^n$; the unit ball of $\mathbb{C}^n$ with respect to an arbitrary norm; the unit ball of a Banach space
$B(p)$	the unit ball of $\mathbb{C}^n$ with respect to a p-norm
$P(a, R)$	open polydisc of center a and polyradius R
P	the unit polydisc of $\mathbb{C}^n$
$\partial_0 P(a, R)$	the distinguished boundary of the polydisc $P(a, R)$
A^t	the transpose of a matrix A
$\overline{A}$	the conjugate of a matrix A
A^*	the conjugate transpose of a matrix A

$\mathcal{U}$	the set of unitary transformations of $\mathbb{C}^n$
$H(\Omega, \Omega')$	the set of holomorphic mappings from a domain $\Omega \subset \mathbb{C}^n$ into another domain $\Omega' \subset \mathbb{C}^m$
$H(\Omega)$	the set of holomorphic mappings from a domain $\Omega \subset \mathbb{C}^n$ into $\mathbb{C}^n$
$J_f(z)$	the complex Jacobian determinant of f at z
$J_r f(z)$	the real Jacobian determinant of f at z
$\|\cdot\|_p$	p-norm
$\mathrm{Aut}(\Omega)$	the set of biholomorphic automorphisms of a domain $\Omega \subset \mathbb{C}^n$
$L(X, Y)$	the space of continuous complex-linear operators from a complex Banach space X into another complex Banach space Y
$\|A\|$	the norm of the operator $A \in L(X, Y)$
I	the identity in $L(X, X)$
X^*	the dual of the complex Banach space X
$T(z)$	$= \{l_z \in X^* : l_z(z) = \|z\|, \|l_z\| = 1\}$
$\mathcal{N}_0$	$= \{f : B \to X : f \in H(B), f(0) = 0, \mathrm{Re}\,[l_z(f(z))] \geq 0, z \in B \setminus \{0\}, l_z \in T(z)\}$
$\mathcal{N}$	$= \{f \in \mathcal{N}_0 : \mathrm{Re}\,[l_z(f(z))] > 0, z \in B \setminus \{0\}, l_z \in T(z)\}$
$\mathcal{M}$	$= \{f \in \mathcal{N} : Df(0) = I\}$
$S(B)$	the set of normalized biholomorphic mappings on B
$S^*(B)$	the set of normalized starlike mappings on B
$K(B)$	the set of normalized convex mappings on B
$V(h)$	the numerical range of h
$|V(h)|$	the numerical radius of h
$\mathcal{G}$	the set of quasi-convex mappings of type A
$\mathcal{F}$	the set of quasi-convex mappings of type B

Symbols in Chapter 7

π_u	the orthogonal projection of $\mathbb{C}^n$ onto the subspace $\mathbb{C}u$
$\widehat{\Omega}$	the convex hull of a subset Ω of $\mathbb{C}^n$
$E_\Omega(z, v)$	the infinitesimal Carathéodory metric
$F_\Omega(z, v)$	the infinitesimal Kobayashi-Royden metric
$\mathcal{D}_1 f(z)$	$= (1 - \|z\|^2)\|[Df(z)]^{-1}\|^{-1}, z \in B$

Symbols in Chapter 8

$\prec$	subordination
$f(z,t)$	subordination chain, Loewner chain
$f_t(z)$	$= f(z,t)$
$Df(z,t)$	differential in z of $f(z,t)$
$v(z,s,t)$	transition mapping for a Loewner chain
$h(z,t)$	a mapping which belongs to $\mathcal{M}$ as a function of $z \in B$ and is measurable in $t \in [0,\infty)$
$\mathcal{M}_g$	$= \{h \in H(B) : h(0) = 0, Dh(0) = I, \langle h(z), z/\|z\|^2 \rangle \in g(U), z \in B \setminus \{0\}\}$
$S_g^0(B)$	the set of mappings in $H(B)$ which have g-parametric representation
$S^0(B)$	the set of mappings in $H(B)$ which have parametric representation
$S^1(B)$	the set of mappings in $H(B)$ which can be embedded as the first element of a Loewner chain
ACL	absolutely continuous on lines

Symbols in Chapter 9

$\mathcal{B}$	the set of Bloch mappings from the unit ball B of $\mathbb{C}^n$ into $\mathbb{C}^n$
$m(f)$	$= \sup\{(1 - \|z\|^2)\|Df(z)\| : z \in B\}$
$\|f\|_0$	$= \sup\{|J_{f \circ \varphi}(0)|^{1/n} : \varphi \in \text{Aut}(B)\}$
$\mathcal{VB}$	$= \{f \in H(B) : \|f\|_0 < \infty\}$
$\mathcal{VB}_0$	$= \{f \in \mathcal{VB} : \|f\|_0 = 1 \text{ and } J_f(0) = 1\}$
$\mathcal{B}_n(k)$	$= \{f : B \to \mathbb{C}^n : f \in H(B), \|f\| \le k\}$
$\mathcal{B}_{n,0}(k)$	$= \{f \in \mathcal{B}_n(k) : J_f(0) = 1\}$
$\mathbf{B}(\mathcal{F})$	the Bloch constant for the set $\mathcal{F}$
$\mathbf{B}(S^*(B))$	the Bloch constant for $S^*(B)$
$\mathbf{B}(K(B))$	the Bloch constant for $K(B)$

Symbols in Chapter 10

L.I.F.	linear-invariant family
$\mathrm{Aut}(B)$	the group of biholomorphic automorphisms of the Euclidean unit ball B
$\mathrm{Aut}(P)$	the group of biholomorphic automorphisms of the unit polydisc P
$\mathcal{LS}(B)$	the set of normalized locally biholomorphic mappings on the Euclidean unit ball B
$\mathcal{LS}(P)$	the set of normalized locally biholomorphic mappings on the unit polydisc P
$\Lambda_\phi(f)$	the Koebe transform of f with respect to $\phi \in \mathrm{Aut}(B)$ (or $\phi \in \mathrm{Aut}(P)$)
$\Lambda[\mathcal{G}]$	the L.I.F. generated by a nonempty subset $\mathcal{G}$ of $\mathcal{LS}(B)$ (or $\mathcal{LS}(P)$)
$\mathrm{ord}\,\mathcal{F}$	the trace order of a L.I.F. $\mathcal{F}$
$\|\mathrm{ord}\|\,\mathcal{F}$	the norm order of a L.I.F. $\mathcal{F}$
$r_0(\mathcal{F})$	the radius of nonvanishing of a L.I.F. $\mathcal{F}$
$r_1(\mathcal{F})$	the radius of univalence of a L.I.F. $\mathcal{F}$

Symbols in Chapter 11

Φ_n	the Roper-Suffridge extension operator
$r^*(\mathcal{F})$	the radius of starlikeness of a nonempty subset $\mathcal{F}$ of $\mathcal{LS}(B)$
$r_c(\mathcal{F})$	the radius of convexity of a nonempty subset $\mathcal{F}$ of $\mathcal{LS}(B)$

Index

Printed in the United States
by Baker & Taylor Publisher Services